T0350171

Genetic Control of Malaria and Dengue

Genetic Control of Malaria and Dengue

Edited by

Zach N. Adelman
Fralin Life Science Institute and Department of Entomology,
Virginia Tech, Blacksburg, VA, USA

AMSTERDAM • BOSTON • HEIDELBERG • LONDON
NEW YORK • OXFORD • PARIS • SAN DIEGO
SAN FRANCISCO • SINGAPORE • SYDNEY • TOKYO
Academic Press is an imprint of Elsevier

Academic Press is an imprint of Elsevier
125, London Wall, EC2Y 5AS.
525 B Street, Suite 1800, San Diego, CA 92101-4495, USA
225 Wyman Street, Waltham, MA 02451, USA
The Boulevard, Langford Lane, Kidlington, Oxford OX5 1GB, UK

Copyright © 2016 Elsevier Inc. All rights reserved.

No part of this publication may be reproduced or transmitted in any form or by any means, electronic
or mechanical, including photocopying, recording, or any information storage and retrieval system,
without permission in writing from the publisher. Details on how to seek permission, further
information about the Publisher's permissions policies and our arrangements with organizations
such as the Copyright Clearance Center and the Copyright Licensing Agency, can be found at our
website: www.elsevier.com/permissions.

This book and the individual contributions contained in it are protected under copyright by the
Publisher (other than as may be noted herein).

Notices
Knowledge and best practice in this field are constantly changing. As new research and experience
broaden our understanding, changes in research methods, professional practices, or medical treatment
may become necessary.

Practitioners and researchers must always rely on their own experience and knowledge in evaluating
and using any information, methods, compounds, or experiments described herein. In using such
information or methods they should be mindful of their own safety and the safety of others, including
parties for whom they have a professional responsibility.

To the fullest extent of the law, neither the Publisher nor the authors, contributors, or editors, assume
any liability for any injury and/or damage to persons or property as a matter of products liability,
negligence or otherwise, or from any use or operation of any methods, products, instructions, or ideas
contained in the material herein.

ISBN: 978-0-12-800246-9

British Library Cataloguing-in-Publication Data
A catalogue record for this book is available from the British Library.

Library of Congress Cataloging-in-Publication Data
A catalog record for this book is available from the Library of Congress.

For Information on all Academic Press publications
visit our website at http://store.elsevier.com/

Publisher: Sara Tenney
Acquisition Editor: Jill Leonard
Editorial Project Manager: Fenton Coulthurst
Production Project Manager: Chris Wortley
Designer: Mark Rogers

Printed and bound in the United States of America

Contents

1. Transgenic Pests and Human Health: A Short Overview of Social, Cultural, and Scientific Considerations

Tim Antonelli, Amanda Clayton, Molly Hartzog, Sophia Webster and Gabriel Zilnik

2. Concept and History of Genetic Control

Maxwell J. Scott and Mark Q. Benedict

3. **Considerations for Disrupting Malaria Transmission in Africa Using Genetically Modified Mosquitoes, Ecology of Anopheline Disease Vectors, and Current Methods of Control**

Mamadou B. Coulibaly, Sekou F. Traoré and Yeya T. Touré

4. **Ecology of Malaria Vectors and Current (Nongenetic) Methods of Control in the Asia Region**

Patchara Sriwichai, Rhea Longley and Jetsumon Sattabongkot

11. Disruption of Mosquito Olfaction

Conor J. McMeniman

12. Disruption of Mosquito Blood Meal
Protein Metabolism

Patricia Y. Scaraffia

13.   Engineering Pathogen Resistance in Mosquitoes

Zach N. Adelman, Sanjay Basu and Kevin M. Myles

14.   Genetic Control of Malaria and Dengue Using
      Wolbachia

Zhiyong Xi and Deepak Joshi

List of Contributors

Zach N. Adelman, Fralin Life Science Institute and Department of Entomology, Virginia Tech, Blacksburg, VA, USA

Omar S. Akbari, Department of Entomology, University of California, Riverside, CA, USA

Tim Antonelli, Department of Mathematics, Worcester State University, Worcester, MA, USA

Roberto Barrera, Entomology and Ecology Activity, Dengue Branch, Centers for Disease Control and Prevention, San Juan, Puerto Rico

Sanjay Basu, Fralin Life Science Institute and Department of Entomology, Virginia Tech, Blacksburg, VA, USA

Mark Q. Benedict, Centers for Disease Control Foundation, Atlanta, GA, USA

James K. Biedler, Department of Biochemistry, Virginia Tech, Blacksburg, VA, USA; Fralin Life Science Institute, Virginia Tech, Blacksburg, VA, USA

Benjamin J. Blumberg, W. Harry Feinstone Department of Molecular Microbiology and Immunology and the Johns Hopkins Malaria Research Institute, Bloomberg School of Public Health, Johns Hopkins University, Baltimore, MD, USA

Margareth L. Capurro, Universidade de São Paulo, São Paulo, Brazil

Danilo O. Carvalho, Universidade de São Paulo, São Paulo, Brazil

Amanda Clayton, Department of Economics, North Carolina State University, Raleigh, NC, USA

Jan E. Conn, The Wadsworth Center, New York State Department of Health, Albany, NY, USA; Biomedical Sciences Department, School of Public Health, State University of New York-Albany, Albany, NY, USA

Mamadou B. Coulibaly, Malaria Research and Training Center, University of Sciences, Techniques and Technologies of Bamako, Bamako, Mali

George Dimopoulos, W. Harry Feinstone Department of Molecular Microbiology and Immunology and the Johns Hopkins Malaria Research Institute, Bloomberg School of Public Health, Johns Hopkins University, Baltimore, MD, USA

Yara A. Halasa, Schneider Institutes for Health Policy, Heller School, Brandeis University, Waltham, MA, USA

Brantley A. Hall, Fralin Life Science Institute, Virginia Tech, Blacksburg, VA, USA; Interdisciplinary PhD Program in Genetics, Bioinformatics, and Computational Biology, Virginia Tech, Blacksburg, VA, USA

Molly Hartzog, Department of Communication, Rhetoric, and Digital Media, North Carolina State University, Raleigh, NC, USA

Anthony A. James, Department of Molecular Biology and Biochemistry, University of California, Irvine, CA, USA; Department of Microbiology & Molecular Genetics, University of California, Irvine, CA, USA

Xiaofang Jiang, Fralin Life Science Institute, Virginia Tech, Blacksburg, VA, USA; Interdisciplinary PhD Program in Genetics, Bioinformatics, and Computational Biology, Virginia Tech, Blacksburg, VA, USA

Deepak Joshi, Department of Microbiology and Molecular Genetics, Michigan State University, East Lansing, MI, USA

Rhea Longley, Faculty of Tropical Medicine, Mahidol Vivax Research Unit, Mahidol University, Bangkok, Thailand; The Walter and Eliza Hall Institute of Medical Research, Melbourne, VIC, Australia

Vanessa Macias, Department of Molecular Biology and Biochemistry, University of California, Irvine, CA, USA

John M. Marshall, Division of Biostatistics, School of Public Health, University of California, Berkeley, CA, USA

Conor J. McMeniman, Department of Molecular Microbiology and Immunology, Malaria Research Institute, Johns Hopkins Bloomberg School of Public Health, Baltimore, MD, USA

Kevin M. Myles, Fralin Life Science Institute and Department of Entomology, Virginia Tech, Blacksburg, VA, USA

Hector Quemada, Biosafety Resource Network, Institute of International Crop Improvement, Donald Danforth Plant Science Center, St. Louis, MI, USA

Paulo E. Ribolla, Departamento de Parasitologia, Instituto de Biociências, Universidade Estadual Paulista "Júlio de Mesquita Neto", Botucatu, São Paulo, Brazil

Jetsumon Sattabongkot, Faculty of Tropical Medicine, Mahidol Vivax Research Unit, Mahidol University, Bangkok, Thailand

Patricia Y. Scaraffia, Department of Tropical Medicine, Vector-Borne Infectious Disease Research Center, School of Public Health and Tropical Medicine, Tulane University, New Orleans, LA, USA

Maxwell J. Scott, Department of Entomology, North Carolina State University, Raleigh, NC, USA

Donald S. Shepard, Schneider Institutes for Health Policy, Heller School, Brandeis University, Waltham, MA, USA

Sarah M. Short, W. Harry Feinstone Department of Molecular Microbiology and Immunology and the Johns Hopkins Malaria Research Institute, Bloomberg School of Public Health, Johns Hopkins University, Baltimore, MD, USA

Robert E. Sinden, The Jenner Institute, The University of Oxford, Oxford, UK; The Department of Life Sciences, Imperial College London, South Kensington, UK

Patchara Sriwichai, Faculty of Tropical Medicine, Department of Medical Entomology, Mahidol University, Bangkok, Thailand

Yeya T. Touré, Malaria Research and Training Center, University of Sciences, Techniques and Technologies of Bamako, Bamako, Mali

Sekou F. Traoré, Malaria Research and Training Center, University of Sciences, Techniques and Technologies of Bamako, Bamako, Mali

Zhijian J. Tu, Department of Biochemistry, Virginia Tech, Blacksburg, VA, USA; Fralin Life Science Institute, Virginia Tech, Blacksburg, VA, USA; Interdisciplinary PhD Program in Genetics, Bioinformatics, and Computational Biology, Virginia Tech, Blacksburg, VA, USA

Eduardo A. Undurraga, Schneider Institutes for Health Policy, Heller School, Brandeis University, Waltham, MA, USA

Sophia Webster, Department of Entomology, North Carolina State University, Raleigh, NC, USA

Zhiyong Xi, Department of Microbiology and Molecular Genetics, Michigan State University, East Lansing, MI, USA

Gabriel Zilnik, Department of Entomology, North Carolina State University, Raleigh, NC, USA

Biography

CHAPTER 1

Tim Antonelli is an IGERT fellow and a PhD student in Biomathematics at North Carolina State University. He is interested in modeling how mosquito populations react to various control strategies, including the release of genetically modified mosquitoes that are unable to transmit disease. He also works on parameter identifiability and parameter estimation for mosquito models using data collected from Iquitos, Peru.

Amanda Clayton received her BA in Economics from Illinois Wesleyan University. She is currently an IGERT fellow and a PhD student in Economics at North Carolina State University. Her research interests include microeconomic development, global health, feminist economics, and the policy and regulation of genetic engineering technologies.

Molly H. Storment is an IGERT fellow and a PhD student in Communication, Rhetoric, and Digital Media at North Carolina State University. Broadly speaking, she studies how language shapes knowledge production in the sciences. Her dissertation explores the influence of digital information technologies, such as the GenBank database, on invention in genetic engineering.

Sophia Webster is an IGERT fellow and a PhD student in Entomology at North Carolina State University. Under the direction of Dr. Max Scott, she works with the dengue mosquito vector, *Aedes aegypti*, to develop gene drive systems for mosquito population suppression and replacement.

Gabriel Zilnik received his BA in Anthropology from Arizona State University where he became interested in the evolutionary effects of agriculture on arthropods. Gabriel is currently an IGERT fellow and a PhD student in Entomology, studying the genetics of adaptation using juvenile-hormone disrupting insecticides as a system under the tutelage of Fred Gould.

CHAPTER 2

Maxwell J. Scott began his career working on the chromatin structure of steroid hormone inducible genes in chickens. A continued interest in epigenetics but a desire to work on a system with better genetic tools led to studies on X chromosome dosage compensation in *Drosophila melanogaster*. This work led to a realization that the sex determination and dosage compensation genetic regulatory systems could potentially be manipulated to make male-only strains of insect pests. His lab has continued to investigate epigenetic regulatory mechanisms in *Drosophila* and to develop conditional female lethal strains of insect pests. The latter has been mostly on the Australian sheep blowfly and the New World screwworm fly, which are major pests of livestock. He is currently a Professor in the Department of Entomology, North Carolina State University, Raleigh, NC, USA.

Mark Q. Benedict began his career working on *Anopheles albimanus* classical and aberration genetics in a laboratory developing tools for genetic control. His interests have continued in this vein and have ranged from developing genetic sexing strains, studying species complexes, germline transformation methods, predicting species distributions in the context of control, molecular biology and biosafety. In addition, he has published papers on mosquito physiology and larval development density effects. He led the vector activities of the Malaria Research and Reference Reagent Resource Center (MR4) and led the mosquito SIT group at the International Atomic Energy Agency (IAEA) where he developed methods, field sites, and equipment for mass-rearing. He is currently a research biologist at the CDC Foundation in Atlanta, GA, USA.

CHAPTER 3

Mamadou B. Coulibaly is a graduate from the University of Notre Dame (2006). Since then, Dr. Coulibaly has been back in Mali where he leads a research laboratory at the Malaria Research and Training Center at the University of Sciences, Techniques and Technologies of Bamako. His lines of research include developing novel genetic approaches for malaria vector control. He is also involved in operational research oriented toward malaria vector control (ITNs/LLINs, IRS) to help making informed decisions by policymakers.

Sekou F. Traoré is currently the Director of the Malaria Research and Training Center Entomology/Mali ICER (International Center for Excellence in Research). He has more than 20 years of experience as a specialist in Entomology and control of vector-borne diseases in developing countries. His current research involves field and laboratory studies on malaria, leishmaniasis, filariasis, and recurrent tick-borne fevers. Additional research lines involve interdisciplinary studies on vector-borne diseases in both urban and rural environments, development and field-testing improved methods for vector control.

Yeya T. Touré led the Malaria Research and Training Center up to 2001 when he joined the Tropical Disease Research program at the World Health Organization (TDR/WHO). "He played an essential role in stimulating research molecular biology, genetics, and genomics of tropical disease vectors. This led to the completion of the genome sequence of the malaria mosquito, the *Anopheles gambiae*, which opened opportunities to better understand vector biology and insecticide resistance mechanisms and develop new tools" (credit; Jamie Guth, WHO senior communication manager, 2014). Prof. Toure retired from TDR/WHO when he was leading the unit for vectors, environment, and society research.

CHAPTER 4

Patchara Sriwichai received her PhD in Tropical Medicine at the Faculty of Tropical Medicine, Mahidol University in 2007. She is currently a lecturer in the Department of Medical Entomology at this faculty. She has experience in medical entomology of Tropical disease. Her research focus is on malaria—vector relationship, vector biology, disease transmission, vector competency and insect immunity specifically related to vector—parasite interaction, as well as vector surveillance, prediction, and evaluation of vector capacity in the endemic areas. She is also interested in malaria elimination programs and utilization of integration of vector control tools. Some of her previous work has involved protein targets that have potential to be malaria transmission blocking vaccines.

Rhea Longley is a postdoctoral researcher at the Mahidol Vivax Research Unit, Mahidol University, Thailand, and the Walter and Eliza Hall Institute of Medical Research, Australia, where she researchers *Plasmodium vivax* malaria. Her current work focuses on understanding the human immune response to *P. vivax*, and how this can be used to eliminate malaria from the Asia Pacific region. She received her PhD in 2014 from the University of Oxford, and was a 2010 Rhodes scholar.

Jetsumon Sattabongkot (Prachumsri) was a senior scientist for more than 25 years at the USA Medical Component, Armed Forces Research Institute of Medical Sciences (AFRIMS), an oversea laboratory of the Walter Reed Army Institute of Research, Bethesda, MD, USA. She has moved to the Faculty of Tropical Medicine, Mahidol University in 2011 to establish the Mahidol Vivax Research Unit, where she is the director of the unit. This unit is under Center of Excellence for Malaria. Her research focus is on malaria transmission and epidemiology, biology of different stages of malaria parasites in human and mosquito vectors, new tools for diagnosis and surveillance, and evaluation of tools for malaria control and elimination such as, but not limited to, integrated vector control, transmission blocking vaccines, antiliver stage compounds, etc. She has international collaboration worldwide and published more than 152 papers under J. Sattabongkot and J. Prachumsri.

CHAPTER 5

Jan E. Conn is a biologist who conducts research on malaria vector adaptation, genetics, and ecology in the Neotropics. She is a Research Scientist at the Wadsworth Center, New York State Department of Health, and Professor in the Biomedical Sciences Department at the School of Public Health at SUNY-Albany.

Paulo E. Ribolla is a biologist with a specialization in Biochemistry and Molecular Biology who conducts research on different aspects of neglected diseases in Brazil. He is an Associate Professor at the University of São Paulo, lecturing in Human Parasitology. His lab develops projects on dengue, malaria, and leishmaniasis.

CHAPTER 6

Roberto Barrera conducted his first studies on the ecology of *Aedes aegypti* in 1977, when he was pursuing the biology degree at the Central University of Venezuela (UCV). He continued investigating the ecology of mosquitoes inhabiting natural and artificial containers and obtained his PhD from the Ecology Program at Pennsylvania State University in 1988. He did a post-doctoral at the Entomology Department, University of Florida (1994–1995). After retiring as a full Professor from UCV, he accepted a position as Entomology and Ecology Activity Chief, Dengue Branch, CDC in 2003. His main research interests are vector ecology and control and eco-epidemiology of vector-borne pathogens, such as dengue.

CHAPTER 7

Robert E. Sinden has for the past 40 years studied the cell biology of malaria parasites, in particular of the sexual stages and biology of transmission—on which he has published approximately 300 papers. He has contributed to global reviews of malaria research activities (MalERA) and remains a strong advocate of the "rediscovered" philosophy of attacking malarial parasites not only to reduce clinical disease in the infected patient, but more importantly to achieve stable/sustainable reductions in transmission between persons in endemic communities. His research team is currently focusing on translating the knowledge acquired from previous basic research into the discovery, development and implementation of effective transmission-blocking drugs and vaccines. He is currently Head of Malaria Cell Biology at the Jenner Institute, the University of Oxford.

CHAPTERS 8, 13

Zach N. Adelman is an associate professor in the Department of Entomology and Fralin Life Science Institute at Virginia Tech. Following earlier work on the generation of pathogen-resistant mosquitoes and the development of novel mosquito promoters, Dr. Adelman's research has more recently focused on the development of novel gene editing/gene replacement approaches for disease vector mosquitoes as well as understanding genetic interactions between arthropod-borne viruses and their mosquito vectors.

Sanjay Basu is a postdoctoral researcher in the Department of Entomology at Virginia Tech. His doctoral research at Keele University focused on the generation of *Plasmodium*-resistant *Anopheles* mosquitoes. More recently, Dr. Basu has focused on the development of gene editing tools such as recombinase-mediated cassette exchange and CRISPR/Cas9 for both *Aedes aegypti* and *Anopheles stephensi*.

Kevin M. Myles is an associate professor in the Department of Entomology and Fralin Life Science Institute at Virginia Tech. Dr. Myles' recent involvement in the development of gene editing/gene replacement technologies stems from a desire to bring more powerful genetic resources to bear on the traditional focus of the Myles laboratory, understanding the role of small RNA pathways in the transmission of mosquito-borne viruses.

CHAPTER 9

John M. Marshall is an assistant professor in the School of Public Health at the University of California, Berkeley, CA, USA. His research has focused on several aspects of mosquito genetic control, including the design of novel gene drive systems and the development of mathematical models for their impact on malaria epidemiology. He has also conducted surveys of public attitudes to transgenic mosquitoes for malaria control in Mali, West Africa, and has contributed to dialogue on their international regulatory issues.

Omar S. Akbari is a postdoctoral scholar in the Division of Biology and Biological Engineering at the California Institute of Technology. His research has focused on the molecular design of novel gene drive systems in mosquitoes and fruit flies. Of note, his research has produced the first synthetic gene drive system capable of confined population replacement in *Drosophila*, and he has conducted pioneering work on the *Aedes aegypti* transcriptome of relevance to engineering gene drive systems in mosquitoes.

CHAPTER 10

James K. Biedler is a research scientist in the Department of Biochemistry and Fralin Life Science Institute at Virginia Tech who is interested in the development of genetic-based mosquito control methods. Biedler has conducted work in *Anopheles stephensi* for the identification of regulatory sequences that enabled maternal delivery of artificial miRNAs and proteins via the ovary to embryos. Such methods may facilitate the development of gene-drive and other mosquito control strategies.

Xiaofang Jiang is a graduate student in the Interdisciplinary Program of Genetics, Bioinformatics, and Computational Biology at Virginia Tech.

Her research focuses on sex-biased genes and dosage compensation in mosquitoes.

Brantley A. Hall is a graduate student in the Interdisciplinary Program of Genetics, Bioinformatics, and Computational Biology at Virginia Tech. His research focuses on finding genes in the Y chromosome and the male-determining locus in mosquitoes.

Zhijian J. Tu is a Professor in the Department of Biochemistry and Fralin Life Science Institute at Virginia Tech. He has a sustained interest in selfish genetic elements in mosquitoes. More recently, his work focuses on using systems biology or functional genomics approaches to study sex-determination and embryonic development in mosquitoes. On the basis of such fundamental information, his group is developing novel genetic applications to control mosquito-borne infectious diseases.

CHAPTER 11

Conor J. McMeniman is a Research Associate at the Laboratory of Neurogenetics and Behavior, The Rockefeller University in New York, NY, USA. Dr. McMeniman received his BSc with first-class honors in Parasitology in 2003, and a PhD in Biological Science in 2009 from The University of Queensland, Australia. In 2009, he moved to Rockefeller supported by a Marie Josée and Henry Kravis Postdoctoral Fellowship from The Rockefeller University, and a Human Frontier Science Program (HFSP) Long-term Postdoctoral Fellowship. His current research focuses on the development and use of genome-engineering methods to study mosquito sensory biology.

CHAPTER 12

Patricia Y. Scaraffia is an assistant professor in the Department of Tropical Medicine at Tulane University. She is also a member of Vector-Borne Infectious Disease Research Center in the School of Public Health and Tropical Medicine at Tulane University. Dr. Scaraffia received her PhD in Cordoba, Argentina. She conducted her postdoctoral training in Dr. Michael Wells' laboratory at the University of Arizona. She is particularly interested in unraveling the physiological, biochemical, and molecular bases underlying the regulation of nitrogen and carbon metabolism in mosquitoes, as well as in discovering new metabolic targets that can be used for the design of better mosquito-control strategies.

CHAPTER 14

Zhiyong Xi is an associate professor in the Department of Microbiology and Molecular Genetics in Michigan State University, and the director of Sun Yat-sen University-Michigan State University Joint Center of Vector Control for Tropical Diseases. Dr. Xi's research is focusing on *Wolbachia*—mosquito interactions and its impact on mosquito vector competence for dengue virus and malaria parasite. As a leader in developing *Wolbachia* symbiosis with mosquito vectors, he is also working on a field trial to test the *Wolbachia*-based strategy for dengue control.

Deepak Joshi is a postdoctoral research associate in the Department of Microbiology and Molecular Genetics in Michigan State University. Dr. Joshi is mainly working on tests of *Wolbachia*-based population replacement and population suppression in laboratory populations of *Anopheles* malaria vectors. His interests also include understanding of *Wolbachia*-associated fitness and the impact of *Wolbachia* on the malaria parasite in *Anopheles* mosquitoes.

CHAPTER 15

George Dimopoulos is a professor in the Department of Molecular Microbiology and Immunology, Bloomberg School of Public Health, Johns Hopkins University. Dr. Dimopoulos has over 20 years experience with medical/molecular entomology of the vectors *Anopheles gambiae* and *Aedes aegypti*. His research has mainly focused on the mosquito's innate immune system and the mosquito midgut microbiota, and how they interact with the human pathogens *Plasmodium falciparum* and the dengue viruses, using genomics, functional genomics, and molecular biology techniques and approaches.

Sarah M. Short is a postdoctoral fellow in the Dimopoulos Group who studies interactions between mosquitoes and their gut microbes. She is interested in understanding determinants of variability in the mosquito gut microbiota and the mechanisms underlying interactions between mosquitoes and specific gut microbes that affect vector competence. More generally, she is interested in the complex nature of insect immunology and the genetic, ecological and evolutionary factors that shape it.

Benjamin J. Blumberg is currently a PhD candidate in the Dimopoulos group who studies tripartite interactions of the mosquito immune system, *Plasmodium* parasites, and the mosquito microbiome. Ben has studied the bacteria- and IMD pathway-independent anti-*Plasmodium* defenses, and the role of filamentous fungi in modulating *Plasmodium* susceptibility in *Anopheles* mosquitoes.

CHAPTER 16

Hector Quemada is the Director of the Biosafety Resource Network at the Donald Danforth Plant Science Center. This project provides regulatory and product development expertise for publicly funded crop development projects. He was the manager of the Biotechnology and Biodiversity Interface grant component of the Program for Biosafety Systems, a project helping to build regulatory capacity in developing countries. He was the founder of Crop Technology Consulting, Inc., a consulting firm conducting technical and biosafety assessment for biotechnology programs in developing countries. He has experience in developing transgenic crop varieties for the private sector.

CHAPTER 17

Donald S. Shepard is a professor at the Schneider Institute for Health Policy at the Heller School, Brandeis University. Director of the Institute's Cost and Value Group, Dr. Shepard is an internationally recognized expert in the field of health economics. His major concentrations are cost and cost-effectiveness analysis in health and health financing, and his research is concerned with health problems of both the United States and developing countries. Dr. Shepard has over two decades of experience in estimating the costs of dengue surveillance and vector control strategies, the economic and disease burden of dengue fever, and cost-effectiveness of control strategies. His research includes Brazil, Cambodia, El Salvador, Guatemala, India, Malaysia, Mexico, Panama, Philippines, Puerto Rico, Thailand, Venezuela, and other countries.

Yara A. Halasa has been working in health policy and economic analyses of health and health-related projects both in the United States and internationally for the past decade. She has a medical background, an MS in health economics, and an MA in health policy. She is a Research Associate at the Schneider Institutes for Health Policy at the Heller School, Brandeis University. She has published in various scholarly journals, including two recent articles about the economic burden of dengue in Puerto Rico, which included a detailed analysis of vector control strategies and costs, and studies on willingness to pay for vector control efforts.

Eduardo A. Undurraga is a Senior Research Associate at the Schneider Institutes for Health Policy at the Heller School, Brandeis University. His current research centers on health economics and impact evaluation, including the economic and disease burden of dengue fever, and the effects of social and economic factors on health among indigenous people of Latin America. Originally trained as an engineer at Universidad Católica de Chile, and later, as a political scientist at Universidad Alberto Hurtado, Dr. Undurraga conducts research on health economics, health metrics, social epidemiology, and inequality. He has been working on dengue fever-related projects in the Americas and Southeast Asia for the past 5 years.

CHAPTER 18

Danilo O. Carvalho is a biologist, specializing in mosquito biology, identification and ecology. He has also experience in molecular biology and genetic manipulation. Since 2010, he has been involved in PAT—the program releasing transgenic male mosquitoes in the field for population suppression as a project manager.

Margareth L. Capurro is a professor at the University of São Paulo. She has experience in the biochemistry and molecular biology of mosquitoes (*Aedes aegypti, Anopheles aquasalis*), gene expression, effector molecules, transgenic insects, dengue and malaria. Currently she serves as coordinator of PAT—Aedes Transgenic Project in Bahia state.

CHAPTER 19

Vanessa Macias is an advanced PhD student in the Department of Molecular Biology and Biochemistry at the University of California, Irvine. She obtained both her Bachelor and Master of Science degrees at the New Mexico State University. She is researching basic problems in mosquito vectors of disease and is interested in small RNA biology in this medically important group of insects.

Anthony A. James is Distinguished Professor of Microbiology & Molecular Genetics (School of Medicine) and Molecular Biology & Biochemistry (School of Biological Sciences) at the University of California, Irvine (UCI). He is a member of the National Academy of Sciences (USA). His research emphasizes the use of genetic and molecular-genetic tools to develop synthetic approaches to interrupting pathogen transmission by mosquitoes.

Help us make our content even better

We are very interested in hearing your feedback about the quality and content of our books. A short survey is all that is required to ensure we continue to deliver the best content in scientific publishing.

Elsevier will make a monthly donation to a selected charity on behalf of readers who have completed our survey. A list of eligible charities from which you may choose will be made available during the survey process and each month the charity obtaining the highest number of votes will receive the donation from Elsevier.

Please follow the link to complete the short online survey and help us to help others.

http://www.elsevier.com/booksfeedback

Chapter 1

Transgenic Pests and Human Health: A Short Overview of Social, Cultural, and Scientific Considerations*

Tim Antonelli[1], Amanda Clayton[2], Molly Hartzog[3], Sophia Webster[4] and Gabriel Zilnik[4]

[1]*Department of Mathematics, Worcester State University, Worcester, MA, USA,*
[2]*Department of Economics, North Carolina State University, Raleigh, NC, USA,* [3]*Department of Communication, Rhetoric, and Digital Media, North Carolina State University, Raleigh, NC, USA,* [4]*Department of Entomology, North Carolina State University, Raleigh, NC, USA*

INTRODUCTION

The global problems of dengue fever and malaria are multifaceted, complex issues that span many disciplines, including human health, ecology, economics and urban development, health and environmental policy, social work, and risk analysis. Effective disease control and prevention therefore requires integrated research from all of these disciplines in order to understand the problem from as many angles as possible and within its social and cultural contexts. This type of interdisciplinary approach that integrates perspectives from the natural sciences, social sciences, and humanities was recently endorsed by the American Academy of Arts and Sciences as critical for developing effective solutions for the world's problems [1]. Adopting this approach, we introduce the ethical, regulatory, social, and economic aspects of control programs for dengue fever and malaria, relating to both currently used control techniques as well as the emerging technologies involving genetically modified organisms (GMOs).

The goal of this chapter is not to offer a definitive stance on whether or not genetically modified (GM) technologies should be used to control mosquito-borne diseases, but rather to offer a cursory look at the complex issues that span multiple disciplines, governmental and nongovernmental

---

* All authors contributed equally to this work.

Genetic Control of Malaria and Dengue.
© 2016 Elsevier Inc. All rights reserved.

organizations, and community interests. In conclusion, we argue that a discussion of whether or not to implement GM technologies should be conducted within the larger discussion of national, regional, and global disease control strategies. These control plans should consider an integration of multiple control strategies and adapt to suit differing social and cultural contexts based on the area under consideration.

CURRENT STATE OF GMOS

In 1996, agriculture experienced a genetic revolution. Before the planting season, the United States Environmental Protection Agency (EPA) had approved the commercial sale of what would become the most widespread transgenic cultivars. Recombinant DNA technology has revolutionized biological sciences with practical impacts in fields ranging from medicine to agriculture [2]. New crops and modified organisms would soon come to be known as GMOs. Crops carried genes from bacteria conferring resistance to Roundup™ (glyphosate) weed killer and to certain insect species. Bacteria were engineered to produce human insulin. With regard to pest management, the impact of transgenics remains acutely felt in agriculture. Entire agricultural systems were constructed around new transgenic cultivars; new industries were born, while old ones failed. Land-grant institutions around the country helped research the impacts of these new varieties. Fields from the applied life sciences produced thousands of articles in biochemistry, molecular biology, conservation biology, ecology, evolution, plant science, weed science, environmental resource management, and many more regarding the efficacy and safety of transgenic cultivars [3]. Yet, this technology has its detractors. Many groups such as Union of Concerned Scientists and Gene Watch point to issues with regulatory systems in assessing safety or environmental concerns related to transgenic organisms.

Controlling pests with transgenic technology is predominantly accomplished with δ-endotoxins (Cry toxins) from strains of *Bacillus thuringiensis*. Commonly known as Bt crops, the plants have a host of attractive features. Most notable is the narrow spectrum of pests that each Cry toxin affects. At the time of this writing, varieties of Bt crops primarily target lepidopteran, dipteran, and coleopteran pests. Furthermore, a single gene encodes each Cry toxin making the combination of toxins, known as stacking, relatively straightforward [4]. Growers have found these crops extremely useful; in 2014, transgenic Bt crops constituted 84% of cotton and 80% of corn grown in the United States [5]. Developing countries such as India, China, South Africa, Brazil, and Argentina have seen explosive growth in transgenic crop adoption. Those five countries accounted for nearly 50% of transgenic crops (including herbicide tolerant cultivars) grown worldwide in 2011 [6]. However, these crops are not without their drawbacks. While the primary pests of these crops have been controlled, a surge of secondary piercing—sucking pests such as stink bugs and aphids has become a problem in some regions of the world [7].

Consequently, the increase in insecticide use to control secondary pests may offset the decreased insecticide applications for the primary pests now controlled by Bt. Thus, detailed knowledge of the pest assemblage is useful when approaching transgenic control through direct modification. Similarly, in thinking about GM mosquitoes to combat dengue and malaria, detailed knowledge of the transmission cycle and host assemblage is required to know how the system might respond to genetic control of a single species.

While transgenic crops are widespread in much of North America and Asia, this is not necessarily the case around much of the globe. For example, many nations in Europe restrict transgenic cultivars and in some cases have even seen a decline in field trials of these cultivars [6]. Concern over the safety of these crops remains intense, but as it stands now, no credible scientific evidence has been presented demonstrating adverse effects associated with consumption of transgenic crops [8]. However, the moral and ethical arguments against transgenic crops seem to have the most traction and these arguments are more difficult to resolve with scientific data alone. Below we discuss some of the ethical implications surrounding transgenic insects, which are derived from literature surrounding transgenic cultivars.

DENGUE FEVER AND MALARIA

The WHO provides fact sheets (available online) on both dengue and malaria that are straightforward and highly informative. Here, we provide a summary and comparison of the two diseases focusing on their global prevalence, symptom severity, and vector characteristics. We also discuss briefly the availability and efficacy of existing treatment and prevention methods for both diseases. Table 1.1 displays a summary of the key facts for both diseases.

Dengue Fever

Dengue is caused by at least four independent viruses that are all transmitted primarily by the mosquito *Aedes aegypti*. The most typical form of the disease is commonly called dengue fever and its symptoms include fever, rash, headache, and joint and retro-orbital pain. The severe form of the disease, called severe dengue or dengue hemorrhagic fever (DHF), can result in vomiting, internal hemorrhaging, and even death [13].

The WHO estimates that there are 50−100 million dengue infections each year, mostly in tropical regions, though a more recent estimate is nearer to 400 million due to the large number of asymptomatic and unreported cases [12]. Despite its high incidence, dengue fever is one of seventeen diseases classified as a neglected tropical disease (NTD) [14]. In terms of human health impact, NTDs are often compared to "the big three": malaria, HIV/AIDS, and tuberculosis, which receive significantly more attention in

TABLE 1.1 Key Facts About Dengue and Malaria [9−12]

	Dengue	Malaria
Vector	*Aedes aegypti, Aedes albopictus* (secondary)	About 20 species from the *Anopheles* genus
Strains	Four virus serotypes from the *Flavivirus* genus	Four parasite species from the *Plasmodium* genus
Severity	Contracting a second serotype results in a higher likelihood of experiencing severe dengue	Prevalence and severity varies with parasite (*Plasmodium falciparum* is the most common and deadly)
Immunity	Contracting one serotype provides permanent immunity to that strain and temporary immunity to the others	Partial immunity is accumulated over time and provides protection against severe disease
Diagnosis	ELISA tests for antigens (IgM & IgG), PCR	Rapid diagnostic tests for antigens, microscopy, PCR
Symptoms	*Classic*: fever, rash, headache, muscle aches, retro-orbital pain, vomiting	*Classic*: fever, headache, chills, vomiting
	Severe: internal hemorrhaging, severe abdominal pain and vomiting, respiratory distress	*Severe*: anemia, respiratory distress, cerebral malaria, organ failure
Mortality	Without treatment: about 20% mortality rate	About 627,000 deaths in 2012
	With treatment: less than 1% mortality rate	About 90% of deaths were from Africa, mostly among children
Global burden	WHO [9]: 50−100 million cases per year	WHO [10]: about 207 (473−789) million cases in 2012
	Bhatt et al. [12]: about 390 million cases per year, including asymptomatic	
Risk groups	Children, elderly, imunocompromised	Children, elderly, imunocompromised, tourists and immigrants
Vaccines	In development but not yet available	In development but not yet available
Treatment	*Classic*: fluids, pain medication, rest	Antimalarial medications (parasite resistance is a continuing issue)
	Severe: fluid replacement therapy, blood transfusion	
Common vector control	Container control, IRS of insecticides, larvicide packets in collected water	LLINs, indoor residual spraying of insecticides

funding, research, and social welfare projects than the 17 NTDs. The disproportionate attention is partly a result of "the big three" being outlined specifically in the Millenium Development Goals, where NTDs are included only in the "other diseases" category. However, while NTDs typically carry a low mortality rate, they are both promoted by and promote poverty, are highly debilitative, and disproportionately impact women and children. Perhaps most importantly, NTDs can increase the severity and prevalence of "the big three" through coinfection and coendemicity [15,16]. As a result, Hotez et al. encourage the development of a global plan for "the big three" that includes control of 17 NTDs as a powerful tool in the process [17].

Malaria

Malaria is a parasitic disease spread by about 20 different species of *Anopheles* mosquitoes in tropical and subtropical climates throughout the world. According to the WHO, malaria is more prevalent in areas with species of *Anopheles* that have longer lifespans or that have breeding habits leading to increased mosquito populations [10]. Because malaria is spread by so many species of mosquitoes, the direct suppression of the mosquito populations responsible for transmission is complicated. However, because all *Anopheles* species bite at night, one of the simplest and most effective forms of malaria prevention is to use long-lasting insecticide-treated bed nets (LLINs) to keep humans from being bitten while sleeping.

Symptoms of malaria include fever, headache, chills, and vomiting. Severe complications can involve anemia, respiratory distress, and cerebral malaria in children and other forms of organ failure in adults. However, as with dengue, severe and fatal complications from malaria are generally avoidable via effective vector control practices and fast access to medical treatment [11]. Antimalarial medications are able to both treat and prevent malaria but parasitic resistance to medications is an ongoing issue [10].

Recent WHO estimates indicate that there were about 207 (with an uncertainty range of 473−789) million cases of malaria in 2012, about 627,000 of which resulted in death [10]. More than 90% of these deaths occurred in Africa, mostly among children where, according to the WHO, "a child dies every minute from malaria" [10]. Although the death toll of malaria is still high globally and especially among children in Africa, the occurrence of malaria cases and deaths has decreased by around 50% since 2000 [10].

Dengue and Malaria Control

Currently, the most commonly used techniques for dengue control include chemical control in the form of either larvicide in water sources or adulticide applied via indoor residual spraying (IRS) or aspiration packs; and cultural control in the form of recruiting communities to empty containers of standing water that may serve as larval-rearing sites. For malaria, cultural control in the form of using LLINs during sleep is the most common form of vector control. This form of control is highly effective for malaria prevention but does not successfully prevent dengue fever because *Ae. aegypti* bite during the day. The other main form of malaria vector control is the implementation of IRS to reduce adult *Anopheles* populations. Antimalarial medications can also be taken preemptively to prevent malaria transmission and their use is recommended by the WHO for travelers, pregnant women, and children under 5 years old in high transmission areas [10].

Emerging techniques to control both diseases using GMOs can be broadly categorized as either population suppression, which seek to reduce the numbers of mosquitoes, or population replacement, which seeks to replace the disease-carrying population with a transgenic strain incapable of transmitting disease (for a more detailed description of these strategies, see Chapter 1 of WHO/TDR, 2014) [18]. GMO technology is controversial, however, especially in the United States and Europe, where a heated debate continues surrounding GM foods.

THINGS TO CONSIDER BEFORE IMPLEMENTING GMO CONTROL METHODS

Table 1.2 lists key issues to consider before implementing GMO control methods. In the following sections, we focus primarily on the simpler system of dengue transmission and control (see also Chapter 6). While malaria is a more complex system (see Chapters 4 and 5), the points we raise should still apply to using GMO methods for the prevention of malaria. Issues of

TABLE 1.2 Considerations for Potential Use of GM Technologies for Disease Control

- Economic burden of dengue or malaria in the area under consideration
- Burden on quality of life
- Burden on healthcare system in time of epidemic
- Local community's perception of disease risk
- Local community's willingness to participate in cultural control
- Local community's values and belief systems regarding environmental protection and care
- Efficacy and public acceptance of currently used control measures, locally and in neighboring areas
- Financial cost, quality of life cost, and ethical cost of candidate technologies for mosquito control
- Benefits of disease prevention over disease treatment in the area under consideration
- Other culturally specific considerations in the area under consideration

regulation (Chapter 17), public opinion (Chapter 19), and ethics pertaining to the use of GMOs as a control technique are likely similar for both diseases, as these issues relate more to the emerging technologies of genetic modification rather than to the specific diseases to which these technologies are applied.

Allocating Resources Between Treatment and Control

It is important to prioritize treatment strategies to mitigate severe health problems resulting from disease transmission. General improvements to healthcare infrastructures along with other forms of economic development would likely decrease instances of malaria and dengue as well as many other communicable diseases. For any control methods implemented to reduce transmission of dengue or malaria, it is important that local communities are consulted and actively engaged in policy decisions and implementation.

Dengue treatment is usually simple and highly effective at reducing death rates if infected individuals are able to obtain timely access to necessary treatment facilities [9]. However, researchers and policymakers working with the virus note that healthcare infrastructures in low- and middle-income countries are often incapable of handling the influx of cases that occurs during an epidemic [9,19]. The response to this problem seems to have often been to push for increased dengue prevention rather than to try and tackle healthcare infrastructure issues directly [20]. Although the control and prevention of dengue is vital to reducing the negative impacts of the virus in the long run, it is important that researchers and policymakers not overlook the immediate importance of ensuring individual access to dengue treatment.

The WHO handbook on dengue management states that "emergency preparedness and response are often overlooked by program managers and policy-makers," and that "while plans have frequently been prepared in dengue-endemic countries, they are seldom validated" [21, pp. 123–124]. However, this problem could be due to a lack of resources rather than a lack of diligence. Because each area will typically only experience an epidemic every few years, it may be difficult to maintain the resources needed to treat high case loads of dengue. We suggest that mobile dengue response units be formed at the international level with neighboring countries pooling resources to maintain effective response teams. This is an area in which NGOs like the Red Cross, NIH, or WHO could step in to provide the necessary resources and expertise to be able to respond to the needs of a larger area.

The effective treatment of dengue will not eliminate disease incidence or transmission. Prevention of the disease will still only be possible through preemptive vector control practices or the development of an effective vaccine. However, the effective control and prevention of dengue is likely to take an extensive amount of time and resources. In the interim, steps should be taken as quickly as possible to ensure that all areas are capable of treating cases of dengue and severe dengue in order to reduce serious health complications and fatalities to the lowest possible levels. There is no justification for accepting high death rates from dengue while long-term solutions are being developed when short-term treatment solutions are currently available for implementation.

Economic Development

Although the risk of dengue transmission is present and possibly increasing in parts of Europe and the United States due to increased global temperatures and the presence of *Aedes* mosquitoes in these regions, the vast majority of countries at the highest risk levels are low- to middle-income countries. While this is likely to be in part due to the fact that many low- and middle-income countries are located in tropical and subtropical climate regions, there are several infrastructural factors that are likely to contribute to a nation's level of dengue risk as well. Improving upon these infrastructural issues would likely not only reduce incidences of dengue but would also have other health benefits for the individuals in the affected areas.

Because *Ae. aegypti* breed in open water containers, one of the main infrastructural obstacles to preventing the spread of dengue lies in poor water and sewage availability. According to the WHO, "Dengue afflicts all levels of society but the burden may be higher among the poorest who grow up in communities with inadequate water supply and solid waste infrastructure, and where conditions are most favorable for multiplication of the main vector, *Ae. aegypti*" [21]. This is because areas without reliable waste disposal or piped water tend to have issues with water drainage and/or are forced to collect water in open containers for household use. These open pools of water then act as viable oviposition sites for the *Ae. aegypti* mosquito [22]. Improving waste disposal and piped water availability would also lead to many health benefits for affected communities that spread far beyond reduced incidences of dengue fever, such as a reduction in hookworm and gastrointestinal diseases that are also prevalent in areas most affected by dengue [23].

As noted earlier, *Ae. aegypti* bite during the day, which makes bed nets an ineffective control method against the mosquitoes. Infrastructural improvements to household construction, particularly regarding the availability of screened windows, air conditioning, and enclosed walls and roofs, would therefore further reduce dengue risk by preventing *Ae. aegypti* from entering households and biting inhabitants during the day [22,24]. Such household improvements are costly however and would necessitate either higher household incomes or subsidization by outside sources like governmental or nongovernmental organizations.

Making permanent improvements to the healthcare infrastructures in dengue-endemic countries would increase the ability of these countries to handle epidemics and minimize severe disease complications and deaths. These infrastructural improvements would also have health benefits extending outside of dengue outcomes by increasing the ability of healthcare infrastructures to treat a wide range of diseases requiring intravenous fluid replacement therapies or other simple medical interventions [25]. Improving transportation infrastructures by building roads and increasing the availability of affordable public transportation would further increase the accessibility of healthcare facilities, thereby reaching a wider spectrum of individuals in need of treatment.

There is ample research linking health outcomes to educational and other economic outcomes [26−28]. Limited healthcare access due to low-income levels leads to worsened health outcomes which can keep individuals from obtaining a formal education or from working, thus leading to even lower incomes. A poverty trap is thus formed wherein low incomes lead to poor health outcomes which contribute to even lower incomes [27]. However, infrastructural improvements and other development programs that increase the access of low-income individuals to healthcare have the potential to stop or even reverse the cycle of these poverty traps since healthier individuals are more likely to be able to obtain a formal education and/or to participate in the labor market [26,28].

Community Engagement

Community engagement for dengue control emerged in the 1980s as a new attempt at sustainable control for the mosquito vector *Ae. aegypti*. It is envisioned as a bottom-up control strategy, one that is carried out by citizens of the community and guided by local leaders rather than government officials. However, especially in the initial stages, collaboration between the government and local leaders is an integral part of community engagement. This strategy is in contrast to a top-down approach, that is, a program run entirely by the government and health officials without input from or expectations of the local community. In some cases, attempts at community engagement end up looking very similar to the traditional government-run control, especially once funding for a trial program has ceased. If implemented correctly, the sense of leadership and ownership in resources and ideas should make a community more responsive and engaged in addressing the dengue problem even after outside support is withdrawn [29].

Community engagement for control of *Anopheles* mosquitoes, which transmit malaria, uses some similar and some different techniques for the mosquito's different behaviors and feeding habits. Rather than focusing on emptying containers with standing water in and around homes, as is done for *Ae. aegypti*, community engagement focuses on distributing and educating about bed nets. Both mosquitoes may be controlled through insecticidal spraying of homes however, especially when it is carried out properly by the household and by the vector control employees.

In the 1980s, the WHO put funding into community engagement trial programs and initially gave funding to Thailand to conduct trials using the new strategy. However, these initial programs were not very successful because they did not involve true community engagement. The program was government-directed, still maintained as top-down control, and participating citizens were simply told what to do, so when the support was withdrawn the community programs fell apart [29].

The trials in the 1980s taught important lessons: community engagement programs will not be sustainable unless there are continued economic incentives and the programs will not receive funding from the limited government health dollars once they have been successful. However, even with continued programmatic incentives and government funding, there are still other factors that underscore the success of a community-based program. If the economic incentives disappear after the external funding ends, the incentive of improved health and fewer cases of dengue should continue to motivate community participation in control programs but it may not be enough, especially during times when dengue is absent from an area and other health concerns take precedence.

While community-based programs are intended to ultimately give control to members of the community, relying solely on the community presents problems itself. Even when incentives are present, members of the community must be convinced that removing larval habitats is in their best interests and that controlling *Ae. aegypti* is a priority. Some reinforcement and involvement from the government must always be maintained to ensure that the community is educated and continuing to implement the control measures. The shift from governmental control to local control will take time because the community may see the task as one for which the government is responsible [29].

Since the 1980s, community-based programs to control dengue have been implemented in many areas around the world. Some of these programs have been successful and sustained over periods of years after funding has ceased, while others continue to look like the initial trials in Thailand where the programs deteriorate after funding and other incentives are removed.

Today, some of the most successful community-based programs are present in Cuba, where local Cubans appear to have truly embraced the cause and believe in suppressing dengue through educating community members, removing larval habitats, and going door-to-door during epidemic periods to check for symptoms of dengue. In 1981, the first and largest DHF epidemic presented itself in the Americas. In response, the Cuban government trained and mobilized over 15,000 workers to go house to house educating citizens about dengue and mosquito vector control, in addition to extensive pesticide application [30].

Initially, the 1980s success of Cuban programs depended on top-down control with enforcement through anti-mosquito breeding laws. People were educated on how to prevent mosquitoes from breeding in and around their homes and were fined by inspectors who were sent to frequently check individual households and enforce the laws [29]. Today, numerous studies suggest that the top-down control is no longer needed even when outside incentives are not present. For example, a 2007 study conducted in Santiago de Cuba focused on the sustainability of a community-based approach for 2 years after external funding was withdrawn [31]. The sustainability was evaluated through direct observation, questionnaires, group interviews, and routine entomological surveys using the breteau and entomological house indices. Two years after the external support was removed, people living in neighborhoods who had received the intervention continued to correctly apply larvicides and store water properly; as a result, larval indices continued to decline. Comparatively, larval and house indices of people living in the control area, who had not received community engagement support and education, increased [31]. This study provides evidence that the community-based approaches in Cuba are sustainable and effective at reducing *Ae. aegypti*.

A study in Taiwan examined the impacts and stress that volunteers in community health experience and suggested that the volunteers experience a lack of support in the role they are expected to fulfill, as well as a lack of proper education and work overload [33]. These types of concerns and considerations are important for community engagement as the volunteers are expected to dedicate time to extra duties outside of their everyday jobs and families needs. Although the participants in this study are volunteers, in other community-based programs the people do not necessarily volunteer to participate, especially during trial periods where governmental or outside support is maintaining that the community follows through with the tasks requested of them. A careful balance between giving the community too much responsibility versus not giving enough is difficult to achieve and what a "successful program" looks like will differ dramatically between countries and even local areas. New techniques, such as the use of transgenic mosquitoes to control vector populations, may help to reduce the burden of community health volunteers. However, before release, it is important to gather the input from public opinion studies in the communities using community engagement to understand the attitudes, concerns, and ideas that people have surrounding transgenic insect releases [34].

Considering the current programs that use traditional nontransgenic approaches, a good community engagement strategy is one that (i) is sustainable over decades and evolves to meet the rising number of dengue cases each year, (ii) empowers citizen to be involved, but does not place too much responsibility on the community so that engagement disappears after incentives or lawful actions are no longer there, and (iii) is widely accepted and does not involve forcing a community to be a part of a control program that goes against its beliefs and values.

If transgenic mosquitoes are released to suppress dengue, the programs are likely to change in terms of the levels of involvement required from the government and community. First, the removal and monitoring of larval habitats by private homeowners would still be useful in reducing cases, but if released mosquitoes are able to suppress wild populations to nontransmissible levels, likely less monitoring would be required. Second, the amount of insecticidal spraying inside and outside of homes would be reduced. This would reduce the need for homeowners to vacate during spraying as well as reduce the risk of chemical exposure. Third, engagement would be more frontloaded in the sense that government and community collaboration would take place before the transgenic mosquitoes are released. GMO educational events and gathering public opinion would be essential to gain a sense of the public acceptance or denial of the GMO technology. After this, the next step would be to understand the reasons why people accept or deny transgenic mosquitoes as a method to suppress dengue. All of this collaboration would be done before the release of mosquitoes, and hence most of the community engagement would be done before the program actually begins and with less community involvement needed after implementation. This is in contrast to the current programs in which community participation is oftentimes required and even increases overtime as governments or outside sources pull funding (Box 1.1).

Values and Ethics of Control Measures

Here, we offer an "informed layperson's" ethical framework for considering the release of GM mosquitoes. Entire careers and numerous volumes have been dedicated to the study of bioethics. We hope this will serve, at the very least, as a spark for further exploration of relevant ethical and social concepts and issues surrounding the use of transgenic mosquitoes for public health. The principles we believe to be most pertinent to the discussion are outlined with some hypothetical examples drawn from literature and adapted to the modification of pest species. Table 1.3 offers a brief overview of the principles, namely: stewardship, animal welfare, justice as fairness, and precaution.

Stewardship

The stewardship principle states that humans are entrusted to care for and promote the good quality of air, water, soil, ecosystems, biodiversity, and the earth as a whole [35]. Illustrating this role, Resnik states a steward is like a property manager and "should ensure that the property is not damaged and should make improvements on the property" [35]. This includes, first

BOX 1.1 Community Engagement

Community engagement, involvement, and development to reduce mosquito-borne diseases all refer to the concept of giving a community leadership and ownership of ideas and resources after education and collaboration from the government. Community engagement (bottom-up/horizontal control) is intended to be a sustainable method that is less costly and more effective at reducing dengue than programs run solely by the government (top-down control).

Ideally, the initial steps of community engagement involve governmental vector control employees educating and collaborating with local communities and then slowly the community takes over responsibility for some of the tasks the government once performed. More than this, the strategy is ideally sustainable because the communities have a *desire* to reduce the mosquitoes and believe in the methods they have been taught. Sustainability is key to the strategy because the initial money the government or outside source of funding had will eventually be used up and the initial incentives to perform the tasks may no longer be present. Thus, the success of trials for community engagement are difficult to assess unless long-term studies are carried out years after the funding and incentives are removed.

To date, the most successful community engagement programs appear to be in Cuba, where Cubans have embraced the techniques needed to suppress dengue such as removing larval habitats, going door-to-door to check for dengue symptoms, and educating other citizens about suppression and control techniques.

and foremost, the residents of the property. Field trials for transgenic insects intended to control disease should be conducted in an area where the disease is a recognized public health concern. Mechanisms should be put in place to protect these residents and especially those who are affected by the disease during the field trial. This protection would include informing the community about the trials and providing free healthcare for the targeted disease. These mechanisms would help to ensure that the benefits significantly outweigh the risks to the community and to the residents' environment [36]. All life depends on environmental resources for survival; thus, stewardship argues that it is no longer defensible to solely consider the natural resource needs of humans. The principle of stewardship moves the ethical discussion toward a biocentric view that nature has its own moral worth [35,37]. Control measures that may conflict with the notions of stewardship include chemical pesticides, environmental management, and transgenic technology.

Long-term damage to natural resources has occurred from short- and long-term applications of chemical insecticides such as DDT to control disease vectors [35,38]. In the years immediately following World War II, unrestricted pesticide use posed dangers to nature and humans, which drew increasing public and academic attention [38–40]. Given the preponderance

TABLE 1.3 Brief Outline of Ethical Principles

Principle	Short Definition	Potential Questions to Address
Stewardship	The environment must be cared for in such a way as to provide natural resources for future generations	What could the future environmental impact look like? How would this technology change the impact of controlling this pest?
Animal welfare	Animals have rights in so much that "because it is an animal" is not an acceptable justification for actions taken against them	Is genetic modification of this species necessary? Does genetic modification unreasonably or unnecessarily interfere with biological drivers of this species?
Justice as fairness	A fair decision is one in which maximizes liberty for all and the distribution of effects follows that the least advantaged individuals in a society will receive the greatest benefits	Who benefits from this technology and who bears the associated risks? How are benefits and risks divided among those directly involved with the technology? Is the distribution of risks and benefits fair (as defined by Rawls)?
Precaution	With the acknowledgment that zero risk is an impossible standard, reasonable risks to the environment, health, and safety of participants must be considered prior to initiating a control program	What are the plausible environmental, health, and safety risks of this technology? Would further research significantly alter the impact or reduce the probability of adverse effects?

of evidence demonstrating the damage unrestricted pesticide use can cause to the environment, one would be hard-pressed to find environmental scientists that would endorse such use. Still, one can strongly argue for judicious application of pesticides if they are used to promote a universally recognized goal of high priority, such as human health [35].

Animal Welfare

Animal biotechnology has often been considered in the light of the modification of vertebrates. Ethicists have developed a number of ethical theories that allowed for the modification of animals under certain conditions [41,42]. Discourse centered predominantly on domesticated animals, their ecological relationships with the natural environment, and their relationships to humans. Mosquitoes, and invertebrates in general, present a new and challenging test for these ethical theories developed in response to the potential of transgenic farm animals.

How should animals be treated? Do humans have "dominion" over life, the right to do as we please? Modern arguments that humans do have to make ethical decisions in treating animals arise from Peter Singer's 1975 utilitarian treatise *Animal Liberation*. Can a dog, cow, or fish suffer? If yes, then in the interest of maximizing happiness, humans are obligated not to inflict upon them any unnecessary suffering [43]. Animals deserve our respect, but how much respect they deserve is still debated.

Finding strict utility insufficient, other authors have attempted to formulate a more deontological—or constrained—view of how we should treat animals. For instance, some have argued that all sentient animals possess intrinsic value, with sentience being defined as the ability to feel [41,44]. By Intrinsic value we mean an animal has value aside from any use or aesthetic value humans derive from it. We should treat animals as if what we do to them matters to them. Violating an animal's intrinsic value is permissible only if a serious animal or human interest (life or death) is threatened and no alternative measures are available [44].

Yet another concept is *telos*, a creature's "end" or "purpose." Aristotle defined *telos* as the full, flourishing development of existence. In application to mosquitoes, it would constitute the nature of the mosquito or, more abstractly, the "mosquitoness" of the mosquito. Animals have needs and interests, and those needs and interests that matter most are inviolable [41]. For instance, it would be wrong to isolate a social animal from social interactions. Contrast this with the previous statement: we should treat animals well *because* what we do to them matters to them. Can *telos* be changed? Should humans manipulate an animal's *telos*? Let us examine a thought experiment we modified from Michael Hauskeller: [45, p. 59]

> *Scientists genetically design a human. This person lacks the possibility to live a fully human life. Traits have been knocked out so that they cannot use their hands the way humans do, their nose the way humans do, their eyes the way humans do, and so on. Simultaneously, the desire to live a fully 'human' existence is removed so they would not know that anything is missing from their existence.*

Is this morally acceptable? Have the scientists caused this person harm? After all, this person does not know or care that these things have been done to them. Yet, "We could still deplore their state and say that harm has been done to them, because we perceive the gap between what they now are and what they are meant to be," says Hauskeller [45, p. 59]. Alternatively, if we would not do this to a human, is it permissible to do this to an animal?

Critics are particularly suspicious of the extensions beyond utilitarian arguments for animal welfare. For instance, if animals have intrinsic value, then what of bacteria, viruses, and other pathogens? [35] If yes, then is it wrong to eradicate tuberculosis, HIV, or dengue? We find this to be an interesting, but fundamentally frail argument. Examination of potential consequences inhibits any ethical discussions, but because a principle may be uncomfortable or have negative outcomes for humanity is not a reason for rejection. Even if pathogens had intrinsic value, eradication could be justifiable because of the level of morbidity and mortality caused by pathogens. In other words, eradication of pathogens is justified *despite* their intrinsic value. In practical application, conflicts will arise between competing ethical principles. Resolving those conflicts reasonably and fairly is part and parcel in deciding what actions to take.

Justice As Fairness

How does one decide what is just or fair with regard to dengue and malaria control? Aristotle outlined the formal definition of justice: equals should treat each other as equals [46]. We adopted Rawls' concept of *justice as fairness*, which outlines two principles: (i) equal liberties for all and (ii) the difference principle [46]. Under the first, everyone in a society would be entitled to maximum liberty insofar as everyone had equal liberty. The difference principle ensures people have equality of opportunities, but restricts social and economic inequalities to ones that benefit the *least advantaged* members of society. Justice as fairness does not only apply to individuals, but it is equally applicable to organizations. Rawls [46, p. 3] wrote, "A theory however elegant and economical must be rejected or revised if it is untrue; likewise laws and institutions no matter how efficient or well-arranged must be reformed or abolished if they are unjust."

Two areas of justice concern control programs. Procedural justice regards the process of making fair societal decisions, and distributive justice seeks a fair distribution of risks and benefits. How we determine the process of making fair decisions includes several underlying principles such as public participation in decisions that affect them, transparency in public decisions, and that people have equal protection under the law. Legal and political systems are responsible for carrying out procedural justice [46]. Determining how risks and benefits are distributed is a complex issue with social and cultural considerations. One way to make that determination is to use "veil of ignorance" thought experiment [46]. Under the veil, members of a group would negotiate the distribution of risks and benefits in a society not knowing what the negotiators' socioeconomic position in the society would be. Rawls argues that this would provide a powerful incentive to formulate an equitable distribution of risks and benefits. For example, one would most likely not place a majority of the risk from insecticide exposure on individuals without access to healthcare because, after the negotiation, they might occupy that place in society.

This has straightforward applications to the control of mosquitoes and vector-borne disease. How should communities and nations organize their infrastructure to treat those affected by dengue or malaria? Access to health-care options needs to be universally promoted. New control strategies need to balance risks and benefits equitably among society. Those that have the least ability to mitigate suffering from disease should receive the greatest benefit from control. Communities have a right to make informed public decisions on control strategies that directly impact their interests.

Precaution

How should a society weigh the risks and benefits of a control measure? Science can only give estimates of probabilities of what may occur based on the best evidence available at the time. This is because science works not by proving hypotheses but by rejecting alternative hypotheses [47]. It arrives at explanations of natural processes through narrowing down probable outcomes until only one or a handful remain. This means that the scientific method does not have the capability to precisely predict what *will* happen in the future only the *likelihood* of what could happen. Still, when weighing the risks of control strategies, the worst advice is to "be careful" [48]. This also happens to be a mischaracterization of the precautionary principle.

There lies some difficulty in defining the precautionary principle due to the often vague language used in official documents [49−51]. The practical approach to the precautionary principle is embodied in Principle 15 of the *Rio Declaration*: [52]

> *In order to protect the environment, the precautionary approach shall be widely applied by States according to their capabilities. Where there are threats of serious or irreversible damage, lack of full scientific certainty shall not be used as a reason for postponing cost-effective measures to prevent environmental degradation.*

While this may still seem vague, a number of authors have attempted to clarify the principle. Scientific evaluation of the probability of risks should be employed to determine to what extent a reasonable risk of environmental harm exists [49]. Individual nations have the right to determine the level of acceptable risks to take, in so far as taking those risks does not unfairly place burdens on neighboring states or individuals least able to bear such risks.

The precautionary principle has engendered a heated debate in the scientific community. On one end, it is seen as irrational, incoherent, and paralyzing to discovery, research and exploration. Conversely, proponents argue that it is a policy tool for making practical decisions in spite of scientific uncertainty. When appropriately defined and applied with careful weighing of reasonable risks, the precautionary principle can allow regulators and policymakers to take informed, rational approaches to protect the environment and human health without paralyzing scientific discovery [35,51].

The precautionary principle will have the most potential to affect proposed systems with gene drives such as homing endonucleases, killer-rescue, and *Wolbachia* [53−55]. These approaches have the potential to spread to areas or countries with bans or moratoriums on transgenic technology. Therefore, any drive system will need to prepare for this encroachment and be able to reverse or cancel the spread of the system into unwanted areas. Furthermore, where the system is designed for eradication of the species there are the possibility of ecological knock-on effects. These effects may not be predictable or even known until the control strategy is underway. While these examples are not reasons in-of-themselves not to pursue a strategy, it would seem reasonable to prepare a method to recall the transgene. This agrees with Macer [34] who states that the precautionary principle makes us "reasonably cautious" because it aligns with the principle of "do no harm," although no human action is completely without risk. Thus, there should be continual reevaluation of the risks involved, and a plan to abort the program should be in place [34]. For GM mosquitoes, this could be accomplished with additional drive systems or a rescue construct [55,56].

An Ethical, Cultural, and Social Framework

The development of any framework is difficult. Stepping back to examine why something *is* done or how it *ought* to be done has eye-opening ramifications. If rushed, the creators of such a framework will have difficulty defending it to the public. This process will certainly yield diverse results depending on the situation. We suggest the principles outlined above as a framework with which to approach questions regarding transgenic pests and human health, fully aware that they may often create situations of conflict. Part of creating a framework is resolving conflict between principles or potentially making a value judgment as to which principle supersedes another. But a given framework should not be viewed as relativistic and thus dismissed as unprincipled; rather, it should produce similar results in similar situations. Thus, in deciding how to resolve conflicts between principles one must outline why one principle must take precedence over another. Ethicists and philosophers could be brought into the conversation as there are many tools within their disciplines in which to deal with conflicting principles in a logically defensible manner. We accept that no two situations or communities are exactly the same, but we argue that making a good-faith effort to utilize a framework will result in a publicly defensible approach to disease mitigation and mosquito control.

Regulation, Deliberation, and Public Communication of Biotechnology

Current US Regulation of Biotechnology

Current regulation of GMOs within the United States is delegated under the Coordinated Framework for the Regulation of Biotechnology (CFRB). This framework employed existing legislation[1] and regulatory organizations[2] to regulate biotechnology. In one sense, this is a result of a definitional question; the regulation of GE plants, microorganisms, food, and animals are divided among the EPA, FDA, and USDA based on whether they are defined as "plant pests," "pesticides," "toxic chemicals," or "investigational new animal drugs" [57] (Box 1.2).

Biotechnology and the Public Sphere

Regulation for biotechnologies like transgenic mosquitoes require a sensitivity to the academic and scientific traditions, or how knowledge is constructed and verified, of the nation at hand. Jasanoff calls these traditions "civic epistemologies" [61]. As Jasanoff defines it, a civic epistemology includes the citizens' and government's ideology regarding the roles that science and technology play in the nation. These frameworks are usually implicit, operating in the background of political and technical discussions. Discussions regarding the common values of a nation and how science and technology further or hinder these values can help bring this epistemological framework to the foreground, helping to bring the governing body to a stronger consensus on how to regulate these technologies.

This philosophy of debate perhaps best aligns with the concept of the public sphere, originally developed by Jürgen Habermas [62]. The Habermasian public sphere is intended to be an inclusive social space where issues are introduced, developed, and debated among a group of equal-standing citizens who are brought together by a common interest. However, it should not be taken for granted that all voices are given an equal and appropriate platform. Governing bodies must give due consideration to the voices that arise not only in governmental institutions but also within common arenas such as public school associations, religious institutions, and other community organizations and NGOs [63]. These organizations are distinct from interest groups and lobbyists [63]. Explicit attention to the civic epistemologies of a nation, especially as it is defined in noninstitutionalized community organizations, can help to bring

---

1. These include the Toxic Substances Control Act (TSCA), Federal Insecticide, Fungicide, and Rodenticide Act (FIFRA), Federal Food Drug and Cosmetic Act (FFDCA), and the Federal Plant Pest Act (FPPA) [57].
2. These include the Food and Drug Administration (FDA), the Environmental Protection Agency (EPA), and the US Department of Agriculture (USDA) [57].

BOX 1.2 Regulatory Controversy and Oxitec

British biotech company, Oxitec, made the first release of GM mosquitoes for dengue control in 2009 on the Caribbean Island of Grand Cayman, with another trial in the summer of 2010 [59]. The following year, Oxitec completed another release in Malaysia, supported by the Malaysian government [59]. While the 80% drop in the *Aedes aegypti* mosquito population in Grand Cayman was considered a major success by Oxitec, these releases have not come without controversy, with *Science* claiming "strained ties" between Oxitec and the Bill and Melinda Gates Foundation [58]. While anti-GMO activists have warned against potential risks of releasing GE mosquitoes, *Nature Biotechnology* cited disagreements having to do with regulatory processes; that is, some disagree with the speed at which Oxitec seemed to conduct releases, and the way in which these releases were communicated with the local communities [60]. Oxitec founder Luke Alphey stated that flyers were distributed in Grand Cayman and government officials went door-to-door answering questions, however many remain critical of the unorthodox press release via a YouTube video at the conclusion of the trials [60]. In addition to Oxitec's arguably sparse public communication efforts, other scientists were critical of Oxitec's procedure of conducting field trials and making information public before going through the peer-review system [60]. Later, the Oxitec mosquitoes were released in Brazil to seemingly little public controversy, while the discussion of possible releases in the Florida Keys in the United States sparked a public petition against the releases [80].

Debates regarding Oxitec's procedure of field releases and public communication strategy has directly spilled over into discussions regarding the regulation of GM pests. Most notably, Guy Reeves et al. offered a critique of the current regulatory system in the United States, Cayman Islands, and Malaysia [81]. They strongly criticized the use of categorical exclusions (CEs) based on the 2008 Environmental Impact Statement (EIS) on GE insects. A CE is a request for exemption from drafting a full environmental assessment (EA), with the argument that a new EA would be redundant and unnecessary. According to Reeves et al., CEs for GM insects are largely granted based on the 2008 EIS [81]. The 2008 EIS covered four species of GE insects: pink bollworm moth (*Pectinophora gossypiella*), Mediterranean fruit fly (*Ceratitis capitata*), Mexican fruit fly (*Anastrepha ludens*), and oriental fruit fly (*Bactrocera dorsalis*). They argue that the 2008-EIS is "scientifically deficient on the basis that (i) most consideration of environmental risk is too generic to be scientifically meaningful; (ii) it relies on unpublished data to establish central scientific points; and (iii) of the approximately 170 scientific publications cited, the endorsement of the majority of novel transgenic approaches is based on just two laboratory studies in only one of the four species covered by the document" [59]. Furthermore, Reeves et al. argue that the continual reliance on the 2008 EIS, especially given what they see as scientific deficiencies, suggests that US regulators fail to acknowledge critical technological differences among different GE insects. In conclusion, they argue for a public engagement approach that includes public access to pre-release materials as well as a "high-quality multidisciplinary approach" in order for these new technologies to succeed [81].

(Continued)

BOX 1.2 (Continued)

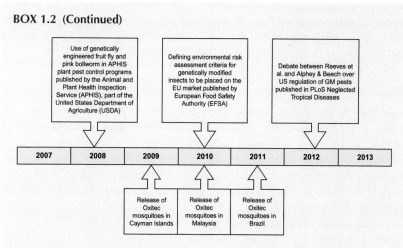

FIGURE 1.1   Timeline of Oxitec mosquito releases and relevant regulation.

In a response, Alphey and Beech pointed to Reeves et al.'s argument for transparency in the pre-release process, stating the "argument has some merit, but needs to be balanced against significant practical difficulties. Technology developers have legitimate rights to protect proprietary information; governments understand this and provide statutory protections" [82]. In addition, Alphey and Beech question Reeves et al.'s basis that regulatory decisions should be based primarily on peer-reviewed scientific works, asserting that this argument "depends on three assumptions: that journal peer review is a superior guarantee of quality than any other method, that no data from any other source can be of adequate quality to warrant consideration, and that regulators themselves are incapable of adequately assessing the quality and significance of data provided to them. Each of these assumptions is naïve at best" [82]. Alphey and Beech believe that regulatory bodies should have access to a wider range of information aside from peer-reviewed studies and that peer review should not be the benchmark for quality regulatory data [82].

As it currently sits, the controversy surrounding the Oxitec field releases seems to primarily circulate around social issues, both the social practices of the scientific community and what is seen as appropriate social interactions among scientists, regulators, and lay communities. As Alphey and Beech point out, Reeves et al. "confuse the concepts of transparency, independence, and scientific quality" [82]. On the other hand, Alphey and Beech seem to reify these concepts themselves, not acknowledging that there may be a diversity of values surrounding each of these concepts, depending on the given situation (Figure 1.1).

attention to marginalized epistemologies, or counterpublics [64], in order to enable a more inclusive form of participatory democracy, leading to a more desirable oversight framework that incorporates differing values, ethics, morals, and concerns of the affected citizens [57].

Models of Public Communication

Sensitivity to the way in which a new technology is presented and discussed within a given community or communities is not merely to quibble over "rhetoric." This idea assumes a definition of "rhetoric" as "mere words" or "mere style" that is added as an extra flourish and has no effect on the core message. However, the academic tradition of rhetoric, stretching back to the work of Aristotle in Ancient Greece, studies the structure and effects of argumentation and persuasion through the three appeals (*ethos*, or credibility and character; *pathos*, or emotion; and *logos*, or logic). While many may initially believe that only *logos* applies to science, many rhetoricians have illustrated how *ethos*, or credibility and character, has been a major driver in scientific communication. *Ethos* plays an especially strong role when scientists are called upon as policy advisors; several historical examples have been documented where scientists were both successful and unsuccessful in developing what was seen as an acceptable *ethos*, including J. Robert Oppenheimer in the case of nuclear warfare, Rachel Carson in the case of pesticides, and the Nongovernmental International Panel on Climate Change [65].[3] When managed carefully and appropriately, a speaker's *ethos* can help to garner trust among public communities [66]. In addition, rhetoricians have shown *pathos*, or emotion, is embedded in science communication, perhaps unconsciously, through arrangement, style, and diction in public communication about biotechnology [67]. This is not to say that *pathos* appeals should be flagged and avoided at all costs, as this would be an impossible task. Rather, a more productive use of rhetoric would be to give due attention to the values and emotions that are implicit in our communication and foreground these in public communication and deliberation. Foregrounding these values and emotions would open up a site of public deliberation that fairly considers the desires of the relevant stakeholders.

However, the model of public communication that has been used for some time by the scientific community has not accounted for this kind of open deliberation. In the past several decades, a deficit model of communication has been used widely by governmental agencies and biotech industries. This model is based on the Shannon–Weaver [68] model of communication that consists of a

3. Walsh [65] dubbed the particular ethos adopted by these, and other, scientists the "prophetic ethos." She argues that the persuasive success of these scientists hinged, in part, on their ability to carefully and appropriately navigate the is/ought divide of science and science policy. That is, demonstrate knowledge of what "is" in the natural world, and firmly connect this with what "ought" to be done in the political and social world.

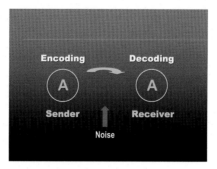

FIGURE 1.2   The deficit model of communication, consisting of a sender, receiver, noise, and feedback loop. *Reproduced with permission from Katz [83].*

sender, receiver, noise, and feedback loop (Figure 1.2). This model comes with a number of problematic assumptions regarding the role of the audience and the role of the communicator. First of all, the traditional model of public communication of science assumes a one-way, linear mode of discourse flowing from the expert speaker to a lay audience, with a singular and transparent means of communication [66]. In this model, the audience is usually perceived as ignorant at best, hostile at worst, and in all cases in need of "simple, clear" information in order to understand reason [69]. This model problematically assumes that pure information can be communicated and received without any alteration to the message; however, language studies scholars have argued that language choice inherently and unavoidable directs attention to certain aspects of a situation, while deflecting others [70]. Appeals to transparency are often made in the deficit model of communication, but the appeal to "transparency" hides the role of language in shaping how we understand science, nature, illness, the human condition, and so on, by ignoring how language unavoidably shapes meaning, ultimately governing what is admitted for discussion into the public sphere (Hartzog and Katz, unpublished manuscript). However, it is not only within the public sphere that style inherently directs what is discussed and how it is discussed. Even in the laboratory, the rhetorical figures of analogy, metaphor, and metonymy[4] have been shown, through ethnographic research, to direct experimental research within the physics laboratory [71].

Given that messages can never truly be sent through a "clear" channel, a model of communication that values multiple interpretations of information will create a more trustworthy, respectful, and productive discursive

---

4. Analogy is a model of argument where one thing is compared to another thing that are essentially dissimilar. For example, discussing a budget in terms of going on a diet. Metaphor is a rhetorical figure that offers a way of "talking about one thing in terms of another." For example, talking about the human brain in terms of a computer. Metonymy is another rhetorical figure that can be defined as "a form of substitution in which something that is associated with x is substituted for x." For example, describing a group of on-duty Secret Service agents as "suits" [70a].

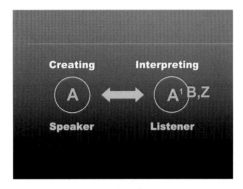

FIGURE 1.3   The rhetorical model of communication, focusing on influence rather than information. *Reproduced with permission from Katz [83].*

landscape for understanding and talking about this technology in regulatory debates. Several alternative models have been proposed by humanists and social scientists, including a rhetorical model of communication [66] (Figure 1.3) and a "genetic" model of communication that understands specialist and public discourses as a "double helix," with one influencing the other [72]. Both the rhetorical and helical model of public communication emphasize a communication that is focused on *influence* rather than *information.* In other words, instead of the one-way information transfer model, these models emphasize recursivity, where scientists and stakeholders use a common language to engage in a back-and-forth dialog that influences one another's positions. In addition, the rhetorical model in particular emphasizes the complexities surrounding the speakers, audiences, languages, histories, and intentions around a given situation. Therefore, a close critique of various texts and spoken exchanges is valued in order to understand what enabled productive/unproductive communication or miscommunication.

Public Opinion of Transgenics

Little research has been done on the public opinion of transgenic mosquitoes for dengue control. However, past research shows that public opinion of genetic modification of living organisms varies [73]. One public opinion survey found that the opinion of a representative sample of the US population varies slightly based on how the technology is presented, whether the technology is described as a "transgenic mosquito," a "genetically modified mosquito," a "genetically engineered mosquito," or "sterile mosquito" [74]. Support for the release of "genetically modified," "genetically engineered" or "transgenic" mosquitoes was generally lower than support for the release of "sterile" mosquitoes. However, respondents exposed to all labels generally indicated that they believed the technology to be safer than insecticides [74].

This change in opinion due to change in the label used is not surprising given past research of this nature. For example, research in H1N1/swine flu coverage illustrates that the latter term generally elicits a more emotional response and connection to the human experience with the flu, while the former distanced the flu from the human experience by emphasizing its scientific aspects [75]. In research concerning the global warming/climate change debate, preference in terminology aligned closely to party preference [76,77]. Sensitivity to the effects of language on public opinion should not be used simply in order to mislead the community into complacency. Rather, future research in public opinion of transgenic technologies should take into account how the technology is presented as one factor that affects public response. This information can then be used create a more accurate assessment of a community's level of accepted risk regarding control measures, which can, in turn, inform regulatory decisions within the community.

In addition to studies of public opinion, future research should take into consideration how transgenic technologies are incorporated into the cultural norms of various communities. In the case of transgenic mosquitoes, for example, this would include an understanding of the community control practices and how these are integrated into the citizens' daily lives. Arnold Pacey [78] offers a critique of what happens when technology is transferred from one culture to another (e.g., from the lab to the field), ultimately arguing for an experiential study of technology in society. That is, a focus on the knowledge and practices of the communities which use and/or are affected by the technologies developed by scientists and engineers. He expresses a deep need to discuss and justify technology using the discourse of these publics, rather than keeping the technologies close to the scientific communities until ready for release.

This cultural sensitivity is especially crucial in many cultural control methods. For example, the removal of vector breeding containers that was proposed to be an effective method of reducing vector populations, but this depends in part on active community involvement [29]. While it had been proposed that community intervention was effective, many of the studies suffered from a lack of methodological rigor (see Community Engagement). It remains disputed if community involvement will decrease over time as enthusiasm for the new control methods fades [31,79]. Careful consideration of the sociological composition in a community is needed to avoid the potential pitfalls of community involvement research that could lead to a loss of trust for implementation of new control measures.

CONCLUSIONS

Every community charged with controlling malaria and/or dengue faces unique challenges regarding its available resources, political infrastructure, and cultural values in implementing any control technique or treatment plan. We fully acknowledge that there can be no single most effective way to control vector-borne disease across so many different scenarios and that any such decision rests entirely within the policymakers and public health officials of the community in question. What we have attempted here is to outline the various considerations that we believe should be taken into account when considering how best to control malaria and dengue, regardless of the decision reached. These considerations include the effectiveness of currently used control measures, the potential use of transgenic control methods, risks and benefits of each, and diagnosis and treatment options including access to clean water and health care, public opinion, and bioethics.

While there remains no cure for dengue, the current treatment option of fluid replacement is simple and its availability should be prioritized, especially because it can benefit the treatment of a myriad of health issues beyond dengue. Likewise, accessibility to currently effective malaria treatment should be prioritized. Alongside the availability of treatment, proper emphasis must be given to informing people on how to correctly diagnose each disease, the timing of which is crucial for proper treatment. With fluid replacement for dengue and antimalarial drugs, combined with early diagnosis, mortality rates from both diseases can be drastically reduced. For overcrowding of hospitals during dengue epidemics, we suggest the development of response units in endemic regions that can be mobilized when necessary to avoid the diversion of hospital resources.

Vector control remains a critical component in the integrated strategy to suppress malaria and dengue; however, current methods should be continually reassessed to monitor their effectiveness while other options, including GMO use, should be considered. We have outlined one ethical framework that we believe to be most widely applicable to the multitude of societies performing or considering vector control. By enumerating several relevant ethical principles (stewardship, animal welfare, justice as fairness, and precaution), we hope to shed light on how to proceed with discussions surrounding this controversial new technology. These principles as well as inclusive public discussions are particularly important considering the gaps in current regulations regarding the use of GMOs. Such discussions across both the expert and nonexpert spheres are required in order for new policy to reflect both the most efficient form of vector control as well as the society's values and concerns.

The fight against malaria and dengue is far from over, but we hope to inform and encourage open discussion regarding the various forms of treatment and vector control, so that communities may take steps toward further reducing disease in the most efficient and responsible means possible.

ACKNOWLEDGMENTS

The authors would like to thank William Klobasa for his contributions in early phases of this project. This paper would not have been possible without the support and feedback of the faculty in the Genetic Engineering and Society program at North Carolina State University, especially Fred Gould, Andrew R. Binder, Matthew Booker, Zachary Brown, Yazmin Cardoza, Nora Haenn, Jennifer Kuzma, William J. Kinsella, Alun Lloyd, Nick Haddad, Marce Lorenzon, and Max Scott. Finally, the invaluable administrative assistance from Karina Todd greatly facilitated the completion of this chapter.

REFERENCES

[1] American Academy of Arts & Sciences. The heart of the matter. Cambridge: American Academy of Arts & Sciences; 2013.

[2] National Research Council (US) Committee on Genetically Modified Pest-Protected Plants. Genetically modified pest-protected plants: science and regulation. Washington, DC: National Academies Press; 2000.

[3] Snell C, Bernheim A, Bergé J-B, et al. Assessment of the health impact of GM plant diets in long-term and multigenerational animal feeding trials: a literature review. Food Chem Toxicol 2012;50:1134−48.

[4] Lereclus D, Delecluse A, Lecadet M-M. Diversity of *Bacillus thuringiensis* toxins and genes. In: Entwistle P, Cory JS, Bailey MJ, Higgs S, editors. Bacillus thuringiensis, an environmental biopesticide: theory and practice. Chichester, UK: John Wiley & Sons Ltd.; 1993. p. 37−69.

[5] USDA. Recent trends in GE adoption. United States Department of Agriculture, <http://www.ers.usda.gov/data-products/adoption-of-genetically-engineered-crops-in-the-us/recent-trends-in-ge-adoption>; 2014 [accessed 30.08.14].

[6] Marshall A. Existing agbiotech traits continue global march. Nat Biotechnol 2012;30:207.

[7] Wang S, Just D, Pinstrup-Andersen P. Bt-cotton and secondary pests. Int J Biotechnol 2008;10:113−21.

[8] Nicolia A, Manzo A, Veronesi F, Rosellini D. An overview of the last 10 years of genetically engineered crop safety research [published online February 17], <http://informahealthcare.com/doi/abs/10.3109/07388551.2013.823595>; 2014 [accessed 30.08.14].

[9] WHO. Dengue and severe dengue: fact sheet no. 117. Geneva: World Health Organization; 2014 <http://www.who.int/mediacentre/factsheets/fs117/en/> [accessed 8.07.14].

[10] WHO. Malaria: fact sheet no. 94. Geneva: World Health Organization; 2014 <http://www.who.int/mediacentre/factsheets/fs094/en/> [accessed 8.07.14].

[11] CDC. Malaria. Centers for Disease Control and Prevention, <http://www.cdc.gov/malaria/index.html>; 2014 [accessed 30.08.14].

[12] Bhatt S, Gething PW, Brady OJ, et al. The global distribution and burden of dengue. Nature 2013;496:504−7.

[13] Kyle JL, Harris E. Global spread and persistence of dengue. Annu Rev Microbiol 2008; 62:71−92.

[14] WHO. Research priorities for the environment, agriculture and infectious diseases of poverty. Geneva: World Health Organization; 2013.

[15] Hotez PJ, Yamey G. The evolving scope of PLoS neglected tropical diseases. PLoS Negl Trop Dis 2009;3:e379.

[16] Hotez PJ. A plan to defeat neglected tropical diseases. Sci Am 2010;302:90−6.

[17] Hotez PJ, Molyneux DH, Fenwick A, Ottesen E, Sachs SE, Sachs JD. Incorporating a rapid-impact package for neglected tropical diseases with programs for HIV/AIDS, tuberculosis, and malaria. PLoS Med 2006;3:e102.

[18] WHO/TDR. Guidance Framework for testing genetically modified mosquitoes. Geneva: World Health Organization; 2014.

[19] Gubler DJ. Epidemic dengue/dengue hemorrhagic fever as a public health, social and economic problem in the 21st century. Trends Microbiol 2002;10:100−3.

[20] Tun-Lin W, Lenhart A, Nam VS, et al. Reducing costs and operational constraints of dengue vector control by targeting productive breeding places: a multi-country non-inferiority cluster randomized trial. Trop Med Int Health 2009;14:1143−53.

[21] WHO. Dengue: guidelines for diagnosis, treatment, prevention and control. Geneva: World Health Organization; 2009.

[22] Morrison AC, Zielinski-Gutierrez E, Scott TW, Rosenberg R. Defining challenges and proposing solutions for control of the virus vector *Aedes aegypti*. PLoS Med 2008;5:e68.

[23] Bleakley H. Disease and development: evidence from hookworm eradication in the American south. Q J Econ 2007;122:73−117.

[24] Reiter P, Lathrop S, Bunning M, et al. Texas lifestyle limits transmission of dengue virus. Emerg Infect Dis 2003;9:86−9.

[25] WHO. Emerging infectious diseases: World Health Day 1997 information kit. Geneva: World Health Organization; 1997.

[26] Bleakley H. Health, human capital, and development. Annu Rev Econom 2010;2: 283−310.

[27] French D. Causation between health and income: a need to panic. Empir Econ 2012;42: 583−601.

[28] Strauss J, Thomas D. Health, nutrition, and economic development. J Econ Lit 2008;36: 766−817.

[29] Gubler DJ, Clark GG. Community involvement in the control of *Aedes aegypti*. Acta Trop 1996;61:169−79.

[30] Armada Gessa JA, Gonzalez RF. Application of environmental management principles in the programme for eradication of *Aedes* (Stegomyia) *aegypti* (Linneus, 1762) in the Republic of Cuba, 1984. New Delhi: WHO Regional Office for South-East Asia, *Dengue Newsletter* 1987; vol. 3. pp. 111−119.

[31] Toldeo Romani ME, Vanlerberghe V. Achieving sustainability of community-based dengue control in Santiago de Cuba. Soc Sci Med 2007;64:976−88.

[32] Deleted in review.

[33] Gau Y-M, Buettner P, Usher K, Stewart L. Burden experienced by community health volunteers in Taiwan: a survey. BMC Public Health 2013;13:491.

[34] Macer D. Ethical, legal and social issues of genetically modified disease vectors in public health. Geneva: World Health Organization; 2003.

[35] Resnik DB. Environmental health ethics. Cambridge: Cambridge University Press; 2012.

[36] Resnik DB. Ethical issues in field trials of genetically modified disease-resistant mosquitoes. Dev World Bioeth 2014;14:37−46.

[37] Worrell R, Appleby MC. Stewardship of natural resources: definition, ethical and practical aspects. J Agric Environ Ethics 2000;12:263−77.

[38] Carson R. Silent spring. New York: Houghton Mifflin Harcourt; 2002.

[39] Van Den Bosch R. The pesticide conspiracy. Los Angeles: University of California Press; 1989.

[40] Perkins JH. Insects, experts, and the insecticide crisis: the quest for new pest management strategies. New York: Plenum Press; 1982.

[41] Rollin BE. The moral status of research animals in psychology. Am Psychol 1985;40: 920−6.

[42] Verhoog H, Matze M, van Bueren EL, Baars T. The role of the concept of the natural (naturalness) in organic farming. J Agric Environ Ethics 2003;16:29–49.

[43] Singer P. Animal liberation: towards an end to man's inhumanity to animals. St Albans, UK: Granada Publishing Ltd; 1977.

[44] Verhoog H. The concept of intrinsic value and transgenic animals. J Agric Environ Ethics 1992;5:147–60.

[45] Hauskeller M. Biotechnology and the integrity of life: taking public fears seriously. Hampshire, England: Ashgate Publishing, Ltd.; 2007.

[46] Rawls J. A theory of justice. Cambridge, MA: Harvard University Press; 1971.

[47] Platt JR. Strong inference: certain systematic methods of scientific thinking may produce much more rapid progress than others. Science 1964;146:347–53.

[48] Comstock G. Ethics and genetically modified foods. In: Gottwald F-T, Igensiep HW, Meinhardt M, editors. Food ethics. Springer Science & Business Media; 2010. p. 49–66.

[49] Resnik DB. Is the precautionary principle unscientific? Stud Hist Philos Biol Biomed Sci 2003;34:329–44.

[50] Peterson M. The precautionary principle is incoherent. Risk Anal 2006;26:595–601.

[51] Dorman P. Evolving knowledge and the precautionary principle. Ecol Econ 2005;53:169–76.

[52] United Nations. Rio declaration on environment and development. Rio de Janeiro: United Nations Environment Programme; 1992.

[53] Windbichler N, Menichelli M, Papathanos PA, Thyme SB, Li H, Ulge UY, et al. A synthetic homing endonuclease-based gene drive system in the human malaria mosquito. Nature 2011;473:212–15.

[54] Gould F, Huang Y, Legros M, Lloyd AL. A killer-rescue system for self-limiting gene drive of anti-pathogen constructs. Proc Biol Sci 2008;275:2823–9.

[55] Sinkins SP, Gould F. Gene drive systems for insect disease vectors. Nat Rev Genet 2006;7:427–35.

[56] Burt A. Site-specific selfish genes as tools for the control and genetic engineering of natural populations. Proc Biol Sci 2003;270:921–8.

[57] Kuzma J, Najmaie P, Larson J. Evaluating oversight systems for emerging technologies: a case study of genetically engineered organisms. J Law Med Ethics 2009;37:546–86.

[58] Enserink M. GM mosquito trial strains ties in Gates-funded project. Science/AAAS News Sci Insid 2010; <http://news.sciencemag.org/2010/11/gm-mosquito-trial-strains-ties-gates-funded-project> [accessed 30.08.14]

[59] Enserink M. GM mosquito release in Malaysia surprises opponents and scientists—again. Science/AAAS News Sci Insid 2011; <http://news.sciencemag.org/asia/2011/01/gm-mosquito-release-malaysia-surprises-opponents-and-scientists—again> [accessed 30.08.14]

[60] Subbaraman N. Science snipes at Oxitec transgenic-mosquito trial. Nat Biotechnol 2011; 29:9–11.

[61] Jasanoff S. Designs on nature: science and democracy in Europe and the United States. Princeton, NJ: Princeton University Press; 2007.

[62] Habermas J. The structural transformation of the public sphere: an inquiry into a category of bourgeois society. Cambridge, MA: MIT Press; 1991.

[63] Hauser GA. Vernacular dialogue and the rhetoricality of public opinion. Commun Monogr 1998;65:83–107.

[64] Fraser N. Rethinking the public sphere: a contribution to the critique of actually existing democracy. Soc Text 1990;25:56–80.

[65] Walsh L. Scientists as prophets: a rhetorical genealogy. New York: Oxford University Press; 2013.

[66] Katz S, Miller C. The low-level radioactive waste siting controversy in North Carolina: toward a rhetorical model of risk communication. In: Herndl CG, Brown SC, editors. Green culture: environmental rhetoric in contemporary America. University of Wisconsin Press; 1996. p. 111–40.

[67] Katz SB. Language and persuasion in biotechnology communication [published online January 24], <http://www.agbioforum.org/v4n2/v4n2a03-katz.htm>; 2002 [accessed 30.08.14].

[68] Shannon CE, Weaver W. The mathematical theory of communication. Urbana, IL: University of Illinois Press; 1967.

[69] Bucchi M. Handbook of public communication of science and technology. New York: Routledge; 2008.

[70] Burke K. Language as symbolic action: essays on life, literature, and method. Berkeley, CA: University of California Press; 1966.

[70a] Jasinski J. Sourcebook on rhetoric: key concepts in contemporary rhetorical studies. Thousand Oaks, CA; 2001.

[71] Graves HB. Rhetoric in(to) science: style as invention in inquiry. Cresskill, NJ: Hampton Press; 2005.

[72] Bucchi M. Can genetics help us rethink communication? Public communication of science as a "double helix.". New Genet Soc 2004;23:269–83.

[73] Brossard D, Shanahan J. The media, the public and agricultural biotechnology. Cambridge, MA: CABI; 2007.

[74] Shipman M. First-ever national survey on genetically engineered mosquitoes shows mixed support; 2012.

[75] Angeli E. Metaphors in the rhetoric of pandemic flu: electronic media coverage of H1N1 and swine flu. J Tech Writ Commun 2012;42(3):203–22.

[76] Schuldt JP, Konrath SH, Schwarz N. "Global warming" or "climate change"?: whether the planet is warming depends on question wording. Public Opin Q 2011;75:115–24.

[77] Whitmarsh L. What's in a name? Commonalities and differences in public understanding of "climate change" and "global warming.". Public Underst Sci 2008;18:401–20.

[78] Pacey A. Meaning in technology. Cambridge, MA: MIT Press; 2001.

[79] Spiegel J, Bennett S, Hattersley L, Hayden MH, Kittayapong P, Nalim S, et al. Barriers and bridges to prevention and control of dengue: the need for a social–ecological approach. Ecohealth 2005;2:273–90.

[80] Maxmen A. Florida abuzz over mosquito plan. Nature 2012;487:286.

[81] Reeves RG, Denton JA, Santucci F, Bryk J, Reed FA. Scientific standards and the regulation of genetically modified insects. PLoS Negl Trop Dis 2012;6:e1502.

[82] Alphey L, Beech C. Appropriate regulation of GM insects. PLoS Negl Trop Dis 2012;6: e1496.

[83] Katz, S.B. Biotechnology and global miscommunication with the public: rhetorical assumptions, stylistic acts, ethical implications. In: Professional Communication Conference, 2005. IPCC 2005. Proceedings. International, pp. 274–280, 10–13 July, 2005. <http://dx.doi.org/10.1109/IPCC.2005.1494186>.

Chapter 2

Concept and History of Genetic Control

Maxwell J. Scott[1] and Mark Q. Benedict[2]

[1]Department of Entomology, North Carolina State University, Raleigh, NC, USA, [2]Centers for Disease Control Foundation, Atlanta, GA, USA

INTRODUCTION

Genetic control of pests is clearly distinguished from other suppression methods that may be familiar to readers. In the context of this chapter, we define "genetic control" as *reducing pest damage using factors that disseminate by mating or inheritance.* This definition encompasses the use of chromosomally inherited genetic factors including rearrangements, mutations and transgenes, radiation- or chemically induced dominant lethal mutations, as well as sexually transmitted symbionts. Genetic control is further distinguished from other methods of control in that it is considered an area-wide control measure, that is, one that consists of *reducing pest damage using measures whose effectiveness depends on application over large expanses.* Area-wide control usually affects continuous populations into which immigration of unaffected individuals is detrimental to effectiveness and is prevented by quarantine and inspection activities, geographic factors, and augmentation by the genetic control itself. Area-wide control methods are seldom effective when applied over small extents into which high levels of immigration occur.

Genetic control is further distinguished from other methods of pest control by being species-specific. By design, it directly affects only one pest. This is a double-edged sword: the scope of effect is limited to what may only be one among many pests; on the other hand, it minimizes environmental impact by the same specificity. This characteristic means that genetic control, as currently envisioned, is most effective when a single species causes most harm in the area of concern.

AGRICULTURAL APPLICATIONS OF GENETIC CONTROL LEAD THE FIELD

The history and development of genetic control of agriculturally important insect pests has been previously covered by several excellent reviews [1–4].

Genetic Control of Malaria and Dengue.
© 2016 Elsevier Inc. All rights reserved.

Thus, the purpose of this chapter is not to provide a comprehensive review but rather a brief introduction to the field and highlight issues that are relevant for the ongoing development of transgenic strains for genetic control of disease vectors, particularly mosquito vectors of human diseases including malaria and dengue. We will begin the story of genetic control of vectors by describing efforts to control agricultural pests, which is the technical foundation, stimulus, and example upon which efforts to control mosquitoes has been based.

Eradication of the New World Screwworm Fly *Cochliomyia hominivorax* Using the Sterile Insect Technique

The first application of the sterile insect technique (SIT), and arguably the most successful, has been in the eradication of *Cochliomyia hominivorax* from all of North and Central America [3,4]. The SIT concept was developed by E.F. Knipling in the 1930s and initially implemented in the 1950s in the successful *C. hominivorax* eradication campaigns in Curacao and Florida. The central idea was to mass rear *C. hominivorax*, sterilize by ionizing radiation, and then release the sterilized flies over the targeted area [5]. A radiation dose was selected that achieved 100% sterility in males and females due to the induction of dominant lethal mutations. Virgin females that mated with a sterile male would not produce viable offspring. Simple mathematical models predicted that with successive releases of an excess of sterile males (Knipling suggested a ninefold excess) that the targeted population could be eradicated. As the population decreases, the ratio of sterilized males to fertile males increases with each generation until the probability that a virgin female will mate with a nonsterilized male is essentially zero. This characteristic of the control becoming more efficient as populations diminish is a unique capacity of SIT.

In his paper outlining the SIT concept [5], Knipling wrote that "this method will be difficult and costly under the most favorable circumstances" but might be useful as an "eradication tool for highly destructive pests" or for preventing the establishment of an infestation of an invasive pest. Thus, Knipling sought to both encourage the use of SIT for some pests and discourage its use for pests that did not meet key criteria:

1. Ability to mass rear the insect economically.
2. The insect must be present in low numbers in the field. This either could occur naturally, as for *C. hominivorax*, or could result from intensive application of other control measures such as insecticides or parasitoids.
3. The sterilization process should not have adverse effects on the males mating behavior or lifespan. That is, sterilization should have a minimal impact on the fitness of the males in the field.
4. Ability to disperse the sterilized males such that they can effectively compete with fertile males for virgin females.

Knipling also thought that it was important that females only mate once but was later persuaded that this is not an issue if the sterile males and their sperm are competitive. After the success in Florida, the *C. hominivorax* SIT program was extended to the remaining infested United States in the 1960s and then subsequently into Mexico and all of Central America. The cost of the over 50 year, multicountry eradication program was high (estimated at USD 1300 million in 2005 [6]). However, the economics of the program have been very positive with total *annual* producer benefits estimated at USD 1300 million [6]. Currently, 31 million *C. hominivorax* are raised per week in a plant in Pacora, Panama. Sterilized flies are distributed in southern Panama and up to 20 nautical miles within the northern interior of Columbia. This acts as a "buffer zone" preventing reinfestation of Central America by flies from South America.

What factors have made the *C. hominivorax* SIT program so successful? It certainly helped that the criteria listed above were met: the fly is naturally present at low densities [5]; economical mass-rearing methods were developed [7] (and continue to be improved [8]); sterilization has only a modest adverse effect on male fitness [9]; and chilled flies can efficiently distributed by air. Other factors that have likely contributed to the success of the program (in no particular order of importance):

- Species identification. Prior to 1933, *C. hominivorax* and the close relative *Cochliomyia macellaria* were thought to be the same species [10]. The adults were difficult to distinguish by morphological means, but the species can now be readily identified at the larval and adult stages. The realization that *C. macellaria* was a separate species was important as it is present in much higher densities in the field and is not an obligate parasite.
- Absence of appreciable mating barriers across the eradication zone. The SIT program could have failed if virgin females in the field would not mate with the released sterilized males. Theoretically, this could have arisen through genetic selection or if the females were from an isolated population that did not interbreed with other populations (i.e., cryptic species). Regarding the latter, Richardson and colleagues presented evidence for the existence of genetically distinct screwworm populations [11]. However, these claims were vigorously rebutted [12] and the success of the program supports the absence of appreciable mating barriers.
- A large monitoring network was established during the eradication period that involved local veterinarians, farmers, high school teachers, etc. Any detected animal infestations were thoroughly treated with insecticides and samples sent to the USDA. Thus, the technology by itself was not seen as a "silver bullet" but must be augmented with other measures on the ground for reducing the insect population.
- Agreements between neighboring countries are essential, as insects will cross national boundaries. The successful eradication of *C. hominivorax*

required formal international agreements between the USDA and respective agencies in Mexico and all of the Central American countries.

- Effective management. Proper execution requires management by a team dedicated solely to the genetic control program.

- Insect distribution. Modern GPS-guided navigation and georeferencing has significantly improved the efficiency of insect distribution and reduced costs [13].

- Input sought from prominent geneticists outside of the program before implementing the strategy. Knipling corresponded with other geneticists well before he published the initial description of SIT. For example, on March 6, 1950 (5 years prior to publication of the first paper on SIT!), Knipling outlined the approach in a letter to H.J. Muller [14], who had first described the mutagenic effects of radiation in *Drosophila* [15]. Muller's reply was enthusiastic about the prospect of using sterilized flies, providing the sterile flies could compete effectively with fertile males in the field.

- Public engagement. From the outset (i.e., Florida campaign), the screw-worm program sought to inform the public and address any concerns.

The failure in a separate program in recent years to eradicate *C. hominivorax* from Jamaica using SIT highlights the importance of some of the factors listed above [16]. For example, the Jamaica program was managed by civil servants who had other primary responsibilities. Further, ground surveillance and treatment of infested animals was poor.

Use of Genetic Sexing Strains in Mediterranean Fruit Fly SIT Programs

In his reply to Knipling's letter, Muller asked if there was some way to separate the sterilized mass-reared males from females prior to release as the "efficiency of the treated males as competitors for the other males would of course be reduced." While a means for separating male and female *C. hominivorax* was never attained, this is routinely achieved in current Mediterranean fruit fly (*Ceratitis capitata*) SIT programs by using so-called genetic sexing strains (GSS) [17]. The impetus for the development of GSS was the problem that sterilized females could puncture the skin of fruit, which could lead to spoilage due to microbial growth. As anticipated by Knipling, it has been shown that male-only releases of *C. capitata* are also three to five times more efficient at population suppression than bisexual releases under field conditions [18]. In the most widely used *C. capitata* GSS, females are homozygous for an autosomal recessive temperature-sensitive lethal (*tsl*) mutation [17]. Males also have one autosome that carries the *tsl* mutation but in addition have a functional copy of the *tsl* gene translocated to the Y chromosome. Thus, only females die when raised at the nonpermissive temperature. Although natural genetic recombination in males is very rare in Diptera, GSS strains are prone to breakdown due to recombination. That is, some female offspring inherit a recombined chromosome carrying the functional *tsl* gene and survive elevated temperatures. These events can be minimized by using suitable chromosome inversions in the GSS and a "filter-rearing system" to remove recombinants [17]. GSS males are also semi-sterile, as some of their offspring do not develop due to aneuploidy. For example, female offspring that inherit the autosome carrying the translocated piece of the Y chromosome die, as they are heterozygous for a large region of the autosome. The GSS is employed in a very large mass-rearing facility in Guatemala to provide up to 3 billion sterile males per week! The sterile males currently are mostly used to prevent the spread of *C. capitata* from Guatemala to Mexico (where it was eliminated by SIT) and are also employed as a preventative measure around ports of entry in Florida and California [1].

The advent of modern methods for genetic manipulation raised the possibility of developing stable sexing strains of *C. capitata*. Indeed, transgenic strains of *C. capitata* have been made that carry dominant tetracycline-repressible female lethal genetic systems. The two systems that have been made both employ the sex-specific intron from the *transformer* (*tra*) gene [19] to obtain sex-specific expression. The single-component system relies on overexpression of the tetracycline transactivator (tTA) and is lethal mostly at the pupal stage [20]. A two-component female embryo lethal system employs an early gene promoter to drive expression of tTA, which in turn regulates expression of a proapoptotic gene that contains the sex-specific *tra* intron [21]. The tetracycline-repressible systems have been successfully transferred/developed for other tephritid fruit pests [22,23]. In contrast, the development of a GSS using classical approaches requires isolation of suitable Y:autosome translocations coupled with selectable mutations for each species of interest. In large cage trials, successive releases of transgenic *C. capitata* female lethal strains were effective in suppression of stable populations of wild-type medfly [24]. However, transgenic female lethal strains remain to be evaluated in open-field tests.

The Australian Sheep Blowfly (*Lucilia cuprina*) Field Female Killing Strains

The success of the screwworm SIT program also encouraged the development of strains for genetic control of another myiasis-causing pest, *Lucilia cuprina*. However, rather than institute an SIT program *per se*, efforts were made to develop a potentially more cost-effective approach. The so-called field female killing (FFK) strains that were developed carried multiple chromosomal translocations involving the Y chromosome [25,26]. FFK females were homozygous for recessive eye color mutations, which were not lethal in the mass-rearing facility but were in the field. Males were not blind as they had normal copies of the eye color genes on the T(Y;5) and T(3;Y;5) chromosomes. The males were also semi-sterile as 75% of their offspring were aneuploid. Modeling suggested that because of the combination of recessive female lethality and male semi-sterility, the FFK approach was more efficient than SIT at low release ratios [27]. Thus, the FFK approach was predicted to be more cost effective than SIT. As the males did not need to be sterilized by radiation, there was no need to have a large central mass-rearing facility with an expensive irradiator. The potential of building several smaller rearing facilities across Australia was attractive given the vast distances of the Australian continent.

In a small island (40 km²) trial in 1985−1986, releases of the FFK strain were successful, achieving very high levels of genetic death and suppressing the *L. cuprina* population to undetectable levels [25]. A subsequent larger island field trial had some initial success but ultimately failed. The program experienced practical difficulties with mass-rearing the strain. For example, the strain was unstable and prone to breakdown due to recombination in males [28]. Potentially, this could have been overcome with the addition of suitable chromosomal inversions to the FFK strain. However, with a decline in the wool price, the *L. cuprina* genetic control program was abandoned.

With the long-term aim of developing stable sexing strains, the Scott lab has developed methods for transformation and isolated and characterized several *L. cuprina* genes and gene promoters, including the *transformer* gene [29,30]. This work led to the development of several transgenic strains of *L. cuprina* that carry a single-component dominant tetracycline-repressible female lethal gene [31]. Lethality is at the pupal stage but with the isolation of *Lucilia* early embryo gene promoters and proapoptotic genes [32], a female embryo lethal genetic system is under development. It is anticipated that the technology can be readily transferred to other calliphorid livestock pests, including *C. hominivorax*. If so, the ongoing SIT program could potentially carry out male-only releases, which has not previously been possible.

The Current State of Using SIT to Control Agricultural Pests

A joint FAO-IAEA program promotes the incorporation of SIT into area-wide integrated pest management of pest species. The program maintains a current directory of SIT facilities (http://nucleus.iaea.org/sites/naipc/dirsit/SitePages/All%20Facilities.aspx). As mentioned earlier, the *C. hominivorax* facility located in Pacora, Panama, provides sterile flies for the "buffer zone" in southern Panama. In addition to Guatemala, several countries including Argentina, Chile, Spain, Australia, and South Africa have *C. capitata* SIT facilities. Other fruit fly SIT programs include the Queensland fruit fly (*Bactrocera tryoni*) facility in Australia and Oriental fruit fly (*Bacterocera dorsalis*) facility in Thailand. Although several lepidopteron species are major agricultural pests, there are fewer SIT facilities due in part to the insensitivity of Lepidoptera to radiation. However, a substerilizing dose of radiation can be employed as the released semi-sterile males pass on the damaged chromosomes to their offspring [33]. As a consequence the F_1 generation are highly sterile. This form of genetic control is referred to as "inherited sterility" or IS [33]. An IS program was successfully employed in the eradication of the painted apple moth from New Zealand [34]. Other successful lepidopteran genetic SIT control programs include the codling moth (*Cydia pomonella*) facility in Canada and the pink bollworm (*Pectinophora gossypiella*) facility in the United States [35].

ATTEMPTS TO EXTEND GENETIC CONTROL TO MOSQUITOES

While the earliest successes of SIT were against agricultural pests, Knipling recognized that its use need not be limited to these insects and could be effective in public health against insects including mosquitoes [36]. All three major genera of mosquitoes, *Anopheles*, *Aedes/Oclerotatus*, and *Culex*, include species that transmit human and animal disease agents and have been the subjects of attempts to develop genetic control by SIT and related methods.

Those developing genetic control technologies from the earliest days were fully aware of the fact that mosquitoes have a characteristic that could prove to be the Achilles heel of their efforts—high, rather than Knipling's recommended low, population densities. Regardless, several genetic methods in addition to SIT were devised and a few have been field-tested. High densities and reproductive rates may be the central reason why so few attempts to control mosquitoes using genetic methods have been effective, and the history of mosquito genetic control consists largely of laboratory research and small exploratory field projects. Highlights of these efforts will be described by the category of technology rather than chronologically. Readers who would like a more detailed listing of early mosquito releases can consult Benedict and Robinson [37] and Dame [38]. Genetic control of mosquitoes has also been discussed in a recent review [39]. We will also describe a few proposed strategies that have not yet reached field testing but which are of contemporary research interest.

Progress Without Modern Biotechnology

Cytoplasmic Incompatibility and Wolbachia
Population Replacement

One of the earliest successful genetic control efforts against mosquitoes employed *Wolbachia* intracellular symbionts to induce sexual sterility. *Wolbachia* are maternally transmitted and often cause uni- or bidirectional sterility in matings of numerous insect species when they do not contain the same type of symbiont [40]. This phenomenon, called cytoplasmic incompatibility or CI, led to the development of an application for population control termed "insect incompatible technique" or IIT [41], an acronym that was later modified into the more mellifluous and descriptive Incompatible Insect Technique (K. Bourtzis, personal communication) and is now again being evaluated for control of mosquitoes [42]. Because released males of a particular type result in infertile matings, they can be used similarly to irradiated or chemically sterilized males but are fertile when mating among their type in the laboratory or production facility. The technology is a variation on SIT but which is expected to result in less loss of vigor than is caused by irradiation (see earlier). Release of females must be prevented as these could merely establish a novel replacement population that would not result in sterile matings with the released males.

Laven used IIT to accomplish elimination of a population of *Culex quinquefasciatus* (referred to as *Culex pipiens* fatigans in the original manuscript) in Myanmar in just one season using male releases [43], signaling an early success for genetic control. As effective as IIT was in this village, it has limitations. In its natural form, there are numerous incompatibility types [44] so target and release strains must be prematched to ensure that population suppression will result. Among mosquitoes, the only species that are widely and naturally infected with CI-inducing *Wolbachia* are found in *Aedes* and *Culex* species. However, the organism has recently been observed in a limited number of field-collected *Anopheles gambiae* [45]. It remains to be seen how common such observations will prove to be, and naturally infected *Anopheles* have not been manipulated for IIT.

Restrictions on presence and compatibility of CI have limited IIT's development as a general control technology, but the method has been revived by creating unnatural combinations of mosquito species and *Wolbachia* strains. Because *Wolbachia* can now be transferred interspecies artificially [46], this has opened up otherwise impossible options for manipulation, for example, in *Aedes aegypti* [47−49], *Aedes albopictus* [40,50−52], *Aedes polynesiensis* [53], *Anopheles stephensi* [54], and *An. gambiae* [55].

In one particular example, the *Wolbachia* causing bidirectional CI was transferred from *Aedes riversi* to *Ae. polynesiensis*, the major vector of lymphatic filariasis in the south Pacific [53]. Releases accomplished population suppression by CI in male-only releases in French Polynesia [56]. Similar technology for *Ae. albopictus* has been given an EPA Experimental User Permit for field testing in the United States (http://www.regulations.gov/#!docketDetail;D = EPA-HQ-OPP-2013-0254).

Using *Wolbachia* for IIT is not the only application that has arisen. It has been observed that *Wolbachia* has the capacity to spread in natural insect populations and replace *Wolbachia*-free individuals at rates that are useful for public health applications [57]. Building upon this observation, a serendipitously discovered facet of an artificial *Wolbachia* infection has been developed into a burgeoning release program against *Ae. aegypti*. "Eliminate Dengue," led by an Australia-based group, originally proposed to transfer *Wolbachia* to shorten the adult life to a degree that development of arbovirus to infectious levels would be reduced [49]. A *Drosophila melanogaster* strain of *Wolbachia pipientis*, *w*MelPop, was adapted via passage in an *Ae. aegypti* cell-line and was transferred to *Ae. aegypti* embryos by microinjection. While shortened lifespan was in fact observed in laboratory experiments, an even more powerful phenotype was observed in a derived line, *w*MelPop-CLA. Such low rates of dengue and chikungunya virus replication occur that the mosquitoes carrying these *Wolbachia* have dramatically reduced vector competence [58].

CI promotes replacement of the native population and the *Wolbachia*-infected mosquitoes have become prevalent where released in Australia [59]. According to the project Web site (www.eliminatedengue.com), releases are planned in several countries in Australasia and South America. This technology depends on IIT to promote replacement but not suppression and represents a good example of novel combinations of strain characteristics that continue to proliferate in mosquito genetic control. Implementing replacement by these refractory mosquitoes requires female release as *Wolbachia* are maternally transmitted, but relatively small numbers can be released beginning early during the seasonal increase in mosquito populations. Regardless of the need for releasing females, the released mosquitoes are expected to pose no increased risk of disease transmission.

Classical SIT

"Classical" SIT, that is, use of chemicals or irradiation to sexually sterilize males, offers significant flexibility and has been discussed along with variations in the context of mosquitoes [60]. Any target population that can be colonized can be sterilized and released without genetic modification; therefore, the initial development time and cost is minimal. One species-specific characteristic allows this to be done more economically—size dimorphism for female elimination. This has allowed mechanical "sexing" in *Ae. aegypti* [61,62], *Ae. albopictus* [63], and *Cx. quinquefasciatus* [64,65]. Size separation is not useful for *Anopheles* including *An. stephensi* and *An. gambiae* s.l. though it has been used for *Anopheles albimanus* [66]. It is possible that combinations of sexing involving a transgene for this purpose [67] and irradiation will be a useful combination. In the absence of physical means of eliminating females, a genetic sexing system using a selectable marker [68–70] or a chemical toxin delivered by bloodfeeding [71,72] is necessary and logistically nontrivial. Only one mosquito sexing strain based on a temperature-sensitive lethal been developed [73].

Because irradiation is typically used for sterilization [74], male sterility can be considered conditional and can be applied at will meaning that no special conditions are required for culturing the mosquitoes. Chemosterilants have also been used and are believed to cause less somatic harm than irradiation [75], but concerns about the predator effects, supported largely by a single publication [76], and broad concern about environmental contamination with chemicals have caused this approach to be abandoned with, to our knowledge, one exception [77]. The assumption that chemosterilants would be prohibited for environmental release might deserve reconsideration, given a large release of chemosterilized sea lampreys in the Great Lakes [78] obtained regulatory approval. Using transgenic approaches for sterilization or female elimination in a hybrid system could also improve classical SIT. While irradiation is often dismissed as causing reductions in male mating competitiveness, and indeed does so as seen in the references below, the same references also demonstrate that the damage may not be unacceptable as has been shown for *Ae. albopictus* [79], *Anopheles arabiensis* [80,81], *An. stephensi*, and *An. gambiae* [82].

Drawing inspiration and technology directly from screwworm, one of the earliest attempts to control *Ae. aegypti* in Pensacola, Florida, USA, began in 1960. It was an ambitious multiyear effort using irradiated males, air transport from Savannah, Georgia, USA, and pupa release [83]. No clear population reductions were observed in spite of high ratios of sterile to wild males ranging from 25:1 to 941:1. Failure was variously attributed to confounding control site variability, movement of containers, and immigration. However, we note that even the minimum irradiation dose of the range used (reported as 11−18,000 roentgens) is more than twice as high as a suitable sterilizing dose [84−86] and undoubtedly resulted in low competitiveness. This effort made a costly and sobering statement that more needed to be understood about sterilization, mating, and population biology of *Ae. aegypti* before one could expect success.

The most convincing and large implementation of classical SIT was performed in the 1970s in El Salvador and was reported in a series of papers that remain models for how one can, with relatively simple methods, evaluate and eliminate isolated populations [66,87−89]. This project used chemosterilants and size separation to produce sterile *An. albimanus* males (∼80% purity). Currently, most projects would not find this level of female release acceptable, but in only one season the local population was eliminated. We recommend that those planning mosquito releases of any kind consult this exemplary series of four papers. The project included sound population surveillance and monitored effects on sympatric *Anopheles pseudopunctipennis* as a control to determine that observed population reductions were restricted to the target species.

Current classical SIT efforts include one against *Ae. albopictus* in Italy that has achieved some success [63] and an active pre-release activity in the Northern State of Sudan [90]. It remains possible that this easily adaptable and widely acceptable technology will prove useful particularly where populations that undergo strong seasonal depression and are delimited by geographic or ecological factors are targeted and suppression or elimination can be sustained. Several challenges to producing the large numbers of mosquitoes have been overcome [91,92] and might be useful to many technologies, but release methods particularly have not advanced beyond merely opening a cage and allowing adults to fly out. Aerial release is widely anticipated as the most versatile release method but implementation has not moved beyond discussion.

Chromosome Rearrangements

Technology to create sexually sterile male mosquitoes was adapted directly particularly from screwworm and the Mediterranean fruit fly SIT programs. Other activities against mosquitoes echoed similar methods to control the sheep blowfly *L. cuprina* (described earlier) using various chromosomal translocation/inversion schemes (e.g., Refs [93–95]). Briefly, these strategies anticipated introducing combinations of translocations and inversions that cause semi-sterility and compound chromosomes in natural populations. The latter can have population replacement effects. The dramatic loss of fitness due to semi-sterility associated with chromosome rearrangements is not necessarily associated with broader losses of vigor [96] and translocation males have often (but not always, e.g., Ref. [97]) been observed to mate competitively [98,99].

Translocation/inversion-bearing mosquitoes have been released numerous times to evaluate competitiveness and less commonly to suppress populations. Surprisingly, the chromosomal aberrations can persist in mosquito populations in spite of their association with semi-sterility. A translocation strain of *Cx. pipiens* was used to successfully reduce—and nearly eliminate—a population in southern France in the village of Notre Dame [91]. Not only was the population suppressed, but the translocation persisted into the next season. Larger subsequent releases had smaller effects on population levels but the translocation persisted for several years without further releases [100]. Persistence was also observed in populations of *Ae. aegypti* in Delhi, India [101] and coastal Kenya [102] but not in another study in Kenya [103], an outcome ascribed at least in part to known deficiencies in vigor.

Unlike the model genetic organism *D. melanogaster*, in which numerous mutants are available and culture of hundreds of lines is routine, few markers were available in mosquitoes with which to manipulate chromosomes for strategies involving chromosome rearrangements, though much valuable genetic analysis was performed and mutations identified [104,105]. These limitations resulted in few strains and rearrangements from which to choose the most suitable ones for release. Mosquitoes were much more difficult than *D. melanogaster* to culture and to manipulate to produce useful strains and this approach was abandoned.

The assessment of the late Chris Curtis still sums up the current opinion of most of these schemes: "Perhaps there was a time in the late 1960s and early 1970s when this led us to start work on some intellectually delightful but impractical schemes" [106]. His retrospective sentiment should lodge in the consciousness of those developing biologically overly simplistic yet arcane schemes today to be cautious about their potential.

Population Replacement Without Transgenesis

A proposal has recently been made to change the genotype frequencies of wild populations by identifying, selecting, and disseminating natural anti-pathogen (low competence) genotypes [107]. This nontransgenic method, called "genetic shifting" was described using, as an example, dengue transmission by *Ae. aegypti* and is based on classical plant/animal breeding. By introgressing laboratory selected low-competence genotypes into wild populations, the authors predict that the transmission potential can be reduced below thresholds necessary for disease for sustained periods and would be reasonably tolerant to immigration, a vulnerability of genetic control strategies such as SIT. We will return to a similar method which proposed a transgenic variation that provides further modeling support for their general approach below.

Those proposing such schemes might benefit from the study of papers in which only markers were released, for example, release of *Silver* mutants of *Ae. aegypti* in Mississippi, USA [108], Delhi, India [101], and Kenya [109]. The persistence of translocations described above is also instructive and, given their clearly reduced fitness due to semi-sterility, provide observational support for testing genetic shifting and similar techniques that do not require a driving mechanism or other positive fitness.

A review of genetic control of mosquitoes would not be complete without highlighting the ambitious and productive bilateral research program conducted from 1969 to 1975 in India. The program had ecological, chemical control activities, and a genetic control component, mainly against *Cx. quinquefasciatus* (called *Cx. pipiens* fatigans in many publications of that era) with a smaller investment in *Ae. aegypti* and *An. stephensi* [110]. The project was led by the Indian Council of Medical Research, National Institute of Medical Research, WHO, and the US Public Health Service. This prolific program developed considerable capacity, published many scientific manuscripts (numbering 104 [111]) and was, in our opinion, a model of an energetic, stimulating, and well-resourced hive of technology innovation in a developing country. Several field trials of various genetic methods were performed, but unfortunately, the considerable achievements were eclipsed by controversy.

Numerous concerns were expressed about the activities of the program, and WHO deferred to the government of India to correct misconceptions.

Among the objections are some similar to those raised today against transgenic mosquitoes: lax regulation in the country where research was being conducted, release of mosquitoes that are not fully sterile, the possibility of mutations in the target populations that create a vector that is more efficient, economically unattractive, and introduction of exotic genotypes. It is possible that greater transparency and responsiveness to criticisms would have been possible if the policies for communications had anticipated and facilitated more rapid responses [112], but the Indian government eventually succumbed to pressure and ended the program. The considerable achievements could easily occupy half of this chapter: they were instead eclipsed by a perception of the program as a public relations failure, an outcome that reflects the difficulties of implementing agile responses in highly bureaucratic structures. Fortunately, the scientists involved in the program continued to be leaders in genetic control within India and elsewhere.

After the end of the WHO/US Public Health Service/India efforts and when UDSA's efforts against *An. albimanus* were completed in the late 1970s [113], field projects contracted and laboratory efforts to modify mosquitoes using classical genetics gradually withered. One by one, programs ended, leaders retired, mutant and chromosomal aberration strains were discarded and genetics laboratories shifted their emphasis to related studies such as population genetics and phylogenetics. Enthusiasm for the popular methods died and a field release hiatus of decades occurred before modern biotechnology appeared in the 1980s, seeming to some, to offer salvation to a dying field.

Modern Biotechnology Attempts to Cover the Achilles Heel of Mosquito SIT

Controlling vast mosquito populations is a formidable challenge that demands either highly organized campaigns [114] that might not be feasible or a clever approach that avoids the seeming impossibility of using inundation of mosquitoes in the face of intimidatingly large population sizes. Essentially, three methods have been proposed to overcome the obstacle of producing and releasing sufficient numbers of mosquitoes: larval competition, use of symbionts, and some form of "drive" to spread useful phenotypes analogous to the method described above with *Wolbachia* population replacement in *Ae. aegypti*. Some of these have now been field-tested. All rely upon making "a better mosquito" that reduces the need for inundative releases. Methods that are being developed will be described only briefly and will be covered elsewhere in this volume, but in keeping with the historical nature of this chapter, we describe one that has actually been implemented in field programs.

Increasing Suppression Leveraging Larval Competition

One approach to enhance conventional SIT is to add a suppressive effect exerted by the progeny of matings of the released transgenic males. Negative density dependent survival of larvae [115] can in principle be leveraged to reduce populations beyond what could be accomplished by SIT [116]. Negative density dependence is being exploited in two technologies that the English company Oxitec is developing. One flavor of Oxitec's RIDL™ transgenic mosquitoes, the bi-sex lethal OX512A strain of *Ae. aegypti*, contains a transgene that is inherited by F_1 progeny but which results in high mortality in heterozygotes, mostly during pupation [116]. By designing the genetic modification so that progeny larvae compete with any remaining wild-type larvae but die before pupation, models predict that population suppression can be accomplished with lower numbers of "sterile" males. The technology has been deployed for population suppressions in Grand Cayman [117] and Brazil [118]. Controlled experiments have not yet been performed to confirm that the anticipated improved population suppression due to larval competition is realized. Another variation of the RIDL technology allows transmission of a gene that prevents female, but not male, flight (female flightless) [119]. Releases of this other type of RIDL mosquitoes resulted in population elimination in large cages in the laboratory [120] but not in field cages [121]. The latter appears to have been largely because of a major mating disadvantage of the OX3604C males and possibly reduced flight activity [122].

Paratransgenesis Using Transgenic Symbionts

Despite considerable effort, there is no report of genetically transformed *Wolbachia*. This is unfortunate because the ability to engineer *Wolbachia* would make it an even more attractive microorganism for disseminating useful genetic factors. On the other hand, the exploding amount of knowledge of the microbiota of mosquitoes is offering opportunities to culture and modify fungal and bacterial symbionts that are associated with mosquitoes [123–125] and manipulation with microbiota is rapidly gaining notice as a method to explore for modification of mosquito phenotypes.

Genetically modified bacterial symbionts of mosquitoes including *Asaia* [126] and *Wickerhamomyces* [127] provide opportunities and several laboratories are investigating their transmission and means to express antipathogen effectors. Only laboratory experiments have been performed to date, but the fact that the modified symbionts invade quickly [126] and are easily manipulated is promising.

What genetic control approaches generally provide—species specificity— has not yet been determined for these organisms and it is not known which of the organisms that can be manipulated are present only in the target species; the presence of some is known to not be specific [123]. Specificity could be accomplished by mosquito-specific expression of transgenes or those that cause no harm in other organisms that might also host them, but whether such a type of specificity is acceptable for regulatory approval has not been explored.

Population Replacement with Factors Detrimental to Mosquitoes

Historically, driving transgenes into populations has been thought to require some factor—an engine if you will—to pull the effector, the cargo. Separation of the driving factor from the effector has long been anticipated to be problematic for one candidate, Type II transposable elements, because transposable elements tend to shed extraneous DNA sequences during the process of spreading [128]. It is expected that this would happen to a modified transposable element as well. There are other options that might be useful engines.

Medea (maternal effect dominant embryonic arrest) is a selfish DNA system discovered in the red flour beetle *Tribolium castaneum* [129] consisting of a maternally encoded toxin and an antidote [130]. Unless progeny embryos receive the antidote, they die. This characteristics provides a driving characteristic that has the potential to spread antipathogen genes in natural populations [131]. Populations of wild *T. castaneum* have been invaded by two forms of *Medea* [132], thus demonstrating drive potential. Unlike the use of *Wolbachia* in *Ae. aegypti* (as implemented by Eliminate Dengue) in which drive and effect are inseparable, *Medea* would require coupling with an effector. Proposals for how this might be accomplished have been made and modeled [131]. What is not clear yet is whether this engine would dump its cargo in order to increase its speed.

Perhaps the most ambitious and counterintuitive approach to suppressing mosquito populations is being developed to use selfish DNAs as means to suppress vector populations [133]. In this technology, rather than asking the engine to pull a cargo, the engine itself *is* the cargo, thus attempting to make spread and benefit inseparable.

The two approaches of most current interest are to spread factors such as homing endonucleases that would shift the sex ratio of populations toward harmless males whose abundance does not limit population sizes [134]. Another potential target is to use a transgene to cause female-fertility gene mutations [133]. If such mutations are recessive, they can increase in frequency in the population until they begin to have a suppressive effect [135,136]. Spreading detrimental genetic factors and simultaneously suppressing the population seems impossible. Indeed, it depends upon use of genetic factors that are inherited at rates higher than is typical of normal alleles and which have acceptable losses of fecundity. Spread has been demonstrated in laboratory tests with *An. gambiae* [137] but in the absence of a deleterious factor. Whether any of these technologies is ever implemented will depend in part upon technological progress, but their use will present challenges of acceptability, regulation, and international treaty implications [138]. None of these technologies has been field-tested.

Population Replacement with Beneficial Factors

When mosquito transgenesis became possible in the late 1990s, those working on genetic control strategies quickly envisioned creating "refractory" mosquitoes that would not be capable of transmitting disease. It was assumed that, because there was natural variation in vector competence [139], it should be possible to produce transgenic mosquitoes that were unable to transmit the parasite. After numerous efforts achieved incompletely refractory modified mosquitoes (e.g., Refs [140–142]), transgenic mosquitoes have been developed that greatly reduce both malaria parasite development [143] and dengue virus [144]. The problem that remains is how to disseminate these effectors. One of the authors (M.Q.B.) has proposed testing them reversibly by inundation [145] and this has now been modeled as a feasible method either alone or in combination with a suppressive transgene [146]. The likelihood that such a transgene would have some fitness cost and slowly disappear from the population provides an element of safety that makes at least preliminary tests attractive but which countervail the economy hoped for by use of gene drive.

While the potential power and economy of genetic control is tantalizing, with few exceptions, it has consisted of more promise, proposals, and programs than control products. However, given the tremendous burden of malaria, filariasis, dengue, yellow fever, West Nile virus, and chikungunya, successfully eliminating transmission of even one of these in a cost-effective program justifies the relatively modest investment that has been made. The emerging success of *Wolbachia* and RIDL™ technologies may be only the first among several more powerful and economical genetic techniques for reducing human and animal mortality and morbidity due to mosquitoes. As demonstrated by the development of genetic control products whose beneficial characteristics were unexpected and the novel combinations that are being devised, there is reason for optimism that technology that successfully reaches the goals will indeed be realized.

ACKNOWLEDGMENTS

We thank Steve Skoda, John Welch, and Pamela Phillips for comments on the chapter. The comments and corrections of an anonymous reviewer are also appreciated.

REFERENCES

[1] Klassen W, Curtis CF. History of the sterile insect technique. In: Dyck VA, Hendrichs J, Robinson AS, editors. Sterile insect technique principles and practice in area-wide integrated pest management. Dordrecht, The Netherlands: Springer; 2005. p. 3–36.

[2] Koyama J, Kakinohana H, Miyatake T. Eradication of the melon fly, *Bactrocera cucurbitae*, in Japan: importance of behavior, ecology, genetics, and evolution. Annu Rev Entomol 2004;49:331–49.

[3] Krafsur ES, Whitten CJ, Novy JE. Screwworm eradication in North and Central America. Parasitol Today 1987;3:131–7.

[4] Wyss JH. Screwworm eradication in the Americas. Ann N Y Acad Sci 2000;916:186–93.

[5] Knipling EF. Possibilities of insect control or eradication through the use of sexually sterile males. J Econ Entomol 1955;48:459–62.

[6] Vargas-Teran M, Hofmann HC, Tweddle NE. Impact of screwworm eradication programmes using the sterile insect technique. In: Dyck VA, Hendrichs J, Robinson AS, editors. Sterile insect technique principles and practice in area-wide integrated pest management. Dordrecht, The Netherlands: Springer; 2005. p. 629–50.

[7] Melvin R, Bushland RC. A method of rearing *Cochliomyia americama* C. & P. on artificial media. USDA Bur Entomol Plant Quar 1936;ET-88.

[8] Chaudhury MF, Skoda SR. A cellulose fiber-based diet for screwworm (Diptera: Calliphoridae) larvae. J Econ Entomol 2007;100:241–5.

[9] Crystal MM. Sterilization of screwworm flies (Diptera: Calliphoridae) with gamma rays: restudy after two decades. J Med Entomol 1979;15:103–8.

[10] Cushing EC, Patton WS. Studies on the higher Diptera of medical and veterinary importance. *Cochliomyia americana* sp. nov., the screw-worm fly of the New World. Ann Trop Med Parasitol 1933;27:539–51.

[11] Richardson RH, Ellison JR, Averhoff WW. Autocidal control of screwworms in North America. Science 1982;215:361–70.

[12] Lachance LE, Bartlett AC, Bram RA, et al. Mating types in screwworm populations? Science 1982;218:1142–3.

[13] Tween G, Rendon P. Current advances in the use of cryogenics and aerial navigation technologies for sterile insect delivery systems. In: Vreysen MJB, Robinson AS, Hendrichs J, editors. Area-wide control of insect pests. Dordrecht, The Netherlands: Springer; 2007. p. 603–15.

[14] Knipling EF, Edward F. Knipling papers, screwworm eradication collection, special collections. National Agricultural Library.

[15] Muller HJ. Artificial transmutation of the gene. Science 1927;66(1699):84–7.

[16] Dyck VA, Reyes Flores J, Vreysen MJB, et al. Management of area-wide integrated pest management programmes that integrate the sterile insect technqiue. In: Dyck VA, Hendrichs J, Robinson AS, editors. Sterile insect technique principles and practice in area-wide integrated pest management. Dordrecht, The Netherlands: Springer; 2005. p. 525–45.

[17] Franz G. Genetic sexing strains in Mediterranean fruit fly, an example for other species amenable to large scale rearing for the sterile insect technique. In: Dyck VA, Hendrichs J, Robinson AS, editors. Sterile insect technique principles and practice in area-wide integrated pest management. Dordrecht, The Netherlands: Springer; 2005. p. 427–51.

[18] Rendon P, McInnis D, Lance D, et al. Medfly (Diptera: Tephritidae) genetic sexing: large-scale field comparison of males-only and bisexual sterile fly releases in Guatemala. J Econ Entomol 2004;97:1547–53.

[19] Pane A, Salvemini M, Delli Bovi P, et al. The transformer gene in *Ceratitis capitata* provides a genetic basis for selecting and remembering the sexual fate. Development 2002;129:3715–25.

[20] Fu G, Condon KC, Epton MJ, et al. Female-specific insect lethality engineered using alternative splicing. Nat Biotechnol 2007;25:353–7.

[21] Ogaugwu CE, Schetelig MF, Wimmer EA. Transgenic sexing system for *Ceratitis capitata* (Diptera: Tephritidae) based on female-specific embryonic lethality. Insect Biochem Mol Biol 2013;43:1−8.

[22] Ant T, Koukidou M, Rempoulakis P, et al. Control of the olive fruit fly using genetics-enhanced sterile insect technique. BMC Biol 2012;10:51.

[23] Schetelig MF, Handler AM. A transgenic embryonic sexing system for *Anastrepha suspensa* (Diptera: Tephritidae). Insect Biochem Mol Biol 2012;42:790−5.

[24] Leftwich PT, Koukidou M, Rempoulakis P, et al. Genetic elimination of field-cage populations of Mediterranean fruit flies. Proc Biol Sci 2014;281. Available from: http://dx.doi.org/10.1098/rspb.2014.1372.

[25] Foster GG, Weller GL, James WJ, et al. Advances in sheep blowfly genetic control in Australia. Management of insect pests: nuclear and related molecular and genetic techniques. Vienna: International Atomic Energy Agency; 1993. p. 299−312.

[26] Black IV WC, Alphey L, James AA. Why RIDL is not SIT. Trends Parasitol 2011;27:362−70.

[27] Foster GG, Vogt WG, Woodburn TL, et al. Computer simulation of genetic control. Comparison of sterile males and field-female killing systems. Theor Appl Genet 1988;76:870−9.

[28] Foster GG, Maddern RH, Mills AT. Genetic instability in mass-rearing colonies of a sex-linked translocation strain of *Lucilia cuprina* (Wiedemann) (Diptera: Calliphoridae) during a field trial of genetic control. Theor Appl Genet 1980;58:169−75.

[29] Sandeman RM, Levot GW, Heath AC, et al. Control of the sheep blowfly in Australia and New Zealand—are we there yet? Int J Parasitol 2014;44:879−91.

[30] Scott MJ. Development and evaluation of male-only strains of the Australian sheep blowfly, *Lucilia cuprina*. BMC Genet 2014;15(Suppl. 2):S3.

[31] Li F, Wantuch HA, Linger RJ, et al. Transgenic sexing system for genetic control of the Australian sheep blow fly *Lucilia cuprina*. Insect Biochem Mol Biol 2014;51:80−8.

[32] Edman RM, Linger RJ, Belikoff EJ, et al. Functional characterization of calliphorid cell death genes and cellularization gene promoters for controlling gene expression and cell viability in early embryos. Insect Mol Biol 2015;24:58−70.

[33] Carpenter JE, Bloem S, Marec F. Inherited sterility in insects. In: Dyck VA, Hendrichs J, Robinson AS, editors. Sterile insect technique principles and practice in area-wide integrated pest management. Dordrecht, The Netherlands: Springer; 2005. p. 115−46.

[34] Suckling DM, Barrington AM, Chhagan A, et al. Eradication of the Australian painted apple moth *Teia anartoides* in New Zealand: trapping, inherited sterility, and male competitiveness. In: Vreysen MJB, Robinson AS, Hendrichs J, editors. Area-wide control of insect pests. Dordrecht, The Netherlands: Springer; 2007. p. 603−15.

[35] Bloem KA, Bloem S, Carpenter JE. Impact of moth supression/eradication programmes using the sterile insect technique or inherited sterility. In: Dyck VA, Hendrichs J, Robinson AS, editors. Sterile insect technique principles and practice in area-wide integrated pest management. Dordrecht, The Netherlands: Springer; 2005. p. 677−700.

[36] Knipling EF, Laven H, Craig GB, et al. Genetic control of insects of public health importance. Bull World Health Org 1968;38:421−38.

[37] Benedict MQ, Robinson AS. The first releases of transgenic mosquitoes: an argument for the sterile insect technique. Trends Parasitol 2003;19:349−55.

[38] Dame DA, Curtis CF, Benedict MQ, et al. Historical applications of induced sterilisation in field populations of mosquitoes. Malar J 2009;8(Suppl. 2):S2.

[39] Alphey L. Genetic control of mosquitoes. Annu Rev Entomol 2014;59:205−24.

[40] Sinkins SP. *Wolbachia* and cytoplasmic incompatibility in mosquitoes. Ins Biochem Mol Biol 2004;34:723–9.

[41] Boller EF. Cytoplasmic incompatibility in *rhagoletis cerasi*. In: Robinson A, Hooper G, editors. Fruit flies, their biology, natural enemies and control. Amsterdam: Elsevier; 1989. p. 39–74.

[42] Hancock PA, Sinkins SP, Godfray HCJ. Strategies for introducing *wolbachia* to reduce transmission of mosquito-borne diseases. PLoS Negl Trop Dis 2011;5:e1024.

[43] Laven H. Eradication of *Culex pipiens fatigans* through cytoplasmic incompatibility. Nature 1967;216:383–4.

[44] Barr AR. Cytoplasmic incompatibility in natural populations of a mosquito, *Culex pipiens* L. Nature 1980;283:71–2.

[45] Baldini F, Segata N, Pompon J, et al. Evidence of natural *Wolbachia* infections in field populations of *Anopheles gambiae*. Nat Commun 2014;5:1–7.

[46] Hughes GL, Rasgon JL. Transinfection: a method to investigate *Wolbachia*–host interactions and control arthropod-borne disease. Insect Mol Biol 2014;23:141–51.

[47] Xi Z, Khoo CCH, Dobson SL. *Wolbachia* establishment and invasion in an *Aedes aegypti* laboratory population. Science 2005;310:326–8.

[48] Ruang-Areerate T, Kitthawee SE. *Wolbachia* transinfection in *Aedes aegypti*: a potential gene driver of dengue vectors. Proc Natl Acad Sci USA 2006;103:12534–9.

[49] McMeniman CJ, Lane RV, Cass BN, et al. Stable introduction of a life-shortening *wolbachia* infection into the mosquito *aedes aegypti*. Science 2009;323:141–4.

[50] Xi Z, Dean JL, Khoo C, et al. Generation of a novel *Wolbachia* infection in *Aedes albopictus* (Asian tiger mosquito) via embryonic microinjection. Insect Biochem Mol Biol 2005;35:903–10.

[51] Xi Z, Khoo C, Dobson SL. Interspecific transfer of *Wolbachia* into the mosquito disease vector *Aedes albopictus*. Proc R Soc Lond B Biol Sci 2006;273:1317–22.

[52] Blagrove MSC, Arias-Goeta C, Failloux AB, et al. *Wolbachia* strain *wMel* induces cytoplasmic incompatibility and blocks dengue transmission in *Aedes albopictus*. Proc Natl Acad Sci USA 2012;109:255–60.

[53] Dean JL, Dobson SL. Characterization of *Wolbachia* infections and interspecific crosses of *Aedes* (Stegomyia) *polynesiensis* and *Ae.* (Stegomyia) *riversi* (Diptera: Culicidae). J Med Entomol 2004;41:894–900.

[54] Bian G, Joshi D, Dong Y, et al. *Wolbachia* invades *Anopheles stephensi* populations and induces refractoriness to *Plasmodium* infection. Science 2013;340:748–51.

[55] Hughes GL, Vega-Rodriguez J, Xue P, et al. *Wolbachia* strain *wAlbB* enhances infection by the rodent malaria parasite *plasmodium berghei* in *Anopheles gambiae* mosquitoes. Appl Environ Microbiol 2012;78:1491–5.

[56] O'Connor L, Plichart C, Sang AC, et al. Open release of male mosquitoes infected with a *wolbachia* biopesticide: field performance and infection containment. PLoS Negl Trop Dis 2012;6:e1797.

[57] Turelli M, Hoffmann AA. Rapid spread of an inherited incompatibility factor in California *Drosophila*. Nature 1991;353:440–2.

[58] Moreira LA, Iturbe-Ormaetxe I, Jeffery JA, et al. A *wolbachia* symbiont in *aedes aegypti* limits infection with dengue, chikungunya, and plasmodium. Cell 2009;139:1268–78.

[59] Hoffmann AA, Montgomery BL, Popovici J. Successful establishment of *Wolbachia* in *Aedes* populations to suppress dengue transmission. Nature 2011;476:454–7.

[60] Alphey L, Benedict M, Bellini R, et al. Sterile-insect methods for control of mosquito-borne diseases: an analysis. Vector Borne Zoonotic Dis 2010;10:295–311.

[61] Fay RW, McCray Jr EM, et al. Mass production of sterilized male *Aedes aegypti*. Mosq News 1963;23:210−14.

[62] Harris AF, Nimmo D, McKemey AR, et al. Field performance of engineered male mosquitoes. Nat Biotechnol 2011;29:1034−7.

[63] Bellini R, Medici A, Puggioli A, et al. Pilot field trials with *aedes albopictus* irradiated sterile males in italian urban areas. J Med Entomol 2013;50:317−25.

[64] Sharma VP, Patterson RS, Ford HR. A device for the rapid separation of male and female mosquito pupae. Bull World Health Org 1972;47:429.

[65] Gerberg EJ, Hopkins TM, Gentry JW. Mass rearing of *Culex pipiens fatigans* under ambient conditions. Mosq News 1969;65(3):382−5.

[66] Dame DA, Lofgren CS, Ford HR, et al. Release of chemosterilized males for the control of *Anopheles albimanus* in El Salvador. II. Methods of rearing, sterilization, and distribution. Am J Trop Med Hyg 1974;23:282−7.

[67] Catteruccia F, Benton J, Crisanti A. An *Anopheles* transgenic sexing strain for vector control. Nat Biotechnol 2005;23:1414−17.

[68] Robinson A. Genetic sexing in *Anopheles stephensi* using dieldrin resistance. J Am Mosq Control Assoc 1986;2:93−5.

[69] Curtis C, Akiyama J, Davidson G. A genetic sexing system in *Anopheles gambiae* species A. Mosq News 1976;36:492−8.

[70] Seawright J, Kaiser P, Suguna S, et al. Genetic sexing strains of *Anopheles albimanus* wiedemann. Mosq News 1981;41:107−14.

[71] Yamada H, Soliban SM, Vreysen MJ, et al. Eliminating female *Anopheles arabiensis* by spiking blood meals with toxicants as a sex separation method in the context of the sterile insect technique. Parasit Vectors 2013;6:197.

[72] Lowe RE, Fowler JEF, Bailey DL, et al. Separation of sexes of adult *Anopheles albimanus* by feeding of insecticide-laden blood. Mosq News 1981;41:634−8.

[73] Baker RH, Sakai RK, Raana K. Genetic sexing for a mosquito sterile-male release. J Hered 1981;72:216−18.

[74] Helinski ME, Parker AG, Knols B. Radiation biology of mosquitoes. Malar J 2009; 8(Suppl. 2):S6.

[75] Schmidt CH, Dame DA, Weidhaas DE. Radiosterilization vs. chemosterilization in house flies and mosquitoes. J Econ Entomol 1964;57:753−6.

[76] Bracken GK, Dondale CD. Fertility and survival of *Achaearanea tepidariorum* (Araneida: Theridiidae) on a diet of chemosterilized mosquitoes. Can Entomol 2003;104:1709−12.

[77] Gato R, Lees RS, Bruzon RY, et al. Large indoor cage study of the suppression of stable *Aedes aegypti* populations by the release of thiotepa-sterilised males. Mem Inst Oswaldo Cruz 2014;109:365−70.

[78] Bergstedt RA, McDonald RB, Twohey MB, et al. Reduction in sea lamprey hatching success due to release of sterilized males. J Great Lakes Res 2003;29(Suppl. 1):435−44.

[79] Balestrino F, Medici A, Candini G, et al. Gamma ray dosimetry and mating capacity studies in the laboratory on *Aedes albopictus* males. J Med Entomol 2010;47:581−91.

[80] Helinski MEH, Knols BGJ. Mating competitiveness of male *Anopheles arabiensis* mosquitoes irradiated with a partially or fully sterilizing dose in small and large laboratory cages. J Med Entomol 2008;45:698−705.

[81] Yamada H, Vreysen MJB, Gilles JRL, et al. The effects of genetic manipulation, dieldrin treatment and irradiation on the mating competitiveness of male *Anopheles arabiensis* in field cages. Malar J 2014;13:318.

[82] Andreasen MH, Curtis CF. Optimal life stage for radiation sterilization of *Anopheles* males and their fitness for release. Med Vet Entomol 2005;19:238−44.

[83] Morlan HB, McCray Jr EM, Kilpatrick JW. Field tests with sexually sterile males for control of *Aedes aegypti*. Mosq News 1962;22:295−300.

[84] Terzian LA, Stahler N. The effect of age and sex ratio on the mating activity of *Anopheles quadrimaculatus* Say. Rep Naval Med Res Inst 1954;Oct:261−8.

[85] Hallinan E, Rai KS. Radiation sterilization of *Aedes aegypti* in nitrogen and implications for sterile male technique. Nature 1973;244:368−9.

[86] Terwedow HA, Asman MA. *Aedes sierrensis*: determination of the optimal radiation dose for competitive sterile-male control. In: Proceedings of 45th Annual Conference on California Mosquito & Vector Control Association 1977;115−18.

[87] Lofgren CS, Dame DA, Breeland SG, et al. Release of chemosterilized males for the control of *Anopheles albimanus* in El Salvador. III. Field methods and population control. Am J Trop Med Hyg 1974;23:288−97.

[88] Breeland SG, Jeffery GM, Lofgren CS, et al. Release of chemosterilized males for the control of *Anopheles albimanus* in El Salvador. I. Characteristics of the test site and the natural population. Am J Trop Med Hyg 1974;23:274−81.

[89] Weidhaas DE, Breeland SG, Lofgren CS, et al. Release of chemosterilized males for the control of *Anopheles albimanus* in El Salvador. IV. Dynamics of the test population. Am J Trop Med Hyg 1974;23:298−308.

[90] Malcolm CA, Sayed el B, Babiker A, Girod R, et al. Field site selection: getting it right first time around. Malar J 2009;8(Suppl. 2):S9.

[91] Balestrino F, Benedict MQ, Gilles JRL. A new larval tray and rack system for improved mosquito mass rearing. J Med Entomol 2012;49:595−605.

[92] Balestrino F, Gilles JRL, Soliban SM, et al. Mosquito mass rearing technology: a cold-water vortex device for continuous unattended separation of *Anopheles arabiensis* pupae from larvae. J Am Mosq Control Assoc 2011;27:227−35.

[93] Laven H, Cousserans J, Guille G. Eradicating mosquitoes using translocations: a first field experiment. Nature 1972;236:456−7.

[94] Kaiser PE, Seawright JA, Benedict MQ, et al. Homozygous translocations in *Anopheles albimanus*. Theor Appl Genet 1983;65:207−11.

[95] Seawright JA, Haile DG, Rabbani MG, et al. Computer simulation of the effectiveness of male-linked translocations for the control of *Anopheles albimanus* Wiedemann. Am J Trop Med Hyg 1979;28:155−60.

[96] Oliva CF, Benedict MQ, Soliban SM, et al. Comparisons of life-history characteristics of a genetic sexing strain with laboratory strains of *Anopheles arabiensis* (Diptera: Culicidae) from Northern Sudan. J Med Entomol 2012;49:1045−51.

[97] Grover KK, Suguna SG, Uppal DK, et al. Field experiments on the competitiveness of three genetic control systems of *Aedes aegypti* (L.). Entomol Exp Appl 1975;20:8−18.

[98] Baker RH, Reisen WK, Sakai RK, et al. *Anopheles culicifacies*: mating behavior and competitiveness in nature of males carrying a complex chromosomal aberration. Ann Entomol Soc Am 1980;73:581−8.

[99] Reisen WK, Baker RH, Sakai RK, et al. *Anopheles culicifacies* Giles: mating behavior and competitiveness in nature of chemosterilized males carrying a genetic sexing system. Ann Entomol Soc Am 1981;74:395−401.

[100] Laven H, Cousserans J, Guille G. Experience de lutte genetique contre *Culex pipiens* dans la region de Montpellier. Bull Biol 1971;108:253−7.

[101] Rai KS, Grover KK, Suguna SG. Genetic manipulation of *Aedes aegypti*: incorporation and maintenance of a genetic marker and a chromosomal translocation in natural populations. Bull World Health Org 1973;48:49−56.

[102] McDonald PT, Hausermann W, Lorimer N. Sterility introduced by release of genetically altered males to a domestic population of *Aedes aegypti* at the Kenya coast. Am J Trop Med Hyg 1977;26:553−61.

[103] Lorimer N, Lounibos LP, Petersen JL. Field trials with a translocation homozygote in *Aedes aegypti* for population replacement. J Econ Entomol 1976;69:405−9.

[104] Kitzmiller JB. Genetics, cytogenetics, and evolution of mosquitoes. Adv Genet 1976;18:1−60.

[105] Kitzmiller JB. Mosquito cytogenetics: a review of the literature 1953−62. Bull World Health Org 1963;29:345−55.

[106] Curtis C. Genetic control of insect pests: growth industry or lead balloon? Biol J Linn Soc Lond 1985;26:359−74.

[107] Powell JR, Tabachnick WJ. Genetic shifting: a novel approach for controlling vector-borne diseases. Trends Parasitol 2014;30:282−8.

[108] Fay RW, Craig GB. Genetically marked Aedes aegypti in studies of field populations. Mosq News 1969;29:121−7.

[109] Lorimer N. Long-term survival of introduced genes in a natural population of *Aedes aegypti* (L.) (Diptera: Culicidae). Bull Entomol Res 1981;71:129−32.

[110] Anonymous. WHO-supported collaborative research projects in India: the facts. WHO Chronicle 1976;30:131−9.

[111] Curtis CF. Destruction in the 1970s of a research unit in India on mosquito control by a sterile male release and a warning for the future. Antenna 2007;31:214−16.

[112] Anonymous. Oh, New Delhi; oh, Geneva. Nature 2004;256:365−7.

[113] Dame DA, Lowe RE, Williamson DL. Assessment of released sterile *Anopheles albimanus* and *Glossina morsitans* morsitans. In: Pal R, Kitzmiller JB, Kanda T, editors. Cytogenetics and genetics of vectors: proceedings of the XVIth International conference of entomology. Amsterdam; 1981. p. 231−48.

[114] Soper FL. The elimination of urban yellow fever in the Americas through the eradication of *Aedes aegypti*. Am J Pub Health 1963;53:7−16.

[115] Southwood TR, Murdie G, Yasuno M, et al. Studies on the life budget of *Aedes aegypti* in Wat Samphaya, Bangkok, Thailand. Bull World Health Org 1972;46:211−26.

[116] Phuc HK, Andreasen MH, Burton RS, et al. Late-acting dominant lethal genetic systems and mosquito control. BMC Biol 2007;5:1−11.

[117] Harris AF, McKemey AR, Nimmo D, et al. Successful suppression of a field mosquito population by sustained release of engineered male mosquitoes. Nat Biotechnol 2012;30:828−30.

[118] Malavasi A. Project *Aedes* transgenic population control in Juazeiro and Jacobina Bahia, Brazil. BMC Proc 2014;8(Suppl. 4): O11.

[119] Fu G, Lees RS, Nimmo D, et al. Female-specific flightless phenotype for mosquito control. Proc Natl Acad Sci USA 2010;107:4550−4.

[120] Wise de Valdez MR, Nimmo D, Betz J, et al. Genetic elimination of dengue vector mosquitoes. Proc Natl Acad Sci USA 2011;108:4772−5.

[121] Facchinelli L, Valerio L, Ramsey JM, et al. Field cage studies and progressive evaluation of genetically-engineered mosquitoes. PLoS Negl Trop Dis 2013;7:e2001.

[122] Bargielowski I, Kaufmann C, Alphey L, et al. Flight performance and teneral energy reserves of two genetically-modified and one wild-type strain of the yellow fever mosquito *Aedes aegypti*. Vector Borne Zoonotic Dis 2012;12:1053−8.

[123] Ricci I, Damiani C, Capone A, et al. Mosquito/microbiota interactions: from complex relationships to biotechnological perspectives. Curr Opin Microbiol 2012;15: 278−84.

[124] Cirimotich CM, Ramirez JL, Dimopoulos G. Native microbiota shape insect vector competence for human pathogens. Cell Host Microbe 2011;10:307−10.

[125] Minard G, Mavingui P, Moro CV. Diversity and function of bacterial microbiota in the mosquito holobiont. Parasit Vectors 2013;6:146.

[126] Favia G, Ricci I, Marzorati M, et al. Bacteria of the genus *Asaia*: a potential paratransgenic weapon against malaria. In: Aksoy S, editor. Transgenesis and the management of vector-borne disease, 627. The Netherlands: Springer; 2008. p. 49−59.

[127] Bandi C, Daffonchio D, Favia G. The yeast *Wickerhamomyces anomalus* (*Pichia anomala*) inhabits the midgut and reproductive system of the Asian malaria vector *Anopheles stephensi*. Environ Microbiol 2011;13:911−21.

[128] Finnegan DJ. Eukaryotic transposable elements and genome evolution. Trends Genet 1989;5:103−7.

[129] Beeman RW, Friesen KS, Denell RE. Maternal-effect selfish genes in flour beetles. Science 1992;256:89−92.

[130] Akbari OS, Chen C-H, Marshall JM, et al. Novel synthetic *medea* selfish genetic elements drive population replacement in *drosophila*; a theoretical exploration of *medea*-dependent population suppression. ACS Synth Biol 2012. Available from: http://dx.doi. org/10.1021/sb300079h.

[131] Hay BA, Chen C-H, Ward CM, et al. Engineering the genomes of wild insect populations: challenges, and opportunities provided by synthetic *Medea* selfish genetic elements. J Insect Physiol 2010;56:1402−13.

[132] Beeman R, Friesen K. Properties and natural occurrence of maternal-effect selfish genes ("*Medea*" factors) in the red flour beetle, *Tribolium castaneum*. Heredity 1999;82:529−34.

[133] Burt A. Site-specific selfish genes as tools for the control and genetic engineering of natural populations. Proc R Soc Lond B Biol Sci 2003;270:921−8.

[134] Galizi R, Doyle LA, Menichelli M, et al. A synthetic sex ratio distortion system for the control of the human malaria mosquito. Nat Commun 2014;5:1−8.

[135] Deredec A, Godfray HCJ, Burt A. Requirements for effective malaria control with homing endonuclease genes. Proc Natl Acad Sci USA 2011;108:874−80.

[136] North A, Burt A, Godfray HCJ. Modelling the spatial spread of a homing endonuclease gene in a mosquito population. J Appl Ecol 2013. Available from: http://dx.doi.org/ 10.1111/1365-2664.12133.

[137] Windbichler N, Menichelli M, Papathanos PA, et al. A synthetic homing endonuclease-based gene drive system in the human malaria mosquito. Nature 2011. Available from: http://dx.doi.org/10.1038/nature09937.

[138] Marshall JM. The Cartagena Protocol and genetically modified mosquitoes. Nat Biotechnol 2010;28:896−7.

[139] Zheng L, Cornel AJ, Wang R, et al. Quantitative trait loci for refractoriness of *Anopheles gambiae* to *Plasmodium cynomolgi* B. Science 1997;276:425−8.

[140] Moreira LA, Ito J, Ghosh A, et al. Bee venom phospholipase inhibits malaria parasite development in transgenic mosquitoes. J Biol Chem 2002;277:40839−43.

[141] Corby-Harris V, Drexler A, de Jong LW, et al. Activation of *Akt* signaling reduces the prevalence and intensity of malaria parasite infection and lifespan in *Anopheles stephensi* mosquitoes. PLoS Pathog 2010;6:e1001003.

[142]   Meredith JM, Basu S, Nimmo DD, et al. Site-specific integration and expression of an anti-malarial gene in transgenic *Anopheles gambiae* significantly reduces *Plasmodium* infections. PLoS One 2011;6:e14587.

[143]   Isaacs AT, Jasinskiene N, Tretiakov M, et al. Transgenic *Anopheles stephensi* coexpressing single-chain antibodies resist *Plasmodium falciparum* development. Proc Natl Acad Sci USA 2012;109:E1922−30.

[144]   Franz AWE, Sanchez-Vargas I, Raban RR, et al. Fitness impact and stability of a transgene conferring resistance to dengue-2 virus following introgression into a genetically diverse *aedes aegypti* strain. PLoS Negl Trop Dis 2014;8:e2833.

[145]   Benedict MQ. Let it snow: field-testing malaria-refractory strains by inundation. Available from: <http://www.malariaworld.org/blog/let-it-snow-field-testing-malaria-refractory-strains-inundation>.

[146]   Okamoto KW, Robert MA, Gould F, et al. Feasible introgression of an anti-pathogen transgene into an urban mosquito population without using gene-drive. PLoS Negl Trop Dis 2014;8:e2827.

Chapter 3

Considerations for Disrupting Malaria Transmission in Africa Using Genetically Modified Mosquitoes, Ecology of Anopheline Disease Vectors, and Current Methods of Control

Mamadou B. Coulibaly, Sekou F. Traoré and Yeya T. Touré
Malaria Research and Training Center, University of Sciences, Techniques and Technologies of Bamako, Bamako, Mali

MALARIA IN AFRICA: CURRENT SITUATION

In 2014, the WHO reported 198 million (with an uncertainty range of 124–283 million) cases of malaria worldwide with 584,000 (with an uncertainty range of 367,000–755,000) deaths occurred in 2013. The majority of the deaths (90%) occur in sub-Saharan Africa and in children under 5 years of age (78%) [1]. According to WHO, a child under 5 years of age dies from malaria every minute in Africa. In addition to the morbidity and the mortality, malaria is responsible for loss of work days, which contributes to the impoverishment of already poor populations. The direct economic losses are estimated to be USD 12 billion per year. The loss of GDP growth per year is estimated to be 1.3% for Africa [2]. These statistics show that malaria still remains a major public health problem mostly in the poorest regions of the world. However, there has been some progress recently in terms of reducing the disease burden. In fact, the WHO reports that globally the mortality due to malaria was reduced by 47% in all age groups and by 53% in children under 5 years of age from 2000 to 2013. In the African region, the mortality due to malaria was reduced by 54% during the same period. According to modeling data, 3.3 million malaria deaths were averted between 2001 and 2012. The

Genetic Control of Malaria and Dengue.
© 2016 Elsevier Inc. All rights reserved.

majority of these lives saved (69%) came from 10 countries that had the highest malaria burden in 2000. Among the averted cases, 90% are children under 5 years of age in sub-Saharan Africa. If the momentum of this progress is maintained, 8 of the 43 countries in the African region where malaria transmission is ongoing will be able to reduce reported malaria incidence or malaria admission rates by 75% or more. These eight countries are Botswana, Cabo Verde, Eritrea, Namibia, Rwanda, Sao Tome and Principe, South Africa and Swaziland, and the island of Zanzibar (United Republic of Tanzania). Zambia and Ethiopia may achieve 50−75% reduction in admission cases by 2015, while Madagascar could reach about 50%. However, the pace of the decrease in malaria incidence and mortality seems to have slowed down between 2011 and 2012 partly due to the fact that the model used for estimating malaria deaths in children less than 5 years old in Africa uses insecticide-treated mosquito net (ITN) coverage as an input, and there was no increase in ITN coverage in 2011−2012 following decreases in funding for malaria control in 2011 [1]. Not only is the pace of the decrease in malaria incidence slowing down but in some regions malaria cases are increasing. In Algeria, northern Africa where the malaria burden is not the highest, the number of reported cases went from 35 in 2000 to 59 in 2012. In some countries, such as Mali, western Africa, despite high bed net coverage rates (80.2% in children less than 5 years old and 75.1% in 15- to 49-year-old women) malaria prevalence still remains as high as 52% in children less than 5 years of age [3]. In some other countries in the African region, such as Rwanda (Central Africa), there is fragile progress. In fact, while malaria cases and mortality rates decreased between 2000 and 2010, the number of confirmed cases increased from 2011 to 2012 [4]. The same pattern has been observed in Sao Tome.

In summary, the African regions [1] show different patterns of progress in malaria control. In the West African region, only in Cabo Verde did malaria cases decline continuously. In Central Africa, considerable reductions are noted; however, recently observed resurgence may hinder the pre-elimination efforts in some countries of this region. In the East and high transmission areas of Southern region, there seems to be a sustained decrease in malaria admission cases. In the low transmission southern Africa, progress has been notable since 2000. The five countries of this region (Botswana, Namibia, South Africa, Swaziland, and Zimbabwe) and three other countries (Angola, Mozambique, and Zambia) signed a malaria elimination initiative (E8) in March 2009. This initiative aims at eliminating malaria from four countries (Botswana, Namibia, South Africa, and Swaziland) by 2015.

MALARIA VECTORS IN THE AFRICAN REGION

In the African region, three dominant malaria vector species are described. They are *Anopheles gambiae* sensu stricto, *Anopheles arabiensis*, and

Anopheles funestus sensu stricto [5]. The two former are both members of the complex *An. gambiae* s.l. which contains five other sibling species (*Anopheles melas, Anopheles merus, Anopheles quadriannulatus* A and B, and *Anopheles bwambae*). *Anopheles quadriannulatus* B, the Ethiopian species (species A has been described in South Africa), is also referred to as *Anopheles amharicus* Hunt, Wilkerson, and Coetzee sp.n. [6].

Anopheles funestus s.s. is a member of the *An. funestus* s.l. complex that comprises at least 11 sibling species. They are *An. funestus* Giles, *Anopheles vaneedeni* Gillies and Coetzee, *Anopheles rivulorum* Leeson, *An. rivulorum-like, Anopheles leesoni* Evans, *Anopheles confusus* Evans and Leeson, *Anopheles parensis* Gillies, *Anopheles brucei* Service, *Anopheles aruni* Sobti, *Anopheles fuscivenosus* Leeson, and an Asian member *Anopheles fluviatilis* James (in Refs [7−9]).

For simplicity, from now on "*An. gambiae*" will be used for *An. gambiae* s.s. and "*An. funestus*" will be used for *An. funestus* s.s. *Anopheles gambiae, An. arabiensis,* and *An. funestus* are the three main malaria vectors in Africa. The three have adapted to specific ecological niches with possible overlaps as described by Sinka et al. in 2012 [5]. Highly influenced by human activities, the distribution of *An. gambiae* and that of *An. arabiensis* are subject to seasonal and spatial variations. In general they are sympatric, however the relative frequencies of each of the two are related to specific ecological setting. For instance in Mali, generally *An. gambiae* has been found in higher frequencies compared to *An. arabiensis* from samples collected in different ecological settings. These data showed that the frequencies of *An. gambiae* are the highest in Southern Sudan savanna zones. The frequencies of *An. arabiensis* are very low in Sudan savanna zones ($\sim 7\%$) but increase from these zones toward Northern Sudan savanna where they can reach 20%. The highest frequencies of *An. arabiensis* are observed in the Northern drier zones where they could reach 60% [10]. In fact, the frequencies of *An. arabiensis* increase with increasing aridity showing a cline [10,11].

The niche partitioning is also observed within *An. gambiae*, which contains different reproductive units defined by chromosomal polymorphisms. These units are referred to as chromosomal forms. They are named Forest, Bissau, Bamako, Mopti, and Savanna. Molecular identifications of these groups resulted in two initial molecular types referred to as M and S molecular forms. The M molecular has recently been raised to a species and referred to as *Anopheles coluzzii* [6]. The frequencies of the Bamako chromosomal form increase from almost undetectable during the dry season to reach a peak at the middle of the rainy season. This unit is believed to have a predilection for rock pools for breeding [10]. The Mopti chromosomal form also referred to as the M molecular form or *An. coluzzii*, is found almost all the year round with higher frequencies at the start of the rainy season. In Mali, it displays a notable ecoplasticity with a distribution ranging from humid savannas to the Sahel and sub-Saharan zones [10]. It has been reported from

West and Central Africa. The Savanna chromosomal form, which together with the Bamako chromosomal form, composes what is called the S molecular form, prefers rain-dependent larval sites though it could also use sites in the most arid area of the savanna belt as *An. arabiensis* does. In contrast in East Africa, the S molecular form due to its 2Rb karyotype has been able to spread to the most arid areas such as Somalia, Ethiopia, and Sudan where *An. arabiensis* predominates. The temporal variation is similar to that of the Bamako chromosomal form.

The Forest chromosomal form, for which a molecular identification is not yet established, is found in rain forest zones and in humid or derived savannas [12]. The Bissau chromosomal form has been described in coastal rice cultivated zones of the Gambia, Southern Senegal (Casamance), Guinea Bissau, and Guinea Conakry. It is salt tolerant and colonizes inland regions [12,13]. In summary, the ecological diversity is higher within *An. gambiae* in West Africa compared to East Africa.

The frequencies of *An. gambiae* are higher in rural areas where typical larval sites are available. The characteristics of such typical sites are but not limited to sunlit, shallow, still water, predominantly human made (these may differ according to localities). *Anopheles gambiae* is also present in urban areas where these typical larval sites may not be easy to find. Therefore, in such settings, *An. gambiae* have adapted to urban areas exploiting water-filled domestic containers or other types of available sites different from the preferred ones [14]. Under some conditions in urban settings, *An. gambiae* will use polluted water bodies as larval sites [15]. *Anopheles gambiae* generally feeds on humans (anthropophilic) indoors (endophagic) though there are recent evidences that a considerable number of individuals is feeding outdoor (exophagic) [16] raising the question of outdoor disease transmission. This has an important implication for vector control strategies. *Anopheles gambiae* also, generally, rests indoors (endophilic) though here too more recent evidences are showing populations resting outdoors (exophilic).

Compared to *An. gambiae* much less attention has been paid to *An. funestus* though it is the second major malaria vector and in some areas it is the major one.

Anopheles funestus s.s. is the most widespread of the complex in the African region. It is composed of two reproductive units referred as Kiribina and Folonzo. As *An. gambiae*, it generally feeds indoors on humans preferably and rests indoors. However, recent studies have shown that Kiribina has a propensity to rest outdoors [17]. This has implications as the two large-scale vector control interventions in use in the African region target indoor feeding and resting populations. *Anopheles funestus* breeds preferably in large, permanent, or semi-permanent fresh water bodies covered with vegetations (e.g., swamps, ponds) but it is known also to be a plastic species. In Mali, for example, it is found in rice cultivation areas [18,19]. Its densities are subject to seasonal variations. In these areas of Mali, its peak densities are observed when those of *An. gambiae* are lowest, hence maintaining a perennial malaria transmission through relay [19].

MALARIA PARASITES AND THEIR PUBLIC HEALTH SIGNIFICANCE

To date, four major malaria parasites infecting humans are described globally. They are *Plasmodium falciparum*, *Plasmodium malaria*, *Plasmodium vivax*, and *Plasmodium ovale*. A fifth, initially known to infect monkeys, has been recently found infecting humans [20]; it is named *Plasmodium knowlesi*. *Plasmodium falciparum* has the largest distribution in the African region and is the most deadly. Its frequencies ranged regionally from 60% to 100% from 2007 to 2012. *Plasmodium vivax* follows with frequencies as high as 40% in Erithrea and Ethiopia [21]. Because it does not affect Duffy negative individuals (most Africans are Duffy negative), *P. vivax* was believed to be a typical Asian parasite until Liu et al. traced its origin to African apes [22]. The other species are present in much lower frequencies. So far, *P. knowlesi* has not been reported from the African region.

CURRENT APPROACHES FOR MALARIA VECTOR CONTROL

(This section provides a general outline of the current vector control approaches and the one based on genetically modification of mosquitoes. Other chapters of the book discuss details and specifics.)

In the malaria endemic regions of Africa, several methods are used by the populations to protect themselves. A panoply of individual and collective protection methods is applied, from traditional and cultural prospects to the modern products. However, the current large-scale interventions that are recommended by the WHO (citation) are (i) the use of ITNs among which the most current are long-lasting insecticidal bed nets (LLINs), (ii) the indoors residual spraying (IRS), and (iii) larval source management.

In the African region, 39 of the 44 countries with ongoing malaria transmission distribute bed nets free of charge [21]. However, bed nets are not free for all age groups in all the countries. In some countries, they will be free of charge for target groups (children under 5 years of age and pregnant women) only. Reports from the national malaria control programs show that ITNs/LLINs are distributed through three main channels: (i) mass distribution campaigns that account for 89%, (ii) antenatal cares that accounts for 7%, and (iii) immunization that accounts for 3%. Other methods represent 1% [1]. According to WHO, the distribution of ITNs/LLINs in sub-Saharan Africa increased from 6 million in 2004 to 145 million in 2010. Then it decreased in 2011 to 92 million and to 70 million in 2012. The projected number for 2013 was estimated to be 136 million. This number is 214 million for 2014 [1]. Many countries are scaling-up the distribution using the universal coverage scheme (one net for 1.8 person). The efficacy of bed net use in malaria control is now thought to be proven.

According to 2013, world malaria report from WHO IRS is recommended for malaria control in 40 countries in Africa. Fifteen of these may use IRS for the control of epidemics. In 31 of the 40 countries where IRS is recommended, it is used in combination with ITNs. Other combinations such as IRS and larval source managements have good potential and should be investigated deeply. Reports from NMCPs (National Malaria Control Programme) around the world claim that 4% (135 million persons) of the global population at malaria risk were protected by IRS in 2012. This declined to 3.5% in 2014 [1]. In the African region, the protection rate varied between 10% and 12% of the populations at risk between 2009 and 2011. In 2012, this rate decreased to 8% and to 7% in 2014 [1].

Larval source and environmental management is recommended by WHO in combination with IRS or the bed nets, in specific ecological setting where mosquito larval sites are few, fixed, and accessible [23]. In 2012, six countries in African region reported to WHO activities related to larval source and environmental management. This intervention needs more investigations to see how best it can be used mostly in combination with existing and or new strategies.

Both ITNs/LLINs and IRS, the two main large-scale malaria vector control interventions currently in use, target mosquitoes that feed and/or rest indoors. Malaria parasites and vectors are developing resistance to drugs and insecticides, respectively. Therefore, developing new approaches to complement recommended methods has become a necessity. Currently, a new vector control tool based on the release of genetically modified mosquitoes (GMMs) is being developed. This novel approach is based on (i) reducing the mosquito population to a level where the disease transmission is no longer maintained (population reduction or suppression) or (ii) making mosquitoes unable to harbor the parasites/develop the parasites (population replacement). Each of these two technologies can be designed so that: (i) the modification persists for a short period (self-limiting) and multiples releases are necessary or (ii) the modification persists for a long time (self-sustaining) and multiple releases will not be necessary.

The release of GMMs will be complementary to both ITNs/LLINs and IRS [24]. Genetically modified (GM) release will target all individual in the target species, regardless whether mosquitoes feed and or rest indoors or outdoors. Therefore, outdoors biting and or resting individuals that will not be affected by either IRS or ITNs/LLINs will be attained using the GM approach. Another important contribution that the GM approach will bring as a complement to these existing strategies is that insecticide-resistant mosquitoes will not be spared. They will be affected as well as the insecticide-susceptible individuals.

Being complementary to IRS and or ITNs/LLINs, the question that might be raised is how will GMMs be affected by the insecticides used in ITNs/LLINs or IRS. A well-designed study might answer this question but one can hypothesize that both transgenic and wild-type mosquitoes will have similar theoretical chances to be exposed to insecticide where insecticides are used in control measures. It is also likely that the outcome will be similar. Another important consideration for the use of GMMs in malaria vector control will be the susceptibility of GMM to usual insecticides. Insecticide-resistant GMM might not be appropriate for the control strategies. In addition, it might complicate mitigation efforts in case of accidental release. However, efforts should be made to investigate the effect of insecticide resistance on GMMs.

GMM could be used in combination with either one of the existing control tools (e.g., GMM + IRS, GMM + ITNs/LLINs, etc.). Once the GM approach is well established for malaria control, operational research should be conducted to assess the efficacy of such combinations. Larval source management could also be a combination element even if it is not as largely implemented as ITNs/LLINs and IRS.

BIOLOGICAL AND ECOLOGICAL CONSIDERATIONS FOR GM RELEASE AND MONITORING

The major malaria vectors in the African region are cryptic species complexes in general. The *An. gambiae* complex (*An. gambiae* s.l.) is composed of seven sibling species. *An. gambiae*, the mostly widely distributed member of the *An. gambiae* complex, is composed of molecular forms named M and S. The M molecular form has recently been raised to species level and is referred to as *An. coluzzii*. *Anopheles funestus* is a complex of at least 11 members. The most dangerous of this complex, *An. funestus*, is composed of two subunits as mentioned above. So far, no chromosomal or molecular forms have been described for *An. arabiensis*, but studies in and out of the Kilombero valley in Tanzania have shown that it, too, could be undergoing genetic differentiations based on ecological preferences of subpopulations [25]. In addition to the ecological differentiation, another important population structuring to take into account is the susceptibility of specific populations to malaria parasites. In fact, in 2011, studies have shown that not only novel ecological units are present (GOUNDRY) but they have different susceptibility level to malaria parasites in Burkina Faso [26].

All the populations described above have preferential larval sites even if it is not excluded that some sites can be found shared by two or more species. Therefore, the ecological niche partitioning should be given attention in designing a GMM-based intervention. Geographical isolation could also play a role in mosquito populations structuring. Baseline investigations could help identify whether or not there are several geographically isolated populations and if they are genetically distinct or not.

This structuring of the mosquito populations has an important implication for vector control tools especially for the genetic approaches based on GMMs. For such control tools, homogeneous mosquito populations would be ideal as single genetic constructs could be used for an entire population. As most of the vector groups are either complexes or present ecological differentiation to some extent, the following question need to be explored: Will multiple constructs be used (a specific construct for a specific population)? At this point, as far as *An. gambiae*, the major vector, is concerned there is a debate on whether the amount of gene flow between populations is sufficient to consider the use of single construct. While it is important to solve this question, it is important to remind that malaria continues to kill hundreds of thousands people among which children under 5 years of age and pregnant women are paying the heaviest tribute. Should we wait until the question is solved? Carefully designed studies and a better understanding of the biology of local mosquito populations will provide orientations in answering this question. Such experiments may take advantage of the fact that one or two groups predominate with mosquito species complexes in the African region. For instance in Mali, *An. coluzzii* could represent about 80% or more of the collected mosquitoes within the *An. gambiae* complex. In other regions, *An. arabiensis* predominates. These predominating populations can first be targeted by the GMM approach. This is one of the orientations scientists currently working on the GMM technology for malaria control are considering. Of course one of the questions that come up is whether the niches left empty by the predominating populations following a successful genetic intervention will be occupied by the minor malaria vector? Or by a population which has never been a vector before? Careful designed preliminary studies can test these hypotheses to provide clear answers. Monitoring interventions on a long term could also provide answers. In conclusion, the GMM technology, in complement with other methods, will play a crucial role in malaria control, elimination, and/or eradication depending on the setting.

Ecological concerns need to be addressed for both the social perception and the safe deployment for the ecosystem. One of the recurrent questions about GM organisms is: what happens if one or more mosquito species are completely eradicated? At this stage of the technology, there might not be much evidence to provide a specific answer. Mosquitoes have reported to be part of some small bats' diet even though they represented less than 1% of the whole diet (http://www.mosquito.org/faq#bats). More recent data have concluded that mosquitoes are important diet items for small bats that may be more likely to be affected by mosquito population reduction [27,28]. Mosquitoes have also been reported to be involved in pollination. John Smith Dexter reported for the first time in 1913 that Orchids were pollinated by mosquitoes although other insects participate [29]. Mosquitoes have been reported participating in pollination of other plants including the goldenrod. In contrast, because mosquitoes are not the only diet items for bats, though could be important in a way, nor are they the only pollinators for plants, there are thoughts that eradicating them might make the ecosystem hiccup but life would go on. These views are debatable as there could be other animals and/or other plants out there for which mosquitoes might be of considerable importance. Therefore, there is an urgent need for more investigation to determine the impact eradicating mosquitoes. Regarding the deployment of GMMs, in particular, there is a need to go back one step and ask if mosquito eradication is the ultimate goal; probably not. Mosquito population reduction/replacement in order to stop disease transmission might be more achievable and more acceptable. But one might wonder whether the genetic modification technologies based on the population reduction would not push mosquito population to extinction even if this is not the aim at the first place. This might be anticipated through monitoring of the mosquito populations throughout the deployment of the GMMs as disease control strategies and plans for corrective actions.

OPERATIONAL CONSIDERATIONS AND CAPACITY BUILDING IN AFRICA

As with every new technology, careful step-by-step approaches should be the rule in implementing the GMM technology for disease control. In most of the cases the technology itself, that is, the protein engineering, is conducted in developed countries outside the disease endemic countries (DECs). Testing should be conducted in laboratory cages (small insectary cages and large cages). These will help in understanding many aspects of the GMM including the mating behavior, survival, safety, and the stability of the gene in the mosquito populations, etc. Following successful results from these assays, the same testing path, as much as possible, should be repeated in DECs before any control efforts is attempted. WHO and several experts have designed a guideline that describes the GMM implementation in four phases similar to the vaccine development scheme [24]. According to this guideline, Phase 1 should be designed to understand the safety and the efficacy in small-scale laboratory studies first and then in large laboratory cages meeting the required confinement standards. This will assess whether the candidate GMM has the desired biological and functional characteristics. GMMs that show success in Phase 1 will be allowed to be tested for Phase 2 where the confinement will be achieved in a more natural setting. WHO and experts suggest two options for Phase 2 testing: (i) testing in large laboratory and field malaria spheres; this is referred to as testing in physical confinement and (ii) testing in small ecologically confined setting, this is releasing in small areas that are geographically or climatically isolated [30–32]. Both or either one can be initiated according to the guideline; however, as this is a new technology and is extremely sensitive in terms of social perceptions, it would probably be advisable to conduct both instead of one. Afterward, when more evidences have

TABLE 3.1 Summary of GMM Testing Phases and Objectives from the WHO 2014 Guidelines

Phase	Objective	Scale
Phase 1: Laboratory studies including cages	Assess biological and functional characteristics toward	Small
	Assess safety and efficacy	
Phase 2: Field trials (physical and/or ecological containment)	Assess biological and functional characteristics toward	Small
	Assess safety and efficacy	
	Assess impact on local mosquitoes	
Phase 3: Staged open-field releases	Assess impact of GMMs on infection and/or disease in human populations	Increasing size and complexity at single or multiple sites
	Assess function and efficacy	
Phase 4: Postimplementation surveillance/monitoring	Assess effectiveness under operational conditions	Increasing size and complexity at single or multiple sites
	Assess human and environmental safety	

been gathered, flexibility of doing one or the other may be allowed. Ultimately, though, the choice of the options will depend on the country regulatory authorities and on what is known on the GMM technology to be used. The WHO guidance describes the situation where a physical confinement testing might not be necessary as in a case where the testing has been conducted in another venue. While this is reasonable, it will be advisable to make sure that the two venues have ecological similarities otherwise repeating the testing in the new venue might be necessary.

When successful results are obtained in Phase 2, the next phase that involves sequential trials on a larger scale can be initiated in a single or multiple sites [24]. The performance under different conditions (e.g., different transmission levels, variations in mosquito densities, the presence of other vectors, etc.) can be assessed. Later in this phase, the impact of the technology on the human epidemiology (i.e., infection and disease) will be assessed. The WHO guidelines suggest a fourth phase after the results from Phase 3 are satisfactory. In this phase, longitudinal surveillance should be initiated to assess the effectiveness in local conditions (Table 3.1 for the phases and the objectives). Monitoring will take place and should be on a long-term to regularly measure the safety for human health by looking at the disease incidences. Monitoring must also be extended to the environmental safety. For example, it will be advisable to monitor the populations of important insects such as bees for horizontal transfer even if the probability for this may be negligible for some technologies.

The GMM technology as other tools will be implemented in DECs. It is essential that adequate infrastructure, equipment, and personnel are in place in these countries. The laboratory receiving the technology should meet the required arthropod containment standards [33]. In the case of malaria control, there seem to be an agreement on ACL2 for self-limiting transgenes (e.g., those that are expected to disappear from the environment). In the African region, one laboratory of this kind was set up in Mali in August 2010 in collaboration with the University of Keele, UK under the leadership of Paul Eggleston, Frederic Tripet, Sekou F. Traore, and Mamadou B. Coulibaly. The project that supported the laboratory was sponsored by Wellcome Trust. To date on the initiative of TARGET MALARIA, a non-profit research consortium (sponsored by FNIH−BMGF), the laboratory in Mali is being refurbished and there are two additional laboratories under construction in Kenya and in Burkina Faso. These laboratories should receive the appropriate and up-to-date equipment to allow the production of solid data. In parallel, the personnel from DECs laboratories should be trained on all the procedures relative to the implementation of the technology from the conceptions to the deployment. The technology transfer must be a reality at all steps so that scientists from DECs can really take ownership and avoid the issue of parachuting technologies to the endemic countries.

Once GMM technologies are deployed in DECs, a number of elements should be monitored. Another essential component of the operational considerations is the community engagement and the regulatory aspects. These will be detailed elsewhere but it is noteworthy to emphasize on the importance of these components. Community engagement should be integrated in the operations from the beginning to the end. A neglected ill perception of the intervention can kill the efforts. Therefore, monitoring the social perception and taking the appropriate corrective actions will be necessary as the opportunity of using the GMM approaches for disease control should not be missed.

REFERENCES

[1]   WHO. World malaria report. *Geneva: World Health Organization*, <http://www.who.int/malaria/publications/world_malaria_report_2014/en/>; 2014.

[2]   Roll Back Malaria. Key malaria facts. <http://wwwrbmwhoint/keyfactshtml>; 2014.

[3]   National Statistics. Demographic and health survey in Mali (EDSM V); 2012−2013.

[4]   Karema C, Aregawi MW, Rukundo A, Kabayiza A, Mulindahabi M, Fall IS, et al. Trends in malaria cases, hospital admissions and deaths following scale-up of anti-malarial interventions, 2000−2010, Rwanda. Malar J 2012;11:236.

[5]   Sinka ME, Michael JB, Sylvie M, Yasmin R-P, Theeraphap C, Maureen C, et al. A global map of dominant malaria vectors. Vector Parasit 2012;5:69.

[6]   Coetzee M, Hunt RH, Wilkerson R, Della Torre A, Coulibaly MB, Besansky NJ. *Anopheles coluzzii* and *Anopheles amharicus*, new members of the *Anopheles gambiae* complex. Zootaxa 2013;3619(3):246−74.

[7]   Green CA, Hunt RH. Interpretations of variation in ovarian polytene chromosomes of *Anopheles funestus* Giles, *An. parensis* Gillies, and *An. aruni*. Genetica 1980;51:187−95.

[8]   Koekemoer LL, Kamau L, Hunt RH, Coetzee M. A cocktail polymerase chain reaction assay to identify members of the Anopheles funestus (Diptera: Culicidae) group. Am J Trop Med Hyg 2002;66:804−11.

[9]   Green CA. Cladistic analysis of mosquito chromosome data (*Anopheles* (Cellia) Myzomyia). J Hered 1980;73:2−11.

[10] Toure YT, Petrarca V, Traore SF, Coulibaly A, Maiga HM, Sankare O, et al. The distribution and inversion polymorphism of chromosomally recognized taxa of the *Anopheles gambiae* complex in Mali, West Africa. Parasitologia 1998;40(4):477−511.

[11] Coluzzi M, Sabatini A, Petrarca V, DiDeco MA. Chromosomal differentiation and adaptation to human environments in the *Anopheles gambiae* complex. R Soc Trop Med Hyg 1979;73 (5):483−97.

[12] Coluzzi M, Petrarca V, DiDeco MA. Chromosomal intergradation and incipient speciation in *Anopheles gambiae*. Boll Zool 1985;52:45−63.

[13] Bryan JH, Di Deco MA, Petrarca V, Coluzzi M. Inversion polymorphism and incipient speciation in *Anopheles gambiae s.str.* in the Gambia, West Africa. Genetica 1982;59: 167−76.

[14] Chinery WA. Effects of ecological changes on the malaria vectors *Anopheles funestus* and the *Anopheles gambiae* complex of mosquitoes in Accra, Ghana. J Trop Med Hyg 1984; 87(2):75−81.

[15] Awolola TS, Oduola AO, Obansa JB, Chukwurar NJ, Unyimadu JP. *Anopheles gambiae* s. s. breeding in polluted water bodies in urban Lagos, southwestern Nigeria. J Vector Borne Dis 2007;44(4):241−4.

[16] Reddy Mr, Overgaard HJ, Abaga S, Reddy VP, Caccone A, Kiszewski AE, et al. Outdoor host seeking behaviour of *Anopheles gambiae* mosquitoes following initiation of malaria vector control on Bioko Island, Equatorial Guinea. Malar J 2011;10:184.

[17] Guelbeogo WM, Sagnon N, Liu F, Besansky NJ, Costantini C. Behavioural divergence of sympatric *Anopheles funestus* populations in Burkina Faso. Malar J 2014;13:65.

[18] Sinka ME, Bangs MJ, Manguin S, Coetzee M, Mbogo CM, Hemingway J, et al. The dominant Anopheles vectors of human malaria in Africa, Europe and the Middle East: occurrence data, distribution maps and bionomic precis. Parasit Vectors 2010;3:117.

[19] Dolo G, Briet OJ, Dao A, Traore SF, Bouare M, Sogoba N, et al. Malaria transmission in relation to rice cultivation in the irrigated Sahel of Mali. Acta Trop 2004;89(2):147−59.

[20] Singh B, Kim Sung L, Matusop A, Radhakrishnan A, Shamsul SS, Cox-Singh J, et al. A large focus of naturally acquired *Plasmodium knowlesi* infections in human beings. Lancet 2004;363(9414):1017−24.

[21] WHO. World malaria report. *Geneva: World Health Organization*, <http://www.who.int/ malaria/publications/world_malaria_report_2013/en/index.html>; 2013.

[22] Liu W, Li Y, Shaw KS, Learn GH, Plenderleith LJ, Malenke JA, et al. African origin of the malaria parasite *Plasmodium vivax*. Nat Commun 2014; 5:3346.

[23] WHO. Interim position statement—the role of larviciding for malaria control in sub-Saharan Africa. Geneva: World Health Organization, <http://www.who.int/malaria/publications/atoz/larviciding_position_statement/en/index.html>; 2012 [accessed 15.10.13].

[24] WHO. Guidance framework for testing of genetically modified mosquitoes. World Health Organization on behalf of the Special Programme for Research and Training in Tropical Diseases; 2014. ISBN 9789241507486. <http://www.who.int/tdr/publications/year/2014/Guidance_framework_mosquitoes.pdf>.

[25] Ng'habi KR, Knols BG, Lee Y, Ferguson HM, Lanzaro GC. Population genetic structure of *Anopheles arabiensis* and *Anopheles gambiae* in a malaria endemic region of southern Tanzania. Malar J 2011;10:289.

[26] Riehle MM, Guelbeogo WM, Gneme A, Eiglmeier K, Holm I, Bischoff E, et al. A cryptic subgroup of *Anopheles gambiae* is highly susceptible to human malaria parasites. Science 2011;331(6017):596—8.

[27] Gonsalves L, Bicknell B, Law B, Webb C, Monamy V. Mosquito consumption by insectivorous bats: does size matter? PLoS One 2013;8(10):e77183.

[28] Gonsalves L, Law B, Webb C, Monamy V. Foraging ranges of insectivorous bats shift relative to changes in mosquito abundance. PLoS One 2013;8(5):e64081.

[29] Dexter JS. Mosquitoes pollinating orchids. Sciences 1913;37(962):867 New Series.

[30] Scott TW, Takken W, Knols BG, Boete C. The ecology of genetically modified mosquitoes. Science 2002;298(5591):117—19.

[31] Benedict M, D'Abbs P, Dobson S, Gottlieb M, Harrington L, Higgs S, et al. Guidance for contained field trials of vector mosquitoes engineered to contain a gene drive system: recommendations of a scientific working group. Vector Borne Zoonotic Dis 2008;8(2):127—66.

[32] Alphey L, Beard CB, Billingsley P, Coetzee M, Crisanti A, Curtis C, et al. Malaria control with genetically manipulated insect vectors. Science 2002;298(5591):119—21.

[33] ACME-ASTMH. Arthropod containment guidelines, version 3.1; December 20, 2001.

Chapter 4

Ecology of Malaria Vectors and Current (Nongenetic) Methods of Control in the Asia Region

Patchara Sriwichai[1], Rhea Longley[2,3] and Jetsumon Sattabongkot[2]

[1]Faculty of Tropical Medicine, Department of Medical Entomology, Mahidol University, Bangkok, Thailand, [2]Faculty of Tropical Medicine, Mahidol Vivax Research Unit, Mahidol University, Bangkok, Thailand, [3]The Walter and Eliza Hall Institute of Medical Research, Melbourne, VIC, Australia

MALARIA RISK IN THE SOUTH EAST ASIA REGION

Approximately 1.4 billion people are at some risk of malaria infection in the South East Asia region (SEAR) from the 10 encompassed countries considered endemic, with 352 million at high risk [1]. According to the WHO, the region has seen remarkable change over the past 13 years, with a drop of more than 75% in the incidence of microscopically confirmed cases in 6 of the 10 endemic countries. Furthermore, Sri Lanka reported zero cases for the first time in 2013. From the remaining 9 countries, 1.5 million cases were reported, with the majority in India (58%), Myanmar (22%), and Indonesia (16%). The 2014 East Asia summit formally endorsed a malaria-free Asia Pacific by 2030, reflecting a global change of strategy from malaria control to elimination [2]. A number of tools will be critical in the success of any elimination or eradication strategy by reducing malaria transmission, namely antimalarial drugs, vaccines, and vector control measures.

The most common and effective malaria vector control strategies currently in use are based on insecticides: indoor residual spraying (IRS) and long-lasting insecticide-treated nets (LLINs). However, these interventions were designed for indoor-biting vectors, and hence are not necessarily effective in all regions. This demonstrates the importance of having a sound understanding of the vector species ecology within regions before embarking on vector control interventions. Within the SEAR, there is a large diversity of vector species with pronounced behavioral plasticity [3], making this

Genetic Control of Malaria and Dengue.
© 2016 Elsevier Inc. All rights reserved.

region one of the most complex in terms of vector ecology and in turn for vector control measures. This chapter aims to (i) describe the ecology of common malaria vectors within Asia, (ii) provide an overview of the current (nongenetic) methods of vector control, and (iii) the success of such methods used within this region. Finally, key challenges facing (nongenetic) vector control methods will be discussed.

ECOLOGY OF COMMON MALARIA VECTORS

Mosquito species that transmit malaria belong to the genus *Anopheles*; globally, there are 41 recognized dominant malaria vectors [4], and 19 are present within the SEAR [3]. Of these 19 species, there are numerous behavioral differences that can have huge impacts on the success of vector control programs such as IRS and LLINs. The species can have preference for biting either humans (anthropophilic) or animals (zoophilic), for indoor (endophagic) or outdoor (exophagic) feeding, and for resting pre- and/or postfeeding either indoors (endophilic) or outdoors (exophilic). A number of reviews have described the major malaria vectors within the SEAR, the most recent by Sinka and colleagues in 2011 [3], Cui and colleagues in 2012 [5], and by Hii and colleagues in 2013 as part of the Mekong Malaria III monograph [6]. Of the 19 dominant malaria vectors described within the SEAR, 12 are primarily anthropophilic (although feeding preference can also change dependent on locality and available food sources). Here, we will briefly review the ecological and behavioral characteristics of only the most important anthropophilic vectors within this region: *Anopheles culicifacies, Anopheles fluviatilis,* and *Anopheles stephensi* across the Indian subcontinent; and *Anopheles dirus* and *Anopheles minimus* within Southeast Asia.

Anopheles culicifacies

Anopheles culicifacies is a major malaria vector within India, the country with the majority of reported malaria cases from the SEAR in 2013. It is actually a species complex containing five species informally designated as A, B, C, D, and E. Of the five species, species E is known as a particularly efficient vector due to its highly anthropophilic and endophilic behavior and susceptibility to both *Plasmodium falciparum* and *Plasmodium vivax* [7]. The remaining species are primarily zoophilic and tend to be considered as playing more minor roles in human malaria transmission in general [8]. This vector species causes malaria transmission primarily in urban areas of the country and generates approximately 65% of total Indian cases annually [9]. Importantly, this species complex appears to have variable ecological habitats [10], but biting is conducted during the night and resting occurs largely indoors [3,8,11].

Anopheles fluviatilis

Anopheles fluviatilis is another complex of species widespread throughout the SEAR (such as Iran, Pakistan, Afghanistan, Nepal, Bangladesh, and Myanmar [3]) and is a particularly important vector in India after *A. culcifacies* [12], at one time accounting for 15% of malaria cases reported annually [9]. The complex consists of at least three sibling species, informally designated as S, T, and U [13]. Species S is highly anthropophilic [14], whereas the others are primarily zoophilic [15]. Much more information is required for this vector complex about the range and locality of each species and behavioral factors, particularly whether it is endo- or exophagic.

Anopheles stephensi

Anopheles stephensi is not a species complex but does contain three ecological variants, with the "type form" considered most important for malaria transmission [8]. It is an important vector in urban areas of India [16] and is generally considered to be both endophagic and endophilic with some behavioral plasticity reported [3]. This species is also found in Iran, Iraq, Bangladesh, China, Myanmar, and Thailand [3].

Anopheles dirus

Anopheles dirus is again widespread throughout the region [17,18] and is one of the major vectors in China, Thailand, Myanmar, Cambodia, and Vietnam [5]. However, this complex contains eight species with variable ecological preferences and epidemiological importance within different countries [8,19], making generalizable statements difficult for this complex. Nevertheless, the habitat is largely in forested areas across the region and a number of species are highly anthropophilic and capable of both indoor and outdoor biting [3,20–22]. This makes vector control of this species complex highly difficult.

Anopheles minimus

Anopheles minimus is considered an important malaria vector throughout forested regions of the SEAR [23], particularly outside of India (namely, China, Thailand, Myanmar, Cambodia, Vietnam, and Laos [5]). There are three species within this complex, with two capable of malaria transmission [24]. Again, this species complex exhibits high levels of plasticity in both ecological and behavioral traits across the region [3], but is highly anthropophilic [25–27] and has recently been shown to have a slight preference for outdoor feeding in Thailand [26].

Other malaria vectors not discussed here that are gaining increasing recognition as important species within this region are the Maculatus [28], Sundaicus [29], Subpictus [3], and Barbirostris [30] groups.

CURRENT (NONGENETIC) METHODS OF VECTOR CONTROL

Indoor Residual Spraying

IRS reduces malaria transmission in areas where the primary vectors feed and/or rest indoors by reducing the survival of mosquitoes that enter houses or sleeping units. It involves spraying an effective dose of insecticide, commonly one to two times per year, on indoor surfaces where malaria vectors are likely to rest after biting. The aim is to reduce human−vector contact through a repellant effect as well as direct killing of the mosquitoes. In order to achieve effective protection, IRS requires a high level of coverage in space and time (more than 80% of living structures within the target region). Globally, an estimated 3.5% of the population at risk was protected by IRS in 2013 [1], which falls far below the threshold required. Nevertheless, IRS has successfully contributed to reducing the incidence of malaria in unstable transmission regions in the past [31].

The WHO Pesticide Evaluation Scheme currently recommends 12 insecticides belonging to 4 chemical classes for IRS (all are nerve poisons). The insecticide used should ideally be rotated annually in order to reduce the risk of resistance (whether behavioral or physiological) and resistance monitoring needs to be undertaken to ensure appropriate insecticides are used. Within the SEAR, the most commonly used insecticide for IRS is the pyrethroids with only one country still reporting the use of DDT [1].

Long-Lasting Insecticide-Treated Nets

LLINs last longer than conventional insecticide-treated bed nets (ITNs), maintaining their biological efficacy for approximately 3 years, but still need to be replaced regularly. LLINs provide both a physical barrier and an insecticidal effect to reduce human−mosquito contact during sleeping hours. In addition to protecting individuals [32], they also have an important community effect as usage by large numbers of community members leads to large-scale killing of mosquitoes. Globally, there is a huge reliance on the pyrethroids, as not only is this class of insecticide used widely for IRS, it is also the only class used for LLINs. In comparison to IRS where the uptake of this intervention in at-risk areas is low, LLINs have been widely adopted with 49% of the at-risk population having access [1].

Source Reduction

Source reduction (SR) is focused on larval control through interventional strategies such as vector habitat modification or manipulation and larviciding or biological control. The WHO position is that SR should only be undertaken in regions where limited and known breeding sites exist that are easy to access and monitor. However, SR has been successful in the past [33].

Integrated Vector Management

The overall approach to vector control should be considered as part of an integrated vector management (IVM) approach. The theory behind IVM is that two or more vector control strategies should be used, and are used in regions most appropriate to each individual strategy. This is supported by both modeling [34,35] and practical evidence [36,37]. In addition to IRS, LLINs and SR, an increase in the use of personal protection measures (other than LLINs) can also be incorporated into IVM. These measures may include window screens and structural changes to prevent entering of mosquitoes into dwellings, insecticide-treated hammocks, use of repellents, and protective clothing. These measures are particularly critical for regions subject to outdoor transmission of malaria as well as areas where the local occupation is related to the forest or staying overnight in the jungle.

PROGRESS OF (NONGENETIC) VECTOR CONTROL IN THE ASIA REGION

According to the WHO, of the 10 malaria endemic countries within the SEAR, either ITNs or LLINs are provided free of charge to all age groups [1], and all national malaria control programs formally recommend IRS. However, actual coverage rates vary dramatically from near 100% of the high-risk population groups in Nepal, Bhutan, and Bangladesh using ITNs to less than 10% in India. IRS was only reported in 6 of the 10 countries, with greater than 50% coverage only in Korea. Only one national malaria control program reported continued use of DDT for IRS; however, only three countries reported monitoring of insecticide resistance.

Evidence of Success

The incidence of malaria within the SEAR has been reduced dramatically over the past 13 years. Vector control strategies must be considered integral to this success, although an abundance of direct evidence is lacking. Sri Lanka is an interesting case study, given zero cases were reported in 2013 for the first time. However, Sri Lanka had previously dramatically reduced malaria incidence in 1963, but suffered a resurgence in cases once control efforts (such as ITNs) and funding were reduced (reviewed in Ref. [38]). *Anopheles culicifacies* species E is the dominant malaria vector in Sri Lanka, which is predominantly indoor biting. This has likely contributed to the success of ITNs in this country. However, Sri Lanka also has a national IVM strategy that includes entomological surveillance and larvicidal activities; thus, it is likely these measures used in combination that were critical to the success of the program.

Scattered evidence also exists demonstrating the efficacy of ITNs or LLINs within the SEAR [39]: in Cambodia, the use of ITNs provided 28%

protective efficacy 10 months after distribution [40]; and two studies in Thailand have demonstrated approximately a 40% reduction in clinical malaria cases following distribution of ITNs [41,42]. A reason for the relatively low levels of protection provided by ITNs is likely that the dominant vector species for malaria in both Cambodia and Thailand are outdoor biting (*An. dirus* and *An. minimus*). In comparison, great success was reported by a study conducted in China during 1985−1987; the use of impregnated nets reduced indoor vector density by up to 93% and monthly malaria incidence by up to 89% [43]. The region of China studied (Guangdong province) at that time had both endo- and exophilic vectors. Further studies are required to determine the efficacy of LLINs and IRS in more regions of the SEAR.

In addition to ITNs, the efficacies of SR measures have also been assessed in some regions of the SEAR. In an area of south India, *Poecilia reticulata* fish were introduced into all the wells and lakes where vector species (*An. culicifacies* and *An. fluviatilis*) were initially present; after only 1 year, no vectors were present and no malaria cases were detected (the village normally had a high annual incidence) [44].

Evidence of Resistance to Insecticides

According to the WHO, of the 10 malaria endemic countries within the SEAR, only three reported insecticide-monitoring activities. It is crucial that such monitoring becomes part of IVM approaches for national malaria control programs, particularly given the reliance on pyrethroids for LLINs and the wide distribution of those nets throughout the region. Within the region, insecticide resistance has been assessed predominantly in India, likely due to the high malaria burden still experienced by this country. Resistance to DDT is widespread throughout the country for the main vector (*An. culicifacies*), and now emerging resistance has also been reported to malathion (used in IRS) and the pyrethroids, but at a much lower level than for DDT [45−47]. In 2012, *An. fluviatilis* was still susceptible to DDT, malathion, and pyrethroids [47].

A network was also set up to monitor resistance in the Mekong region (Cambodia, Laos, Thailand, and Vietnam) [48]. Differences were found across countries and vector species: *An. dirus* and *An. minimus* were found to be susceptible to pyrethroids in all countries except Vietnam, where a degree of resistance was evident for both vector species. However, this study was completed in 2005 and so the current status of resistance in the Mekong is unknown. This is the case for most studies on insecticide resistance within the greater Asia region; they were conducted more than 10 years ago and the information is hence out of date [49−51]. The exception is in Thailand where a more recent study demonstrated susceptibility of both *An. dirus* and *An. minimus* to malathion and pyrethroids [52].

KEY CHALLENGES

The malaria eradication research agenda consultative group on vector control identified three major challenges for this strategy: (i) identifying new classes of insecticides for use in LLINs and IRS, (ii) developing control measures for outdoor-biting vectors, and (iii) determining how to permanently reduce vectorial capacity in high transmission regions (i.e., within Africa) [53]. The first two are also key challenges for vector control programs within the SEAR.

Insecticide Resistance Monitoring

The WHO established the global plan for insecticide resistance management in malaria vectors, encompassing five strategies: (i) resistance monitoring, (ii) management strategies, (iii) identifying mechanisms of resistance, (iv) developing new control tools, and (v) to obtain enabling mechanisms in place. The aim is to track resistance globally, and this is one area where the SEAR needs to improve. Insecticide resistance is not currently being effectively monitored, and few published studies exist. From those that do, insecticide resistance does not yet seem to be a major problem outside of India; this may be a reflection of the lack of monitoring, or due to the outdoor-biting preference of vector species within other SEAR countries (and hence a lack of exposure to IRS/LLINs).

Outdoor Biting

One of the biggest challenges for this region is the high proportion of vector species that exhibit outdoor-biting behavior; outdoor biting is estimated to be responsible for half of the outdoor transmission within this region, excluding *An. culicifacies* [54]. This highlights why traditional methods of vector control such as IRS and LLINs are unlikely to be sufficient to reach elimination targets set by the 2014 East Asia summit. One alternative strategy that has been trialed and was partially successful was the use of long-lasting insecticidal hammocks in both Cambodia (46% protection from *An. minimus* biting) [55] and Vietnam (1.6-fold reduction in malaria prevalence) [56]. Another interesting strategy that could be an alternative to LLINs and IRS in outdoor-biting regions is the use of natural or artificial sugar sources treated with insecticides. As all vectors are relied upon sugar for a food source (and perhaps even more so when infected with *Plasmodium* spp. [57]), this method has been quite effective in the small number of studies undertaken so far [58−60].

Behavioral Changes of Vectors

Another challenge for vector control in this region is the relative plasticity of behavioral phenotypes of multiple vector species. This is also a challenge for endemic settings with indoor-biting vectors as IRS/LLINs have introduced behavior resistance or evolution in these vectors [61]. It will be interesting to see whether the introduction of such vector control measures concentrated on indoor biting may have the same effect in the SEAR, directing more of the vector species toward a primarily exophagic phenotype. Furthermore, given the known behavioral plasticity, more effort needs to be addressed to determine the effects of climate and other environmental changes on the ecology of malaria vectors within the region [28,62,63].

CONCLUSIONS

1. Malaria vectors in Asia are species complexes. Within the complex, host preference behavior, habitat, and biting time can be different and thus impact on their vector capacity and control.
2. There is limited research in evaluating tools that have been used for control of the vectors in this region, due to the above issues. Certain control tools such as the LLINs can be used effectively if the vectors' biting time is the same as the sleeping time of the population.
3. Ideally, coverage and usage of LLINs should be 100% in endemic areas and this is key for success of this vector control method. Likewise, community acceptance of IRS is key to its success. This is the case in Asia where the coverage and usage is not 100% in most endemic areas.
4. SR or biological control of larvae can be used only if larval breeding places are identified and the habitats are suitable for the selected tools; thus, there is limited impact on certain breeding places such as rice fields or forests, for example.
5. Although there are many nongenetic tools for vector control, the limitation is the research-based evaluation of efficiency of those tools in many areas.
6. The limitation for research of malaria vectors in Asia is due to the high complexity of many vector species and the difficulty to maintain most of the important vectors in the laboratory, as it is very labor intensive.
7. In many places, unidentified malaria vectors can play important roles on transmission maintenance when the major vectors have been controlled. Therefore, there should be more studies on potential secondary vectors in the same endemic areas.
8. Insecticide resistance has been monitored sporadically in many areas; however, this was limited to vector species that were abundant in the area or by the availability of colonized species in laboratory, which often do not accurately represent wild caught species. Innovative protocols to determine insecticide resistance in areas that have many vector species with different peaks should be considered.
9. Finally, the challenges in applying genetic control methods to malaria vectors in Asia will be similar to the other nongenetic tools that have

been used to control vectors in this region: species complexity, multiple vectors, and potential vectors in each endemic area and only a few species that can be successfully maintained in the laboratory. However, if genetic tools can be used to control any species in an endemic area, it is likely that this can be applied to other areas in the region where the same species exist, given this region shares many common vectors and species complexes.

REFERENCES

[1] World Malaria Report 2014. Geneva: World Health Organization; 2014.

[2] Alonso PL, et al. A research agenda to underpin malaria eradication. PLoS Med 2011;8: e1000406.

[3] Sinka ME, et al. The dominant *Anopheles* vectors of human malaria in the Asia-Pacific region: occurrence data, distribution maps and bionomic precis. Parasit Vectors 2011; 4:89.

[4] Hay SI, et al. Developing global maps of the dominant anopheles vectors of human malaria. PLoS Med 2010;7:e1000209.

[5] Cui L, et al. Malaria in the greater mekong subregion: heterogeneity and complexity. Acta Trop 2012;121:227.

[6] Singhasivanon P. Towards malaria elimination in the greater Mekong subregion. Southeast Asian J Trop Med Public Health 2013;44(Suppl. 1):iii.

[7] Kar I, et al. Evidence for a new malaria vector species, species E, within the *Anopheles culicifacies* complex (Diptera: Culicidae). J Med Entomol 1999;36:595.

[8] Dev V, Sharma VP. The dominant mosquito vectors of human 2. Malaria in India. In: Manguin S, editor. *Anopheles mosquitoes – New insights into malaria vectors*. Croatia: INTECH Publications; 2013. p. 239–71.

[9] Sharma VP. Re-emergence of malaria in India. Indian J Med Res 1996;103:26.

[10] Surendran SN, Ramasamy R. Some characteristics of the larval breeding sites of *Anopheles culicifacies* species B and E in Sri Lanka. J Vector Borne Dis 2005;42:39.

[11] Basseri HR, Abai MR, Raeisi A, Shahandeh K. Community sleeping pattern and anopheline biting in southeastern Iran: a country earmarked for malaria elimination. Am J Trop Med Hyg 2012;87:499.

[12] Singh N, et al. Dynamics of forest malaria transmission in Balaghat district, Madhya Pradesh, India. PLoS One 2013;8:e73730.

[13] Nanda N, et al. Population cytogenetic and molecular evidence for existence of a new species in *Anopheles fluviatilis* complex (Diptera: Culicidae). Infect Genet Evol 2013;13:218.

[14] Tripathy A, et al. Distribution of sibling species of *Anopheles culicifacies* s.l. and *Anopheles fluviatilis* s.l. and their vectorial capacity in eight different malaria endemic districts of Orissa, India. Mem Inst Oswaldo Cruz 2010;105:981.

[15] Nanda N, et al. Anopheles fluviatilis complex: host feeding patterns of species S, T, and U. J Am Mosq Control Assoc 1996;12:147.

[16] Sumodan PK, Kumar A, Yadav RS. Resting behavior and malaria vector incrimination of *Anopheles stephensi* in Goa, India. J Am Mosq Control Assoc 2004;20:317.

[17] Ngo CT, et al. Diversity of Anopheles mosquitoes in Binh Phuoc and Dak Nong Provinces of Vietnam and their relation to disease. Parasit Vectors 2014;7:316.

[18] Obsomer V, Defourny P, Coosemans M. Predicted distribution of major malaria vectors belonging to the *Anopheles dirus* complex in Asia: ecological niche and environmental influences. PLoS One 2012;7:e50475.

[19] Tananchai C, et al. Species diversity and biting activity of *Anopheles dirus* and *Anopheles baimaii* (Diptera: Culicidae) in a malaria prone area of western Thailand. Parasit Vectors 2012;5:211.

[20] Ritthison W, Tainchum K, Manguin S, Bangs MJ, Chareonviriyaphap T. Biting patterns and host preference of *Anopheles epiroticus* in Chang Island, Trat Province, eastern Thailand. J Vector Ecol 2014;39:361.

[21] Obsomer V, Defourny P, Coosemans M. The *Anopheles dirus* complex: spatial distribution and environmental drivers. Malar J 2007;6:26.

[22] Vythilingam I, et al. The prevalence of Anopheles (Diptera: Culicidae) mosquitoes in Sekong Province, Lao PDR in relation to malaria transmission. Trop Med Int Health 2003;8:525.

[23] Kongmee M, et al. Seasonal abundance and distribution of Anopheles larvae in a riparian malaria endemic area of western Thailand. Southeast Asian J Trop Med Public Health 2012;43:601.

[24] Garros C, Van Bortel W, Trung HD, Coosemans M, Manguin S. Review of the minimus complex of *Anopheles*, main malaria vector in Southeast Asia: from taxonomic issues to vector control strategies. Trop Med Int Health 2006;11:102.

[25] Yu G, et al. The *Anopheles* community and the role of *Anopheles minimus* on malaria transmission on the China-Myanmar border. Parasit Vectors 2013;6:264.

[26] Tisgratog R, et al. Host feeding patterns and preference of *Anopheles minimus* (Diptera: Culicidae) in a malaria endemic area of western Thailand: baseline site description. Parasit Vectors 2012;5:114.

[27] Bashar K, Tuno N, Ahmed TU, Howlader AJ. Blood-feeding patterns of *Anopheles* mosquitoes in a malaria-endemic area of Bangladesh. Parasit Vectors 2012;5:39.

[28] Suwonkerd W, et al. Changes in malaria vector densities over a twenty-three year period in Mae Hong Son Province, northern Thailand. Southeast Asian J Trop Med Public Health 2004;35:316.

[29] Linton YM, et al. *Anopheles* (Cellia) *epiroticus* (Diptera: Culicidae), a new malaria vector species in the Southeast Asian Sundaicus Complex. Bull Entomol Res 2005;95:329.

[30] Durnez L, et al. False positive circumsporozoite protein ELISA: a challenge for the estimation of the entomological inoculation rate of malaria and for vector incrimination. Malar J 2011;10:195.

[31] Pluess B, Tanser FC, Lengeler C, Sharp BL. Indoor residual spraying for preventing malaria. Cochrane Database Syst Rev 2010;CD006657.

[32] Lengeler C. Insecticide-treated bed nets and curtains for preventing malaria. Cochrane Database Syst Rev 2004;CD000363.

[33] Tusting LS, et al. Mosquito larval source management for controlling malaria. Cochrane Database Syst Rev 2013;8:CD008923.

[34] Lutambi AM, Chitnis N, Briet OJ, Smith TA, Penny MA. Clustering of vector control interventions has important consequences for their effectiveness: a modelling study. PLoS One 2014;9:e97065.

[35] Killeen GF, et al. The potential impact of integrated malaria transmission control on entomologic inoculation rate in highly endemic areas. Am J Trop Med Hyg 2000;62:545.

[36]   Okech BA, et al. Use of integrated malaria management reduces malaria in Kenya. PLoS One 2008;3:e4050.

[37]   Fillinger U, Ndenga B, Githeko A, Lindsay SW. Integrated malaria vector control with microbial larvicides and insecticide-treated nets in western Kenya: a controlled trial. Bull World Health Organ 2009;87:655.

[38]   Abeyasinghe RR, Galappaththy GN, Smith Gueye C, Kahn JG, Feachem RG. Malaria control and elimination in Sri Lanka: documenting progress and success factors in a conflict setting. PLoS One 2012;7:e43162.

[39]   Sexton JD. Impregnated bed nets for malaria control: biological success and social responsibility. Am J Trop Med Hyg 1994;50:72.

[40]   Sochantha T, et al. Insecticide-treated bednets for the prevention of *Plasmodium falciparum* malaria in Cambodia: a cluster-randomized trial. Trop Med Int Health 2006;11:1166.

[41]   Luxemburger C, et al. Permethrin-impregnated bed nets for the prevention of malaria in schoolchildren on the Thai-Burmese border. Trans R Soc Trop Med Hyg 1994;88:155.

[42]   Kamol-Ratanakul P, Prasittisuk C. The effectiveness of permethrin-impregnated bed nets against malaria for migrant workers in eastern Thailand. Am J Trop Med Hyg 1992;47: 305.

[43]   Li ZZ, et al. Trial of deltamethrin impregnated bed nets for the control of malaria transmitted by *Anopheles sinensis* and *Anopheles anthropophagus*. Am J Trop Med Hyg 1989;40:356.

[44]   Ghosh SK, et al. Larvivorous fish in wells target the malaria vector sibling species of the *Anopheles culicifacies* complex in villages in Karnataka, India. Trans R Soc Trop Med Hyg 2005;99:101.

[45]   Mishra AK, Chand SK, Barik TK, Dua VK, Raghavendra K. Insecticide resistance status in *Anopheles culicifacies* in Madhya Pradesh, central India. J Vector Borne Dis 2012; 49:39.

[46]   Raghavendra K, et al. A note on the insecticide susceptibility status of principal malaria vector *Anopheles culicifacies* in four states of India. J Vector Borne Dis 2014;51:230.

[47]   Sahu SS, et al. Response of malaria vectors to conventional insecticides in the southern districts of Odisha State, India. Indian J Med Res 2014;139:294.

[48]   Van Bortel W, et al. The insecticide resistance status of malaria vectors in the Mekong region. Malar J 2008;7:102.

[49]   Prapanthadara L, et al. Correlation of glutathione *S*-transferase and DDT dehydrochlorinase activities with DDT susceptibility in *Anopheles* and *Culex* mosquitos from northern Thailand. Southeast Asian J Trop Med Public Health 2000;31(Suppl. 1):111.

[50]   Cui F, Raymond M, Qiao CL. Insecticide resistance in vector mosquitoes in China. Pest Manag Sci 2006;62:1013.

[51]   Chareonviriyahpap T, Aum-aung B, Ratanatham S. Current insecticide resistance patterns in mosquito vectors in Thailand. Southeast Asian J Trop Med Public Health 1999;30:184.

[52]   Pemo D, Komalamisra N, Sungvornyothin S, Attrapadung S. Efficacy of three insecticides against *Anopheles dirus* and *Anopheles minimus*, the major malaria vectors, in Kanchanaburi Province, Thailand. Southeast Asian J Trop Med Public Health 2012;43: 1339.

[53]   Alonso PL, Besansky NJ, Collins FH, Hemingway J, James AA, Lengeler C, et al. A research agenda for malaria eradication: vector control. PLoS Med 8: e1000401, 2011.

[54]   Killeen GF. Characterizing, controlling and eliminating residual malaria transmission. Malar J 2014;13:330.

[55]   Sochantha T, et al. Personal protection by long-lasting insecticidal hammocks against the bites of forest malaria vectors. Trop Med Int Health 2010;15:336.

[56]   Thang ND, et al. Long-lasting insecticidal hammocks for controlling forest malaria: a community-based trial in a rural area of central Vietnam. PLoS One 2009;4:e7369.

[57]   Nyasembe VO, et al. *Plasmodium falciparum* infection increases *Anopheles gambiae* attraction to nectar sources and sugar uptake. Curr Biol 2014;24:217.

[58]   Muller GC, Schlein Y. Efficacy of toxic sugar baits against adult cistern-dwelling *Anopheles claviger*. Trans R Soc Trop Med Hyg 2008;102:480.

[59]   Muller GC, et al. Successful field trial of attractive toxic sugar bait (ATSB) plant-spraying methods against malaria vectors in the *Anopheles gambiae* complex in Mali, West Africa. Malar J 2010;9:210.

[60]   Beier JC, Muller GC, Gu W, Arheart KL, Schlein Y. Attractive toxic sugar bait (ATSB) methods decimate populations of *Anopheles* malaria vectors in arid environments regardless of the local availability of favoured sugar-source blossoms. Malar J 2012;11:31.

[61]   Reddy MR, et al. Outdoor host seeking behaviour of *Anopheles gambiae* mosquitoes following initiation of malaria vector control on Bioko Island, Equatorial Guinea. Malar J 2011;10:184.

[62]   Martens WJ, Niessen LW, Rotmans J, Jetten TH, McMichael AJ. Potential impact of global climate change on malaria risk. Environ Health Perspect 1995;103:458.

[63]   Tanser FC, Sharp B, le Sueur D. Potential effect of climate change on malaria transmission in Africa. Lancet 2003;362:1792.

Chapter 5

Ecology of *Anopheles darlingi*, the Primary Malaria Vector in the Americas and Current Nongenetic Methods of Vector Control

Jan E. Conn[1,2] and Paulo E. Ribolla[3]

[1]*The Wadsworth Center, New York State Department of Health, Albany, NY, USA,* [2]*Biomedical Sciences Department, School of Public Health, State University of New York-Albany, Albany, NY, USA,* [3]*Departamento de Parasitologia, Instituto de Biociências, Universidade Estadual Paulista "Júlio de Mesquita Neto", Botucatu, São Paulo, Brazil*

INTRODUCTION

The Latin American countries with the highest malaria burden are Brazil, Bolivia, Colombia, Ecuador, Guyana, Peru, Suriname, and Venezuela [1,2]. Suriname should now be excluded from this list (Section 6). Overall in the Americas, the most common malaria parasite is *Plasmodium vivax*, usually comprising between 60% and 70% of the human cases, with *Plasmodium falciparum* responsible for the remaining 30−40% [2]. The main vector in these countries, considered to be the most effective vector in the Americas, particularly in the Amazon basin, is *Anopheles darlingi*. This species is broadly but unevenly distributed (see Insert, Figure 5.1) from southern Mexico (Chiapas state) to Nicaragua. There is a gap in Nicaragua, Costa Rica and most of Panama where it is reportedly absent, although it was recently detected in southern Panama for the first time [3]. In South America, it is found from Colombia and Venezuela to northern Argentina and from east of the Andes to the Atlantic coast [3−6]. It has been incriminated as a vector in many localities and regions (Figure 5.1) [5,7,8].

The impact of *An. darlingi* is facilitated by opportunistic feeding (depending mainly on host availability and environmental conditions such as rainfall), plasticity of biting behavior (inside or outside houses, temporal variation), a

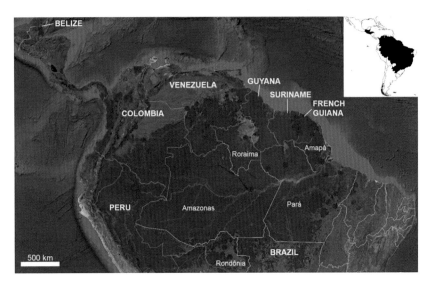

FIGURE 5.1 Localities where *An. darlingi* has been reported infected with *Plasmodium* by ELISA or RT-PCR methods from 1995 to 2014.

range of breeding site types associated with rivers or flooded forest [9−11], and rapid adaptation to novel environments provided by whole ecosystem alteration via factors such as changes in temperature and humidity associated with the fragmentation of continuous forest [12−14]. Such adaptation can lead to increased vectorial capacity and higher malaria transmission risk [15,16]. *Anopheles darlingi* also utilizes artificial water collection depressions, which are by-products of gold-mining, brick-making, and/or road construction, as well as fish ponds, wells, and cisterns, for breeding [12,17,18] and is increasingly involved in transmission in peri-urban and urban settings [17,19,20].

There is evidence in *An. darlingi* for divergence in the nuclear [3,21], but not in the complete mitochondrial genome [22]. A better understanding of the implications of these results should now be possible because of the recent publication of the genome [23]. There is ample documentation of population structure within *An. darlingi* [18,21,24−26], and the effects of distance are evident in some studies [27,28] but in others, geographic/biogeographic barriers [24] or local environmental conditions [18] are the likely drivers of structure and divergence. These population differences in *An. darlingi* were tested using microsatellite loci [21], which detected (i) a large division between Central America plus northeastern Colombia and Venezuela (genotype 2), and an Amazonian region (genotype 1); (ii) within the Amazon region, additional divisions include northeastern Amazonian Brazil; Central and Western Brazil; and Amazonian Peru (Figure 5.2). Pairwise F_{ST} comparisons among 1,376 samples from Belize, Brazil, Guatemala, and Peru ranged

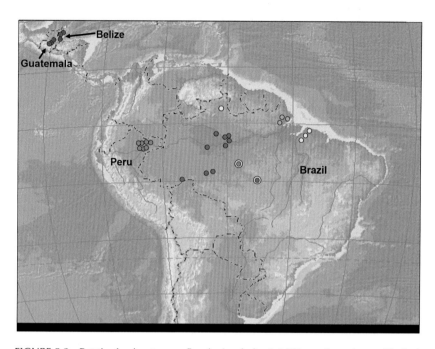

FIGURE 5.2 Putative barriers to gene flow in *An. darlingi:* 1,354 samples, microsatellite loci, analyzed by STRUCTURE, resulting in five groups: [21] Fuchsia—Belize + Guatemala; Orange—Peru; Lime Green—NE Amazonian Brazil; Blue—western & central Amazonian Brazil; Yellow—southeastern Amazonia and central Brazil.

from -0.0005 to 0.3901, with the largest values corresponding to comparisons between genotypes 1 and 2 ($0.1859-0.3901$), all significant at $P < 0.05$ after Bonferroni correction [21]. Additional support for the major north–south division is the presence of five fixed mutation differences between them in sequences of the nuclear *white* gene in 286 samples from Belize, Bolivia, Brazil, Colombia, French Guinea, Guatemala, Panama, Peru, and Venezuela (Conn and collaborators, unpublished). Analysis using a fragment of the mtDNA cytochrome oxidase I (*COI*) gene and additional southern Brazilian samples detected five groups of *An. darlingi* in Brazil: Northeast 1; Northeast 2; Central; South; and Southeast (Figure 5.3) [24,25]. Together these data lend support to a new study using SNPs that detected three incipient species in Brazil [124]. These species may differ in ecological and/or behavioral traits that could be exploited for more effective vector control.

Patterns of heterogeneous transmission in localities where *An. darlingi* has been incriminated have been documented in the Neotropics [7,29–31], even though homogeneous transmission is a basic assumption of the earliest [32] and most subsequent epidemiological models. The urgent need for a novel synthesis to advance the theory of mosquito-borne pathogen

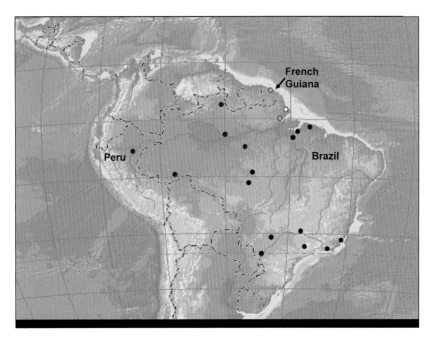

FIGURE 5.3   Putative barriers to gene flow in *An. darlingi* within Brazil: fragment of *COI* gene, analyzed by SAMOVA, resulting in five groups: [24] Red—Central; Lilac—Northeast 2; White—Northeast 1; Dark Royal Blue—South; Black—Southeast.

transmission acknowledging the ecological context of mosquito blood feeding and its quantitative impact on transmission was recently advocated [33]. Such information combined with detailed, accurate distribution data of vectors, *Plasmodium* (as in Ref. [34]), and human cases, could help focus genetic modification (GM) control efforts on malaria hot spots in the ongoing global effort to eliminate malaria.

ECOLOGY OF *ANOPHELES DARLINGI*

Distinct from Africa, where malaria endemic regions comprise a variety of different ecological niches from savanna to humid forest, in Central and South America, such endemic regions are mainly tropical forest, including the Amazon in South America and tropical and subtropical forests in Central America. Especially in the Amazon, forest cover level profoundly impacts the human-biting rate (HBR) of *An. darlingi*, and thus the risk of malaria transmission [35,36]. This difference in primary niche availability may reflect the diversity of malaria vectors in these two continents, that is, Africa is home to several major vectors, *Anopheles coluzzi*, *Anopheles gambiae*, *Anopheles arabiensis*, and *Anopheles funestus* s.l., whereas in the Americas, the primary vector diversity is reduced, with *An. darlingi* as the dominant vector in South America, where malaria is most prevalent, and *Anopheles albimanus* the most important in Central America [5,37,38] and along the Pacific coast in Colombia and northern Peru [7].

Larvae

Anopheles darlingi breeding sites generally are characterized as clear, shaded water bodies with floating or emergent vegetation [11,39,40]. Sometimes during the rainy seasons, immatures have been observed in small puddles and other shallow, temporary water collections but, because this mosquito has high sensitivity to low levels of moisture, it disappears in such environments as rainfall ceases. Thus, under natural conditions, habitats of immature forms are most common in the backwaters of rivers and streams of various sizes [4,41,42]. Streams in forest patches are not necessarily adequate to sustain immature forms of *An. darlingi*, since in such environments pH and temperature vary and shaded areas are uncommon [11]. The persistent perception that breeding sites are mainly along the margins of water collections is not supported by more recent data [11,43], although larvae are frequently associated with detritus patches [44] and submerged tree/shrub root hairs or macrophytes [9,42]. In larger ponds and pools that support *An. darlingi* breeding, larvae are found in deeper areas, away from the edges [6]. Collections that focus only around the edges of breeding sites likely introduce sampling error when larval abundance is measured by number of specimens collected.

Adult Behavior

Anopheles darlingi is notable for considerable plasticity in biting behavior, time, and host source, due to adaptation to local environmental conditions. For example, studies have documented a predominance of exophagy (reviewed in Refs [6,15,45]), endo-exophagy [14,46−48], and mono-, bi-, or trimodal peak biting times [6,29,49]. Two hypotheses could explain such diversity which is related to the interaction between mosquitoes and humans, that is, (i) mosquito genetic differences accounting for behavioral diversity and (ii) environment differences shaping mosquito behavior. These hypotheses are not mutually exclusive and it is probable that both contribute to anopheline plasticity.

The population density of *An. darlingi* and other neotropical anophelines varies dramatically seasonally, and often peaks near the end of the rainy season or during the transition between rainy and dry seasons [5,6,50], but in one substantial longitudinal study Amapá state, Brazil, the highest abundance occurred during the dry season [51]. Based on new findings of seasonally differentiated subpopulations in Rondônia, Brazil [18], two genetically distinct populations of *An. darlingi* may exist in Loreto Department,

Amazonian Peru, where marked changes in seasonal biting times are common [52]. There is evidence of *An. darlingi* resting indoors (endophily) [53,54] and outdoors (exophily) [10] but exophily predominates in many parts of its vast distribution [6]. In Brazil, there is less endophilic behavior of *An. darlingi* following the prolonged use of IRS [55].

Blood meal source depends on host availability: for example in Amapá state, Brazil, in three malaria endemic villages, *An. darlingi* fed primarily on cattle and pigs (~35% each), and less than 10% on humans [10]. This surprising discovery in a mosquito considered to be mainly anthropophilic is important because of the implications for estimating vector biology metrics such as the human blood index and the entomological inoculation rate (EIR) that are used to compare the importance of vectors in human malaria transmission. Considered to be a vital metric in estimating the relative importance of each species in malaria transmission, and in monitoring the effects of interventions, the EIR for *An. darlingi* ranges widely from 0 to 360 (Table 5.1). On average in the endemic regions in the Americas, *pf*EIRs are much lower (~1) than most in Africa and some in Asia [56]. However, there are many communities (in Colombia, French Guiana, Peru, Suriname, and Venezuela) where the *pf*EIR of *An. darlingi* >1 has been recorded (Table 5.1), lending support to the recommended focus on local hyperendemic hot spots [16,57]. Within the context of a global goal of malaria eradication, it has been proposed that the aim should be an EIR of less than one infectious bite per person [58]. This goal may be difficult to achieve in the Neotropics even if insecticide-treated nets (ITNs)/long-lasting insecticide-treated nets (LLINs) coverage is high because of the exophagic/exophilic behavior that predominates in most populations of *An. darlingi*.

ANOPHELES ALBIMANUS AND *ANOPHELES AQUASALIS*: REGIONALLY IMPORTANT VECTORS

Another significant vector in Central America, northern South America, and the Caribbean is *An. albimanus*, characterized as mainly zoophilic, crepuscular, exophagic, and exophilic [7,8]. *Anopheles albimanus* populations in South America are susceptible to *P. vivax* and *P. falciparum*, although the few infectivity indices reported are low, for example, 0.062–0.066% in the Pacific coast of Colombia [74]. In Las Varas, a fishing village in southern coastal Colombia, the EIR of *An. albimanus* was 1.4, but its density was low, and it was not the major local vector [75]. There is variability in host feeding, insecticide resistance, and *Plasmodium* susceptibility throughout its distribution and evidence of population structure, but not multiple species [76].

An interesting link between El Niño and the increase in malaria cases transmitted by *An. albimanus* along the Pacific coast of Colombia was detected [77] related to its geographic distribution that coincides with areas of greater than 1000 mm rainfall and two rainy seasons per year [7]. Adult abundance is

TABLE 5.1 Vector Metrics of *An. darlingi* Incriminated in Malaria Transmission in Localities Across Latin America from 1995 to 2015 (See Figure 5.1 for Locations)

Location	Human-Biting Rate (HBR)	Entomological Inoculation Rate (EIR)	Annual EIR	References
Boa Vista, Roraima, Brazil	0–6.2 bites/person/month	0–0.42 bites/person/month	0–5.04	da Silva-Vasconcelos et al. [59]
Boa Vista, Roraima, Brazil	0.03–0.92 bites/person/h	–	–	Povoa et al. [60]
Matapí, Amapá, Brazil	53.9–837.7 bites/person/night	0.040–0.988 bites/person/night	14.6–360.62	Galardo et al. [61]
Rorainópolis, Roraima, Brazil	0.6–6.24 bites/person/night	–	–	de Barros et al. [20]
Careiro, Amazonas, Brazil	0–15.6 bites/person/h	–	–	Martins-Campos et al. [62]
Córdoba & Antioquia, Colombia	0.04–0.163 bites/person/night	–	–	Gutiérrez et al. [63]
Villavicencio, Meta, Colombia	2.2–55.5 bites/person/night	2.9 bites/person/year	2.9	Ahumada et al. [47]
Puerto Carreño, Vichada, Colombia	5.1 bites/person/night	1.6 bites/person/year	1.6	Jimenez et al. [19]
UCS region, Colombia	0.03–15.1 bites/person/night	3.7 bites/person/year	3.7	Naranjo-Diaz et al. [64]
San José del Guaviare, Colombia	5.2 bites/person/night	2.6 bites/person/year	2.6	Jimenez et al. [48]
Dibulla, La Guajira, Colombia	12.24 bites/person/night	15.8 bites/person/year	15.8	Herrera-Varela et al. [50]

(Continued)

TABLE 5.1 (Continued)

Location	Human-Biting Rate (HBR)	Entomological Inoculation Rate (EIR)	Annual EIR	References
Upper Maroni, French Guiana	255.5 bites/person/night	14.4–27.4 bites/person/year	14.4–27.4	Girod et al. [65]
Maroni region, French Guiana	0.67–5.66 bites/person/h	5–10 bites/person/year	5–10	Fouque et al. [66]
French Guiana	1.3–38.9 bites/person/night	5.7–8.7 bites/person/year	5.7–8.7	Girod et al. [67]
Suriname/French Guiana border	3.5–19.0 bites/person/night	8.7–66.4 bites/person/year	8.7–66.4	Hiwat et al. [68]
Suriname	1.09 (indoor); 1.43 (outdoor) bites/person/h	0.8–3.7 bites/person/year	0.8–3.7	Hiwat et al. [69]
Iquitos-Nauta road, Peru	0.03–8.33 bites/person/6 h	0.1–38 bites/person/year	0.1–38	Vittor et al. [36]
Loreto Department, Peru	–	0–12 bites/person/month	0–144	Reinbold-Wasson et al. [70]
Loreto Department, Peru	0.102–41.13 bites/person/h	0.02–5.3 bites/person/night	–	Parker et al. [16]
Upper Orinoco River, Southern Venezuela	–	128.6 bites/person/year	128.6	Magris et al. [71]
Southern Venezuela	1.02–10.94 bites/person/night	–	–	Moreno et al. [72]
Southern Venezuela	0.74 bites/person/night	2.21 bites/person/year	2.21	Moreno et al. [73]
Lower Caura River Basin, Venezuela	38.21 bites/person/night	0–20.85 bites/person/year	0–20.85	Rubio-Palis et al. [14]

Annual EIR, number of infective bites/person/year.

closely tied to precipitation, although this can also vary. Based on sequences of the *COI* and *white* genes, there are three lineages in *An. albimanus*: one includes mosquitoes from Nicaragua, Costa Rica, and western Panama; the second one includes those from central and eastern Panama as well as the Caribbean coast of Colombia; the third includes those found along the Pacific coasts of Colombia and Ecuador [78]. These lineages display limited gene flow, and the barriers between them date from the Pleistocene.

The distribution of the vector *Anopheles aquasalis* is limited to coastal parts of the Caribbean, along the Atlantic to southern Brazil, and the Pacific as far south as Ecuador [5]. It becomes a locally important vector when the densities are high, as along some marshy coastal malaria endemic regions, such as eastern Sucre state, Venezuela [79], and in districts of the city of Belém, Brazil, where it was detected infected with *P. falciparum* at 1.15% and *P. vivax* at 0.77% [80]. In coastal eastern Panama, its HBR reaches 9.3 bites/person/night in the indigenous Kuna Yala Comarca where it is presumed to be involved in transmission of *Plasmodium* [38]. *Anopheles aquasalis* is also responsible for sporadic malaria outbreaks around large port cities, for example, in Trinidad in the early 1990s where introduced *P. vivax* was responsible for several cases [81].

COLONIZATION

In a PUBMED search for "anopheles" and "colonization," the two oldest results are articles documenting *An. darlingi* colonization [82,83]. Despite these early efforts, sustained *An. darlingi* colonization was not achieved until recently [84]. During this gap of 67 years, many laboratory experiments have been limited by the absence of a reference strain. Even the published *An. darlingi* genome [23] was conducted with a wild population collected at multiple time points over 3 years from near Manaus, Amazonas state, Brazil, and, for this reason, has some limitations.

Several other Neotropical anophelines have been colonized, including *An. albimanus* [85−87], *Anopheles albitarsis* [88], *An. aquasalis* [89], and *Anopheles pseudopunctipennis* [90,91]. One of the most significant experiments to examine Neotropical anopheline susceptibility to *P. vivax* compared colonies of *An. albimanus* and *An. pseudopunctipennis* in Mexico [92]. These species are in different subgenera, *Nyssorhynchus* and *Anopheles*, respectively, and occupy very distinctive ecological niches such that they do not co-occur spatially or temporally. Laboratory colony feeding experiments determined that the three local *P. vivax* populations were most compatible with their respective sympatric anopheline species. The most significant finding was that reciprocal selection between malaria parasites and mosquito vectors has led to local adaptation of *P. vivax* [92].

Different methodologies have been used to overcome the main bottleneck for Neotropical anopheline colonization, selection for the ability to mate in a confined space (stenogamy) [93]. With *An. albitarsis* this was solved by an artificial-mating protocol during the initial generations [88], although this laborious technique is not practical for use in large colonies for long periods of time. Fortunately, artificial mating could be suspended after several generations as the colony adapted to natural mating [88]. In contrast, *An. pseudopunctipennis* and *An. darlingi* colonies were established by inducing mating with flashing light [84,90,91]. Either stroboscopic blue light or white light was used during the dark period of a 12:12 LD cycle to induce mating. As in artificial mating, the induction was required only during the initial period of colonization (6 generations for *An. pseudopunctipennis*, 12 generations for *An. darlingi*) and, subsequently, mating occurred without external stimuli. Crossing experiments between colonized and free-living *An. albitarsis* s.s. from the same initial source population showed that faster-male evolution of mating ability may have taken place during the colonization process, which, for this species, was only established after 30 generations of artificial mating [94].

COEVOLUTION OF *ANOPHELES* AND *PLASMODIUM*

Malaria needs to be understood in the context of relationships among three organisms: *Plasmodium*, anophelines, and humans. The interaction between humans and parasite begins at the moment of infection. Humans, through their immune response, shape parasite diversity and quantity during infection. Moreover, parasites may or may not cause symptoms in humans via pathogenic mechanisms. The interaction between humans and vectors is initiated at the time of a blood meal and is correlated with the availability of human–vector contact. The latter interaction has not been explored in as much detail as the one between parasites and humans, but it is critically important.

During sexual reproduction inside the vector, new parasite haplotypes may be generated. Population bottlenecks could occur in the digestive tract, where a small number of oocysts are formed, and later, in the salivary glands, when a small number of sporozoites are inoculated into humans at the time of a blood meal [95]. Parasites are likely under selection pressure by as-yet unknown vector genetic factors.

The presence of asymptomatic malaria in humans in the Amazon region has been documented for several years [57,96]. Asymptomatic malaria cases seem to be associated with small isolated riverine human populations with little mobility [97–99]. In such individuals, the continuing challenge of a small range of strains of *P. vivax* and/or *P. falciparum* leads to an immune system response in adult life, resulting in low parasitemia infections, often imperceptible to traditional diagnosis by microscopy. These individuals generally have no symptoms for several months but can infect vectors as the latter blood feed [100].

The population structure of the two main human species of *Plasmodium* in the Americas displays little local variability. The introduction of *P. falciparum* into the Americas from Africa, for which many lines of indirect evidence exist, may explain this [101]. In contrast, the combined effects of geographical structure and low incidence of *P. vivax* malaria in the Americas have resulted in low diversity at a local geographic scale, but high diversity at a global scale [102]. For this scenario to occur, isolated human, parasite, and anopheline populations must have had opportunities to interact.

Each *Plasmodium* species has a different interaction with *An. darlingi*. Although the interaction of *An. darlingi*–*P. vivax* appears to be older, the interaction of this vector (and other Neotropical vectors) with *P. falciparum* is considered to be very recent [101]. *Plasmodium falciparum* has had a long history of interaction with Old World anophelines, which are members of the subgenus *Cellia*, and it has had to adapt considerably to each of the three Neotropical subgenera (*Anopheles*, *Nyssorhynchus*, and *Kerteszia*) that include competent malaria vectors. Old World and Neotropical anopheline subgenera are estimated to have split approximately 100 million years ago [22]. Genetic differentiation between Old and New World anopheline subgenera is about twice that between the genera *Aedes* and *Culex*. It is therefore somewhat surprising that neotropical anophelines are vectors of *P. falciparum* although they are notably less competent than African vectors such as *An. gambiae* s.s. There are no available data on simultaneous infection experiments carried out with New and Old World vectors but historical information is suggestive. In 1930, *An. gambiae* was accidentally introduced from Africa into the port city of Natal, Brazil. This introduction led to a local malaria epidemic with high mortality rates [103]. Although F_1 from wild *An. darlingi* in the laboratory are susceptible to both *P. falciparum* [104] and *P. vivax* [105], infection rates with *P. falciparum* seem to be higher compared to those with *P. vivax* in endemic areas, where human infections are approximately 80% *P. vivax* and 20% *P. falciparum* (Gil, personal communication) [18]. *Anopheles darlingi* can be infected by *P. falciparum*, but is not especially efficient in human transmission. In fact, *An. darlingi* infection rates are highly variable across different studies, ranging from less than 1% [17,106,107] to as high as 4–5% [18,59,108]. Aside from differences in *Plasmodium* detection method, which vary from direct observation of oocysts and sporozoites to ELISA, PCR, and multiplex RT-PCR, if there is no correlation between low and high infection rates and methodology used, the variability of *An. darlingi* infection rates is likely real and correlated with mosquito genetic and environmental factors.

CURRENT CONTROL OF *ANOPHELES DARLINGI*

Adult Stage

Current control measures across Latin America for *An. darlingi* and other local vectors have mostly been very focal, and often consist of indoor residual spraying (IRS) of insecticides in reaction to local malaria outbreaks. However, the majority of Neotropical vectors are mainly exophagic and exophilic, thus IRS is quite limited. The effectiveness of ITNs, or LLINs depends on the biting behavior and peak biting time of the local vectors. For example, in the small community of San Jose del Guaviare, Guaviare, Colombia, where *An. darlingi* is the primary vector, its peak indoor biting time is 12:00−01:00 h, so ITNs in this village would be effective against this species [48]. In contrast, in Puerto Carreño, a small city in Vichada, western Colombia, *An. darlingi* collected indoors and outdoors showed only early evening and early morning biting peaks, when people are not generally protected under ITNs. Further underscoring the limitations of ITNs in Puerto Carreno, another vector in this city, *Anopheles marajoara*, bites only in the early evening, outdoors, and indoors [19].

The impact of ITNs on reducing vector populations has not been evaluated in Latin America. For example, even in villages in the peri-Iquitos region where ITNs were freely distributed by the PAMAFRO program, supported by the Global Fund from 2006 to 2009, more houses report using local nets that are not treated with insecticides than ITNs (Moreno and collaborators, unpublished). The latter may be hotter, and it remains uncertain how long ITNs last under local conditions.

Nevertheless, ITNs may be the best option in certain situations. For example, a 2-year study that collected baseline data prior to intervention in Venezuelan Yanomami communities found that ITNs prevented 56% of new malaria cases and reduced parasitemia prevalence by 83% during the high transmission season [109]. The conclusion of this study was that ITNs are the most feasible control in this forested community where the villages are widely dispersed and interventions such as IRS are either not culturally acceptable or ineffective because of type of housing and/or vector behavior.

Few populations of malaria vectors in Latin America have been tested for insecticide resistance, a neglected area of research. The first detection of resistance in *An. darlingi* was to DDT in Quibdo, Colombia [110]. It has subsequently been shown to be resistant to pyrethroids in the coastal state of Choco, Colombia and to lambda-cyhalothrin in Ame-Bete, Colombia [111]. However, in two villages in Caquetá and two villages in Santander, Colombia, *An. darlingi* populations were fully susceptible to pyrethroids, organophosphates, and to DDT [112].

There have been efforts to reduce the cost of repellents so they could be more feasibly become part of integrated vector control strategies. Field tests in a site in Guatemala where the main vector is *An. albimanus* and in Peru where *An. darlingi* predominates, successfully compared controls (no repellent), 15% *N,N*-Diethyl-*meta*-toluamide (DEET), and two inexpensive repellents, *para*-methane-diol (PMD) and lemongrass oil (LG). In both localities, the PMD/LG repellent was effective, providing protection of 98% and 95% to *An. albimanus* and *An. darlingi*, respectively, up to 6 h after application [113]. Another study combined the use of ITNs and a plant-based insect repellent, *Eucalyptus maculate* citriodon, in peri-urban and rural areas around Riberalta and Guayaramerin, Amazonian Bolivia, where early evening biting behavior by *An. darlingi* was common. This controlled trial showed that insect repellent applied to the skin had a significant epidemiological effect in reducing malaria incidence [114].

Guyana, Suriname, and French Guiana together have the highest numbers and concentrations of *P. falciparum* in the Americas [2]. Malaria in Suriname has been concentrated mainly in the interior where the 63,000 inhabitants at risk consist of Amerindians and Maroons living in tribal communities, mostly along rivers, and migrants, many crossing the French Guiana border, who live and work small-scale gold mines in the forest. A series of strategies undertaken between 2006 and 2009 were successful in reducing national malaria case numbers from 8,618 in 2005 to 1,509 in 2009, the latter number considered to be at a pre-elimination level. The strategy included IRS, distribution of LLINs, case management (testing for correct diagnosis and treatment), an aggressive active case detection campaign, a media awareness campaign, strengthening the epidemic detection and response system, and standardization of monitoring and evaluation methods [115]. The other factor that decreased malaria transmission was the collapse of *An. darlingi* populations due to a combination of climatic events and ITNs such that the HBR decreased from 1.43 bites/human/h outdoor and 1.09 indoor, to zero [69]. The reduction in malaria cases in Suriname is an outstanding example of a successful series of interventions against malaria, and luck, and in some smaller countries, such as Guyana or French Guiana, or endemic regions where malaria vectors are limited by their geographical distribution (e.g., coastal *An. aquasalis*), this may be an appropriate model. Even in the massive Amazon Basin, integrated interventions, perhaps even malaria elimination, could be successful at a local scale, if appropriately tailored to local mosquito vector ecology and behavior.

Larval Stage

Several studies have tested the feasibility of using *Bacillus sphaericus* (Vectolex CG) to treat breeding sites to reduce local populations of primary vectors such as *An. albimanus* in coastal Buenaventura, Colombia [116], *An. darlingi* in Iquitos, Peru [117], *An. aquasalis* in coastal southeastern Venezuela [118] and various vectors in Bolivar state, Venezuela [119]. One of the many hazards of gold-mining operations in Brazil is the likelihood that the pools used and abandoned will be colonized by *An. darlingi* larvae. An important field experiment to reduce populations of larvae and adults involved the use of low concentrations of *B. sphaericus* over a 52-week period in Calçoene, Amapá, Brazil [120]. In treated pools, late-stage larvae were reduced by 78% and pupae by 93% compared with control, untreated pools. Adult *An. darlingi* was reduced by 53% during the rainy season, the last 5 months of this study.

All of these studies focused on *B. sphaericus* for its persistence, effectiveness, and low nontarget toxicity, and all showed fairly high success in reducing aquatic anopheline stages and subsequently adults. The only caveat is that none determined the effectiveness in reducing local malaria prevalence (before and after the treatments of *B. sphaericus*). Nevertheless, focal larval treatments could be incorporated in combination with ITNs and other interventions in integrated malaria control efforts, particularly where breeding sites are accessible, and relatively limited in size. Although the interventions were not evaluated prior to treatment, a malaria outbreak in Dibulla, La Guajira, Colombia in 2008−2009, which reported 1,617 cases, was controlled by a combination of IRS, ITN, and application of *B. sphaericus* such that the number of cases in Dibulla was reduced to 148 in 2013. The main vector in Dibulla is *An. darlingi* [50].

SCENARIOS POST LOCAL ELIMINATION OF *ANOPHELES DARLINGI*

If GM and other *An. darlingi* elimination strategies are able to reduce EIRs below one where *An. darlingi* is the most abundant anopheline species, how might that ecological diorama change? This will depend on the diversity of local vector species, which is turn depends on habitat and geological/tectonic history. In the forested interior of Suriname, anopheline species diversity is low, another factor which may aid in the maintenance of a zone free of *An. darlingi* [69]. Around the city of Macapá, Amapá state, Brazil, where *An. darlingi* was very abundant [121], populations dwindled and were effectively replaced by the vector *An. marajoara* due to a combination of deforestation, urbanization, human migration, and increased biomass (cattle). Unfortunately, this major decrease in *An. darlingi* abundance did nothing to reduce malaria transmission or local caseload because *An. marajoara* is an effective vector [61,122]. Around the port city of Belém, Pará state, Brazil, *An. darlingi* was eradicated in 1968, but it, and local malaria transmission, subsequently made a comeback, possibly as a result of the continuing expansion of Belém and illegal settlements into the surrounding forest in the 1990s [80].

Another interesting ecological situation is the unprecedented invasion by *An. darlingi* of the Loreto Department in northern Amazonian Peru, resulting in thousands of malaria cases [123]. Now, *An. darlingi* is the most abundant regional anopheline [16,70], but prior to the early 1990s another less effective, less abundant species, *Anopheles benarrochi* B, was the local vector. There is still substantial regional species diversity [70], but there are also hyperendemic pockets where *An. darlingi* is essentially the only species of consequence. In summary, for *An. darlingi*, successful elimination may depend mainly on local habitat and species diversity.

CONCLUSIONS

Malaria reduction, and perhaps elimination, can be achieved in the Americas but this will require much better coordination of patient diagnosis and treatment, including asymptomatic patients, with information on the behavior and ecology of the local malaria vectors, whether or not they are *An. darlingi*. Both *An. darlingi* and *An. albimanus* are very adaptable, have detectable barriers to gene flow across their distributions, and are susceptible to *P. vivax* and *P. falciparum*. IRS and ITN interventions are generally ineffective in reducing human–vector contact with exophagic and exophilic anophelines, although there are populations of *An. darlingi* and *An. albimanus* that are mainly endophagic where such strategies could be effective. The very low level of insecticide resistance in *An. darlingi* is encouraging, and efforts toward more integrated methods, such as plant-based repellents combined with ITNs, IRS, and focal larval reduction where vectors are mostly early evening outdoor biters, could be highly effective.

ACKNOWLEDGMENTS

We thank Sara Bickersmith, New York State Department of Health, Albany, NY, USA, for maps and editorial comments. Financial support was provided by US National Institutes of Health (NIH) ICEMR grant U19 AI089681 to JMV and NIH grant R01AI54139 to JEC.

REFERENCES

[1] Corbel V, N'Guessan R. Distribution, mechanisms, impact and management of insecticide resistance in malaria vectors. In: Manguin S, editor. A pragmatic review, anopheles mosquitoes—new insights into malaria vectors. Rijeka, Croatia: InTech; 2013.

[2] World Malaria Report 2012. Geneva, Switzerland; 2012.

[3] Loaiza J, Scott M, Bermingham E, Rovira J, Sanjur O, Conn JE. *Anopheles darlingi* (Diptera: Culicidae) in Panama. Am J Trop Med Hyg 2009;81(1):23–6.

[4] Forattini OP. Culicidologia médica: identificação, biologia, epidemiologia/Medical culicidology: ID, biology, epidemiology. EDUSP; 2002.

[5] Sinka ME, Rubio-Palis Y, Manguin S, et al. The dominant *Anopheles* vectors of human malaria in the Americas: occurrence data, distribution maps and bionomic precis. Parasit Vectors 2010;3:72.

[6] Hiwat H, Bretas G. Ecology of *Anopheles darlingi* Root with respect to vector importance: a review. Parasit Vectors 2011;4:177.

[7] Montoya-Lerma J, Solarte YA, Giraldo-Calderon GI, et al. Malaria vector species in Colombia: a review. Mem Inst Oswaldo Cruz 2011;106(Suppl. 1):223–38.

[8] Conn JE, Quiñones ML, Póvoa MM. Phylogeography, vectors, and transmission in Latin America. In: Manguin S, editor. *Anopheles* mosquitoes—new insights into malaria vectors. Rijeka, Croatia: Intech Open; 2013.

[9] Rejmankova E, Rubio-Palis Y, Villegas L. Larval habitats of anopheline mosquitoes in the Upper Orinoco, Venezuela. J Vector Ecol 1999;24(2):130–7.

[10] Zimmerman RH, Galardo AK, Lounibos LP, Arruda M, Wirtz R. Bloodmeal hosts of *Anopheles* species (Diptera: Culicidae) in a malaria-endemic area of the Brazilian Amazon. J Med Entomol 2006;43(5):947–56.

[11] de Barros FS, Arruda ME, Gurgel HC, Honorio NA. Spatial clustering and longitudinal variation of *Anopheles darlingi* (Diptera: Culicidae) larvae in a river of the Amazon: the importance of the forest fringe and of obstructions to flow in frontier malaria. Bull Entomol Res 2011;101(6):643–58.

[12] Vittor AY, Pan W, Gilman RH, et al. Linking deforestation to malaria in the Amazon: characterization of the breeding habitat of the principal malaria vector, *Anopheles darlingi*. Am J Trop Med Hyg 2009;81:5–12.

[13] Pardini R, Bueno Ade A, Gardner TA, Prado PI, Metzger JP. Beyond the fragmentation threshold hypothesis: regime shifts in biodiversity across fragmented landscapes. PLoS One 2010;5(10):e13666.

[14] Rubio-Palis Y, Bevilacqua M, Medina DA, et al. Malaria entomological risk factors in relation to land cover in the Lower Caura River Basin, Venezuela. Mem Inst Oswaldo Cruz 2013;108(2):220–8.

[15] Moutinho PR, Gil LH, Cruz RB, Ribolla PE. Population dynamics, structure and behavior of *Anopheles darlingi* in a rural settlement in the Amazon rainforest of Acre, Brazil. Malar J 2011;10:174.

[16] Parker BS, Paredes Olortegui M, Penataro Yori P, et al. Hyperendemic malaria transmission in areas of occupation-related travel in the Peruvian Amazon. Malar J 2013;12:178.

[17] Gil LH, Tada MS, Katsuragawa TH, Ribolla PE, da Silva LH. Urban and suburban malaria in Rondonia (Brazilian Western Amazon) II. Perennial transmissions with high anopheline densities are associated with human environmental changes. Mem Inst Oswaldo Cruz 2007;102(3):271–6.

[18] Angella AF, Salgueiro P, Gil LH, Vicente JL, Pinto J, Ribolla PE. Seasonal genetic partitioning in the neotropical malaria vector, *Anopheles darlingi*. Malar J 2014;13:203.

[19] Jimenez P, Conn JE, Wirtz R, Brochero H. [*Anopheles* (Diptera: Culicidae) vectors of malaria in Puerto Carreno municipality, Vichada, Colombia]. Biomedica 2012;32(Suppl. 1): 13–21.

[20] de Barros FS, Honorio NA, Arruda ME. Survivorship of *Anopheles darlingi* (Diptera: Culicidae) in relation with malaria incidence in the Brazilian Amazon. PloS One 2011;6 (8):e22388.

[21] Mirabello L, Vineis JH, Yanoviak SP, et al. Microsatellite data suggest significant population structure and differentiation within the malaria vector *Anopheles darlingi* in Central and South America. BMC Ecol 2008;8:3.

[22] Moreno M, Marinotti O, Krzywinski J, et al. Complete mtDNA genomes of *Anopheles darlingi* and an approach to anopheline divergence time. Malar J 2010;9:127.

[23] Marinotti O, Cerqucira GC, De Almeida LGP, et al. The Genome of *Anopheles darlingi*, the main neotropical malaria vector. Nucleic Acids Res 2013;41(15):7387–400.

[24] Pedro PM, Sallum MAM. Spatial expansion and population structure of the neotropical malaria vector, *Anopheles darlingi* (Diptera: Culicidae). Biol J Linn Soc 2009;97(4):854–66.

[25] Pedro PM, Uezu A, Sallum MA. Concordant phylogeographies of 2 malaria vectors attest to common spatial and demographic histories. J Hered 2010;101(5):618–27.

[26] Santos LM, Gama RA, Eiras AE, Fonseca CG. Genetic differences based on AFLP markers in the mosquito species *Anopheles darlingi* collected in versus near houses in the region of Porto Velho, RO, Brazil. Genet Mol Res 2010;9(4):2254–62.

[27] Conn JE, Rosa-Freitas MG, Luz SL, Momen H. Molecular population genetics of the primary neotropical malaria vector *Anopheles darlingi* using mtDNA. J Am Mosq Control Assoc 1999;15(4):468−74.

[28] Scarpassa VM, Conn JE. Population genetic structure of the major malaria vector *Anopheles darlingi* (Diptera: Culicidae) from the Brazilian Amazon, using microsatellite markers. Mem Inst Oswaldo Cruz 2007;102(3):319−27.

[29] Zimmerman RH, Lounibos LP, Nishimura N, Galardo AK, Galardo CD, Arruda ME. Nightly biting cycles of malaria vectors in a heterogeneous transmission area of eastern Amazonian Brazil. Malar J 2013;12:262.

[30] Junior SG, Pamplona VM, Corvelo TC, Ramos EM. Quality of life and the risk of contracting malaria by multivariate analysis in the Brazilian Amazon region. Malar J 2014;13:86.

[31] Grillet ME, El Souki M, Laguna F, Leon JR. The periodicity of *Plasmodium vivax* and *Plasmodium falciparum* in Venezuela. Acta Trop 2013;130c:58−66.

[32] Macdonald G. Epidemiological basis of malaria control. Bull World Health Organ 1956;15(3−5):613−26.

[33] Smith DL, Perkins TA, Reiner Jr. RC, et al. Recasting the theory of mosquito-borne pathogen transmission dynamics and control. Trans R Soc Trop Med Hyg 2014;108(4): 185−97.

[34] Foley DH, Linton YM, Ruiz-Lopez JF, et al. Geographic distribution, evolution, and disease importance of species within the Neotropical *Anopheles albitarsis* Group (Diptera, Culicidae). J Vector Ecol 2014;39(1):168−81.

[35] Guerra CA, Snow RW, Hay SI. A global assessment of closed forests, deforestation and malaria risk. Ann Trop Med Parasitol 2006;100(3):189−204.

[36] Vittor AY, Gilman RH, Tielsch J, et al. The effect of deforestation on the human-biting rate of *Anopheles darlingi*, the primary vector of falciparum malaria in the Peruvian Amazon. Am J Trop Med Hyg 2006;74:3−11.

[37] Molina-Cruz A, de Merida AM, Mills K, et al. Gene flow among *Anopheles albimanus* populations in Central America, South America, and the Caribbean assessed by microsatellites and mitochondrial DNA. Am J Trop Med Hyg 2004;71(3):350−9.

[38] Loaiza JR, Bermingham E, Scott ME, Rovira JR, Conn JE. Species composition and distribution of adult *Anopheles* (Diptera: Culicidae) in Panama. J Med Entomol 2008;45 (5):841−51.

[39] Forattini OP. Entomologia medica, ol. 1. São Paulo, Brazil: Faculdade de Higiene e Sáude Publica; 1962.

[40] Deane L, Causey O, Deane M. An illustrated key by adult female characteristics for identification of thirty-five species of Anophelini from the Northeast and Amazon regions of Brazil, with notes on the malaria vectors (Diptera: Culicidae). Am J Hyg Mono Ser 1946; 18:1−18.

[41] Consoli RA, Lourenco-de-Oliveira R. Principais mosquitos de importância sanitária no Brasil. Fundação Oswaldo Cruz: Editora Fiocruz; 1994.

[42] Manguin S, Roberts DR, Andre RG, Rejmankova E, Hakre S. Characterization of *Anopheles darlingi* (Diptera: Culicidae) larval habitats in Belize, Central America. J Med Entomol 1996;33(2):205−11.

[43] Morais SA, Urbinatti PR, Sallum MA, et al. Brazilian mosquito (Diptera: Culicidae) fauna: I. *Anopheles* species from Porto Velho, Rondonia state, western Amazon, Brazil. Rev Inst Med Trop Sao Paulo 2012;54(6):331−5.

[44] Achee NL, Grieco JP, Rejmankova E, et al. Biting patterns and seasonal densities of *Anopheles* mosquitoes in the Cayo District, Belize, Central America with emphasis on *Anopheles darlingi*. J Vector Ecol 2006;31(1):45−57.

[45] Santos RL, Padilha A, Costa MD, Costa EM, Dantas-Filho Hde C, Povoa MM. Malaria vectors in two indigenous reserves of the Brazilian Amazon. Rev Saude Publica 2009;43 (5):859−68.

[46] Achee NL, Korves CT, Bangs MJ, et al. *Plasmodium vivax* polymorphs and *Plasmodium falciparum* circumsporozoite proteins in *Anopheles* (Diptera: Culicidae) from Belize, Central America. J Vector Ecol 2000;25(2):203−11.

[47] Ahumada ML, Pareja PX, Buitrago LS, Quinones ML. [Biting behavior of *Anopheles darlingi* Root, 1926 (Diptera: Culicidae) and its association with malaria transmission in Villavicencio (Meta, Colombia)]. Biomedica 2013;33(2):241−50.

[48] Jiménez IP, Conn JE, Brochero H. Malaria vectors in San José del Guaviare, Orinoquia, Colombia. J Am Mosq Control Assoc 2014;30(2):91−8.

[49] Voorham J. Intra-population plasticity of *Anopheles darlingi*'s (Diptera, Culicidae) biting activity patterns in the state of Amapa, Brazil. Rev Saude Publica 2002;36(1):75−80.

[50] Herrera-Varela M, Orjuela LI, Penalver C, Conn JE, Quinones ML. *Anopheles* species composition explains differences in *Plasmodium* transmission in La Guajira, northern Colombia. Mem Inst Oswaldo Cruz 2014;109(7):955−9.

[51] Barbosa LMC, Souto RNP, Ferreira RMD, Scarpassa VM. Composition, abundance and aspects of temporal variation in the distribution of *Anopheles* species in an area of Eastern Amazonia. Rev Soc Bras Med Trop 2014;47(3):313−20.

[52] León W, Valle J, Naupay R, Tineo E, Rosas A, Palomino M. Comportamiento estacional del *Anopheles* (*Nyssorhynchus*) *darlingi* Root 1926 en localidades de Loretao y Madre de Dios, Peru 1999−2000. Rev Peru Med Exp Salud Pública 2003;20:22−7.

[53] Roberts DR, Alecrim WD, Tavares AM, Radke MG. The house-frequenting, host-seeking and resting behavior of *Anopheles darlingi* in southeastern Amazonas, Brazil. J Am Mosq Control Assoc 1987;3(3):433−41.

[54] Quinones ML, Suarez MF. Indoor resting heights of some anophelines in Colombia. J Am Mosq Control Assoc 1990;6(4):602−4.

[55] Gil LH, Alves FP, Zieler H, et al. Seasonal malaria transmission and variation of anopheline density in two distinct endemic areas in Brazilian Amazonia. J Med Entomol 2003;40 (5):636−41.

[56] Gething PW, Patil AP, Smith DL, et al. A new world malaria map: plasmodium falciparum endemicity in 2010. Malar J 2011;10:378.

[57] da Silva-Nunes M, Moreno M, Conn JE, et al. Amazonian malaria: asymptomatic human reservoirs, diagnostic challenges, environmentally driven changes in mosquito vector populations, and the mandate for sustainable control strategies. Acta Trop 2012;12: 281−91.

[58] Ulrich JN, Naranjo DP, Alimi TO, Muller GC, Beier JC. How much vector control is needed to achieve malaria elimination? Trends Parasitol 2013;29(3):104−9.

[59] da Silva-Vasconcelos A, Kato MY, Mourao EN, et al. Biting indices, host-seeking activity and natural infection rates of anopheline species in Boa Vista, Roraima, Brazil from 1996 to 1998. Mem Inst Oswaldo Cruz 2002;97(2):151−61.

[60] Povoa MM, de Souza RT, Lacerda RN, et al. The importance of *Anopheles albitarsis* E and *An. darlingi* in human malaria transmission in Boa Vista, state of Roraima, Brazil. Mem Inst Oswaldo Cruz 2006;101(2):163−8.

[61] Galardo AK, Arruda M, D'Almeida Couto AA, Wirtz R, Lounibos LP, Zimmerman RH. Malaria vector incrimination in three rural riverine villages in the Brazilian Amazon. Am J Trop Med Hyg 2007;76(3):461–9.

[62] Martins-Campos KM, Pinheiro WD, Vitor-Silva S, et al. Integrated vector management targeting *Anopheles darlingi* populations decreases malaria incidence in an unstable transmission area, in the rural Brazilian Amazon. Malar J 2012;11:351.

[63] Gutierrez LA, Gonzalez JJ, Gomez GF, et al. Species composition and natural infectivity of anthropophilic *Anopheles* (Diptera: Culicidae) in the states of Cordoba and Antioquia, Northwestern Colombia. Mem Inst Oswaldo Cruz 2009;104(8):1117–24.

[64] Naranjo-Diaz N, Rosero DA, Rua-Uribe G, Luckhart S, Correa MM. Abundance, behavior and entomological inoculation rates of anthropophilic anophelines from a primary Colombian malaria endemic area. Parasit Vectors 2013;6:61.

[65] Girod R, Gaborit P, Carinci R, Issaly J, Fouque F. *Anopheles darlingi* bionomics and transmission of *Plasmodium falciparum*, *Plasmodium vivax* and *Plasmodium malariae* in Amerindian villages of the Upper-Maroni Amazonian forest, French Guiana. Mem Inst Oswaldo Cruz 2008;103(7):702–10.

[66] Fouque F, Gaborit P, Carinci R, Issaly J, Girod R. Annual variations in the number of malaria cases related to two different patterns of *Anopheles darlingi* transmission potential in the Maroni area of French Guiana. Malar J 2010;9:80.

[67] Girod R, Roux E, Berger F, et al. Unravelling the relationships between *Anopheles darlingi* (Diptera: Culicidae) densities, environmental factors and malaria incidence: understanding the variable patterns of malarial transmission in French Guiana (South America). Ann Trop Med Parasitol 2011;105(2):107–22.

[68] Hiwat H, Issaly J, Gaborit P, et al. Behavioral heterogeneity of *Anopheles darlingi* (Diptera: Culicidae) and malaria transmission dynamics along the Maroni River, Suriname, French Guiana. Trans R Soc Trop Med Hyg 2010;104(3):207–13.

[69] Hiwat H, Mitro S, Samjhawan A, Sardjoe P, Soekhoe T, Takken W. Collapse of *Anopheles darlingi* populations in Suriname after introduction of insecticide-treated nets (ITNs); malaria down to near elimination level. Am J Trop Med Hyg 2012;86(4):649–55.

[70] Reinbold-Wasson DD, Sardelis MR, Jones JW, et al. Determinants of *Anopheles* seasonal distribution patterns across a forest to periurban gradient near Iquitos, Peru. Am J Trop Med Hyg 2012;86(3):459–63.

[71] Magris M, Rubio-Palis Y, Menares C, Villegas L. Vector bionomics and malaria transmission in the Upper Orinoco River, Southern Venezuela. Mem Inst Oswaldo Cruz 2007;102 (3):303–11.

[72] Moreno JE, Rubio-Palis Y, Paez E, Perez E, Sanchez V. Abundance, biting behaviour and parous rate of anopheline mosquito species in relation to malaria incidence in gold-mining areas of southern Venezuela. Med Vet Entomol 2007;21(4):339–49.

[73] Moreno JE, Rubio-Palis Y, Paez E, Perez E, Sanchez V, Vaccari E. Malaria entomological inoculation rates in gold mining areas of Southern Venezuela. Mem Inst Oswaldo Cruz 2009;104(5):764–8.

[74] Gutierrez LA, Naranjo N, Jaramillo LM, et al. Natural infectivity of *Anopheles* species from the Pacific and Atlantic Regions of Colombia. Acta Trop 2008;107(2):99–105.

[75] Escovar JE, Gonzalez R, Quinones ML. Anthropophilic biting behaviour of *Anopheles (Kerteszia) neivai* Howard, Dyar & Knab associated with Fishermen's activities in a malaria-endemic area in the Colombian Pacific. Mem Inst Oswaldo Cruz 2013;108(8): 1057–64.

[76] Loaiza JR, Scott ME, Bermingham E, Rovira J, Conn JE. Evidence for pleistocene population divergence and expansion of *Anopheles albimanus* in southern central America. Am J Trop Med Hyg 2010;82(1):156−64.

[77] Poveda G, Rojas W, Quinones ML, et al. Coupling between annual and ENSO timescales in the malaria-climate association in Colombia. Environ Health Perspect 2001;109(5):489−93.

[78] Loaiza JR, Scott ME, Bermingham E, et al. Late Pleistocene environmental changes lead to unstable demography and population divergence of *Anopheles albimanus* in the northern Neotropics. Mol Phylogenet Evol 2010;57(3):1341−6.

[79] Grillet ME, Barrera R, Martinez JE, Berti J, Fortin MJ. Disentangling the effect of local and global spatial variation on a mosquito-borne infection in a neotropical heterogeneous environment. Am J Trop Med Hyg 2010;82(2):194−201.

[80] Povoa MM, Conn JE, Schlichting CD, et al. Malaria vectors, epidemiology, and the re-emergence of *Anopheles darlingi* in Belem, Para, Brazil. J Med Entomol 2003;40(4): 379−86.

[81] Chadee DD, Kitron U. Spatial and temporal patterns of imported malaria cases and local transmission in Trinidad. Am J Trop Med Hyg 1999;61(4):513−17.

[82] Bates M. The laboratory colonization of *Anopheles darlingi*. J Natl Malar Soc (US) 1947; 6(3):155−8.

[83] Gigioli G. Laboratory colony of *Anopheles darlingi*. J Natl Malar Soc 1947;6(3):159−64.

[84] Moreno M, Tong C, Guzman M, et al. Infection of laboratory-colonized *Anopheles darlingi* mosquitoes by plasmodium vivax. Am J Trop Med Hyg 2014;90(4):612−16.

[85] Rozebbom LE. The life cycle of laboratory-bred *Anopheles albimanus* Wiedemann. Ann Entomol Soc Am 1936;29(3):480−9.

[86] Bailey DL, Dame DA, Munroe WL, Thomas JA. Colony maintenance of *Anopheles-albimanus* Wiedmann by feeding preserved blood through natural membrane. Mosq News 1978;38(3):403−8.

[87] Carrillo MP, Suarez MF, Morales A, Espinal C. Colonización y mantenimiento de una cepa Colombina de *Anopheles albimanus* Wiedemann, 1820 (Diptera: Culicidae). Biomedica 1981;1:64−6.

[88] Horosko S, Lima JBP, Brandolini MB. Establishment of a free-mating colony of *Anopheles albitarsis* from Brazil. J Am Mosq Control Assoc 1997;13(1):95−6.

[89] Da Silva ANM, Dos Santos CCB, Lacerda RNL, et al. Laboratory colonization of *Anopheles aquasalis* (Diptera: Culicidae) in Belem, Para, Brazil. J Med Entomol 2006;43 (1):107−9.

[90] Villarreal C, Arredondo-Jimenez JI, Rodriguez MH, Ulloa A. Colonization of *Anopheles pseudopunctipennis* from Mexico. J Am Mosq Control Assoc 1998;14(4):369−72.

[91] Lardeux F, Quispe V, Tejerina R, et al. Laboratory colonization of *Anopheles pseudopunctipennis* (Diptera: Culicidae) without forced mating. C R Biol 2007;330(8):571−5.

[92] Joy DA, Gonzalez-Ceron L, Carlton JM, et al. Local adaptation and vector-mediated population structure in *Plasmodium vivax* malaria. Mol Biol Evol 2008;25(6):1245−52.

[93] Lounibos LP, Lima DC, Lourenco-de-Oliveira R. Prompt mating of released *Anopheles darlingi* in western Amazonian Brazil. J Am Mosq Control Assoc 1998;14(2):210−13.

[94] Lima JBP, Valle D, Peixoto AA. Adaptation of a South American malaria vector to laboratory colonization suggests faster-male evolution for mating ability. BMC Evol Biol 2004;4.

[95] Wang SB, Jacobs-Lorena M. Genetic approaches to interfere with malaria transmission by vector mosquitoes. Trends Biotechnol 2013;31(3):185−93.

[96] Alves FP, Durlacher RR, Menezes MJ, Krieger H, Silva LH, Camargo EP. High prevalence of asymptomatic *Plasmodium vivax* and *Plasmodium falciparum* infections in native Amazonian populations. Am J Trop Med Hyg 2002;66(6):641−8.

[97] Marcano TJ, Morgado A, Tosta CE, Coura JR. Cross-sectional study defines difference in malaria morbidity in two Yanomami communities on Amazonian boundary between Brazil and Venezuela. Mem Inst Oswaldo Cruz 2004;99(4):369−76.

[98] Ladeia-Andrade S, Ferreira MU, de Carvalho ME, Curado I, Coura JR. Age-dependent acquisition of protective immunity to malaria in riverine populations of the Amazon Basin of Brazil. Am J Trop Med Hyg 2009;80(3):452−9.

[99] Katsuragawa TH, Gil LH, Tada MS, et al. The dynamics of transmission and spatial distribution of malaria in riverside areas of Porto Velho, Rondonia, in the Amazon region of Brazil. PLoS One 2010;5(2):e9245.

[100] Alves FP, Gil LH, Marrelli MT, Ribolla PE, Camargo EP, Da Silva LH. Asymptomatic carriers of *Plasmodium* spp. as infection source for malaria vector mosquitoes in the Brazilian Amazon. J Med Entomol 2005;42(5):777−9.

[101] Molina-Cruz A, Barillas-Mury C. The remarkable journey of adaptation of the *Plasmodium falciparum* malaria parasite to New World anopheline mosquitoes. Mem Inst Oswaldo Cruz 2014;109(5):662−7.

[102] Taylor JE, Pacheco MA, Bacon DJ, et al. The evolutionary history of *Plasmodium vivax* as inferred from mitochondrial genomes: parasite genetic diversity in the Americas. Mol Biol Evol 2013;30(9):2050−64.

[103] Parmakelis A, Russello MA, Caccone A, et al. Historical analysis of a near disaster: *Anopheles gambiae* in Brazil. Am J Trop Med Hyg 2008;78(1):176−8.

[104] Klein TA, Lima JBP, Tada MS. Comparative susceptibility of anopheline mosquitos to *Plasmodium falciparum* in Rondonia, Brazil. Am J Trop Med Hyg 1991;44(6):598−603.

[105] Klein TA, Lima JBP, Tada MS, Miller R. Comparative susceptibility of anopheline mosquitos in Rondonia, Brazil to infection by *Plasmodium vivax*. Am J Trop Med Hyg 1991;45(4):463−70.

[106] Tadei WP, Dossantos JMM, Costa WLD, Scarpassa VM. Biology of amazonian anophelines.12. Species of anopheles, transmission dynamics and control of malaria in the urban area of ariquemes (Rondonia, Brazil). Rev Inst Med Trop Sao Paulo 1988;30(3):221−51.

[107] de Oliveira-Ferreira J, Lourenco-de-Oliveira R, Teva A, Deane LM, Daniel-Ribeiro CT. Natural malaria infections in anophelines in Rondonia State, Brazilian Amazon. Am J Trop Med Hyg 1990;43(1):6−10.

[108] de Arruda M, Carvalho MB, Nussenzweig RS, Maracic M, Ferreira AW, Cochrane AH. Potential vectors of malaria and their different susceptibility to *Plasmodium falciparum* and *Plasmodium vivax* in northern Brazil identified by immunoassay. Am J Trop Med Hyg 1986;35(5):873−81.

[109] Magris M, Rubio-Palis Y, Alexander N, et al. Community-randomized trial of lambdacyhalothrin-treated hammock nets for malaria control in Yanomami communities in the Amazon region of Venezuela. Trop Med Int Health 2007;12(3):392−403.

[110] Suarez MF, Quinones ML, Palacios JD, Carrillo A. First record of DDT resistance in *Anopheles darlingi*. J Am Mosq Control Assoc 1990;6(1):72−4.

[111] Fonseca-Gonzalez I, Quinones ML, McAllister J, Brogdon WG. Mixed-function oxidases and esterases associated with cross-resistance between DDT and lambda-cyhalothrin in *Anopheles darlingi* Root 1926 populations from Colombia. Mem Inst Oswaldo Cruz 2009; 104(1):18−26.

[112] Santacoloma L, Tibaduiza T, Gutierrrez M, Brochero H. Susceptibility to insecticides of *Anopheles darlingi* Root 1840, in two locations of the departments of Santander and Caqueta. Biomedica 2012;32:22−8.

[113] Moore SJ, Darling ST, Sihuincha M, Padilla N, Devine GJ. A low-cost repellent for malaria vectors in the Americas: results of two field trials in Guatemala and Peru. Malar J 2007;6.

[114] Hill N, Lenglet A, Arnez AM, Carneiro I. Plant based insect repellent and insecticide treated bed nets to protect against malaria in areas of early evening biting vectors: double blind randomised placebo controlled clinical trial in the Bolivian Amazon. BMJ 2007; 335(7628):1023−5.

[115] Hiwat H, Hardjopawiro LS, Takken W, Villegas L. Novel strategies lead to pre-elimination of malaria in previously high-risk areas in Suriname, South America. Malar J 2012;11.

[116] Suarez MF, Morales C. Impact of *Bacillus sphaericus* (Vectolex® CG) in the control of *Anonpheles albimanus* and *Culex spp.* in Buenaventura, Colombia. J Am Mosq Contr 1999;16:16.

[117] Berrocal E, Carey C, Rodriguez L, Calampa C, Valdivia L. Residual effect of *Bacillus sphaericus*—Vectolex® CG for the control of *Anopheles darlingi* in Iquitos, Peru. J Am Mosq Contr 2000;16:305.

[118] Berti MJ, Gonzales JE. Evaluacion de la efectividad y persistencia de una nueva formulacion de Bacillus sphaericus contra larvae de *Anopheles aquasalis* Curry (Diptera: Culicidae) en criaderos naturales del estado Sucre, Venezuela. Bol Mal Saneamiento Amb 2004;44:21−7.

[119] Moreno JE, Acevedo P, MArtinez A, Sanchez V, Petterson L. Evaluacion de la persistencia de una formulacion commercial de *Bacillus sphaericus* en criaderos naturales de anofelinos vectores de malaria en estado Bolivar, Venezuela. Bol Mal Saneamiento Amb 2010;50:109−17.

[120] Galardo AKR, Zimmerman R, Galardo CD. Larval control of *Anopheles* (*Nyssorhynchus*) *darlingi* using granular formulation of *Bacillus sphaericus* in abandoned gold-miners excavation pools in the Brazilian Amazon Rainforest. Rev Soc Bras Med Trop 2013;46(2):172−7.

[121] Deane LM, Causey OR, Deane MP. Notas sobre a distribuição e a biologia dos anofelinos das Regiões Nordestina e Amazônica do Brasil. Rev Serv Esp Saúde Pública 1948;1: 827−965.

[122] Conn JE, Wilkerson RC, Segura MN, et al. Emergence of a new neotropical malaria vector facilitated by human migration and changes in land use. Am J Trop Med Hyg 2002;66(1):18−22.

[123] Aramburu Guarda J, Ramal Asayag C, Witzig R. Malaria reemergence in the Peruvian Amazon region. Emerg Infect Dis 1999;5(2):209−15.

[124] Emerson KJ, Conn JE, Bergo ES, Randel MA, Sallum MAM. Brazilian *Anopheles darlingi* Root (Diptera: Culicidae) Clusters by major biogeographical region. PLoS One 2015;10(7).

Chapter 6

Considerations for Disrupting Dengue Virus Transmission; Ecology of *Aedes aegypti* and Current (Nongenetic) Methods of Control

Roberto Barrera

Entomology and Ecology Activity, Dengue Branch, Centers for Disease Control and Prevention, San Juan, Puerto Rico

CURRENT BURDEN OF DENGUE

Dengue virus (DENV) incidence has increased 30-fold in the last five decades [1]. A recent estimate indicates that there were between 284 and 528 million DENV infections in 2010, of which 96 million exhibited some sort of symptom or apparent infection [2]. Among these, 70% occurred in Asia, 16% in Africa, 13.8% in the Americas, and 0.2% in Oceania. For example, annual incidence of severe dengue in Indonesia increased from 0.05/100,000 in 1968 to 35–40/100,000 in 2013 [3]. There is evidence pointing at increased DENV transmission in Africa [1], but because dengue is markedly underreported and underdiagnosed, it is difficult to establish its true extent [4]. Dengue also seems to be emerging or expanding its distribution in the Eastern Mediterranean Region (Saudi Arabia, Pakistan, Yemen, Sudan, Djibouti, and Somalia) [1,5].

The more recent global expansion of *Aedes albopictus* [6] has been responsible for the reemergence of local dengue transmission in the United States [7] and Europe (Croatia, France) [8]. Well-established populations of *Aedes aegypti* have been associated with repeated autochthonous transmission in the southern USA (Florida, TX) [9,10]. It is interesting to note that in spite of the recent geographical expansion and increased incidence of dengue, this disease used to have a wider geographical distribution than presently [11]. Among the areas that have had contractions in dengue distribution

Genetic Control of Malaria and Dengue.
2016 Published by Elsevier Inc.

103

are the southern United States, Europe and Japan, Australia, China and South Africa [11]. This review focuses on the ecology and nongenetic control of the domestic variety of *Ae. aegypti*, the main vector of dengue viruses worldwide, and on some of the difficulties that genetic approaches would face to control this mosquito.

CAUSES OF WIDESPREAD DENGUE TRANSMISSION

The main causes for the unprecedented surge of dengue are human population growth, urbanization, globalization, human movement, changing life styles, and insufficient vector control [12]. These variables are interdependent and reflect a complex environment, where more people live closer together in unplanned urban areas with deficient public services (piped water supply, domestic garbage collection, and sewerage) [13]. As a result, water-storage vessels, trashed containers (disposable food and drink plastics, tires, etc.), and even septic tanks provide aquatic habitats where dengue vectors undergo immature development and adult mosquito emergence. Substandard housing in a tropical climate, where air-conditioning is not affordable and window/door screens are absent, propitiate a high frequency of host−vector contacts. Increased international commerce and air transportation have caused unprecedented movement of both vectors and viruses [12,14]. The current, prevalent approach to DENV vector control is mostly reactive, seeking to control epidemics. Unfortunately, there is no hard evidence showing that widespread dengue epidemics can be controlled [15]. Lack of organized, well-funded vector control programs, trained personnel, and local operational research limits the efficiency of dengue vector control. The ever increasing disease incidence suggests that vector control is not controlling dengue, yet there is no evaluation of what the dengue figures would be in the absence of current vector control efforts. Excessive reliance on the use of insecticides has caused the evolution of resistance, although mainly in *Ae. aegypti* [16−18]. Improvements in vector surveillance are needed to evaluate the impact of vector control measures, as well as new tools for dengue vector control [19].

MAJOR DENGUE VECTORS WORLDWIDE

The main mosquito vector of DENV worldwide is *Ae. aegypti*, followed in importance by *Ae. albopictus*. People supply *Ae. aegypti* with water-holding containers for immature development, harborage, or refuge inside buildings (where this mosquito prefers to rest and feed), and sources of blood (people and domestic animals). Because *Ae. albopictus* is less prone to rest indoors, humans mainly provide them with outdoor containers and blood. Although the latter species is highly anthropophilic, it also feeds on a variety of vertebrates other than humans that do not amplify dengue viruses [20]. These two mosquito species also differ in terrestrial habitats. The domestic variety of *Ae. aegypti* can occupy many terrestrial habitats as long as they are inhabited by people, such as highly urbanized areas without vegetation, suburban, and rural areas. Even though *Ae. albopictus* is highly anthropophilic, it is more dependent on vegetation and the resources associated with it (sources of sugar, vertebrates) and it is thus more common in suburban and rural areas. As a result, *Ae. aegypti* has a higher frequency of contacts with people, which explains why this mosquito is the main vector of dengue viruses.

HOW DOES THE ECOLOGY OF *AEDES AEGYPTI* COMPLICATE CONTROL EFFORTS?

Domestication

Aedes aegypti takes refuge in dark recesses indoors (endophilic) and uses a variety of containers both indoors (e.g., water-storage containers, flower pots) and around the home (e.g., wells, discarded containers, plant saucers, animal drinking pans). This behavior enormously complicates its control. Because people may be out at work or deny entry to health inspectors, a variable number of premises in urban areas go without vector surveillance and control activities. It has been shown that efforts to eliminate *Ae. aegypti* through source reduction (elimination of containers), focal (larviciding), and perifocal (application of residual insecticide around containers to eliminate landing mosquitoes) treatments per house failed in an urban area where 32% of the houses were closed in comparison with another area where only 6% of the houses were closed [21]. The reason for failure was most likely that the local population of *Ae. aegypti* could repopulate the area from the houses that could not be treated.

Part of the difficulty in controlling *Ae. aegypti* lies in the fact that many types of containers cannot be eliminated through source reduction, such as water-storage vessels, potted plants and trivets, animal drinking pans, paint trays, toys, pails, septic tanks, and cavities in structures (fence poles, bricks, uneven floors and roofs, roof gutters, air-conditioned trays). Discarded containers that can be eliminated through source reduction may reappear on backyards in a few weeks. Even some people accumulate many objects and containers that would appear useless but refuse giving them up (e.g., hoarding).

Density-Dependent Population Regulation

Aedes aegypti as well as other mosquitoes that undergo immature development in natural or artificial containers are low-density mosquitoes [22]. Notwithstanding, dengue transmission may occur at relatively low mosquito densities. For example, a model revealed that dengue transmission would be possible above threshold levels of 0.5−1.5 *Ae. aegypti* pupae per person at

28°C and 0−67% dengue seroprevalence [23]. This is equivalent to 0.1−0.4 female adult mosquitoes per person after considering that a pupa can last 2 days and that half of the pupae are females. Additionally, DENV transmission has occurred in areas with low values of immature mosquito indicators such as the Breteau Index [24].

It has been shown that *Ae. aegypti* is commonly exposed to food limitation and competition for limited resources in the larval stage [25−27]. The limited source of energy (e.g., decomposing plant materials) and usual lack of primary producers (e.g., algae) in containers are thought to be causes why aquatic predators are uncommon in containers. Aquatic predators may also be uncommon in containers of highly urbanized areas as a result of isolation from natural areas where they occupy natural containers, such as tree holes and bamboo internodes. Additionally, some urban predators that exploit containers, such as the mosquito *Toxorhynchites* spp., are restricted to well-shaded, vegetated areas and do not disperse over open areas [28]. By comparison, mosquito species that develop in bodies of water on the ground (flooded areas, ponds, margins of rivers, etc.), such as many *Aedes*, *Culex*, and *Anopheles* species are locally abundant and less limited by food; instead they are mainly regulated by predation and hydrologic processes (flooding, desiccation) [29]. Direct density-dependence determines that the population of *Ae. aegypti* would not have limitations to realize its maximum reproductive capacity when its population is at low density and that its potential for increase becomes limiting at higher densities. Because of this process, *Ae. aegypti* may quickly recover from vector control measures and other disturbances that temporarily reduce its populations. It has also been proposed that *Ae. aegypti* is regulated in a density-dependent way in the adult stage by the defensive behavior of people [30]. That is, when several females are attempting to bite a person, the presence of the mosquitoes is noticed, eliciting a person's defensive behavior, but if only one or two females are around, they may go unnoticed. This is possible because *Ae. aegypti* tends to bite on the ankles and elbows in a seclusive way. Although its bite is painless, many people develop an inflammatory reaction and itching soon after the bite, so that the presence of just a few *Ae. aegypti* can be detected by people and evoke their defensive behavior. Yet, more field investigations are needed to understand better the nature of density-dependence in *Ae. aegypti* [31] and its relationship with control measures.

Spatial Heterogeneity and Super-Producer Aquatic Habitats of *Aedes aegypti*

The spatial distribution of *Ae. aegypti* is highly heterogeneous or overdispersed, with many places having few mosquitoes and a few places having many mosquitoes. Several studies have found that a few types of containers produce most *Ae. aegypti* pupae [32,33]. The contribution of container types to mosquito density is the product of the number of pupae per type of container and the number of containers of each type. There are also variations in mosquito productivity among houses. For example, a few houses (6%) with large yards, many trees, and containers with a large volume of water produced more than 60% of all *Ae. aegypti* pupae in a locality in Puerto Rico [33]. These common findings have led to the recommendation that by targeting the most productive containers the number of mosquitoes could be brought down below threshold levels for dengue transmission [23]. The distribution of *Ae. aegypti* adults has also been shown to be highly heterogeneous, with clustering of adult mosquitoes up to 30 m [34], but other studies have found local clustering at the block level, possibly reflecting the contribution of highly productive households within blocks [35]. There have been attempts to characterize the aspect of highly productive premises to facilitate vector control, such as the Premise Condition Index where shade and untidiness of yards were significantly associated with the proportion of positive premises and the number of positive containers per premise [36]. Places with elevated densities of both mosquitoes and people may disproportionally contribute to a rapid epidemic spread of dengue viruses [37]. It has also been suggested that dengue reduction would be greater in areas with higher human density after removing super-producer containers [37].

Cryptic Aquatic Habitats

There are increasing reports of *Ae. aegypti* developing in nonconventional containers or in containers that cannot be easily located by visual inspections alone, such as wells, storm drains, sumps, roof gutters, elevated water tanks, and even septic tanks [38,39]. Some of these containers can produce more *Ae. aegypti* than the traditional containers that can be visually located (e.g., discarded tires and miscellaneous containers, plant pots, jars and barrels for water storage) [38,40−42]. For example, one septic tank was producing 1,449 *Ae. aegypti* adults per day in southern Puerto Rico. The average number of adult *Ae. aegypti* captured in the main room of houses with backpack aspirators in a community with septic tanks was 3.6 (max = 234), whereas in a nearby community without septic tanks it was 1.4 (max = 34) [38]. Conventional vector control operations targeting surface containers and even spatial spraying of insecticides would not affect these underground populations of *Ae. aegypti*. The presence of subterranean aquatic habitats (wells, manholes) of *Ae. aegypti* has been linked to DENV transmission in Australia [43].

Egg Quiescence

Eggs of *Ae. aegypti* are laid on the walls of water-holding containers where they can withstand desiccation for several months (aka "egg bank"), depending on temperature and humidity [44,45]. Although possibly not representative of most populations, some *Ae. aegypti* eggs hatched and mosquitoes eventually reproduced after being kept dry in a laboratory for 21 months [46]. The capacity to withstand desiccation for several months is important in seasonal tropical environments where the dry season can last several months [45]. Additionally, this adaptation confers *Ae. aegypti* high resilience or the capacity of the population to recover after disturbances (e.g., drought, vector control, storms). For example, if all larvae, pupae, and adults were eliminated from a locality at once, *Ae. aegypti* could still recolonize the area from its eggs as soon as they are flooded with water (rain, humans). This fact has important implications for mosquito and dengue control.

There are two ways to eliminate dormant eggs as a factor that would repopulate treated areas: (i) by destroying the eggs or killing the pharate larvae in them or (ii) by killing the larvae when they hatch sometime later. Unfortunately, there are no commercial ovicides against *Ae. aegypti* or other container mosquitoes. Although not registered as an insecticide, chlorine bleach (NaOCl) sprayed on the inner walls of containers is effective against eggs (20 or 10 ppt + smectite clay solution) [47] and immature stages (2 tablespoons per 5 L of water) of *Ae. aegypti* [48]. If larvicides lasted longer than egg viability, then hatching larvae after flooding caused by rains or the addition of water by people would be eliminated. Current larvicides do not last but a few days to weeks, therefore unless larvicides are repeatedly applied, the area is bound to be repopulated from the egg bank. Elimination of the egg bank along with conventional vector control measures in containers that cannot be eliminated (e.g., wash basins) [49] may result in a more substantial or sustained impact on the vector population [50].

POPULATION DYNAMICS OF *AEDES AEGYPTI* AND DENGUE

The population dynamics of *Ae. aegypti* is strongly influenced by latitude/elevation, weather, and human behavior. This mosquito can establish persistent populations at latitudes below the 10°C isotherm in the northern and southern hemispheres and below 2,000 m of elevation [44]. However, provided timely introductions, *Ae. aegypti* can establish transient populations at higher latitudes in the temperate zone during the summer, which in the past were responsible for epidemics of yellow fever and dengue in the United States [51,52]. The length of the season with reproductively active mosquitoes is mainly constraint by temperature in subtropical areas [53−55]. The abundance of *Ae. aegypti* is mainly driven by rainfall in tropical areas and larger populations exist during the wet seasons [56−63]. However, there is also an important effect of human behavior on the temporal variation of mosquito abundance, which consists in keeping containers with water that are used by *Ae. aegypti* to complete its immature development. Individual water storage resulting from substandard piped water supply determines that this mosquito can thrive even during long dry seasons [13,57,58]. There are many types of containers kept with water by people during critical parts of the year, which seem to be important enough to sustain DENV transmission and collectively, are the main cause for DENV endemicity in seasonal tropical areas [59].

Dengue outbreaks usually occur during the hot and wet parts of the year in both hemispheres [64]. Typically, peaks in mosquito density that are driven by seasonal rainfall precede maximum dengue incidence [59,64−69], although dengue outbreaks have been reported during the dry season as a result of extensive water storage [70]. Lack of association between *Ae. aegypti* populations and dengue has been described in areas where most aquatic habitats were kept with water by people [71]. The relationship between vectors and dengue in areas near the Equator where weather conditions are suitable for dengue transmission year-round are not so straightforward [72]. There are some scenarios that confound interpretations of the influence of mosquito abundance on dengue transmission, which may cast doubts on the usefulness of vector surveillance. For example, there may be a large number of vectors but no dengue epidemic even in the presence of local DENV transmission. This phenomenon has been observed in Puerto Rico where the first peak of *Ae. aegypti* adults, which typically occurs after the first peak of rainfall in May, is not associated with dengue outbreaks. However, a similar peak of mosquitoes that results from the second peak of rainfall in late summer is associated with dengue outbreaks [59]. It has also been observed that relatively large densities of *Ae. aegypti* do not result in major outbreaks when major epidemics had just happened [67]. It is also possible to observe large mosquito populations and no disease outbreaks if dengue viruses have not yet arrived to the area.

MAIN CURRENT LIMITATIONS TO CONTROL DENGUE

A main limitation to control dengue is dealing with four different viruses (DEN-1, DEN-2, DEN-3, and DEN-4) that are disseminated worldwide and cocirculate in most large urban centers in tropical regions [12]. If it were just one virus, built herd immunity following an outbreak would prevent another outbreak to happen in the following years until enough susceptible people were recruited (immigration, births), such as it has been shown for yellow fever [73]. Traditional dengue control is not known for targeting particular dengue viruses in endemic areas; the approach has been to try to control all at the same time, following a similar approach to the development of a tetravalent dengue vaccine [74]. Targeting specific DENV serotypes may prove valuable if most control efforts would concentrate in preventing the spread of a new invading serotype (virus containment). There are successful examples of DENV containment through vector control, although this requires active and efficient virus and vector surveillance [75]. Currently, the widespread dissemination of dengue viruses along with air transportation determine frequent reintroductions to endemic and nonendemic areas, which constantly challenge dengue control [76].

Another important limitation to controlling dengue is the prevalent approach to controlling dengue epidemics instead of preventing them. There is not much evidence showing that major dengue epidemics can actually be controlled [15]. "Realizing the rapidity with which the epidemic evolved, the inevitable delay in responding, and the uncertain ability to control the epidemic once it had begun, the Puerto Rico experience suggests that a better strategy may lie with prevention rather than firefighting efforts" [77]. There is a major element involved with the type of approach assumed to control dengue (preventive vs reacting), which lies in the kind of organization that each one requires. A common argument for not adopting a preventive strategy is lack of funds, but it is surprising to observe how funds usually become available when a major epidemic strikes [78].

The ensuing question is how best to prevent dengue epidemics. Before discussing possibilities, it is necessary to point out the lack of planned investigations on how to prevent dengue epidemics in large urban centers, most likely because of the scale at which they occur, which makes controlled interventions difficult to achieve. Another limitation is that it is not ethically acceptable to leave areas that might be impacted by dengue as "nonintervention" areas to compare with intervention areas.

The prevention of dengue epidemics requires, at least, considering that dengue transmission is heterogeneous in space and time, where some areas are more frequently or more intensely hit than others and that dengue epidemics may form many months before they appear. For example, it has been shown that certain neighborhoods in a dengue endemic city had higher incidence and persistence of dengue (hot spots), where a small fraction of all neighborhoods presented most of the cases year after year [79]. Hot spots are areas with appropriate environmental conditions with stable and elevated mosquito populations [80]. Dengue viruses and mosquitoes can be disseminated out of hot spots into other neighborhoods. Preemptive vector control in hot spots should cause a reduction of dengue incidence not only in hot spots but also in other areas of the city. This approach is being tested for malaria control [81].

It has been shown that dengue epidemics can be anticipated several months in advance, which seems to indicate that the proper conditions for the formation of an epidemic are not necessarily those that are observed at peak transmission. For example, a simple early warning system (endemic channel) consists of observing if the number of suspected dengue cases is above the recent, nonepidemic years' 75th percentile for two consecutive weeks (CDC, unpublished). In 2010, alarm signs of an epidemic were detected on epidemiological weeks 4−5 and the major epidemic ever reported in Puerto Rico [82] peaked on weeks 32−34, which represented a lead time of 7 months. Because historically dengue cases reach minimum values in January−February in Puerto Rico, it would appear that conditions for the development of an epidemic were already happening at that time (e.g., early, widespread interepidemic transmission). It has been shown in theory that high, initial virus reproduction rates determine the magnitude of the epidemic several months later [83]. The ideal strategy would be to keep sustained, preventive vector control in order to prevent dengue epidemics [84], but if it is not possible, then vector control interventions should be intensified early in the season (e.g., before the rainy season) because they cause extended reductions in vector abundance and DENV transmission [72,85,86]. Thus, applying preemptive vector control measures to hot spots early in the mosquito season seems to be a plausible hypothesis for preventing dengue epidemics.

Lack of evaluation of the efficacy of vector control measures is perhaps one of the most important reasons why it is so difficult to control dengue. The application of ineffective vector control measures negatively impacts the much needed country's support to understand how best to control *Ae. aegypti* and dengue. Local, operative research is needed to understand the environmental and social variables involved in the ecology of *Ae. aegypti* and its interaction with people [87]. Lack of highly trained entomological personnel and support to conduct local research are not the only reasons for lack of success in dengue vector control.

Traditionally, it has been very difficult to monitor the numbers of *Ae. aegypti* mosquitoes in order to evaluate the impact of control measures on vector and DENV transmission [88]. Indicators of immature presence (*Stegomyia indices*) do not inform how many mosquitoes there are, but they also face the limitations of visual localization of container habitats and lack of access to many of the buildings producing *Ae. aegypti*. Pupal surveys were recommended as an improvement for vector surveillance, because enumerating pupae is a better proxy for the number of adult mosquitoes being produced [89]. The main limitation with estimating pupae abundance is the large variance associated with it, which implies having to take very large sample sizes (e.g., number of houses) [90−92]. Nevertheless, pupal surveys are important to indicate what containers are producing most *Ae. aegypti* and design targeted control interventions [32,33]. The major reason why vector surveillance of dengue vectors have concentrated on the immature stages has been the lack of efficient traps to capture adult mosquitoes. Fortunately, key recent developments in trapping adults of container *Aedes* make it possible to obtain reliable estimations of their relative abundance and understand better its relation to virus transmission [61,93]. As vector control programs start using the new vector surveillance tools, it will be easier to determine the effectiveness of various vector control measures.

CURRENT INTERVENTIONS TO CONTROL MOSQUITO VECTORS

Current vector control interventions consist of immature mosquito control through source reduction and larviciding and insecticide spraying to eliminate adult mosquitoes. Immature mosquito control faces the limitations of (i) restricted access to the places where mosquitoes are being produced (closed houses, refusing treatments) [21], (ii) existence of cryptic aquatic habitats that are missed [38,94], and (iii) resistance to the most common larvicide (organophosphate insecticide temefos) [16,95]. Among these three factors, the latter is more easily solved thanks to the existence of several larvicides with very different modes of action, such as those from microorganism origin (*Bacillus thuringiensis* israelensis, spinosad), juvenile hormone analogs (methoprene, pyriproxyfen), chitin synthesis inhibitors (diflubenzuron, novaluron), and monomolecular films and oils. Biological control, particularly through the use of copepods [96], source reduction, and ample community participation have been successful in Vietnam to eliminate container *Aedes* in large water-storage vessels [97]. However, many of the predators available for use to control *Ae. aegypti*, including cyclopoid copepods, predatory mosquitoes (*Toxorhynchites* spp.), and larvivorous fish cannot easily get established in containers that periodically dry out (i.e., small containers). Entomopathogenic fungi, such as *Beauveria bassiana* and *Metarhizium anisopliae*, have potential as biocontrol agents against container *Aedes* mosquitoes. Mosquitoes are infected by tarsal contact with spores that penetrate the cuticle, grow, and produce toxins that kill the mosquito [98]. Another development involving biological control is the release of *Wolbachia*-infected males of container *Aedes* mosquitoes, which after mating with uninfected females produce a sterilizing effect through cytoplasmic incompatibility [99−101].

Controlling adult *Ae. aegypti* is most commonly done by spraying concentrated insecticides at ultralow volume [102]. The purpose is eliminating flying or resting adult mosquitoes that may be participating in the DENV transmission cycle. This technique transiently reduces the number of adult mosquitoes but has negligible impact on mosquito population growth [22]. When space spraying is the only vector control method employed, it has been ineffective at reducing mosquito populations [103]. Indoor spatial spraying of nonresidual insecticides is more effective because most *Ae. aegypti* mosquitoes tend to rest indoors. Outdoor spatial spraying is expected to be more effective on *Ae. albopictus* because it is mostly an exophilic species [104]. Recent observations suggest that frequent indoor spraying has reduced DENV infections in Iquitos [72]. Residual insecticide spraying, the key element in the past successful eradication of *Ae. aegypti* [105], is almost inexistent in contemporary dengue vector control programs. Indoor residual spraying of insecticides is used in some control programs for the focal control of mosquitoes around introduced dengue cases [106].

There are a number of novel approaches to controlling DENV vectors that target the adult stages of the mosquito, such as new adulticides [107], toxic baits [108], insecticide impregnated materials (curtains [109], bed nets [110], covers for water-storage containers [111], and ovitraps [112]), sticky gravid traps [61,113], autodissemination of insect growth regulators by contaminated (pyriproxyfen) [114], or fungus-infected [115] adult mosquitoes, and the release of genetically modified (GM) [116] and *Wolbachia*-infected mosquitoes [101]. Controlling the adult stages of DENV vectors is much needed to complement immature mosquito control, which by itself has not been able to solve the problem. Furthermore, rather than looking for a "silver bullet" solution to controlling *Aedes*, what it is most likely to happen is that successful control will be achieved by a combination of partially effective tools. For example, successful elimination of *Ae. aegypti* from a Caribbean island was achieved by simultaneously applying indoor and outdoor insecticide spraying, residual insecticide spraying of mosquito resting places (indoor, around containers), source reduction, and larviciding, whereas failure to eliminate this mosquito in another island was likely a result of not concurrently applying indoor residual spraying [105].

IMPROVING MOSQUITO CONTROL

Local Aedes Populations Should be Managed in Space and Time Using an Area-Wide Approach

Vector control measures should "maintain pest populations at low, nondamaging, densities ... this means creating a stable and low equilibrium" [117]. In this case, it would mean keeping the vector population below DENV transmission thresholds [23]. For example, the deployment of sticky, autocidal ovitraps that are serviced every 2 months has kept the *Ae. aegypti* population at sustained low densities (60–80%) for over 3 years in southern Puerto Rico [118]. Whether these densities are low enough to prevent dengue or chikungunya virus transmission is being investigated. Managing the population of *Ae. aegypti* may be best accomplished by working out relatively small areas at a time (e.g., neighborhoods), although it depends on the control method and program resources.

Effective Control Agent

Control agents are used to kill mosquitoes (e.g., pesticides, predators, parasites), suppress their reproductive capacity (e.g., sterile insect techniques, ovitraps), or both (e.g., autocidal gravid traps). It is required to verify how effective is the control agent before field implementation (e.g., insecticide resistance, male competitiveness). It is also important to assess for how long the agent keeps its effectiveness and the frequency of reapplications. For example, most larvicides do not retain full effectiveness but for a few weeks. Comparatively, the organochlorine insecticide DDT has been shown to have a residual effect far in excess of 6 months on some natural populations of *Ae. aegypti* [119]. Extended control means that larger areas can be treated with the same personnel.

Efficient Delivery System of the Control Agent

Traditional delivery systems search for the locations of mosquitoes to administer specific control agents, such as by visiting houses to apply source reduction and larviciding or spraying insecticides. Calibration of insecticide sprayers, training of personnel, and field supervision are essential components of this type of delivery system. Many of the most recent developments targeting adult DENV vectors do not require the active search of mosquitoes. In the case of vector control by genetically modified mosquitoes (GMM) or *Wolbachia*-infected mosquitoes, released males are both control agents and delivery systems that are expected to contact their targets on their own. Other control methods rely on luring adult mosquitoes to a variety of devices such as autodissemination stations, toxic baits, ovitraps, gravid traps, and insecticide-treated surfaces (such as covers for water-storage vessels, bed nets, curtains, etc.).

Sufficient Coverage

Coverage is related to the extent at which the vector population is exposed to the control agent. For example, outdoor spraying of insecticides from aircraft or truck-mounted equipment can in theory cover all of the urban area that needs vector control. However, because of the endophilic behavior of *Ae. aegypti*, droplet penetration is insufficient to reach a large fraction of the indoor resting mosquito population. Thus, although the control agent is delivered everywhere in the treatment area, it perhaps does not reach but a small fraction of the extant adult *Ae. aegypti* population. Coverage limitations are expected while conducting door-to-door control of immature *Aedes* if there are cryptic aquatic habitats. Methods relying on luring adult mosquitoes or releasing adult mosquitoes need to consider the flight range of *Ae. aegypti* to distribute their efforts as to cover most of the population in space and time. Some questions pertaining coverage include how many sterile males need to be released, how many release stations need to be used, how frequently should males be released, how many ovitraps need to be deployed, how many dissemination/bait stations are required, what percentage of houses need to be treated with insecticide impregnated materials, etc.

Evaluation of Impact

The effects of control measures on the *Ae. aegypti* population should evaluate its extent or degree of reduction and duration. This is usually accomplished by comparing vector densities before and after the control intervention in areas with and without treatments (longitudinal studies) [118] or by contrasting "comparable" treated and untreated areas (cluster randomized trials) [120]. A common problem that should be avoided is mosquito immigration from nearby, untreated areas [121,122]. For the proof of concept of a new control tool in field trials, only the best available tools for capturing *Ae. aegypti* should be used, which include indoor aspiration techniques and some of the newer passive or active adult traps.

PERSPECTIVES OF SUCCESS USING GENETICALLY MODIFIED MOSQUITOES TO CONTROL DENGUE VECTORS

The sterile insect technique (SIT) has been successfully used for eliminating or controlling several important agricultural pests [123]. The use of GM or *Wolbachia*-infected adult mosquitoes to control *Ae. aegypti* is unique in that target (mosquito), control agent (modified gene, bacteria), and delivery system (mating) are the same. The major advantage is that once it has been approved by the community, mosquitoes can be released from public areas, thus avoiding the need to request individual householders' permission to gain access to their premises. As explained before, this is one of the major limitations to current *Ae. aegypti* control approaches. The effectiveness of the GMM technique will depend on the efficacy of the control agent (genetic modification) to suppress the reproduction of extant females (e.g., degree of sterility) or the survival of cohorts inheriting modified genes (e.g., lethal genes). It is also important that released mosquitoes can effectively deliver the genes through adequate male competitiveness, longevity, and flight capacity. The higher the quality of the specimens the fewer ones that would need to be released [124]. Finally, there is the issue of coverage that includes releasing sufficient numbers of modified males, adopting adequate release frequencies, and appropriate spatial coverage of the release stations [125].

Two aspects are discussed in relation to the use of GMM to suppress *Ae. aegypti*: some limitations for successful implementation and the integration of this technique with other control methods.

Self-Limiting Population Suppression

1. There are several features of the biology and ecology of *Ae. aegypti* that complicate a genetics-based, suppression control strategy:
    a. The limited flight range of *Ae. aegypti* (a few hundred meters) dictates that release stations should be closely spaced. For example, a spatial model predicted a minimum of 2.2 release stations per hectare and that the presence of gaps resulted in rapid area recolonization [125]. By comparison, the new world screwworm (*Cochliomyia hominivorax*) can disperse more than 280 km, so fewer release stations can cover larger areas.
    b. The spatial pattern of dispersal of *Ae. aegypti* tend to be clumped or aggregated [35,92], which is expected to require greater release rates

as compared to a randomly dispersed population [126]. Results of a cell automata model even suggested that spatial heterogeneity of oviposition sites may make it very difficult or impossible to suppress *Ae. aegypti* using SIT [127]. Yet, because the spatial dispersal of vectors varies seasonally or as a result of vector control [35], temporal changes in aggregation should be monitored. It is also important to have uniform control of the population. It has been hypothesized that reducing all the target population by 90% in each generation is more effective than reducing 90% of the population by 99% but leaving 10% of the population untreated [128].

c. The capacity of the eggs of *Ae. aegypti* to withstand desiccation and remain viable inside the containers for months [45] provides an element of vector population resistance and resilience that requires sustained control efforts, possibly spanning many months.

d. Previous observations with GM *Ae. aegypti* (chromosomal translocations) showed that an induced sterility of 60−70% did not cause any significant changes in the mosquito population [129,130]. The authors concluded that induced sterility was at least similar to density-dependent mortality in the larval stages. Thus, the levels of population suppression would need to reach much higher values before starting noticing significant reductions in the mosquito population. Some animal populations show inverse density-dependence and an Allee effect, whereby the population goes extinct below a critical density threshold [131]. Because *Ae. aegypti* seems to be under direct density-dependence regulation, lowering its density may have no impact or even increase population numbers [132].

e. Like with other vector control methods, mosquito immigration into treated areas may require either the use of buffer control zones around the treated area or significantly increased release efforts [133].

2. Integration of GMM with other vector control methods. Although the sterile insect technique has been used to successfully eliminate some insect pests, it has more commonly been used as a complementary pest suppression technique [123,134]. The efficiency of SIT in terms of the ease of further suppression is greater when the pest population is smaller [135]. Thus, a number of vector control measures that would not harm the GM male mosquitoes can be applied in advance to reduce the population of *Ae. aegypti*, such as source reduction and larviciding, biological control, and mosquito traps. Some control measures could be applied concomitantly with genetic control. For example, the elimination of already mated females along with the sterilization of unmated females can cause a much greater impact [123]. So, control measures directed at gravid females, such as using sticky gravid ovitraps, may make GMM approaches more feasible and economic.

Self-Sustaining Vector Population Replacement

An alternate method to controlling dengue is by temporally or permanently replacing a DENV competent population of *Ae. aegypti* with one that cannot transmit the virus. Several genetic modification approaches are being explored, although none of them have yet reached the stage of open-field tests [136]. The objective of this approach is introducing a self-sustaining (heritable) genetic modification that makes the mosquito incapable of transmitting DENV. Any genetic modification of this sort needs to have a small fitness cost to the mosquito in order to persist in nature [130,136].

A similar approach has been adopted using strains of *Wolbachia* that disrupt DENV transmission by *Ae. aegypti* [137]. These studies have shown rapid dissemination of the bacteria in urban populations of *Ae. aegypti* in Australia, which was accomplished by releasing modest numbers of infected mosquitoes (10,000−22,000 per week) in a relatively short time (3 months). The initial implementation of this technique was preceded by source reduction and followed by adaptive releases of infected mosquitoes in areas showing low infection coverage [137]. Thus, efforts to replace a vector population with nonvector individuals seem to be considerably smaller than those required to suppress or eliminate a vector population. Long-term evaluation of the effectiveness of the *Wolbachia* strain to block DENV transmission by field mosquitoes will be required to determine if any evolutionary changes occur.

REFERENCES

[1] WHO. Global strategy for dengue prevention and control, 2012−2020. WHO/HTM/NTD/VEM/2012.5. Geneva: World Health Organization; 2012.

[2] Bhatt S, Gething PW, Brady OJ, et al. The global distribution and burden of dengue. Nature 2013;496(7446):504−7.

[3] Karyanti MR, Uiterwaal CS, Kusriastuti R, et al. The changing incidence of dengue haemorrhagic fever in Indonesia: a 45-year registry-based analysis. BMC Infect Dis 2014;14(1):412.

[4] Amarasinghe A, Kuritsk JN, Letson GW, Margolis HS. Dengue virus infection in Africa. Emerg Infect Dis 2011;17(8):1349−54.

[5] Aziz AT, Al-Shami SA, Mahyoub JA, Hatabbi M, Ahmad AH, Rawi CS. An update on the incidence of dengue gaining strength in Saudi Arabia and current control approaches for its vector mosquito. Parasit Vectors 2014;7(1):258.

[6] Benedict MQ, Levine RS, Hawley WA, Lounibos LP. Spread of the tiger: global risk of invasion by the mosquito *Aedes albopictus*. Vector Borne Zoonotic Dis 2007;7(1):76−85.

[7] ProMED. Dengue/DHF update (95): Americas; 2013.

[8] Tomasello D, Schlagenhauf P. Chikungunya and dengue autochthonous cases in Europe, 2007−2012. Travel Med Infect Dis 2013;11(5):274−84.

[9] Clark GG. Dengue and dengue hemorrhagic fever in northern Mexico and south Texas: do they really respect the border? Am J Trop Med Hyg 2008;78(3):361−2.

[10] Radke EG, Gregory CJ, Kintziger KW, et al. Dengue outbreak in Key West, Florida, USA, 2009. Emerg Infect Dis 2012;18(1):135−7.

[11] Van Kleef E, Bambrick HJ, Hales S. The geographic distribution of dengue fever and the potential influence of global climate change. TropIKAnet 2010;1–22.

[12] Gubler DJ. Dengue, urbanization and globalization: the unholy trinity of the 21st century. Trop Med Health 2011;39(Suppl. 4):3–11.

[13] Barrera R, Navarro JC, Mora JD, Dominguez D, Gonzalez J. Public service deficiencies and *Aedes aegypti* breeding sites in Venezuela. Bull Pan Am Health Organ 1995;29 (3):193–205.

[14] Huang Z, Huang A, Das Y, Qiu A. Web-based GIS: the vector-borne disease airline importation risk (VBD-AIR) tool. Int J Health Geogr 2012;11(1):33.

[15] World Health Organization. Dengue guidelines for diagnosis, treatment, prevention and control: new edition. Geneva: World Health Organization; 2009.

[16] Ranson H, Burhani J, Lumjuan N, Black IV WC. Insecticide resistance in dengue vectors. TropIKAnet 2010;1.

[17] Vontas J, Kioulos E, Pavlidi N, Morou E, della Torre A, Ranson H. Insecticide resistance in the major dengue vectors *Aedes albopictus* and *Aedes aegypti*. Pestic Biochem Physiol 2012;104(2):126–31.

[18] Marcombe S, Farajollahi A, Healy SP, Clark GG, Fonseca DM. Insecticide resistance status of United States populations of *Aedes albopictus* and mechanisms involved. PLoS One 2014;9(7):e101992.

[19] Morrison AC, Zielinski-Gutierrez E, Scott TW, Rosenberg R. Defining challenges and proposing solutions for control of the virus vector *Aedes aegypti*. PLoS Med 2008;5 (3):362–6.

[20] Hawley WA, Hawley WA. The biology of *Aedes albopictus*. J Am Mosq Control Assoc Suppl 1988;1:1–39.

[21] Chadee DD, Chadee DD. Effects of 'closed' houses on the *Aedes aegypti* eradication programme in Trinidad. Med Vet Entomol 1988;2(2):193–8.

[22] Reiter P, Gubler DJ. Surveillance and control or urban dengue vectors. In: Gubler D, Kuno G, editors. Dengue and dengue hemorrhagic fever. UK: Cab International; 1997. p. 425–62.

[23] Focks DA, Brenner RJ, Hayes J, Daniels E. Transmission thresholds for dengue in terms of *Aedes aegypti* pupae per person with discussion of their utility in source reduction efforts. Am J Trop Med Hyg 2000;62(1):11–18.

[24] Sanchez L, Vanlerberghe V, Alfonso L, et al. *Aedes aegypti* larval indices and risk for dengue epidemics. Emerg Infect Dis 2006;12(5):800–6.

[25] Southwood TR, Murdie G, Yasuno M, Tonn RJ, Reader PM. Studies on the life budget of *Aedes aegypti* in Wat Samphaya, Bangkok, Thailand. Bull World Health Organ 1972;46 (2):211–26.

[26] Arrivillaga J, Barrera R. Food as a limiting factor for *Aedes aegypti* in water-storage containers. J Vector Ecol 2004;29(1):11–20.

[27] Barrera R, Amador M, Clark GG. Ecological factors influencing *Aedes aegypti* (Diptera: Culicidae) productivity in artificial containers in Salinas, Puerto Rico. J Med Entomol 2006;43(3):484–92.

[28] Focks D. Toxorhynchites as biocontrol agents. J Am Mosq Control Assoc 2007;23 (Suppl. 2):118–27.

[29] Barrera R, Medialdea V. Development time and resistance to starvation of mosquito larvae. J Nat Hist 1996;30(3):447–58.

[30] Schofield CJ. Vector population responses to control interventions. Ann Soc Belg Med Trop 1991;71(Suppl. 1):201–17.

[31]   Legros M, Lloyd AL, Huang Y, Gould F. Density-dependent intraspecific competition in the larval stage of *Aedes aegypti* (Diptera: Culicidae): revisiting the current paradigm. J Med Entomol 2009;46(3):409–19.

[32]   Focks D, Alexander N, Villegas E, et al. Multicountry study of *Aedes aegypti* pupal productivity survey methodology: findings and recommendations. Geneva, Switzerland: TDR/WHO; 2006.

[33]   Barrera R, Amador M, Clark GG. Use of the pupal survey technique for measuring *Aedes aegypti* (Diptera: Culicidae) productivity in Puerto Rico. Am J Trop Med Hyg 2006;74 (2):290–302.

[34]   Getis A, Morrison AC, Gray K, Scott TW. Characteristics of the spatial pattern of the dengue vector, *Aedes aegypti*, in Iquitos, Peru. Am J Trop Med Hyg 2003;69(5):494–505.

[35]   Barrera R. Spatial stability of adult *Aedes aegypti* populations. Am J Trop Med Hyg 2011;85(6):1087–92.

[36]   TunLin W, Kay BH, Barnes A. The premise condition index: a tool for streamlining surveys of *Aedes aegypti*. Am J Trop Med Hyg 1995;53(6):591–4.

[37]   Padmanabha H, Durham D, Correa F, Diuk-Wasser M, Galvani A. The interactive roles of *Aedes aegypti* super-production and human density in dengue transmission. PLoS Negl Trop Dis 2012;6(8):e1799. Available from: http://dx.doi.org/10.1371/journal.pntd.0001799.

[38]   Barrera R, Amador M, Diaz A, Smith J, Munoz-Jordan JL, Rosario Y. Unusual productivity of *Aedes aegypti* in septic tanks and its implications for dengue control. Med Vet Entomol 2008;22(1):62–9.

[39]   Arana-Guardia R, Baak-Baak CM, Lorono-Pino MA, et al. Stormwater drains and catch basins as sources for production of *Aedes aegypti* and *Culex quinquefasciatus*. Acta Trop 2014;134:33–42.

[40]   Gonzalez R, Gamboa R, Perafan O, Suarez MF, Montoya J. Experience of an entomological analysis of the breeding sites of *Aedes aegypti* and *Culex quinquefasciatus* in Cali, Colombia. Rev Colomb Entomol 2007;33:18–156.

[41]   Mackay AJ, Amador M, Diaz A, Smith J, Barrera R. Dynamics of *Aedes aegypti* and *Culex quinquefasciatus* in septic tanks. J Am Mosq Control Assoc 2009;25(4): 409–16.

[42]   Kay BH, Ryan PA, Russell BM, Holt JS, Lyons SA, Foley PN. The importance of subterranean mosquito habitat to arbovirus vector control strategies in north Queensland, Australia. J Med Entomol 2000;37(6):846–53.

[43]   Russell BM, McBride WJ, Mullner H, et al. Epidemiological significance of subterranean *Aedes aegypti* (Diptera: Culicidae) breeding sites to dengue virus infection in Charters Towers, 1993. J Med Entomol 2002;39(1):143–5.

[44]   Christophers SR. *Aedes aegypti*, the yellow fever mosquito: its life history, bionomics, and structure. London, Great Britain: Cambridge University Press; 1960.

[45]   Trpis M, Trpis M. Dry season survival of *Aedes aegypti* eggs in various breeding sites in the Dar es Salaam area, Tanzania. Bull World Health Organ 1972;47(3):433–7.

[46]   Fox I. Viability of Puerto Rican *Aedes aegypti* eggs after long periods of storage. Mosq News 1974;34(3):274–5.

[47]   Mackay AJ, Amador M, Felix G, Acevedo V, Barrera R. Evaluation of household bleach formulations as ovicides for the control of Aedes aegypti. J Am Mosq Control Assoc 2015;31(1):77–84.

[48]   Barrera R, Amador M, Clark GG. The use of household bleach to control *Aedes aegypti*. J Am Mosq Control Assoc 2004;20(4):444–8.

[49] Fernandez EA, Leontsini E, Sherman C, et al. Trial of a community-based intervention to decrease infestation of *Aedes aegypti* mosquitoes in cement washbasins in El Progreso, Honduras. Acta Trop 1998;70(2):171−83.

[50] Machaca J, Llontop F, Pasapera F, Castañeda C. Eliminación mecánica de huevos del *Aedes aegypti* para la erradicación del Dengue urbano. Localidad de Sechura—Piura, Abril—Diciembre 2001. Rev Peruana Epidemiol 2001;10:1−5.

[51] Shope R. Global climate change and infectious diseases. Environ Health Perspect 1991;96:171−4.

[52] Soper FL, Soper FL. Dynamics of *Aedes aegypti* distribution and density. Seasonal fluctuations in the Americas. Bull World Health Organ 1967;36(4):536−8.

[53] Micieli MV, Garcia JJ, Achinelly MF, Marti GA. Population dynamics of the immature stages of *Aedes aegypti* (Diptera: Culicidae), vector of dengue: a longitudinal study (1996−2000). Rev Biol Trop 2006;54:979−83.

[54] Vezzani D, Velazquez SM, Schweigmann N. Seasonal pattern of abundance of *Aedes aegypti* (Diptera: Culicidae) in Buenos Aires city, Argentina. Mem Inst Oswaldo Cruz 2004;99(4):351−6.

[55] Estallo EL, Carbajo AE, Grech MG, et al. Spatio-temporal dynamics of dengue 2009 outbreak in Cordoba City, Argentina. Acta Trop 2014;136:129−36.

[56] Regis LN, Acioli RV, Silveira Jr. JC, et al. Characterization of the spatial and temporal dynamics of the dengue vector population established in urban areas of Fernando de Noronha, a Brazilian oceanic island. Acta Trop 2014;137C:80−7.

[57] Barrera R, Avila J, Navarro JC. Population dynamics of *Aedes aegypti* (L.) in urban areas with deficient supply of potable water. Acta Biol Venez 1996;16:23−35.

[58] Stewart-Ibarra AM, Lowe R. Climate and non-climate drivers of dengue epidemics in Southern Coastal Ecuador. Am J Trop Med Hyg 2013;88(5):971−81.

[59] Barrera R, Amador M, MacKay AJ. Population dynamics of *Aedes aegypti* and dengue as influenced by weather and human behavior in San Juan, Puerto Rico. PLoS Negl Trop Dis 2011;5(12):e1378.

[60] Codeco CT, Honorio NA, Rios-Velasquez CM, et al. Seasonal dynamics of *Aedes aegypti* (Diptera: Culicidae) in the northernmost state of Brazil: a likely port-of-entry for dengue virus 4. Mem Inst Oswaldo Cruz 2009;104(4):614−20.

[61] Barrera R, Amador M, Acevedo V, Caban B, Felix G, Mackay A. Use of the CDC autocidal gravid ovitrap to control and prevent outbreaks of *Aedes aegypti* (Diptera: Culicidae). J Med Entomol 2014;51(1):145−54.

[62] Schultz GW. Seasonal abundance of dengue vectors in Manila, republic of the Philippines. Southeast Asian J Trop Med Public Health 1993;24(2):369−75.

[63] Halstead SB. Dengue virus−mosquito interactions. Annu Rev Entomol 2008;53:273−91.

[64] Kuno G. Review of the factors modulating dengue transmission. Epidemiol Rev 1995;17 (2):321−35.

[65] Moore CG, Cline BL, Ruiz-Tiben E, Lee D, Romney-Joseph H, Rivera-Correa E. *Aedes aegypti* in Puerto Rico: environmental determinants of larval abundance and relation to dengue virus transmission. Am J Trop Med Hyg 1978;27(6):1225−31.

[66] de Souza SS, da Silva IG, da Silva HHG. Association between dengue incidence, rainfall and larval density of *Aedes aegypti*, in the State of Goias. Rev Soc Bras Med Trop 2010;43(2):152−5.

[67] Chadee DD, Shivnauth B, Rawlins SC, Chen AA. Climate, mosquito indices and the epidemiology of dengue fever in Trinidad (2002−2004). Ann Trop Med Parasitol 2007;101(1):69−77.

[68] Dibo MR, Chierotti AP, Ferrari MS, Mendonca AL, Chiaravalloti Neto F. Study of the relationship between *Aedes (Stegomyia) aegypti* egg and adult densities, dengue fever and climate in Mirassol, state of Sao Paulo, Brazil. Mem Inst Oswaldo Cruz 2008;103(6):554–60.

[69] Barbosa GL, Lourenco RW. [Analysis on the spatial-temporal distribution of dengue and larval infestation in the municipality of Tupa, State of Sao Paulo]. Rev Soc Bras Med Trop 2010;43(2):145–51.

[70] Eamchan P, Nisalak A, Foy HM, Chareonsook OA. Epidemiology and control of dengue virus infections in Thai villages in 1987. Am J Trop Med Hyg 1989;41:95–101.

[71] Tonn RJ, Sheppard PM, Macdonald WW, et al. Replicate surveys of larval habitats of *Aedes aegypti* in relation to dengue haemorrhagic fever in Bangkok, Thailand. Bull World Health Organ 1969;40(6):819–29.

[72] Stoddard ST, Wearing HJ, Reiner Jr. RC, et al. Long-term and seasonal dynamics of dengue in Iquitos, Peru. PLoS Negl Trop Dis 2014;8(7):e3003.

[73] Carter HR. Yellow fever. An epidemiologic and historical study of its place of origin. Baltimore, MD: Williams & Wilkins Co; 1931.

[74] Cassetti MC, Halstead SB. Consultation on dengue vaccines: progress in understanding protection, 26–28 June 2013, Rockville, Maryland. Vaccine 2014;32(26):3115–21.

[75] Vazquez-Prokopec GM, Kitron U, Montgomery B, Horne P, Ritchie SA. Quantifying the spatial dimension of dengue virus epidemic spread within a tropical urban environment. PLoS Negl Trop Dis 2010 Dec 21;4(12):e920. Available from: http://dx.doi.org/10.1371/journal.pntd.0000920.

[76] Gardner L, Sarkar S. A global airport-based risk model for the spread of dengue infection via the air transport network. PLoS One 2013 Aug 29;8(8):e72129. Available from: http://dx.doi.org/10.1371/journal.pone.0072129.

[77] Morens DM, Rigau-Perez JG, Lopez-Correa RH. Dengue in Puerto Rico, 1977: public health response to characterize and control an epidemic of multiple serotypes. Am J Trop Med Hyg 1986;35(1):197–211.

[78] Stahl HC, Butenschoen VM, Tran HT, et al. Cost of dengue outbreaks: literature review and country case studies. BMC Public Health 2013 Nov 6;13:1048. Available from: http://dx.doi.org/10.1186/1471-2458-13-1048.

[79] Barrera R, Delgado N, Jimenez M, Villalobos I, Romero I. [Stratification of a hyperendemic city in hemorrhagic dengue]. Rev Panam Salud Publ 2000;8(4):225–33.

[80] Barrera R, Delgado N, Jimenez M, Valero S. Eco-epidemiological factors associated with hyperendemic dengue haemorrhagic fever in Maracay City, Venezuela. Dengue Bull 2002;26:84–95.

[81] Bousema T, Stevenson J, Baidjoe A, et al. The impact of hotspot-targeted interventions on malaria transmission: study protocol for a cluster-randomized controlled trial. Trials 2013;14:36.

[82] Sharp TM, Rivera A, Rodriguez-Acosta R, et al. An island-wide dengue epidemic— Puerto Rico, 2010. Am J Trop Med Hyg 2011;1:400–1.

[83] Focks D, Barrera R. Dengue transmission dynamics: assessment and implications for control. Geneva, Switzerland: World Health Organization on behalf of the Special Programme for Research and Training in Tropical Diseases; 2007.

[84] Pontes RJS, Freeman J, Oliveira-Lima JW, Hodgson JC, Spielman A. Vector densities that potentiate Dengue outbreaks in a Brazilian city. Am J Trop Med Hyg 2000;62(3):378–83.

[85] Chadee DD. Impact of pre-seasonal focal treatment on population densities of the mosquito Aedes aegypti in Trinidad, West Indies: a preliminary study. Acta Trop 2009;109 (3):236–40.

[86] Kay BH, Ryan PA, Lyons SA, Foley PN, Pandeya N, Purdie D. Winter intervention against Aedes aegypti (Diptera: Culicidae) larvae in subterranean habitats slows surface recolonization in summer. J Med Entomol 2002;39(2):356−61.

[87] Barr KL, Focks DA, Messenger AM, Leal A. Cryptic breeding: a potential cause of local dengue transmission in Key West, Florida. Am J Trop Med Hyg 2011;1:380.

[88] Focks D. A review of entomological sampling methods and indicators for dengue vectors. Geneva, Switzerland: Special Programme for Research and Training in Tropical Diseases. World Health Organization; 2003.

[89] Nathan MB, Focks DA. Pupal/demographic surveys to inform dengue-vector control. Ann Trop Med Parasitol 2006;100:S1−3.

[90] Reuben R, Das PK, Samuel D, Brooks GD. Estimation of daily emergence of Aedes aegypti (Diptera: Culicidae) in Sonepat, India. J Med Entomol 1978;14:705−14.

[91] Barrera R, Amador M, Clark GG. Sample-size requirements for developing strategies, based on the pupal/demographic survey, for the targeted control of dengue. Ann Trop Med Parasitol 2006;100(Suppl. 1):S33−43.

[92] Barrera R. Simplified pupal surveys of Aedes aegypti (L.) for entomologic surveillance and dengue control. Am J Trop Med Hyg 2009;81(1):100−7.

[93] Cohnstaedt LW, Rochon K, Duehl AJ, et al. Arthropod surveillance programs: basic components, strategies, and analysis. Ann Entomol Soc Am 2012;105(2):135−49.

[94] Pilger D, Lenhart A, Manrique-Saide P, Siqueira JB, da Rocha WT, Kroeger A. Is routine dengue vector surveillance in central Brazil able to accurately monitor the Aedes aegypti population? Results from a pupal productivity survey. Trop Med Int Health 2011;16(9):1143−50.

[95] Grisales N, Poupardin R, Gomez S, Fonseca-Gonzalez I, Ranson H, Lenhart A. Temephos resistance in Aedes aegypti in colombia compromises dengue vector control. PLoS Negl Trop Dis 2013 Sep 19;7(9):e2438. Available from: http://dx.doi.org/10.1371/journal.pntd.0002438.

[96] Marten GG, Bordes ES, Nguyen M. Use of cyclopoid copepods for mosquito control. Hydrobiologia 1994;292/293:491−6.

[97] Nam VS, Yen NT, Kay BH, Marten GG, Reid JW. Eradication of Aedes aegypti from a village in Vietnam, using copepods and community participation. Am J Trop Med Hyg 1998;59(4):657−60.

[98] Scholte EJ, Knols BGJ, Samson RA, Takken W. Entomopathogenic fungi for mosquito control: a review. J Insect Sci 2004;4:19.

[99] O'Connor L, Plichart C, Sang AC, Brelsfoard CL, Bossin HC, Dobson SL. Open release of male mosquitoes infected with a wolbachia biopesticide: field performance and infection containment. PLoS Negl Trop Dis 2012;6(11):e1797.

[100] Dobson SL. Reversing Wolbachia-based population replacement. Trends Parasitol 2003;19(3):128−33.

[101] Slatko BE, Luck AN, Dobson SL, Foster JM. Wolbachia endosymbionts and human disease control. Mol Biochem Parasitol 2014 Jul;195(2):88−95. Available from: http://dx.doi.org/10.1016/j.molbiopara.2014.07.004.

[102] Reiter P, Nathan MB. Guidelines for assessing the efficacy of insecticidal space sprays for control of the dengue vector Aedes aegypti. Geneva: World Health Organization; 2001.

[103] Pilger D, De Maesschalck M, Horstick O, San Martin JL. Dengue outbreak response: documented effective interventions and evidence gaps. TropIKA. net Journal 2010;1(1), <http://journal.tropika.net/scielo.php?script=sci_arttext&pid=S2078-6062010000100002&lng=en&nrm=iso&tlng=en>.

[104] Farajollahi A, Healy S, Unlu I, Gaugler R, Fonseca D. Effectiveness of ultra-low volume nighttime applications of an adulticide against diurnal *Aedes albopictus*, a critical vector of dengue and chikungunya viruses. PLoS One 2012;7(11):e49181.

[105] Giglioli MEC. *Aedes aegypti* programs in the Caribbean and emergency measures against the dengue pandemic of 1977–1978; a critical review. In: Anon, Dengue in the Caribbean, 1977: proceedings of a workshop held in Montego Bay, Jamaica (8–11 May 1978), Washington, D.C: Pan American Health Organization; 1979. p. 133–52.

[106] Ritchie SA, Hanna JN, Hills SL, et al. Dengue control in North Queensland, Australia: case recognition and selective indoor residual spraying. Dengue Bull 2002;26:7–13.

[107] Hemingway J. The role of vector control in stopping the transmission of malaria: threats and opportunities. Philos Trans R Soc Lond B Biol Sci 2014;369(1645):20130431.

[108] Naranjo D, Qualls W, Mueller G, et al. Evaluation of boric acid sugar baits against *Aedes albopictus* (Diptera: Culicidae) in tropical environments. Parasitol Res 2013;112 (4):1583–7.

[109] Lorono-Pino MA, Garcia-Rejon JE, Machain-Williams C, et al. Towards a casa segura: a consumer product study of the effect of insecticide-treated curtains on *Aedes aegypti* and dengue virus infections in the home. Am J Trop Med Hyg 2013;89(2):385–97.

[110] Lenhart A, Orelus N, Alexander N, Streit T, McCall PJ. Insecticide treated bednets for the control of dengue vectors in Haiti. Am J Trop Med Hyg 2006;75(5):200.

[111] Kroeger A, Lenhart A, Ochoa M, et al. Effective control of dengue vectors with curtains and water container covers treated with insecticide in Mexico and Venezuela: cluster randomised trials. BMJ, <http://onlinelibrary.wiley.com/o/cochrane/clcentral/articles/679/CN-00556679/frame.html>; 2006.

[112] Kittayapong P, Yoksan S, Chansang U, Chansang C, Bhumiratana A. Suppression of dengue transmission by application of integrated vector control strategies at sero-positive GIS-based foci. Am J Trop Med Hyg 2008;78:70–6.

[113] Eiras AE, Buhagiar TS, Ritchie SA. Development of the gravid aedes trap for the capture of adult female container-exploiting mosquitoes (Diptera: Culicidae). J Med Entomol 2014;51(1):200–9.

[114] Caputo B, Ienco A, Cianci D, et al. The "auto-dissemination" approach: a novel concept to fight Aedes albopictus in urban areas. PLoS Negl Trop Dis 2012;6(8):e1793.

[115] Snetselaar J, Andriessen R, Suer RA, Osinga AJ, Knols BGJ, Farenhorst M. Development and evaluation of a novel contamination device that targets multiple life-stages of *Aedes aegypti*. Parasit Vectors 2014;7.

[116] Burt A. Heritable strategies for controlling insect vectors of disease. Philos Trans R Soc Lond B Biol Sci 2014;369(1645):20130432.

[117] Benyman AA. Population theory and pest management. In: Pimentel D, editor. Encyclopedia of pest management. New York, NY: Taylor & Francis Group; 2002. p. 642–4.

[118] Barrera R, Amador M, Acevedo V, Hemme RR, Félix G. Sustained, area-wide control of *Aedes aegypti* using CDC autocidal gravid ovitraps. Am J Trop Med Hyg 2014;91:1269–76.

[119] Giglioli G. Residual effect of DDT in a controlled area of British Guiana tested by the continued release of *Anopheles darlingi* and *Aedes aegypti:* a practical technique for the standardized evaluation of over-all residual efficiency under field conditions. Trans R Soc Trop Med Hyg 1954;48(6):506–21.

[120] Vanlerberghe V, Toledo ME, Rodríguez M, et al. Community involvement in dengue vector control: cluster randomised trial. BMJ, <http://onlinelibrary.wiley.com/o/cochrane/clcentral/articles/285/CN-00705285/frame.html>; 2009.

[121] Bang YH, Gratz N, Pant CP, Bang YH, Gratz N, Pant CP. Suppression of a field population of *Aedes aegypti* by malathion thermal fogs and abate larvicide. Bull World Health Organ 1972;46(4):554−8.

[122] Koenraadt CJM, Aldstadi J, Kijchalao U, Kengluecha A, Jones JW, Scott TW. Spatial and temporal patterns in the recovery of *Aedes aegypti* (Diptera: Culicidae) populations after insecticide treatment. J Med Entomol 2007;44(1):65−71.

[123] Klassen W, Curtis CF. History of the sterile insect technique. In: Dyck VA, Hendrichs J, Robinson AS, editors. Sterile insect technique principles and practice in area-wide integrated pest management. The Netherlands: Springer; 2005. p. 3−36.

[124] Ito Y, Yamamura K. Role of population and behavioural ecology in the sterile insect technique. In: Dyck VA, Hendrichs J, Robinson AS, editors. Sterile insect technique principles and practice in area-wide integrated pest management. The Netherlands: Springer; 2005. p. 177−208.

[125] Oleron Evans TP, Bishop SR. A spatial model with pulsed releases to compare strategies for the sterile insect technique applied to the mosquito *Aedes aegypti*. Math Biosci 2014;254:6−27.

[126] Barclay HJ. Modeling the effects of population aggregation on the efficiency of insect pest-control. Res Popul Ecol 1992;34(1):131−41.

[127] Ferreira CP, Yang HM, Esteva L. Assessing the suitability of sterile insect technique applied to *Aedes aegypti*. J Biol Syst 2008;16(4):565−77.

[128] Knipling EF. Entomology and the management of man's environment. J Aust Entomol Soc 1972;11:153−67.

[129] McDonald PT, Hausermann W, Lorimer N, McDonald PT, Hausermann W, Lorimer N. Sterility introduced by release of genetically altered males to a domestic population of *Aedes aegypti* at the Kenya coast. Am J Trop Med Hyg 1977;26(3):553−61.

[130] Lounibos LP. Genetic control trials and the ecology of *Aedes aegypti* at the Kenya coast. In: Takken W, Scott TW, editors. Ecological aspects for application of genetically modified mosquitoes. Dordrecht, The Netherlands: Kluwer Academic; 2003. p. 33−43.

[131] Courchamp F, Clutton-Brock T, Grenfell B. Inverse density dependence and the allee effect. Trends Ecol Evol 1999;14(10):405−10.

[132] Agudelo-Silva F, Spielman A. Paradoxical effects of simulated larviciding on production of adult mosquitoes. Am J Trop Med Hyg 1984;33(6):1267−9.

[133] Benedict MQ, Robinson AS. The first releases of transgenic mosquitoes: an argument for the sterile insect technique. Trends Parasitol 2003;19(8):349−55.

[134] Mangan RL. Population suppression in support of the sterile insect technique. In: Dyck VA, Hendrichs J, Robinson AS, editors. Sterile insect technique principles and practice in area-wide integrated pest management. The Netherlands: Springer; 2005. p. 407−25.

[135] Klassen W. Area-wide integrated pest management and the sterile insect technique. In: Dyck VA, Hendrichs J, Robinson AS, editors. Sterile insect technique principles and practice in area-wide integrated pest management. The Netherlands: Springer; 2005. p. 39−68.

[136] Alphey L. Genetic control of mosquitoes. Annu Rev Entomol 2014;59:205−24.

[137] Hoffmann AA, Montgomery BL, Popovici J, et al. Successful establishment of *Wolbachia* in *Aedes* populations to suppress dengue transmission. Nature 2011;476 (7361):454−7.

Chapter 7

The Challenge of Disrupting Vectorial Capacity

Robert E. Sinden[1,2]
[1]The Jenner Institute, The University of Oxford, Oxford, UK, [2]The Department of Life Sciences, Imperial College London, South Kensington, UK

PURPOSE

In the past decade, global reviews of the malaria research agenda have identified the urgent need to understand the biology of malaria transmission and to develop methods for its reduction in endemic communities [1,2]. The collection of articles in this book is highly relevant to this new malaria elimination/eradication agenda, which itself appreciates that malarial intervention must not only address the long-standing humanitarian/clinical prerogative to save the life of every infected human host but also reduce the number of new cases resulting from any infected person. Simply stated, the second prerogative requires that the parasite's basic reproductive rate (R_o, or R_c in areas under control measures) is reduced by intervention to <1 in the target population. It is recognized by key stakeholders that unless this is achieved malaria control campaigns would need to be maintained in perpetuity, which is clearly not a sustainable option. Though when seeing the difficulties being experienced in the campaign to eliminate poliovirus (where R_o rarely exceeds 2), we must recognize the prospects to eliminate malaria in any locality, or indeed to eradicate the six species infecting man (for which R_o values might exceed 1000 [3]) are daunting indeed. Despite the enormity of the task, unless we develop new measures to attempt to reduce transmission directly malaria is unlikely to succumb to current or future interventions.

The title I have been given for this contribution "The Challenge of Disrupting Vectorial Capacity" is challenging in itself because I will suggest this is not the objective we should focus upon! I will propose that in the formulation of antimalarial strategies it is the ultimate consequences of intervention upon the human population and not the "intermediate" consequences upon the vector population that needs to be studied and reported.

Genetic Control of Malaria and Dengue.
© 2016 Elsevier Inc. All rights reserved.

Ever since the discovery of mosquito-dependent transmission of malaria [4], important and basic scientific questions are still being asked on the biology, evolution, and taxonomy of mosquitoes. The large majority of the funding put into basic research on mosquito biology presumes that one consequence of the study is the improvement of our understanding of the complex parasite−vector and vector−host biologies, on the assumption that this knowledge will contribute to the control of major pathogens of man and other animals of economic importance. As such, it is important to realize that the key outputs required are not more data, however fascinating, on mosquito biology *per se*, but information on how we may change vector biology to reduce the ability of the insects to transmit disease pathogens, and thus how any proposed interventions reduce infection in man. In the eyes of the writer nowhere has this been more effectively summarized than in the early mathematical excursions of Ross [4], their development by MacDonald in the 1950s [5−7], and subsequently by numerous authors in the recent decades [3,8−10].

$$\text{Ross/MacDonald formula} \quad R_0 = \frac{ma^2 bp^n}{-r \log_e p} \tag{1}$$

where m is the density of vectors in relation to density of hosts; a is the proportion of vectors feeding on a host divided by the length of the gonotrophic cycle in days; b is the proportion of mosquitoes that are infectious; p is the daily survival of vectors; n is the extrinsic incubation period in days; and r is the daily proportion of infected people who become noninfectious to the mosquito.

"Embedded" within the Ross−MacDonald formula lays the mathematical basis for vectorial capacity, that is, the daily rate at which future inoculations arise from a currently infectious case. Vectorial capacity can be described by the following formula

$$\text{Vectorial capacity} \quad C = \frac{ma^2 VP^n}{-\log_e P} \tag{2}$$

where C is the number of infective bites received daily by a single host; m is the density of vectors in relation to density of hosts; a is the proportion of vectors feeding on a host divided by the length of the gonotrophic cycle in days; V is the vector competence, where $V = bh$ (b is the proportion of bitten humans infected; and h is the proportion of biting mosquitoes infected); P is the daily survival of vectors; and n is the extrinsic incubation period in days.

With respect to the current objective of malaria elimination and eradication, what the calculation of vectorial capacity fails to convey is the critical relationship between the infective biting rate and the resultant prevalence of

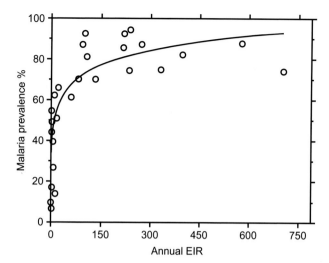

FIGURE 7.1 The saturating relationship between the EIR and parasite prevalence rate in endemic populations. *Figure from Ref. [12].*

infected persons [11]. This is perhaps best illustrated by the work of John Beier who described the saturating relationship between the *E*ntomological *I*noculation *R*ate (EIR) and prevalence of human infection (Figure 7.1) [12]. When progressing the development of a genetically modified mosquito [http://www.who.int/tdr/news/2012/GMM_Guidance_2012.pdf?ua = 1]   to meet the current objectives of malaria elimination and eradication, I will suggest that perhaps the most appropriate parameter to measure when determining the potential impact of an intervention is the *effect size* on parasite transmission (see Section Measuring impact) [3].

In looking at Equation (2), if we are to attempt to disrupt vectorial capacity we have a finite number of parameters we can attack, some of which have been the corner-stone of previous antimalarial campaigns: namely "m" which can be attacked by insecticide/biocide-based campaigns, against adult and larval vectors; "a" the mosquito biting rate, which is directly attacked by the use of bed nets/diversionary blood sources/personal protection regimens, and is a more powerful (squared) parameter because of the necessity of a mosquito to bite twice to complete transmission, but by far the most influential parameter, identified in early campaigns, is "p," the daily survival of infected mosquitoes, which is susceptible to the sustained exposure to insecticides and biocides for the "n" days of the parasites extrinsic cycle. Of these biological control strategies, one of the more exciting is the use of "slow-killing" fungi, for example, *Beauveria* [13] which, if delivered before or during the malaria-infected blood meal, may kill the infected female after she has laid the majority of her eggs but before completion of the extrinsic incubation period and thus before she can transmit the infection onward to a secondary host. This cleverly constrains the very strong selective pressure that would otherwise exist for the vector to develop resistance to the fungus [14]. This of course presumes that the fungal infection did not itself significantly suppress oogenesis.

While it is the powered terms (a and p) in Eqs (1) and (2) that remain the most successful targets to reduce parasite transmission, this does not mean that we should not investigate methods by which to disrupt the weaker, vector-dependent terms (b, or V in Eqs 1 and 2, respectively)—methods that will be complemented by parallel efforts to attack the term r in Eq. (1) by the development of transmission-blocking drugs [15] and vaccines [16].

STRATEGY: HOW DO WE USE THE VECTOR TO ACHIEVE REDUCTIONS IN PARASITE TRANSMISSION?

To achieve reductions in vectorial capacity, we can target either the vector *per se*, or the parasite within the vector, potential interventions for the latter include the antiparasitic transmission-blocking vaccines [16] and drugs [15] targeting the sporogonic stages of development, and genetically modified (GM) suppression of parasite development, for example, by enhancing natural vector antiparasite immunity [17,18], by introducing heterologous refractory mechanisms [19,20], or by deleting/blocking vector gene products essential to parasite survival [21].

Diverse mechanisms exist for the delivery of interventions that aim to reduce the number of mosquitoes that survive beyond the extrinsic incubation period of the parasite and deliver an infectious bite. When attacking the vector *per se*, exposure may be direct or indirect. Examples of the former include insecticides, odor-based intervention, and biological control; examples of the latter include GM technologies to reduce the absolute number of vectors [22] or the number able to find an infectious blood meal [23]. It has been suggested that current bacterial control agents, such as *Bacillus thuringiensis israelensis* [24], may be complemented with other pathogens such as, *Nosema* [25], *Vavraia* [26], and *Bauveria* [13]. An interesting concept (paratransgenesis) to modify genetically these pathogens (or other symbionts, e.g., *Wolbachia* [27]) has been the subject of a recent interesting and very comprehensive reviews [28,29].

Exciting technologies for such GM concepts are now available [30−33], for which it has proved impossible for some not to speculate how these discoveries will translate into revolutionary new intervention strategies; however, in formulating such hypotheses, it is essential to remain aware as to the practical limitations that might be experienced in their development.

LESSONS LEARNT?

Lessons we (may) have learnt from long-established intervention campaigns, include the need for:

1. Effective delivery of the intervention to the appropriate target population
2. Maintenance of the quality of the intervention
3. Sustained delivery as the original drivers for intervention (e.g., case prevalence) decline
4. Design of polyvalent interventions.

Focussing on GM technologies, how do these imperatives impact upon the relevant approaches? Two properties of the global dynamics of malaria transmission biology complicate our abilities to predict the efficacy of interventions. The first is the biological/genetic diversity of all interacting species, man, parasite, and mosquito. The second is the simple abundance of both vectors and parasites, such that they might be considered "supersaturated" systems.

Effective Delivery

Considering just the enormous diversity of vector species (Figure 7.2A), a problem uniquely faced by vector GM technologies is the species-specificity of the intervention and drive mechanism. It is clear that substantial bodies of work still require to be done to understand, in multispecies scenarios (Figure 7.2B and C), how the introduction of one or multiple species of GM-vector can reduce *sustainably* parasite prevalence in the human population, that is, in population suppression strategies, will a "vacated" ecological niche be occupied by an alternate population of vectors? and in population replacement strategies, will the original niche be occupied fully/adequately by the GM line?

When considering the size of the target population being attacked, it has long puzzled the author as to why *in the context of control/elimination/eradication* we would continue to design our interventions to attack the parasites at times when they have the maximal population size and hence potential for evasion, for example, the asexual parasites in the infected vertebrate host. Surely, we increase our chance of success by attacking the bottleneck populations in the targets' life cycles, that is, when *Plasmodium* passes through the ookinete/oocyst population in the mosquito vector and the pre-erythrocytic population in the infected vertebrate—this strategy alone if implemented successfully would reduce the target parasite population by as much as 10^9-fold (Figure 7.3A). In a similar vein, if were we to attack the vector population in the dry season reportedly, we might reduce the target (vector) population 100-fold (Figure 7.3B). Finally, as is well established in insecticide campaigns, we can focus efforts on mosquito breeding sites which may give perhaps 10-fold efficiencies in resource allocation [35] (Figure 7.3C), and simultaneously will, in heterogeneous

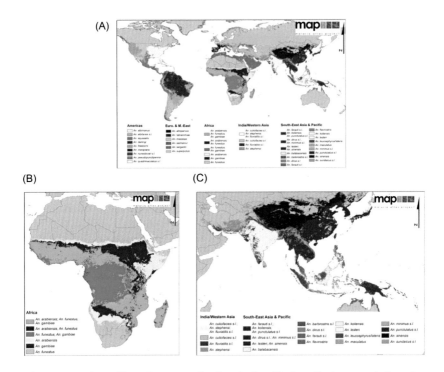

FIGURE 7.2 The problem of vector species-diversity in achieving effective transmission reductions by GM-vector technologies. (A) The global map of reported major vectors of malaria. (B) Vector diversity in Africa illustrating those areas where multiple vectors are found. (C) Vector diversity in Asia illustrating the higher complexity of vector population composition [34].

populations, critically be focussing the attack upon those individuals who are being bitten most [3]. Thus, irrespective of the intervention developed simple consideration of when and where to attack has the potential to produce very significant efficiency savings, and by reducing the genetic resource of the target population, potentially increase the "life-expectancy" of any intervention. When considering GM-vector interventions, the ability to exploit these bottleneck populations is obvious: the gametes and early sporogonic forms of the parasite are prime targets; temporally, the dry season nadir in vector numbers might offer unique opportunities for the introduction of replacement populations; and spatially, efforts can clearly be focussed on the interface between human and vector populations adjacent to mosquito breeding sites.

Maintenance of the Quality of the Intervention

An as-yet unknown issue that will need study is that of the evolutionary stability of GM technologies, malaria eradication is foreseen to be a 20- to 50-year

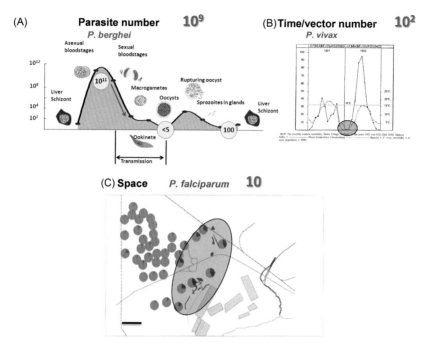

FIGURE 7.3   Attacking transmission bottlenecks can improve the efficacy of intervention. (A) The relative parasite population size at sequential points in the life cycle, the ookinete−oocyst, and sporozoite−pre-erythrocytic schizont represent the nadirs in parasite numbers some 10^9- and 10^8-fold smaller that the peak asexual parasitemia—the current point of attack of schizonticidal compounds. Figure from Ref. [36]. (B) Seasonal fluctuations in mosquito number/case number during a malaria control program in the Netherlands. The overwintering vector population is 10^2-fold lower than that in the peak transmission season. (C) Efficiencies potentially achieved by targeting mosquito breeding sites suggest 10-fold reductions in treatment areas should be achieved. Diagram from Ref. [37].

campaign, depending upon the biology of the modified organism this may equate to hundreds or indeed countless generations of the "intervention." It will be appropriate therefore to at least make effort to ensure the stability of the construct, or at least the innocuous nature of any future mutations that might emerge. Nevertheless, it might be hoped that one of the major advantages of GM manipulation of either the vector, or a symbiotic bacterium, for example, *Wolbachia*, is that, barring genetic disruption of the introduced genetic modification, the quality of the intervention might be sustained at no "additional cost" to the program. Similarly, the quantitative aspects of the intervention within each insect might similarly be sustained throughout the population. Where the latter benefit may not hold is in the model of paratransgenesis as applied to acquired infections, for example, *Pantoea*/DNA densovirus [28,29], as opposed to inherited symbionts (*Wolbachia/Asaia*) [27,38]. The problem will be that following the "release" of the GM agent there is no control over either the number of organisms delivering the inhibitor (and therefore the inhibitor concentration) or the duration of exposure to the GM product. This lack of control mirrors major issues that have plagued the effective and sustainable use of antimalarials, issues which are believed to have contributed to the very rapid selection of resistance against otherwise effective measures [39].

Sustaining Effective Delivery

The decision as to whether to introduce specific GM technologies into concerted programs of disease reduction raises common issues such as the comparative efficacy of the technology with respect to other interventions, and unique issues including the national/local acceptance of the underlying scientific principles. Here, perhaps one of the major strengths of the technology comes to the fore in the later stages of any elimination/eradication program, and that is the cost of maintaining a self-replicating/-sustaining intervention against a background of declining disease prevalence (increasing cost/case). Of course we must be aware, in the early stages of the application of GM technologies, of the need for the acceptance of affected communities, and at least consider the ability to "recall" the intervention if found lacking in any significant way. However, if safe and effective, a useful property of the intervention may prove to be its ability to spread into appropriate biological niches that transcend purely administrative boundaries. On the assumption that elimination/eradication policies will be first applied to areas of low transmission, where there is the highest probability of providing the essential early success (public relations "victory"), it will be essential to prevent the ingress of parasites from the neighboring areas with higher transmission. It is in these situations that GM vectors may prove of particular utility in maintaining cordons-sanitaire around areas where elimination has been achieved, problems that will be exacerbated if the interventions are applied in a scattergun approach. The contribution of stable and effective global authorities in achieving coordination of integrated multinational and sustainable programs of GM intervention cannot be underestimated; in this context, it is reassuring to see the summary of the recent Victoria Falls Declaration (http://www.rollbackmalaria.org/globaladvocacy/pr2014-07-06.html), which overtly recognizes this need.

Polyvalent Intervention

A significant lesson learnt from the use of many biocides, be it antibiotic, insecticide, or antimalarial drug, is that if an intervention is to have a useful life span it needs to be coformulated with agents targeting different molecules. Although this is now frequently, if not invariably, addressed in the formulation of new vaccines or drugs, there remains a large body of the published work seeking to develop GM-based antimalarial strategies that proposes delivery of single mutant genes or monospecific inhibitors, for example, monoclonal antibody, peptide, or drug. It is essential that the enormous intellectual originality, effort, and cost expended in the discovery of these new reagents is not squandered by failing to appreciate a key lesson already learnt. Not only is it imperative to develop combinations of "inhibitors" but to realize the need to be able to sustain the composition of the mixtures delivered to the parasite/vector ecosystem.

GM INTERVENTIONS: HOW DO WE MANAGE THE EXPECTATIONS AND MEASURE IMPACT

Managing Expectation

From experiences in the development of antimalarial transmission-blocking drugs and vaccines, it would be prudent to recognize that the critical measures of efficacy of new GM technologies for parasite control relate to the prevalence and intensity of *parasite infection in the human population*, and not the impact of the intervention on properties of the vector population. Without providing direct evidence of the former, experience has shown that it may prove difficult to attract the approval of the regulatory authorities for any previously unproven intervention. We therefore need to recognize that parameters that are scientifically informative and regularly used by the entomological community may have perhaps what others consider obscure relationships with malaria case prevalence. As examples from the parasite side of the arena: Figure 7.4 illustrates the precursor−product relationships between each successive developmental transition of *Plasmodium berghei* in the mosquito vector. All four relationships clearly saturate at higher parasite densities; thus, it is clear that as an intervention reduces the precursor abundance that the product population may not fall by the same amount (if at all), that is, the fall in precursor number may not be a particularly useful measure of impact of the intervention upon transmission to the vertebrate host. The final, and key, relationship illustrated in the figure is that between the number of sporozoites in the glands and the probability of infecting the next host still eludes effective analysis. Without this knowledge, we must take care that having made "normal" single point observations on some indirect estimate of efficacy (e.g., in the case given, the oocyst intensity or prevalence) we do not raise inappropriate expectations on the potential impact of an intervention upon parasite prevalence in the human population. Simply stated we need to understand the shape of the "reaction profile" of case prevalence in the human population against the parameter being attacked by each specific intervention.

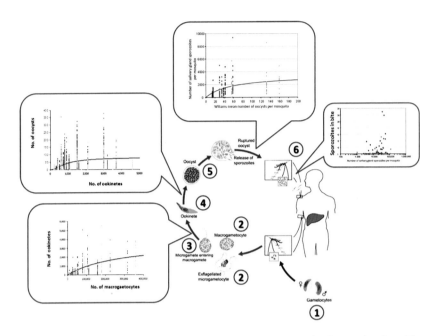

FIGURE 7.4 Managing the expectation of impact of interventions in diverse situations. The figure illustrates why point observations on the reduction in the numbers of any stage of parasite development in the mosquito (points 1−6) cannot be used as a direct estimate on the potential impact on overall transmission of the parasite from the mosquito to the vertebrate host. *Data from Ref. [40].*

Measuring Impact

In the examination of the antiparasitic transmission-blocking measures (both vaccines and drugs), a wide range of assays and diverse-reporting endpoints (e.g., 1, 2, 4, and 5 in Figure 7.4) have been published, none of which have clear and proven predictive value on the reduction in parasite transmission to the human host. Equivalent endpoints used by the entomological community might include reductions in mosquito abundance, EIR, the force of infection, etc. Many of these indices suffer from the fact that the impact of the intervention varies with the local intensity of transmission. Smith et al. [3] described a measure "the effect size" that has direct utility in estimating the impact of an intervention, acting at any point in the parasite life cycle, upon the case prevalence in study populations of differing transmission intensities (see also Ref. [41]). The design of lab-based assays that permit the calculation of this parameter for antivector measures has been described by Blagborough et al. [42] who, as proof-of-principle, demonstrated how the antimalarial drug atovaquone, which has a point efficacy of just 57% (reduction in oocyst intensity at a mean control infection of 50 oocysts), could

eliminate the parasite from the entire laboratory population in just two transmission cycles when mice are bitten by one or two potentially infected mos-
quitoes/cycle, *but failed to have any measurable impact*, even after five
transmission cycles, when bitten by three or more mosquitoes/cycle
(Figure 7.5). The calculated effect size of the intervention was $\sim 20\%$. An
intervention with this effect size, if used in the same way (i.e., at 100% coverage) would result in parasite elimination in endemic areas where R_0 is less
than 1.255 (this embraces 77% of all malaria-infected populations and 63%

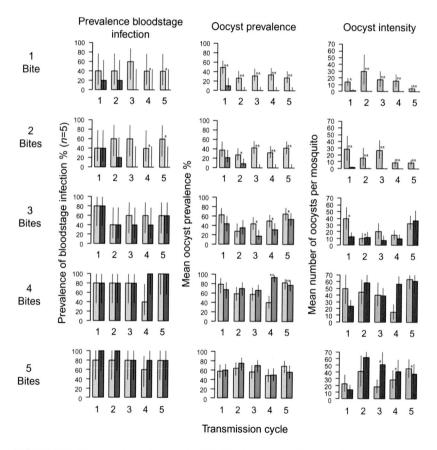

FIGURE 7.5 The impact of a transmission-blocking intervention (the drug Atovaquone) on the
transmission of *Plasmodium berghei* in laboratory populations of *Anopheles stephensi* and *Mus
musculus* (C57Bl6). Observations on the prevalence of malaria infection in vertebrate and vector
populations in control (light bars) and treated (dark bars) populations at each of five sequential
cycles of infection, at five different mosquito biting rates (MBR: 1−5 per cycle). Note this intervention which has a point efficacy of 57% reduction in oocyst number (at a control intensity of
50 oocysts/mosquito) eliminates the parasite from the populations in two generations if the
MBR <3, but that no *measurable* changes occur at higher MBRs [42].

of malaria-infected areas). *Conversely, and of equal importance to the design of appropriate development pipelines, knowing the effect size of a single or combination of interventions, it is then feasible to ensure that when attempting to determine the efficacy of the strategy the trial is conducted in endemic communities where the local transmission intensity is low enough that it actually permits a measurement of impact to be made!*

To be able to integrate the GM technologies into the overall portfolio of transmission-blocking interventions, it might be expedient to harmonize the measurement of impact of new antivector technologies with those of existing antiparasite measures. Despite the complexity of data collection, this might most efficiently be achieved by providing outputs from the different strategies in terms of effect size.

REFERENCES

[1] Alonso PL, Brown G, Arevalo-Herrera M, et al. A research agenda to underpin malaria eradication. PLoS Med 2011;8(1):e1000406.

[2] Feachem RGA, Phillips AA, Hwang J, et al. Shrinking the malaria map: progress and prospects. *Lancet* 2010;376(9752):1566−78.

[3] Smith DL, McKenzie FE, Snow RW, Hay SI. Revisiting the basic reproductive number for malaria and its implications for malaria control. PLoS Biol 2007;5(3):e42. 1−12.

[4] Ross R. The prevention of malaria. London: John Murray; 1910.

[5] MacDonald G. Theory of the eradication of malaria. Bull World Health Organ 1956;15:369−87.

[6] MacDonald G. Epidemiological basis of malaria control. Bull World Health Organ 1956;15:613−26.

[7] MacDonald G. The measurement of malaria transmission. Proc R Soc Med 1955;48(4):295−301.

[8] Chitnis N, Schapira A, Smith DL, Smith T, Hay SI, Steketee R. Mathematical modelling to support malaria control and elimination. RBM/WHO progress and impact series, vol. 5; 2010.

[9] Griffin JT, Hollingsworth TD, Okell LC, et al. Reducing *Plasmodium falciparum* malaria transmission in Africa: a model-based evaluation of intervention strategies. PLoS Med 2010;7(8):e1000324.

[10] Smith T. Imperfect vaccines and imperfect models. Trends Ecol Evol 2002;17:154−6.

[11] Garrett-Jones C, Shidrawi GR. Malaria vectorial capacity of a population of *Anopheles gambiae*: an exercise in epidemiological entomology. Bull World Health Organ 1969;40(4):531−45.

[12] Ulrich JN, Naranjo DP, Alimi TO, Müller GC, Beier JC. How much vector control is needed to achieve malaria elimination? Trends Parasitol 2013;29(3):104−9.

[13] Kikankie CK, Brooke BD, Knols BG, et al. The infectivity of the entomopathogenic fungus *Beauveria bassiana* to insecticide-resistant and susceptible *Anopheles arabiensis* mosquitoes at two different temperatures. Malar J 2010;9:71.

[14] Read AF, Lynch PA, Thomas MB. How to make evolution-proof insecticides for malaria control. PLoS Biol 2009;7(3):e1000058.

[15] Delves M, Plouffe D, Scheurer C, et al. The activities of current antimalarial drugs on the life cycle stages of *Plasmodium*: a comparative study with human and rodent parasites. PLoS Med 2012;9(2):e1001169.

[16] Wu Y, Sinden RE, Churcher TS, Tsuboi T, Yusibov V. Development of malaria transmission blocking vaccines: from concept to product. Adv Parasitol 2015;89:109−52.

[17] Dong Y, Das S, Cirimotich C, Souza-Neto JA, McLean KJ, Dimopoulos G. Engineered anopheles immunity to plasmodium infection. PLoS Pathog 2011;7(12):e1002458.

[18] Corby-Harris V, Drexler A, Watkins de Jong L, et al. Activation of *Akt* signaling reduces the prevalence and intensity of malaria parasite infection and lifespan in *Anopheles stephensi* mosquitoes. PLoS Pathog 2010;6(7):e1001003.

[19] Yoshida S, Shimada Y, Kondoh D, et al. Hemolytic C-type lectin CEL-III from sea cucumber expressed in transgenic mosquitoes impairs malaria parasite development. PLoS Pathog 2007;3(12):e192.

[20] de Lara Cupurro M, Coleman J, Beerntsen BT, et al. Virus-expressed, recombinant single-chain antibody blocks sporozoite infection of salivary glands in *Plasmodium gallinaceum*-infected *Aedes aegypti*. Am J Trop Med Hyg 2000;62:427−32.

[21] Armistead JS, Morlais I, Mathias DK, et al. Antibodies to a single, conserved epitope in *Anopheles* APN1 inhibit universal transmission of falciparum and vivax malaria. Infect Immun 2013;82(2):818−29.

[22] Alphey L, Beard CB, Billingsley P, et al. Malaria control with genetically manipulated insect vectors. Science 2002;298(5591):119−21.

[23] Liu C, Pitts RJ, Bohbot JD, Jones PL, Wang G, Zwiebel LG. Distinct olfactory signaling mechanisms in the malaria vector mosquito. *Anopheles gambiae*. PLoS Biology 2010;8(8): e1000467.

[24] Kroeger A, Horstick O, Riedl C, Kaiser A, Becker N. The potential for malaria control with the biological larvicide *Bacillus thuringiensis israelensis* (Bti) in Peru and Ecuador. Acta Trop 1995;60:47−57.

[25] Hulls RH. The adverse effects of a microsporidian on the sporogony and infectivity of *Plasmodium berghei*. Trans R Sco Trop Med Hyg 1971;65:421−2.

[26] Bargielowski I, Koella JC. A possible mechanism for the suppression of *Plasmodium berghei* development in the mosquito *Anopheles gambiae* by the microsporidian *Vavraia culicis*. PLoS One 2009;4(3):e4676.

[27] Bian G, Joshi D, Dong Y, et al. *Wolbachia* invades *Anopheles stephensi* populations and induces refractoriness to *Plasmodium* infection. Science 2013;340(6133):748−51.

[28] Wang S, Ghosh AK, Bongio N, Stebbings KA, Lampe DJ, Jacobs-Lorena M. Fighting malaria with engineered symbiotic bacteria from vector mosquitoes. Proc Natl Acad Sci USA 2012;109(31):12734−9.

[29] Wang S, Jacobs-Lorena M. Genetic approaches to interfere with malaria transmission by vector mosquitoes: special issue: celebrating 30 years of biotechnology. Trends Biotechnol 2013;31(3):185−93.

[30] James AA. Gene drive systems in mosquitoes: rules of the road. Trends Parasitol 2005;21 (2):64−7.

[31] Deredec A, Godfray HCJ, Burt A. Requirements for effective malaria control with homing endonuclease genes. Proc Natl Acad Sci USA 2011;108(43):E874−80.

[32] Chen CH, Huang H, Ward CM, et al. A synthetic maternal-effect selfish genetic element drives population replacement in *Drosophila*. Science 2007;316(5824):597−600.

[33] Windbichler N, Menichelli M, Papathanos PA, et al. A synthetic homing endonuclease-based gene drive system in the human malaria mosquito. Nature 2011;473(7346):212−15.

[34]   Sinka M, Bangs M, Manguin S, et al. A global map of dominant malaria vectors. Parasit Vectors 2012;5(1):69.

[35]   Coluzzi M. Malaria vector analysis and control. Parasitol Today 1992;8:113−18.

[36]   Sinden RE. A biologist's perspective on malaria vaccine development. Hum Vaccin 2010;6(1):3−11.

[37]   Thompson R, Begtrup K, Cuamba N, et al. The Matola malaria project: a temporal and spatial study of malaria transmission and disease in a suburban area of Maputo, Mozambique. Am J Trop Med Hyg 1997;57:550−9.

[38]   Favia G, Ricci I, Damiani C, et al. Bacteria of the genus *Asaia* stably associate with *Anopheles stephensi*, an asian malarial mosquito vector. Proc Natl Acad Sci USA 2007;104(21):9047−51.

[39]   White NJ, Olliaro PL. Strategies for the prevention of antimalarial drug resistance: rationale for combination chemotherapy for malaria. Parasitol Today 1996;12:399−401.

[40]   Sinden RE, Dawes EJ, Alavi Y, et al. Progression of *Plasmodium berghei* through *Anopheles stephensi* is density-dependent. PLoS Pathog 2007;3(12):e195.

[41]   Gething PW, Smith DL, Patil AP, Tatem AJ, Snow RW, Hay SI. Climate change and the global malaria recession. Nature 2010;465(7296):342−5.

[42]   Blagborough AM, Churcher TS, Upton LM, Ghani AC, Gething PW, Sinden RE. Transmission-blocking interventions eliminate malaria from laboratory populations. Nat Commun 2013;4:1812.

Chapter 8

Gene Insertion and Deletion in Mosquitoes

Zach N. Adelman, Sanjay Basu and Kevin M. Myles
Fralin Life Science Institute and Department of Entomology, Virginia Tech, Blacksburg, VA, USA

RANDOM INSERTION OF GENETIC ELEMENTS

Transposons

Introduction. The poetical notion that the genome of an organism was a tranquil arrangement of cooperative genes working in harmony was forever destroyed with the fundamental discovery by Barbara McClintock of the existence of selfish genetic elements [1]. These transposable elements (TEs), or transposons, do not sit still and do not necessarily share the goals of the genome in which they reside, being capable of moving around the genome in an attempt to increase the frequency in which they are inherited. Class I transposons, descendent from ancient retroviruses, generate an RNA intermediate, which is turned back into DNA in order to reintegrate into a new location in the genome. Class II transposons catalyze their own excision and reintegration; repair by the host cell machinery can restore the first copy resulting in a net gain in transposon number. Both types are highly mutagenic; the integration of a transposon into a new location in the genome can disrupt any gene or regulatory region present therein. Both transposon classes have been highly successful throughout evolutionary history and these sequences now make up a majority of the genome of many organisms, including humans. The mutagenic power of TEs is now widely recognized as a major driving force in evolution (reviewed in Refs [2,3]). As mentioned in Chapter 2, initial efforts at genetic control in mosquitoes utilized chromosomal translocations induced by irradiation. Following the successful transposon-based transformation of *Drosophila melanogaster* using a transposon known as the *P* element [4], a new era where much more targeted genome manipulation was possible was born. In a relatively short timeframe, the efforts of several groups to use *P*-element-based genetic transformation on malaria and dengue mosquitoes resulted in some rare recombination events, but no true transposon-mediated integration [5−7]. Finally, the *P* element, so successful in *Drosophila*, was abandoned, and a search for new transposons to use in mosquito transformation experiments began. It was not until 10 years later that the barrier in mosquito transformation was broken with the successful transposon-based genetic modification of *Aedes aegypti* with the *Mos1* [8], *Hermes* [9], and *piggyBac* [10] elements, and the transformation of *Anopheles stephensi* [11] and *Anopheles gambiae* [12], with the transposons *minos* and *piggyBac*, respectively.

How they work. All current insect transposon-based transformation systems are based on class II DNA elements. These simple, selfish genetic elements consist of a pair of inverted repeats flanking a single open reading frame encoding the transposase protein. Following transcription and translation directed by the host cell machinery, the transpose binds selectively to its own inverted repeat sequences and catalyzes the excision and reintegration of the entire element (reviewed in Ref. [2]). This mechanism, referred to as cut-and-paste transposition, results in the movement of the transposon from one genomic location to another. For insect transformation, this single genetic unit (transposon + inverted repeats) is split into a bipartite system, with the inverted repeats present in one plasmid DNA molecule, and the transpose protein (without inverted repeats) encoded separately (either on a separate plasmid or as *in vitro* transcribed RNA); the use of such transposons to modify mosquito genomes has been the subject of a number of excellent reviews [13−15]. Thus, any genetic segment of interest to the investigator can be placed in between the inverted repeats and subsequently be mobilized by the transpose from the plasmid and into the mosquito genome (Figure 8.1).

Strengths. Transposons have been the workhorse of the mosquito transformation field for more than 15 years now. They are easy to manipulate using standard molecular techniques and are consistent in their rates of

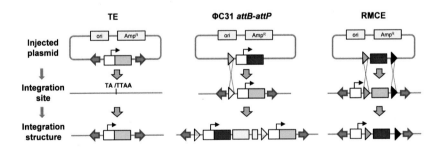

FIGURE 8.1 Methods of germline transformation. Blue arrows indicate the L and R arms of the initial TE-based transposon. Green and red boxes indicate fluorescent marker genes, while yellow represents bacterial sequences such as antibiotic resistance genes or an origin of replication. White/orange arrowheads indicate *attP-attB* sites, while gray/black arrowheads indicate heterospecific *lox* sites. Dotted lines indicate crossing-over events during recombination.

transformation (typically ranging from 1% to 15%, depending on the mosquito/element combination). Transposon-based transformation allows the investigator to sample multiple locations of the genome due to the ubiquitous distribution of preferred target sites (typically the nucleotides TA or TTAA). This sampling of the genome can be used to identify new enhancer elements or genes [16], to find a genetic locus where insertion of the transposon yields the least effect on mosquito fitness [17], or to identify the most suitable location for expression of the transgene of interest [18−20]. Because the transposons used do not rely on any host factors, they can be easily adapted for use in new insect systems with confidence.

Weaknesses. Although a reliable means of inserting genes into the mosquito genome, transposons suffer from a number of weaknesses that have consequences for both studies on basic biology and applied strategies to develop mosquitoes to be used in genetic control programs (Figure 8.2). Certainly, the primary weakness is also their greatest strength—the ability to insert virtually anywhere in the genome. While a desired trait for forward genetic experiments, the lack of an ability to control where a transgene inserts severely complicates the development and characterization of transgenic strains for genetic control programs. The local chromatin structure or the presence of nearby enhancers/repressors can dramatically alter the expression pattern of the transgene, requiring investigators to characterize many insertion sites (typically 5−10) in order to find those that are most favorable [18−20]. Nearby TEs (or the inverted repeats of the transposon used itself) may induce epigenetic changes, resulting in substantial variation between individuals containing the same insertion [21]. Transcriptional initiation on the opposite strand by either recognized or cryptic promoters may induce the formation of dsRNA and result in silencing of the transgene [22].

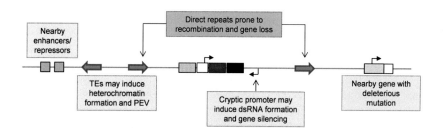

FIGURE 8.2 Position-based effects on transgene expression. A number of factors can influence the stability and/or expression of transgenes. Nearby enhancers/repressors (purple boxes) can reduce, increase, or change the spatial expression of the transgene. Nearby transposable elements (TEs) may induce heterochromatin formation and position effect variegation (PEV). Direct repeats may collapse following random DNA breaks, resulting in loss of the transgene. The presence of cryptic (or recognized) promoters may generate double-stranded RNA (dsRNA), resulting in gene silencing. The transgene may be linked to a deleterious mutation; selection for the transgene may then increase the frequency of this allele in the population, lowering its fitness.

Transgene insertions that occur nearby recessive deleterious mutations can suffer substantial fitness effects when made homozygous for the transgene [23,24]. Concerning long-term stability, the presence of short direct repeats sequences (common in mosquito genomes, particularly in *Ae. aegypti*) can result in recombination and loss of the entire transgene. Transposons are also limited in the cargo that they can carry, with maximum sizes of 10−12 kb. Early on, one of the biggest concerns regarding the use of transposons to develop transgenic mosquitoes was their ability to be remobilized by native elements in the mosquito genome. While such remobilization is indeed efficient in *An. stephensi* [16,25], in mosquito such as *Ae. aegypti*, even intentional remobilization has proven to be extraordinarily difficult [26−28], suggesting for at least this species that this is not a serious concern. However, the biological basis for this lack of remobilization is still unknown, and could certainly vary in the field. Thus, methods for permanently stabilizing transgenes have been developed, which remove one of the inverted terminal repeats [29] or both [30−32]. Finally, once favorable chromosomal integration sites are discovered and characterized, using transposon-based systems alone it is not possible to reuse the same integration site.

Docking-Site-Based Integration

Introduction. The large amount of work needed to characterize transposon-based integration sites in the mosquito genome for characteristics such as transgene expression and variation, fitness and stability has driven the implementation of site-specific, docking-site-based integration methods based on the bacteriophage ΦC31 recombinase [33]. First adapted to human cells [34] and then *Drosophila* [35], the ΦC31 system was first utilized in *Ae. aegypti* as a way to reuse favorable transposon insertions [36]; with subsequent success also in *Aedes albopictus* [37], *An. stephensi* [17], and *An. gambiae* [38,39].

How they work. In its native environment, the bacteriophage ΦC31 encodes a recombinase that recognizes a short DNA sequence in the phage genome (termed *attP*, for attachment-phage) and a short DNA sequence in the host bacterial genome (termed *attB*, for attachment-bacteria). Once both sites are bound, the recombinase catalyzes the integration of the phage genome into the bacterial genome; a process that is unidirectional. As these attachment sites are small and easy to manipulate, they were very readily converted into gene modification tools [34,35]. For typical experiments involving manipulating the genome of mosquitoes, the *attP* site is included in a normal transposon-based construct that integrates randomly into the mosquito genome. Once all of the insertion sites are characterized, the most favorable sites can be reused by encoding an *attB* site into a new target plasmid and introducing it into mosquito embryos while providing a source of recombinase (Figure 8.1).

Strengths. The ability to reuse well-characterized transposon-based insertion sites has made ΦC31-based recombination the current state-of-the-art practice for developing and testing genetically modified mosquito (GMM) strains for genetic control programs [40]. Reusing the same integration site allows a panel of related effector molecules to be evaluated together with the contributions of chromosomal position strictly controlled for. Overall rates of integration into *attP* target sites are similar to overall rates of transposon-based transposition but do not require the extensive downstream characterization of the site after the initial evaluation.

Weaknesses. Docking-site-based systems such as ΦC31 still depend on the random integration of a transposon and may also require additional steps to stabilize the transposon insertion. These insertions are thus restricted to a very small portion of the mosquito genome and are still ultimately descended from a single transformation event where loss of heterozygosity around the insertion site can depress fitness. While screening enough sites can identify insertions free of such effects, it remains to be seen whether the same will hold true under field conditions. ΦC31 integration results in the incorporation of an entire bacterial plasmid, including any antibiotic resistance genes, into the recipient genome. Removing this unwanted material thus requires additional downstream steps prior to inclusion of any modified strains in a genetic control program. Alternatively, including two *attP* target sites in the initial transposon allows the ability to perform recombinase-mediated cassette exchange [41], though this has not been reported in mosquitoes at the time of this writing (Figure 8.1).

Other Recombinases

Isolated from the phage P1, *cre* (causes recombination) recombinase is commonly used to catalyze recombination events in a wide variety of organisms [42−46], though this system is not commonly used in mosquitoes. *Cre*-mediated recombination occurs between two short *loxP* target sites, with each *loxP* site consisting of perfect 13-bp inverted repeats flanking an 8-bp asymmetric spacer. When two such *loxP* sites flank a gene of interest *in cis*, *cre*-mediated recombination results in the effective excision of the gene, leaving behind only a single *loxP* site. While *cre* recombinase has been shown to be highly efficient at catalyzing the excision of DNA sequences in *Ae. aegypti* [47], excision is so thermodynamically favored that transgene insertion events were never recovered [36]. To overcome this obstacle, heterospecific *lox* sites consisting of one wild-type *loxP* site and a second *lox* site containing one or more substitutions in the spacer region have been developed [48−50]. As *cre*-binding specificity is denoted entirely by the inverted repeats, these modified *lox* sequences bind *cre* normally and thus can recombine with themselves, but can no longer recombine with the wild-type *loxP* sequence. Such sites have been used to perform site-specific recombinase-mediated cassette exchange in systems such as *Drosophila* [44], thus it is possible that this system could also be adapted to mosquitoes. The FLP/FRT recombination system derived from the baker's yeast, *Saccharomyces cerevisiae*, is also heavily utilized for genomic manipulation in *D. melanogaster* (reviewed in Ref. [51]), and more recently, it has been shown to be effective in the silkworm [52]. Though this system was able to catalyze interplasmid recombination in *Ae. aegypti* embryos [53], it failed to produce any recombination events with the mosquito chromosome [47].

SITE-SPECIFIC NUCLEASES

Homing Endonucleases

Introduction. A new type of selfish DNA element that was first identified in yeast in the 1970s (reviewed in Refs [54,55]), homing endonucleases have been highly successful in invading eukaryotic, bacterial, archaeal, and even phage genomes despite their simple structure—a single open reading frame. Many (but not all) homing endonucleases identified to date recognize highly specific DNA target sequences found in self-splicing introns. Others, known as inteins, are translated along with a host protein and are posttranslationally processed in a manner that frees the homing endonuclease but leaves the host protein intact. Both strategies have the advantage of limiting detrimental effects to the host cell and its genome, enabling the spread of these selfish genes. Following the identification and biochemical characterization of many homing endonucleases in the 1980s and 1990s, researchers started to repurpose these simple elements for various genome manipulation applications. For example, the introduction of the target sequence recognized by the homing endonuclease I-*Sce*I into the *Drosophila* genome enabled studies of DNA break repair and homologous recombination [56−59]. Homing endonucleases were relatively unknown to the vector biology community at this time, their adaptation beginning when Burt [60] suggested that the primitive form of genome invasion used by homing endonucleases may be well-suited for genetic strategies to control malaria or other vector-borne diseases. Indeed, several homing endonucleases have since been used effectively to introduce site-specific DNA breaks in *An. gambiae* [61,62] and to excise genes from *Ae. aegypti* [63,64]. Through biochemical redesign, the homing endonucleases I-*Cre*I and I-*Ani*I have been modified to recognize targets in the *An. gambiae* genome directly [65]. Transgenic expression of wild-type I-*Ppo*I during spermatogenesis completely sterilized *An. gambiae* mosquitoes [62]; large cage trials using these transgenic sterile mosquitoes indicate effectiveness at crashing populations of this mosquito [66]. The expression of modified versions of I-*Ppo*I in the male testes of *An. gambiae* has led to the development of sex-distorter strains that produce almost all males [39], while I-*Sce*I was shown to successfully invade large cage populations [65].

How they work. The phrase "homing" refers to their ability as mature proteins to return to the site of their mRNA "birth" and introduce a double-stranded break on the homologous chromosome lacking the homing endonuclease gene. Unlike TEs and recombinases, homing endonucleases do not catalyze any reactions beyond DNA cleavage. This extraordinarily simplistic mode of action thus relies entirely on the cellular DNA repair machinery to generate a duplicate version of the homing endonuclease gene, by using the original gene as a repair template. Homing endonucleases can be categorized by their mode of catalyzing DNA cleavage into at least four completely independent families, and an extensive body of literature is available concerning the structural basis for DNA-binding specificity and double-stranded DNA break formation (reviewed in Ref. [54]). Of particular interest is that the target sequences for most homing endonucleases are 18 bp or longer, meaning that even in large eukaryotic genomes the probability of finding an endogenous site is very small. For example, the target sequence of I-*Sce*I is 18 bp, and the corresponding chance of finding an exact match in a random sequence is 1 in 6.87×10^{10}. Given the size of the *An. gambiae* (2.7×10^8 bp) and *Ae. aegypti* (1.3×10^9 bp) genomes, the corresponding chances of randomly finding an I-*Sce*I target site are approximately 1:250 and 1:50, respectively.

Strengths. Homing endonuclease genes are relatively small, with dimeric nucleases such as I-*Cre*I and I-*Ppo*I both less than 500 bp; the larger monomeric nucleases such as I-*Sce*I are still less than 1 kb. Thus, these genes and their resultant proteins are easy to manipulate, express, and purify. When integrated into a mosquito genome, their small size may also reduce the risks of acquiring deleterious mutations that can affect their function. Their extreme specificity can be expected to reduce or virtually eliminate the impacts of off-target cutting; the transgenic expression of I-*Sce*I is well-tolerated by both *Drosophila* and *An. gambiae* [56,65]. Homing endonucleases are a mature technology that have been used to generate the first experimentally validated gene drive system in mosquitoes [65] as well as transgenic sex-distorter strains ready to be deployed in SIT-type programs [39].

Weaknesses. The extreme specificity of homing endonucleases due to extensive protein−DNA contacts at the target site also makes reengineering these molecules to recognize new target sequences extremely difficult. While such modifications are possible [39,67−69], the cost (primarily human capital) associated with modifying homing endonucleases to recognize new and diverse target sites is thus the primary barrier to their widespread use as gene editing and modification tools. While large-scale sequencing of new microbial genomes and comparative genomic analyses have accelerated the discovery of new homing endonuclease genes [70], many of these have reduced or completely lost catalytic activity, meaning that each new nuclease must still be verified experimentally.

Chimeric Nucleases (ZFNs and TALENs)

Instead of attempting to identify naturally occurring nucleases to cut a target DNA sequence using homing endonucleases, or modifying a homing endonuclease to recognize a related sequence, other research groups have attempted to create *de novo* targeted nucleases by fusing a custom DNA-binding domain to a nonspecific nuclease domain. Some of the most well-characterized DNA-binding domains at the time were considered to be the zinc-finger motifs found in a family of transcription factors. Fusions of several such zinc-finger motifs to the nuclease domain of the type IIS restriction enzyme *Fok*I (isolated from the bacteria *Flavobacterium okeanokoites*) resulted in the first zinc-finger nucleases (ZFNs) [71]. As each zinc finger was responsible for mediating interactions with three nucleotides, increasing the number of fingers included in the chimeric protein served to increase specificity of the resultant nuclease. At the same time, cleavage by the *Fok*I domain relied on dimerization of the nuclease domain, meaning two independent ZFNs were required—one to bind upstream of the target on one strand and one to bind downstream of the target on the opposing strand. ZFNs have since been used successfully to edit the genome of a number organisms (reviewed in Refs [72,73]), including inducing gene correction in human cells [74]. It was not until 2013 that ZFNs were first used in mosquitoes, when DeGennaro et al. [75] generated null mutant alleles for the *Ae. aegypti orco* gene, a key component of the insect olfactory system. *Orco*-deficient mosquitoes were unable to differentiate human from animal subjects and could not be repelled by *N,N*-Diethyl-*meta*-toluamide (DEET), though in the presence of CO_2 *orco*-mutant mosquitoes were still attracted to vertebrates. Liesch et al. [76] used ZFNs to knockout a neuropeptide Y-like receptor in *Ae. aegypti* thought to be involved in shutting off host-seeking behavior after a female mosquito obtains a blood meal. Finally, ZFNs were used to knockout a critical component of the *Ae. aegypti* CO_2 receptor complex, rendering these mosquitoes unable to detect this critical host cue [77].

Following the decoding of how transcription-activator-like elements (TALEs) from *Xanthomonas* species bacteria bind specific DNA sequences in the nuclear genomes of the plants they infect [78,79], chimeric nucleases based on this platform (TALE nucleases, or TALENs) were rapidly developed as an alternative to ZFNs [80]. The DNA-binding region of each TALE was found to consist of a series of almost identical 34 amino acid repeats, the only changes being a two amino acid variable region-each unique combination specified binding to one of each of the four nucleotide bases A, C, G, and T. By simply stacking these repeats, it was possible to spell out the desired DNA-binding region, and efficient methods for building new TALENs were developed rapidly [81−85]. In mosquitoes, Aryan et al. [86] demonstrated that TALENs could induce gene editing rates of >30% in *Ae. aegypti* when targeting a gene involved in eye pigmentation. Smidler et al. [87] adopted a different strategy, generating transgenic strains of *An. gambiae* each expressing one half of a TALEN pair targeting the *tep1* gene. No mutations were observed until the two strains were crossed with each other, at which point a fully active TALEN capable of heterodimerization and DNA cleavage was produced. Despite these successes, the cost and difficulty associated with generating new nucleases is likely to preclude their adoption as common laboratory reagents for mosquito genome modification considering the clustered regularly interspaced short palindromic repeat (CRISPR) revolution (see CRISPR/Cas9).

Strengths. At the time when each was developed, both ZFNs and TALENs were the leading technology for genome editing applications, and as described earlier, both platforms have been used successfully to edit the genomes of the most important vectors of malaria and dengue. In the laboratory, their expression can be easily controlled via plasmid-based expression, and when inherited directly [87], appear to be stable in the mosquito genome.

Weaknesses. Both ZFNs and TALENs are relatively difficult to engineer, particularly for smaller academic labs. ZFNs suffered from the fact that the rules that govern protein−DNA contacts are not straightforward and are heavily influenced by context. While several prediction platforms and validation schemes were developed for both ZFNs and TALENs [82−84,88,89], the cost and required expertise associated with these methods has prevented their widespread adoption. While TALENs possess a far simpler set of rules for DNA binding, these molecules suffer from the large and repetitive nature of the recombinant genes used to produce them. Both ZFNs and TALENs rely on the *Fok*I nuclease, which in its wild-type form can form homodimers as well as heterodimers, resulting in cutting at off-target sites [90]. Efforts at structure-based redesign of the dimerization plane decreased such off-target activity [91−93] but also decreased on-target activity in *Ae. aegypti* [94]. Also, the repetitive nature of the ZF or TALE repeats may promote instability if these nucleases are inserted directly into the mosquito genome [95].

CRISPR/Cas9

Introduction. Found in the genome of some bacteria, CRISPRs represent a novel adaptive, heritable immune defense mechanism against foreign DNA [96,97]. These CRISPR loci contain short (30−40 nt) stretches of variable sequences, interspersed by direct repeats. Short RNA transcripts termed CRISPR RNAs bind one or more CRISPR-associated (Cas) proteins, in association with a short transactivating-RNA (trRNA) to direct the Cas endonuclease complex to a given target site. A minimalist system characterized from *Streptococcus pyogenes* involving a single Cas protein (Cas9) is sufficient to act as an RNA-directed dsDNA endonuclease in combination with the appropriate CRISPR short RNAs [97]. Due to the simplicity in designing new CRISPR RNAs, the CRISPR/Cas9 system has competed extremely well against other more established technologies and has been successfully adapted to a range of organisms (reviewed in Ref. [98]). Indeed, both Basu et al. [94] and Kistler et al. [99] reported that the CRISPR/Cas9 system was highly efficient in *Ae. aegypti*, with editing rates as high as 90%. Substantially lower editing rates were reported by Dong et al. [100], though with only a limited number of guide RNAs.

How it works. In a natural system, CRISPRs are associated with trRNA that guides the Cas9 endonuclease to the complimentary sequence of the CRISPR, resulting in cleavage and indel events in the target DNA. Expression of a fusion RNA whereby the CRISPR RNA and transactivating RNA are represented by a single molecule reduces the system to a two component reprogrammable nuclease (one protein, one RNA) [97,101,102]. Following hybridization of the CRISPR RNA, the complimentary strand of DNA is cleaved by an HNH domain of Cas9, whereas the noncomplimentary strand is cleaved by a RuvC-like domain generating a blunt-ended double-stranded break 3 bp upstream from the so-called protospacer-adjacent motif (PAM, a trinucleotide sequence represented by the bases NGG for *S. pyogenes* Cas9). As with other site-specific engineered endonucleases, repair of the double-stranded DNA break is mediated by the host's DNA damage response.

Strengths. The power of the CRISPR system is that developing new nucleases does not involve generating a new protein or a new complicated gene, but simply a new short RNA guide that can be synthesized cheaply using existing technology. This not only enables both high-throughput, large-scale editing screens [103], but also puts editing reagents in the hands of even small laboratories that could not afford homing endonuclease redesign, ZFNs or TALENs. Initial experiments in mosquitoes have used Cas9 endonuclease generated via *in vitro* transcribed mRNA or purified protein [94,99,100]. Other strategies such as transgenic driver lines should be readily achievable as well [104], though multiple groups were unable to obtain editing when using plasmid-expressed Cas9 in mosquitoes [99,100]. DNA plasmid constructs can also carry RNA Pol III-based U6 promoters to drive expression of the chimeric sgRNA, thereby facilitating ease-of-cloning and transfection/injection methods. The efficiency of the system in mosquitoes may be further increased by inclusion of multiple sgRNAs targeting the same or different genes within a single experiment, known as multiplexing, making it possible for multiple-gene knockout investigations [105].

Weaknesses. At present, the primary weaknesses associated with CRISPR/Cas9 center around off-target effects and specificity [106]. Several new analytical methods have been developed to identify such off-target effects [107,108], and it is clear that there is substantial variability in the potential for off-target cutting between different guide RNAs. In the context of mosquito genome modification, gene drive or GMMs, any such off-target effects could result in decreased fitness due to the introduction of indels or excessive chromosomal breakage. While no off-target effects were observed in initial experiments [94], these analyses were limited to a few computationally predicted sites. Such prediction algorithms have recently been found to perform poorly compared to whole genome analysis methods [107,108]. Thus, the full range of off-target effects should be clarified experimentally prior to field-testing of active CRISPR RNAs. The current generation of CRISPR/Cas9 reagents is also restricted by the requirements of the PAM, any thus guide RNAs are restricted to the sequence $N_{18}GG$. As additional CRISPR reagents are developed with different PAM requirements [109], the number of targetable loci in a particular genome should increase further.

TARGETED INSERTION OF GENETIC ELEMENTS THROUGH HOMOLOGOUS RECOMBINATION

As noted above, the insertion of material into the mosquito genome via TEs is a random process. Without the ability to control the landing site, the types of manipulations that can be performed to generate useful phenotypes in mosquitoes are relatively limited. Genetic cargo typically consists of two or more expression cassettes (one marker, one or more effectors) that are expected to act relatively independent of the landing site. With the advent of reprogrammable site-specific nucleases, it becomes possible to generate double-stranded DNA breaks more strategically in the mosquito genome: if these breaks are repaired using a homology-driven process it becomes possible to introduce genetic material in a site-specific manner (Figure 8.3). Liesch et al. [76] used a ZFN along with a donor vector containing 1.3/1.5 kbp of homologous sequence flanking the target site to integrate a fluorescent reporter construct into an *Ae. aegypti* neuropeptide Y-like receptor gene. Similarly, McMeniman et al. [77] used a ZFN targeting the *Ae. aegypti Gr3* CO_2 receptor along with a donor vector containing 0.8/0.8 kbp of homologous sequence to integrate a similar reporter into the *Ae. aegypti* genome. In the malaria mosquito, Bernardini et al. [110] used homologous recombination to insert a ΦC31 docking site onto the Y-chromosome. Taking advantage of a fortuitous transposon insertion, these researchers used the homing endonuclease I-*Sce*I along with a donor vector containing 0.6/2.1 kbp of homologous sequence on each side of the target site. In all these cases, the efficiency of transgene insertion via homology-dependent repair was at least an order of magnitude lower than that of transposon-based systems. This appears to be due largely to the preference by mosquitoes for using end-joining repair to correct double-stranded DNA breaks, a preference shared by most higher eukaryotes. Manipulating the DNA damage response thus represents a critical bottleneck to targeting engineering of the mosquito genome [111]. Altering the ability of an insect to use end-joining factors such as Ligase 4 in *Drosophila* [112] or Ku70 in silkmoth [113] has been shown to increase rates of homology-dependent gene insertion. Basu et al. [94] adapted this concept to *Ae. aegypti*, where

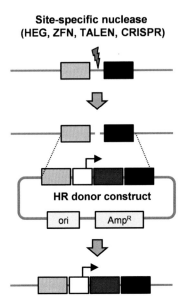

FIGURE 8.3   Homology-dependent repair for transgene integration. Following treatment with a site-specific nuclease (blue lightning), homologous sequence upstream (gray box), and downstream (black box) of the break site can be included in a donor vector to stimulate homology-dependent repair. Only the sequence located between the homology arms is incorporated into the genome, but not the bacterial origin of replication (ori) or antibiotic selection cassette (Amp^R).

silencing of the *ku70* ortholog at the time of embryonic injection resulted in a significant boost to rates of homology-dependent repair. These experiments used homology arms of 1.7/2.2 and 1.1/1.7 kb to introduce an EGFP reporter following TALEN or CRISPR/Cas9-based DNA break induction. The ability to introduce genetic material at any location in the mosquito genome substantially increases the type of beneficial phenotypes that can be engineered into a mosquito population. For example, changes to native promoters or splice sites may alter the timing or expression pattern of genes to better respond to pathogen infection. Even the introduction of single base changes in key positions in the genome could have a dramatic beneficial effect, such as reverting insecticide-resistant mosquitoes back to a susceptible state.

CONTROLLING TRANSGENE EXPRESSION; THE CURRENT STATE OF THE MOSQUITO TOOLBOX

Promoters

Markers. Early mosquito transformation experiments used an eye pigmentation-based rescue to identify a successful transgene insertion. This was based on the *Drosophila cinnabar* gene, which rescues an orthologous defect in a mosquito enzyme known as kyurenine hydroxylase [114]. Depending on the insertion site, eye pigmentation was restored from white to yellow, orange, red or purple depending on the expression of the transgene as dictated by the local chromosomal environment. Despite the simplicity of this system, which did not require fluorescence microscopy, this method had a number of drawbacks. The appropriate white-eyed mutation was only available for *Ae. aegypti*, and was difficult to rear due to substantial inbreeding which reduced both lifespan and the production of progeny. Also, many of the eye-color changes were only visible in adults, requiring massive numbers of progeny to be reared to adulthood. Fortunately, markers based on the expression of one or more fluorescent proteins quickly replaced eye pigment-based rescue. These markers work well in all mosquito species tested, with screening able to be performed in the early larval stages, substantially reducing the labor required. The synthetic eye-specific promoter $3 \times P3$ based on the Pax-6 enhancer and the hsp70 core promoter is the most commonly used [115], along with more general promoters such as the *Drosophila* Act5C [19,116], the baculovirus IE-1 [12,117], and the *Ae. aegypti polyubiquitin* [118]. These promoters have the advantage of driving gene expression in a more robust fashion, enabling identification of transgenic individuals at earlier developmental stages and lower magnifications.

Genes of interest. Following the establishment of transgenic technology in mosquitoes, considerable effort was expended by the vector biology community to characterize promoter elements (for simplicity, the term "promoter" as used refers to both the actual promoter where transcription is initiated and any enhancer/repressor elements required for controlling proper expression) capable of driving the expression of transgenes in the most biologically relevant tissues. These efforts were highly successful, as for both dengue and malaria vectors, a catalog of validated promoters is available for driving gene expression in the mosquito gut, fat body, salivary glands, ovaries, and testes (Figure 8.4, Table 8.1). With the exception of the Vg promoter, which has been extensively dissected [119,120], little is known about the functional elements that contribute to the activity of each promoter. Identification of specific regulatory sequences that control tissue and temporal specificity would be valuable to the development of synthetic promoters with novel expression patterns (e.g., robust expression in both midgut and salivary glands). Some large-scale computational analyses have identified some conserved regulatory sequences [121,122]; the accuracy of such high-throughput approaches is likely to improve as RNA sequencing data helps refine gene models across the various mosquito genomes. Nevertheless, the current cadre of available promoters represents a sufficient entry point for most studies.

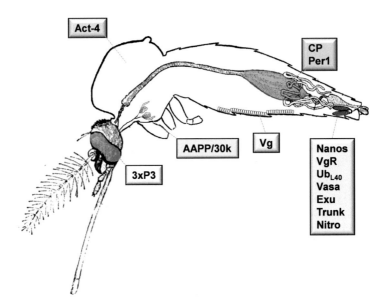

FIGURE 8.4   Tissue-specific promoters for driving transgene expression in mosquitoes. Control elements have been characterized to allow transgene expression in a number of tissues in the mosquito body, including the eyes (green), salivary glands (yellow), midgut (red), fat body (gray), ovaries (blue), and flight muscle (purple). See Table 8.1 for a complete list of validated promoters.

Identifying New Promoters, a Cautionary Tale

While many promoters have been developed for use in transgenic experiments, many others did not behave as expected when used to control transgenic constructs. While model organisms such as *Drosophila* have compact genomes that largely constrain the location of transcriptional regulatory regions such as enhancers/repressors to a few kilobases in most instances, this is not the case for the mosquito genomes, which have expanded substantially due to the proliferation of TEs and repetitive sequences. For example, the average intergenic region is $3 \times$ longer in *An. gambiae* (~ 17 kb), and $10 \times$ longer in *Ae. aegypti* (~ 56 kb) as compared to Drosophila (~ 6 kb) [147]. Most promoter studies will also include the entire 5'UTR and first intron of the donor gene in order to include any functional elements present therein. In both *Drosophila* and *Anopheles*, the average intron size is a manageable ~ 1 kb, while in *Ae. aegypti* introns on average are $5 \times$ longer (~ 5 kb) [147]. It is not uncommon to find introns spanning $20-50$ kb in mosquito genes, severely complicating efforts to include all possible functional elements in transgenic constructs. These increased distances are almost entirely due to the interspersed nature of repetitive sequences and/or TEs in and around protein-coding genes.

TABLE 8.1 Promoter Fragments Used to Drive Transgene Expression in Mosquitoes

Promoter	Mosquito	Sex	Spatial Distribution	Temporal Distribution	Activity in Other Species	References
Mall	Ae. aegypti	♀	Proximal lateral lobe of salivary glands	Constitutive		[123]
Apy	Ae. aegypti	♀	Distal lateral and medial lobes of salivary glands	Constitutive		[123]
AeVg	Ae. aegypti	♀	Fat body	Blood meal induced	An. stephensi	[124]
AeCPA	Ae. aegypti	♀	Midgut	Blood meal induced	An. gambiae	[125,126]
AaVgR	Ae. aegypti	♀	Vitellogenin receptor	Ovary, blood-meal induced		[127]
β2-tub	Ae. aegypti	♂	Testes	Developmental		[128]
Nanos	Ae. aegypti	♀	Ovaries/embryos	Developmental		[129]
30K	Ae. aegypti	♀	Distal lateral lobes of salivary glands	Constitutive		[130]
Ae-Act4	Ae. aegypti	♀	Indirect flight muscles	Pupae	Ae. albopictus	[117]
AeUb$_{L40}$	Ae. aegypti	♀, ♂	Ovary/early larval	Constitutive		[118]
AePUb	Ae. aegypti	♀, ♂	Whole body	Constitutive		[118]
AaHsp70Aa	Ae. aegypti	♀	Head, midgut, salivary gland, and ovary	Constitutive and induced		[131]

(Continued)

TABLE 8.1 (Continued)

Promoter	Mosquito	Sex	Spatial Distribution	Temporal Distribution	Activity in Other Species	References
AaHsp70Bb	*Ae. aegypti*	♀	Head, midgut, salivary gland, and ovary	Constitutive and induced		[131]
Aal-Act4	*Ae. albopictus*	♀	Indirect flight muscles	Pupae	*Ae. aegypti*	[18]
Hex1.2	*Ae. atropalpus*	♀	Larval fat body	Developmental	*Ae. aegypti*	[132]
exu	*Ae. aegypti*	♀	Ovary	Developmental		[133]
trunk	*Ae. aegypti*	♀	Ovary	Developmental		[133]
nitro	*Ae. aegypti*	♀	Ovary	Developmental		[133]
Agper1	*An. gambiae*	♀	Midgut	Constitutive	*An. stephensi, Ae. fluviatilis*	[134]
β2-tub	*An. gambiae*	♂	Testes	Constitutive	*An. stephensi*	[135]
AsVg1	*An. stephensi*	♀	Fat body	Blood meal induced		[136]
AsAAPP	*An. stephensi*	♀	Salivary glands	Constitutive (high) + Blood meal induced		[137]
AgVgt2	*An. gambiae*	♀	Fat body	Blood meal induced		[138]
AgApy	*An. gambiae*	♀	Salivary glands	Constitutive (low)		[139]
AgVasa	*An. gambiae*	♀, ♂	Gonads	Constitutive		[140]

Promoter	Source	Sex	Tissue/Location	Expression	Test species	Reference
AgG12	*An. gambiae*	♀	Midgut	Blood meal induced	*An. stephensi*	[141]
Antryp1	*An. gambiae*	♀	Midgut	Nonresponsive to blood meal	*An. stephensi*	[141]
α1-tub1b	*An. gambiae*	♀	Head, ventral nerve cord, and chordontal organs	Constitutive		[142]
AcCP	*An. culicifacies A*	♀	Midgut	Blood meal induced		[143]
Act88F	*D. melanogaster*	♀, ♂	Pupae and adult flight muscles	Developmental	*Cx. quinquefasciatus* Say	[144]
3 × P3	*Synthetic*	♀, ♂	Neural specific, eye, anal papillae	Constitutive	Many insect species	[115]
Actin5c	*D. melanogaster*	♀, ♂	Muscle	Constitutive	*An. stephensi*, *Ae. aegypti*	[135], [19,116,145]
IE-1	*Baculovirus*	♀, ♂	Whole body	Constitutive	*An. stephensi*, *Ae. aegypti*, *An. gambiae*	[50,146]

Practically speaking then, experimenters should proceed with caution when trying to identify additional promoters for driving transgene expression in mosquitoes. RNA-seq data, now abundantly available for both *An. gambiae* and *Ae. aegypti* can be used to identify multiple candidate genes with the desired expression pattern [133,141,148]. The candidates can be compared using parameters such as (i) distance to nearest gene (particularly if said gene/s have a different expression pattern); (ii) length of first intron (so 5′-UTR sequence up to the initiation codon can be included in test constructs); and (iii) presence/distribution of both unique and repetitive sequences upstream of transcription initiation. Ideally, a candidate test construct can be chosen for situations where functional elements are constrained to the genomic DNA fragment under investigation. For example, the *Ae. aegypti* 30 K [130] and Hsp70 [131] as well as the *An. gambiae* trpy1 [141] gene promoter fragments tested were constrained by the presence of an upstream gene in close proximity.

Bipartite Tet Systems

Introduction. Bipartite systems have been developed so that two independent events are required to control the expression of a transgene. As the name suggests, bipartite systems are comprised of two genetic components: a "driver" that expresses a specific exogenous transcription factor and a "responder" that employs a promoter that is exclusively activated only by that particular transcription factor.

How it works. The most common and well-established bipartite system in mosquitoes involves the activation of gene expression when the antibiotic tetracycline (or its analog, doxycycline) is withheld; this system is known as Tet-Off. Conversely, the opposing Tet-On system specifically activates gene expression in the presence of tetracycline (Tc). Both the Tet-Off and Tet-On systems take advantage of the naturally occurring negative-regulation of resistance-mediating genes by the tetracycline repressor (TetR) found in the bacteria *Escherichia coli*, specifically the regulatory components of the Tn10 tetracycline-resistance operon [149]. It was found that the TetR bound tetracycline more tightly than a previously used lacR complexed with IPTG of the lacR system. This allowed for employment of very low, nontoxic intracellular concentrations of tetracycline that still functioned efficiently. The TetR also shows very specific binding to the Tet operator (TetO) sequence, resulting in a reduction of leaky expression. Doxycycline also has better cell and tissue penetrative properties than IPTG, the inducer for the lacR system [150]. In the Tet-Off system, the TetR is fused with the activation domain of VP16, a protein encoded by herpes simplex virus involved in the triggering of the lytic phase of the viral infection from the latent, dormant state. The resultant chimeric tetracycline transactivation protein, referred to as tTA, binds to the TetO sequence, resulting in the recruitment of the transcriptional machinery and substantially increasing the expression of the gene under its control. When tetracycline is present in the diet of the mosquito, it binds to the tTa protein, thereby preventing DNA binding and gene activation. In the Tet-On system, the reverse tetracycline transactivation protein can only recognize and activate the TetO sequence in the presence of doxycycline. Initially used in mosquitoes by Lycett et al. [151], the Tet-Off system has been utilized as a key component of the transgenic strains used by Oxitec Limited for the genetic control of *Ae. aegypti*[1]. Phuc et al. [19] demonstrated that the employment of the Tet-Off system linked to lethal expression of tTAV (an optimized variant of tTA sequence) that serves as both a transactivator and an effector can result in late-acting lethality [19]. By killing mosquitoes at the fourth instar larval/pupal boundary, this strategy retains density-dependent selection while preventing the emergence of new adult mosquitoes. Functionally, this system uses Tc to prevent tTAV from activating transcription from TetO that would otherwise consequently express more tTAV in a positive feedback loop that would proceed uncontrolled until the death of the mosquito (Figure 8.5A). While transgenic females must be

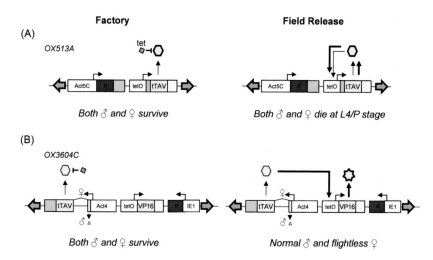

FIGURE 8.5 Tetracycline-repressible control of transgene expression in OX531A and OX3604C mosquitoes. While rearing in the factory, the presence of tetracycline (tet) or its analog doxycyline binds to tTAV and prevents gene expression from TetO. Once released, the tTAV is free to bind TetO, activating expression of either additional tTAV or the VP16 transactivator, the accumulation of which is toxic to cells.

---

1. Any discussion of the transgenic constructs developed and used by Oxitec Limited for the genetic control of dengue vectors must first and foremost acknowledge that this company has voluntarily disclosed via publication all of the experimental details concerning how each mosquito strain was modified, though there is no legal requirement that they do so. This level of transparency is in line with recent WHO recommendations [152] for testing genetically modified mosquitoes and allows the scientific community and public at large the opportunity to evaluate each transgenic mosquito strain on its own merits.

removed manually prior to field releases, the resultant strain (referred to as LA513 in Ref. [19], OX513A in Refs [153,154]) was the first transgenic mosquito used in an open-field release [153].

The transgene construct employed in another transgenic strain of *Ae. aegypti*, OX3604C, incorporates sex-specific splicing, under control of the *Ae-Act4* promoter [117] in combination with the tTAV/Tc system to control a female-specific flightless phenotype. In the presence or absence of Tc, the tTAV is not produced in males due to a combination of minimal promoter activity and an in-frame termination codon in the male splice form (Figure 8.5B). However, in females strong sex-specific expression from the promoter combined with correct splicing of the tTAV mRNA results in sufficient expression of the VP16 transactivator in the indirect flight muscles to render them nonfunctional [117]. OX3604C was successfully tested in large cage trials [155] but has not yet been used in open release. Similar flightless female transgenic strains have been developed and validated for the Asian tiger mosquito, *Ae. albopictus* [18] and the malaria vector *An. stephensi* [156]. Both of these strategies illustrate the potential for differential control of transgene expression and are particularly suited to cases where the expression of the transgene is expected to be detrimental to the health and/or survival of the mosquito.

Strengths. The nature of this approach allows for not only the ability to induce gene expression but also to limit its expression to specific temporal and spatial profiles. A distinct advantage of this approach allows for the scrutiny of genes whose expression may result in lethality, sterility, or a high fitness effects, as these effects will manifest only when both components are present. Another degree of control is afforded by systems that include an exogenous chemical component acting upon the expressed transgenic transactivation factor.

Weaknesses. As demonstrated by Phuc et al. [19], Fu et al. [117], and Labbe et al. [18] many transgenic lines must be generated and evaluated to obtain a transgenic strain that possesses the correct blend of restricted expression in the presence of Tc and sufficient activation upon removal of Tc to achieve the desired phenotypic switch. Once established, there is also the economic cost associated with treating large volumes of larval rearing water with Tc/doxycycline when transgenic mosquitoes are reared at large scale.

Other Bipartite Expression Systems

Other bipartite systems such as the yeast GAL4−UAS and the Neurospora Q systems may prove to be extremely valuable as a means to restrict transgene expression in candidate strains for genetic control programs. As opposed to the Tet-On/Off system, neither of these systems depends upon an exogenous chemical to allow for the repression/suppression of the switch. The promoter-dependent expression of the Gal4 yeast transactivator forms the driver line component whereas the responder lines are generated to include a gene of interest under the transcriptional control of upstream activation sequences (UAS). When both components are present, Gal4 protein binds to the UAS permitting transcription of the candidate gene. Furthermore a natural suppressor of Gal4, Gal80, can provide an inhibitory effect allowing for more precise or complex investigations. This system has been successfully used in a number of different mosquito species including *An. gambiae* [157], *An. stephensi* [16], and *Ae. aegypti* [158,159]. The Q system involves genes identified from filamentous fungi *Neurospora crassa* whose *qa* gene cluster catabolizes quinic acid when glucose levels are limited. Essentially, quinic acid controls expression of the genes required for its own catabolism. The adapted system (similar in component nature to the Tet-On/Off system) utilizes the transcription factor, QF and its repressor, QS. In normal conditions, when glucose is abundant, QF is inhibited by QS thereby arresting expression of the *qa* gene cluster. However, when glucose is limited, the *qa* gene cluster expresses quinic acid that then binds QS in turn allowing QF to promote expression from the *qa* genes. Further, coupling together both the Gal4 and the Q system can attain even more precise control [160].

CONSIDERATIONS AND OUTLOOK

In the near term, first (transposon-based) and second (transposon + ΦC31) generation transgenic approaches will likely remain the dominant technologies used in genetic control strategies for both malaria and dengue. Indeed, all current and proposed genetically-modified mosquito trials involve these manipulations, and due to the amount of resources already invested, these will likely continue. However, with the advent of affordable site-specific gene editing of mosquito genomes, a range of manipulations not previously considered become manageable. For example, the entire need for promoter validation experiments as described above almost becomes unnecessary when a gene of interest can be recombined into the genome to be directly controlled by the native promoter. This can be done in a manner that is destructive to the original gene (replacement) or nondestructive (by recombining into an intron with a competing splice acceptor site). Site-specific recombination could also be used to insert or remove enhancer/repressor elements, thus changing the expression pattern of native mosquito genes in a beneficial manner without introducing any non-native DNA. Site-specific genome editing technologies provide new and powerful methods for stabilizing transgenes, through the generation of micro- or macroinversions [161], as well as new methods for performing gene drive [162,163]. Thus, while TE-based approaches will likely remain useful for the foreseeable future as a basic research tool, the precision offered by CRISPR-induced homologous recombination is unrivaled and may well-dominate the field of genetics-based control of diseases such as malaria and dengue in the coming years (Figure 8.6).

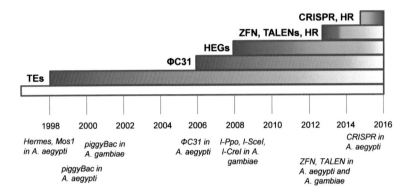

FIGURE 8.6 Timeline of genome modification and manipulation tools in mosquitoes. Since the initial reports of cut-and-paste transposition in mosquitoes in the late 1990s, an increasing number of genetic tools have been validated including the integrase ΦC31, homing endonucleases (HEGs), zinc-finger nucleases (ZFNs), transcription-activator-like nucleases (TALENs), homologous recombination (HR), and CRISPR/Cas based gene editing.

REFERENCES

[1] McClintock B. The origin and behavior of mutable loci in maize. Proc Natl Acad Sci USA 1950;36(6):344−55.

[2] Feschotte C, Pritham EJ. DNA transposons and the evolution of eukaryotic genomes. Annu Rev Genet 2007;41:331−68.

[3] Oliver KR, McComb JA, Greene WK. Transposable elements: powerful contributors to angiosperm evolution and diversity. Genome Biol Evol 2013;5(10):1886−901.

[4] Rubin GM, Spradling AC. Genetic transformation of *Drosophila* with transposable element vectors. Science 1982;218(4570):348−53.

[5] Miller LH, Sakai RK, Romans P, Gwadz RW, Kantoff P, Coon HG. Stable integration and expression of a bacterial gene in the mosquito *Anopheles gambiae*. Science 1987;237 (4816):779−81.

[6] Morris AC, Eggleston P, Crampton JM. Genetic transformation of the mosquito *Aedes aegypti* by micro-injection of DNA. Med Vet Entomol 1989;3(1):1−7.

[7] McGrane V, Carlson JO, Miller BR, Beaty BJ. Microinjection of DNA into *Aedes triseriatus* ova and detection of integration. Am J Trop Med Hyg 1988;39(5):502−10.

[8] Coates CJ, Jasinskiene N, Miyashiro L, James AA. Mariner transposition and transformation of the yellow fever mosquito, *Aedes aegypti*. Proc Natl Acad Sci USA 1998;95 (7):3748−51.

[9] Jasinskiene N, Coates CJ, Benedict MQ, et al. Stable transformation of the yellow fever mosquito, *Aedes aegypti*, with the *Hermes* element from the housefly. Proc Natl Acad Sci USA 1998;95(7):3743−7.

[10] Kokoza V, Ahmed A, Wimmer EA, Raikhel AS. Efficient transformation of the yellow fever mosquito *Aedes aegypti* using the *piggyBac* transposable element vector pBac [3 × P3-EGFP afm]. Insect Biochem Mol Biol 2001;31(12):1137−43.

[11] Catteruccia F, Nolan T, Loukeris TG, et al. Stable germline transformation of the malaria mosquito *Anopheles stephensi*. Nature 2000;405(6789):959−62.

[12] Grossman GL, Rafferty CS, Clayton JR, Stevens TK, Mukabayire O, Benedict MQ. Germline transformation of the malaria vector, *Anopheles gambiae*, with the *piggyBac* transposable element. Insect Mol Biol 2001;10(6):597−604.

[13] Atkinson PW, James AA. Germline transformants spreading out to many insect species. Adv Genet 2002;47:49−86.

[14] Handler AM, Atkinson PW. Insect transgenesis: mechanisms, applications, and ecological safety. Biotechnol Genet Eng Rev 2006;23:129−56.

[15] O'Brochta DA, Handler AM. Perspectives on the state of insect transgenics. Adv Exp Med Biol 2008;627:1−18.

[16] O'Brochta DA, Pilitt KL, Harrell II RA, Aluvihare C, Alford RT. Gal4-based enhancer-trapping in the malaria mosquito *Anopheles stephensi*. G3 2012;2(11):1305−15.

[17] Amenya DA, Bonizzoni M, Isaacs AT, et al. Comparative fitness assessment of *Anopheles stephensi* transgenic lines receptive to site-specific integration. Insect Mol Biol 2010;19(2):263−9.

[18] Labbe GM, Scaife S, Morgan SA, Curtis ZH, Alphey L. Female-specific flightless (fsRIDL) phenotype for control of *Aedes albopictus*. PLoS Negl Trop Dis 2012;6(7): e1724.

[19] Phuc HK, Andreasen MH, Burton RS, et al. Late-acting dominant lethal genetic systems and mosquito control. BMC Biol 2007;5:11.

[20] Franz AW, Jasinskiene N, Sanchez-Vargas I, et al. Comparison of transgene expression in *Aedes aegypti* generated by mariner *Mos*1 transposition and PhiC31 site-directed recombination. Insect Mol Biol 2011;20(5):587−98.

[21] Elgin SC, Reuter G. Position-effect variegation, heterochromatin formation, and gene silencing in *Drosophila*. Cold Spring Harb Perspect Biol 2013;5(8):a017780.

[22] Gullerova M, Proudfoot NJ. Convergent transcription induces transcriptional gene silencing in fission yeast and mammalian cells. Nat Struct Mol Biol 2012;19(11):1193−201.

[23] Irvin N, Hoddle MS, O'Brochta DA, Carey B, Atkinson PW. Assessing fitness costs for transgenic *Aedes aegypti* expressing the GFP marker and transposase genes. Proc Natl Acad Sci USA 2004;101(3):891−6.

[24] Catteruccia F, Godfray HC, Crisanti A. Impact of genetic manipulation on the fitness of *Anopheles stephensi* mosquitoes. Science 2003;299(5610):1225−7.

[25] O'Brochta DA, Alford RT, Pilitt KL, Aluvihare CU, Harrell II RA. *piggyBac* transposon remobilization and enhancer detection in *Anopheles* mosquitoes. Proc Natl Acad Sci USA 2011;108(39):16339−44.

[26] Palavesam A, Esnault C, O'Brochta DA. Post-integration silencing of *piggyBac* transposable elements in *Aedes aegypti*. PLoS One 2013;8(7):e68454.

[27] Smith RC, Atkinson PW. Mobility properties of the *Hermes* transposable element in transgenic lines of *Aedes aegypti*. Genetica 2011;139(1):7−22.

[28] Sethuraman N, Fraser Jr. MJ, Eggleston P, O'Brochta DA. Post-integration stability of *piggyBac* in *Aedes aegypti*. Insect Biochem Mol Biol 2007;37(9):941−51.

[29] Handler AM, Zimowska GJ, Horn C. Post-integration stabilization of a transposon vector by terminal sequence deletion in *Drosophila melanogaster*. Nat Biotechnol 2004;22 (9):1150−4.

[30] Dafa'alla TH, Condon GC, Condon KC, et al. Transposon-free insertions for insect genetic engineering. Nat Biotechnol 2006;24(7):820−1.

[31] Schetelig MF, Scolari F, Handler AM, Kittelmann S, Gasperi G, Wimmer EA. Site-specific recombination for the modification of transgenic strains of the Mediterranean fruit fly *Ceratitis capitata*. Proc Natl Acad Sci USA 2009;106(43):18171−6.

[32] Tkachuk A, Kim M, Kravchuk O, Savitsky M. A new powerful method for site-specific transgene stabilization based on chromosomal double-strand break repair. PLoS One 2011;6(10):e26422.

[33] Thorpe HM, Smith MC. *In vitro* site-specific integration of bacteriophage DNA catalyzed by a recombinase of the resolvase/invertase family. Proc Natl Acad Sci USA 1998;95 (10):5505–10.

[34] Groth AC, Olivares EC, Thyagarajan B, Calos MP. A phage integrase directs efficient site-specific integration in human cells. Proc Natl Acad Sci USA 2000;97(11):5995–6000.

[35] Groth AC, Fish M, Nusse R, Calos MP. Construction of transgenic *Drosophila* by using the site-specific integrase from phage phiC31. Genetics 2004;166(4):1775–82.

[36] Nimmo DD, Alphey L, Meredith JM, Eggleston P. High efficiency site-specific genetic engineering of the mosquito genome. Insect Mol Biol 2006;15(2):129–36.

[37] Labbe GM, Nimmo DD, Alphey L. piggybac- and PhiC31-mediated genetic transformation of the Asian tiger mosquito, *Aedes albopictus* (Skuse). PLoS Negl Trop Dis 2010;4 (8):e788.

[38] Meredith JM, Basu S, Nimmo DD, et al. Site-specific integration and expression of an anti-malarial gene in transgenic *Anopheles gambiae* significantly reduces *Plasmodium* infections. PLoS One 2011;6(1):e14587.

[39] Galizi R, Doyle LA, Menichelli M, et al. A synthetic sex ratio distortion system for the control of the human malaria mosquito. Nat Commun 2014;5:3977.

[40] Pondeville E, Puchot N, Meredith JM, et al. Efficient PhiC31 integrase-mediated site-specific germline transformation of *Anopheles gambiae*. Nat Protoc 2014;9(7):1698–712.

[41] Bateman JR, Lee AM, Wu CT. Site-specific transformation of *Drosophila* via phiC31 integrase-mediated cassette exchange. Genetics 2006;173(2):769–77.

[42] Torres R, Garcia A, Paya M, Ramirez JC. Non-integrative lentivirus drives high-frequency cre-mediated cassette exchange in human cells. PLoS One 2011;6(5):e19794.

[43] Osterwalder M, Galli A, Rosen B, Skarnes WC, Zeller R, Lopez-Rios J. Dual RMCE for efficient re-engineering of mouse mutant alleles. Nat Methods 2010;7(11):893–5.

[44] Oberstein A, Pare A, Kaplan L, Small S. Site-specific transgenesis by Cre-mediated recombination in *Drosophila*. Nat Methods 2005;2(8):583–5.

[45] Watson AT, Garcia V, Bone N, Carr AM, Armstrong J. Gene tagging and gene replacement using recombinase-mediated cassette exchange in *Schizosaccharomyces pombe*. Gene 2008;407(1–2):63–74.

[46] Louwerse JD, van Lier MC, van der Steen DM, de Vlaam CM, Hooykaas PJ, Vergunst AC. Stable recombinase-mediated cassette exchange in *Arabidopsis* using *Agrobacterium tumefaciens*. Plant Physiol 2007;145(4):1282–93.

[47] Jasinskiene N, Coates CJ, Ashikyan A, James AA. High efficiency, site-specific excision of a marker gene by the phage P1 cre-loxP system in the yellow fever mosquito, *Aedes aegypti*. Nucleic Acids Res 2003;31(22):e147.

[48] Lee G, Saito I. Role of nucleotide sequences of loxP spacer region in Cre-mediated recombination. Gene 1998;216(1):55–65.

[49] Siegel RW, Jain R, Bradbury A. Using an in vivo phagemid system to identify non-compatible loxP sequences. FEBS Lett 2001;505(3):467–73.

[50] Langer SJ, Ghafoori AP, Byrd M, Leinwand L. A genetic screen identifies novel non-compatible loxP sites. Nucleic Acids Res 2002;30(14):3067–77.

[51] Venken KJ, Bellen HJ. Genome-wide manipulations of *Drosophila melanogaster* with transposons, Flp recombinase, and PhiC31 integrase. Methods Mol Biol 2012;859: 203–28.

[52] Long DP, Zhao AC, Chen XJ, et al. FLP recombinase-mediated site-specific recombination in silkworm, *Bombyx mori*. PLoS One 2012;7(6).

[53] Morris AC, Schaub TL, James AA. FLP-mediated recombination in the vector mosquito, *Aedes aegypti*. Nucleic Acids Res 1991;19(21):5895−900.

[54] Stoddard BL. Homing endonucleases: from microbial genetic invaders to reagents for targeted DNA modification. Structure 2011;19(1):7−15.

[55] Stoddard BL. Homing endonucleases from mobile group I introns: discovery to genome engineering. Mob DNA 2014;5(1):7.

[56] Bellaiche Y, Mogila V, Perrimon N. I-SceI endonuclease, a new tool for studying DNA double-strand break repair mechanisms in *Drosophila*. Genetics 1999;152(3):1037−44.

[57] Rong YS, Golic KG. Gene targeting by homologous recombination in Drosophila. Science 2000;288(5473):2013−18.

[58] Rong YS, Golic KG. A targeted gene knockout in *Drosophila*. Genetics 2001;157 (3):1307−12.

[59] Rong YS, Golic KG. The homologous chromosome is an effective template for the repair of mitotic DNA double-strand breaks in *Drosophila*. Genetics 2003;165(4):1831−42.

[60] Burt A. Site-specific selfish genes as tools for the control and genetic engineering of natural populations. Proc R Soc Lond B Biol Sci 2003;270(1518):921−8.

[61] Windbichler N, Papathanos PA, Catteruccia F, Ranson H, Burt A, Crisanti A. Homing endonuclease mediated gene targeting in *Anopheles gambiae* cells and embryos. Nucleic Acids Res 2007;35(17):5922−33.

[62] Windbichler N, Papathanos PA, Crisanti A. Targeting the X chromosome during spermatogenesis induces Y chromosome transmission ratio distortion and early dominant embryo lethality in *Anopheles gambiae*. PLoS Genet 2008;4(12):e1000291.

[63] Traver BE, Anderson MA, Adelman ZN. Homing endonucleases catalyze double-stranded DNA breaks and somatic transgene excision in *Aedes aegypti*. Insect Mol Biol 2009;18 (5):623−33.

[64] Aryan A, Anderson MA, Myles KM, Adelman ZN. Germline excision of transgenes in *Aedes aegypti* by homing endonucleases. Sci Rep 2013;3:1603.

[65] Windbichler N, Menichelli M, Papathanos PA, et al. A synthetic homing endonuclease-based gene drive system in the human malaria mosquito. Nature 2011;473(7346):212−15.

[66] Klein TA, Windbichler N, Deredec A, Burt A, Benedict MQ. Infertility resulting from transgenic I-PpoI male *Anopheles gambiae* in large cage trials. Pathog Glob Health 2012;106(1):20−31.

[67] Ashworth J, Taylor GK, Havranek JJ, Quadri SA, Stoddard BL, Baker D. Computational reprogramming of homing endonuclease specificity at multiple adjacent base pairs. Nucleic Acids Res 2010;38(16):5601−8.

[68] Takeuchi R, Choi M, Stoddard BL. Redesign of extensive protein-DNA interfaces of meganucleases using iterative cycles of *in vitro* compartmentalization. Proc Natl Acad Sci USA 2014;111(11):4061−6.

[69] Taylor GK, Petrucci LH, Lambert AR, Baxter SK, Jarjour J, Stoddard BL. LAHEDES: the LAGLIDADG homing endonuclease database and engineering server. Nucleic Acids Res 2012;40(Web Server issue):W110−16.

[70] Takeuchi R, Lambert AR, Mak AN, et al. Tapping natural reservoirs of homing endonucleases for targeted gene modification. Proc Natl Acad Sci USA 2011;108(32):13077−82.

[71] Kim YG, Cha J, Chandrasegaran S. Hybrid restriction enzymes: zinc finger fusions to Fok I cleavage domain. Proc Natl Acad Sci USA 1996;93(3):1156−60.

[72] Porteus MH, Carroll D. Gene targeting using zinc finger nucleases. Nat Biotechnol 2005;23(8):967–73.

[73] Gaj T, Gersbach CA, Barbas III CF. ZFN, TALEN, and CRISPR/Cas-based methods for genome engineering. Trends Biotechnol 2013;31(7):397–405.

[74] Urnov FD, Miller JC, Lee YL, et al. Highly efficient endogenous human gene correction using designed zinc-finger nucleases. Nature 2005;435(7042):646–51.

[75] DeGennaro M, McBride CS, Seeholzer L, et al. *Orco* mutant mosquitoes lose strong preference for humans and are not repelled by volatile DEET. Nature 2013;498(7455):487–91.

[76] Liesch J, Bellani LL, Vosshall LB. Functional and genetic characterization of neuropeptide Y-like receptors in *Aedes aegypti*. PLoS Negl Trop Dis 2013;7(10):e2486.

[77] McMeniman CJ, Corfas RA, Matthews BJ, Ritchie SA, Vosshall LB. Multimodal integration of carbon dioxide and other sensory cues drives mosquito attraction to humans. Cell 2014;156(5):1060–71.

[78] Moscou MJ, Bogdanove AJ. A simple cipher governs DNA recognition by TAL effectors. Science 2009;326(5959):1501.

[79] Boch J, Scholze H, Schornack S, et al. Breaking the code of DNA binding specificity of TAL-type III effectors. Science 2009;326(5959):1509–12.

[80] Christian M, Cermak T, Doyle EL, et al. Targeting DNA double-strand breaks with TAL effector nucleases. Genetics 2010;186(2):757–61.

[81] Briggs AW, Rios X, Chari R, et al. Iterative capped assembly: rapid and scalable synthesis of repeat-module DNA such as TAL effectors from individual monomers. Nucleic Acids Res 2012;40(15):e117.

[82] Li L, Piatek MJ, Atef A, et al. Rapid and highly efficient construction of TALE-based transcriptional regulators and nucleases for genome modification. Plant Mol Biol 2012;78 (4–5):407–16.

[83] Reyon D, Khayter C, Regan MR, Joung JK, Sander JD. Engineering designer transcription activator-like effector nucleases (TALENs) by REAL or REAL-Fast assembly. Curr Protoc Mol Biol 2012; [chapter 12: unit 125].

[84] Reyon D, Tsai SQ, Khayter C, Foden JA, Sander JD, Joung JK. FLASH assembly of TALENs for high-throughput genome editing. Nat Biotechnol 2012;30(5):460–5.

[85] Schmid-Burgk JL, Schmidt T, Kaiser V, Honing K, Hornung V. A ligation-independent cloning technique for high-throughput assembly of transcription activator-like effector genes. Nat Biotechnol 2012;31(1):76–81.

[86] Aryan A, Anderson MA, Myles KM, Adelman ZN. TALEN-based gene disruption in the dengue vector *Aedes aegypti*. PLoS One 2013;8(3):e60082.

[87] Smidler AL, Terenzi O, Soichot J, Levashina EA, Marois E. Targeted mutagenesis in the malaria mosquito using TALE nucleases. PLoS One 2013;8(8):e74511.

[88] Sander JD, Reyon D, Maeder ML, et al. Predicting success of oligomerized pool engineering (OPEN) for zinc finger target site sequences. BMC Bioinformatics 2010;11:543.

[89] Neff KL, Argue DP, Ma AC, Lee HB, Clark KJ, Ekker SC. Mojo Hand, a TALEN design tool for genome editing applications. BMC Bioinformatics 2013;14:1.

[90] Miller JC, Holmes MC, Wang J, et al. An improved zinc-finger nuclease architecture for highly specific genome editing. Nat Biotechnol 2007;25(7):778–85.

[91] Szczepek M, Brondani V, Buchel J, Serrano L, Segal DJ, Cathomen T. Structure-based redesign of the dimerization interface reduces the toxicity of zinc-finger nucleases. Nat Biotechnol 2007;25(7):786–93.

[92] Sollu C, Pars K, Cornu TI, et al. Autonomous zinc-finger nuclease pairs for targeted chromosomal deletion. Nucleic Acids Res 2010;38(22):8269–76.

[93]   Doyon Y, Vo TD, Mendel MC, et al. Enhancing zinc-finger-nuclease activity with improved obligate heterodimeric architectures. Nat Methods 2011;8(1):74−9.

[94]   Basu S, Aryan A, Overcash JM, et al. Silencing of end-joining repair for efficient site-specific gene insertion after TALEN/CRISPR mutagenesis in *Aedes aegypti*. Proc Natl Acad Sci USA 2015;112(13):4038−43.

[95]   Simoni A, Siniscalchi C, Chan YS, et al. Development of synthetic selfish elements based on modular nucleases in *Drosophila melanogaster*. Nucleic Acids Res 2014;42 (11):7461−72.

[96]   Marraffini LA, Sontheimer EJ. CRISPR interference: RNA-directed adaptive immunity in bacteria and archaea. Nat Rev Genet 2010;11(3):181−90.

[97]   Jinek M, Chylinski K, Fonfara I, Hauer M, Doudna JA, Charpentier E. A programmable dual-RNA-guided DNA endonuclease in adaptive bacterial immunity. Science 2012;337 (6096):816−21.

[98]   Doudna JA, Charpentier E. Genome editing. The new frontier of genome engineering with CRISPR-Cas9. Science 2014;346(6213):1258096.

[99]   Kistler KE, Vosshall LB, Matthews BJ. Genome engineering with CRISPR-Cas9 in the mosquito *Aedes aegypti*. Cell Rep 2015;11(1):51−60.

[100] Dong S, Lin J, Held NL, Clem RJ, Passarelli AL, Franz AW. Heritable CRISPR/Cas9-mediated genome editing in the yellow fever mosquito, *Aedes aegypti*. PLoS One 2015;10(3):e0122353.

[101] Cong L, Ran FA, Cox D, et al. Multiplex genome engineering using CRISPR/Cas systems. Science 2013;339(6121):819−23.

[102] Mali P, Yang L, Esvelt KM, et al. RNA-guided human genome engineering via Cas9. Science 2013;339(6121):823−6.

[103] Wang T, Wei JJ, Sabatini DM, Lander ES. Genetic screens in human cells using the CRISPR-Cas9 system. Science 2014;343(6166):80−4.

[104] Ren X, Sun J, Housden BE, et al. Optimized gene editing technology for *Drosophila melanogaster* using germ line-specific Cas9. Proc Natl Acad Sci USA 2013;110 (47):19012−17.

[105] Wang H, Yang H, Shivalila CS, et al. One-step generation of mice carrying mutations in multiple genes by CRISPR/Cas-mediated genome engineering. Cell 2013;153 (4):910−18.

[106] Wu X, Kriz AJ, Sharp PA. Target specificity of the CRISPR-Cas9 system. Quant Biol 2014;2(2):59−70.

[107] Kim D, Bae S, Park J, et al. Digenome-seq: genome-wide profiling of CRISPR-Cas9 off-target effects in human cells. Nat Methods 2015;12(3):237−43.

[108] Tsai SQ, Zheng Z, Nguyen NT, et al. GUIDE-seq enables genome-wide profiling of off-target cleavage by CRISPR-Cas nucleases. Nat Biotechnol 2015;33(2):187−97.

[109] Ran FA, Cong L, Yan WX, et al. *In vivo* genome editing using *Staphylococcus aureus* Cas9. Nature 2015;520(7546):186−91.

[110] Bernardini F, Galizi R, Menichelli M, et al. Site-specific genetic engineering of the *Anopheles gambiae* Y chromosome. Proc Natl Acad Sci USA 2014;111(21):7600−5.

[111] Overcash JM, Aryan A, Myles KM, Adelman ZN. Understanding the DNA damage response in order to achieve desired gene editing outcomes in mosquitoes. Chromosome Res 2015;23(1):31−42.

[112] Beumer KJ, Trautman JK, Bozas A, et al. Efficient gene targeting in *Drosophila* by direct embryo injection with zinc-finger nucleases. Proc Natl Acad Sci USA 2008;105 (50):19821−6.

[113] Ma S, Chang J, Wang X, et al. CRISPR/Cas9 mediated multiplex genome editing and heritable mutagenesis of BmKu70 in *Bombyx mori*. Sci Rep 2014;4:4489.

[114] Cornel AJ, Benedict MQ, Rafferty CS, Howells AJ, Collins FH. Transient expression of the *Drosophila melanogaster* cinnabar gene rescues eye color in the white eye (WE) strain of *Aedes aegypti*. Insect Biochem Mol Biol 1997;27(12):993−7.

[115] Horn C, Wimmer EA. A versatile vector set for animal transgenesis. Dev Genes Evol 2000;210(12):630−7.

[116] Pinkerton AC, Michel K, O'Brochta DA, Atkinson PW. Green fluorescent protein as a genetic marker in transgenic *Aedes aegypti*. Insect Mol Biol 2000;9(1):1−10.

[117] Fu G, Lees RS, Nimmo D, et al. Female-specific flightless phenotype for mosquito control. Proc Natl Acad Sci USA 2010;107(10):4550−4.

[118] Anderson MA, Gross TL, Myles KM, Adelman ZN. Validation of novel promoter sequences derived from two endogenous ubiquitin genes in transgenic *Aedes aegypti*. Insect Mol Biol 2010;19(4):441−9.

[119] Kokoza VA, Martin D, Mienaltowski MJ, Ahmed A, Morton CM, Raikhel AS. Transcriptional regulation of the mosquito vitellogenin gene via a blood meal-triggered cascade. Gene 2001;274(1−2):47−65.

[120] Zhu J, Chen L, Raikhel AS. Distinct roles of Broad isoforms in regulation of the 20-hydroxyecdysone effector gene, Vitellogenin, in the mosquito *Aedes aegypti*. Mol Cell Endocrinol 2007;267(1−2):97−105.

[121] Sieglaff DH, Dunn WA, Xie XS, Megy K, Marinotti O, James AA. Comparative genomics allows the discovery of cis-regulatory elements in mosquitoes. Proc Natl Acad Sci USA 2009;106(9):3053−8.

[122] Biedler JK, Hu W, Tae H, Tu Z. Identification of early zygotic genes in the yellow fever mosquito *Aedes aegypti* and discovery of a motif involved in early zygotic genome activation. PLoS One 2012;7(3):e33933.

[123] Coates CJ, Jasinskiene N, Pott GB, James AA. Promoter-directed expression of recombinant fire-fly luciferase in the salivary glands of *Hermes*-transformed *Aedes aegypti*. Gene 1999;226(2):317−25.

[124] Kokoza V, Ahmed A, Cho WL, Jasinskiene N, James AA, Raikhel A. Engineering blood meal-activated systemic immunity in the yellow fever mosquito, *Aedes aegypti*. Proc Natl Acad Sci USA 2000;97(16):9144−9.

[125] Moreira LA, Edwards MJ, Adhami F, Jasinskiene N, James AA, Jacobs-Lorena M. Robust gut-specific gene expression in transgenic *Aedes aegypti* mosquitoes. Proc Natl Acad Sci USA 2000;97(20):10895−8.

[126] Kim W, Koo H, Richman AM, et al. Ectopic expression of a cecropin transgene in the human malaria vector mosquito *Anopheles gambiae* (Diptera: Culicidae): effects on susceptibility to *Plasmodium*. J Med Entomol 2004;41(3):447−55.

[127] Cho KH, Cheon HM, Kokoza V, Raikhel AS. Regulatory region of the vitellogenin receptor gene sufficient for high-level, germ line cell-specific ovarian expression in transgenic *Aedes aegypti* mosquitoes. Insect Biochem Mol Biol 2006;36(4):273−81.

[128] Smith RC, Walter MF, Hice RH, O'Brochta DA, Atkinson PW. Testis-specific expression of the beta2 tubulin promoter of *Aedes aegypti* and its application as a genetic sex-separation marker. Insect Mol Biol 2007;16(1):61−71.

[129] Adelman ZN, Jasinskiene N, Onal S, et al. *Nanos* gene control DNA mediates developmentally regulated transposition in the yellow fever mosquito *Aedes aegypti*. Proc Natl Acad Sci USA 2007;104(24):9970−5.

[130] Mathur G, Sanchez-Vargas I, Alvarez D, Olson KE, Marinotti O, James AA. Transgene-mediated suppression of dengue viruses in the salivary glands of the yellow fever mosquito, *Aedes aegypti*. Insect Mol Biol 2010;19(6):753−63.

[131] Carpenetti TL, Aryan A, Myles KM, Adelman ZN. Robust heat-inducible gene expression by two endogenous hsp70-derived promoters in transgenic *Aedes aegypti*. Insect Mol Biol 2012;21(1):97−106.

[132] Totten DC, Vuong M, Litvinova OV, et al. Targeting gene expression to the female larval fat body of transgenic *Aedes aegypti* mosquitoes. Insect Mol Biol 2013;22 (1):18−30.

[133] Akbari OS, Papathanos PA, Sandler JE, Kennedy K, Hay BA. Identification of germline transcriptional regulatory elements in *Aedes aegypti*. Sci Rep 2014;4:3954.

[134] Abraham EG, Donnelly-Doman M, Fujioka H, Ghosh A, Moreira L, Jacobs-Lorena M. Driving midgut-specific expression and secretion of a foreign protein in transgenic mosquitoes with AgAper1 regulatory elements. Insect Mol Biol 2005;14(3):271−9.

[135] Catteruccia F, Benton JP, Crisanti A. An *Anopheles* transgenic sexing strain for vector control. Nat Biotechnol 2005;23(11):1414−17.

[136] Nirmala X, Marinotti O, Sandoval JM, et al. Functional characterization of the promoter of the vitellogenin gene, AsVg1, of the malaria vector, *Anopheles stephensi*. Insect Biochem Mol Biol 2006;36(9):694−700.

[137] Yoshida S, Watanabe H. Robust salivary gland-specific transgene expression in *Anopheles stephensi* mosquito. Insect Mol Biol 2006;15(4):403−10.

[138] Chen XG, Marinotti O, Whitman L, Jasinskiene N, James AA, Romans P. The *Anopheles gambiae* vitellogenin gene (VGT2) promoter directs persistent accumulation of a reporter gene product in transgenic *Anopheles stephensi* following multiple blood-meals. Am J Trop Med Hyg 2007;76(6):1118−24.

[139] Lombardo F, Lycett GJ, Lanfrancotti A, Coluzzi M, Arca B. Analysis of apyrase 5′ upstream region validates improved *Anopheles gambiae* transformation technique. BMC Res Notes 2009;2:24.

[140] Papathanos PA, Windbichler N, Menichelli M, Burt A, Crisanti A. The *vasa* regulatory region mediates germline expression and maternal transmission of proteins in the malaria mosquito *Anopheles gambiae*: a versatile tool for genetic control strategies. BMC Mol Biol 2009;10:65.

[141] Nolan T, Petris E, Muller HM, Cronin A, Catteruccia F, Crisanti A. Analysis of two novel midgut-specific promoters driving transgene expression in *Anopheles stephensi* mosquitoes. PLoS One 2011;6(2):e16471.

[142] Lycett GJ, Amenya D, Lynd A. The *Anopheles gambiae* alpha-tubulin-1b promoter directs neuronal, testes and developing imaginal tissue specific expression and is a sensitive enhancer detector. Insect Mol Biol 2012;21(1):79−88.

[143] Kumar A, Sharma A, Sharma R, Gakhar SK. Identification, characterization and analysis of expression of gene encoding carboxypeptidase A in *Anopheles culicifacies* A (Diptera: culicidae). Acta Trop 2014;139:123−30.

[144] Allen ML, Christensen BM. Flight muscle-specific expression of act88F: GFP in transgenic *Culex quinquefasciatus* Say (Diptera: Culicidae). Parasitol Int 2004;53(4):307−14.

[145] Catteruccia F, Nolan T, Blass C, et al. Toward *Anopheles* transformation: Minos element activity in anopheline cells and embryos. Proc Natl Acad Sci USA 2000;97(5):2157−62.

[146] Meredith JM, Underhill A, McArthur CC, Eggleston P. Next-generation site-directed transgenesis in the malaria vector mosquito *Anopheles gambiae*: self-docking strains expressing germline-specific phiC31 integrase. PLoS One 2013;8(3):e59264.

[147] Nene V, Wortman JR, Lawson D, et al. Genome sequence of *Aedes aegypti*, a major arbovirus vector. Science 2007;316(5832):1718−23.

[148] Biedler JK, Qi Y, Pledger D, James AA, Tu Z. Maternal germline-specific genes in the Asian malaria mosquito *Anopheles stephensi*: characterization and application for disease control. G3 2014;5(2):157−66.

[149] Gossen M, Bujard H. Tight control of gene expression in mammalian cells by tetracycline-responsive promoters. Proc Natl Acad Sci USA 1992;89(12):5547−51.

[150] Gossen M, Bujard H. Studying gene function in eukaryotes by conditional gene inactivation. Annu Rev Genet 2002;36:153−73.

[151] Lycett GJ, Kafatos FC, Loukeris TG. Conditional expression in the malaria mosquito *Anopheles stephensi* with tet-on and tet-off systems. Genetics 2004;167(4):1781−90.

[152] Guidance framework for testing of genetically modified mosquitoes. WHO; 2014.

[153] Harris AF, Nimmo D, McKemey AR, et al. Field performance of engineered male mosquitoes. Nat Biotechnol 2011;29(11):1034−7.

[154] Harris AF, McKemey AR, Nimmo D, et al. Successful suppression of a field mosquito population by sustained release of engineered male mosquitoes. Nat Biotechnol 2012;30 (9):828−30.

[155] Wise de Valdez MR, Nimmo D, Betz J, et al. Genetic elimination of dengue vector mosquitoes. Proc Natl Acad Sci USA 2011;108(12):4772−5.

[156] Marinotti O, Jasinskiene N, Fazekas A, et al. Development of a population suppression strain of the human malaria vector mosquito, *Anopheles stephensi*. Malar J 2013;12:142.

[157] Lynd A, Lycett GJ. Development of the bi-partite Gal4-UAS system in the African malaria mosquito, *Anopheles gambiae*. PLoS One 2012;7(2):e31552.

[158] Kokoza VA, Raikhel AS. Targeted gene expression in the transgenic *Aedes aegypti* using the binary Gal4-UAS system. Insect Biochem Mol Biol 2011;41(8):637−44.

[159] Zhao B, Kokoza VA, Saha TT, Wang S, Roy S, Raikhel AS. Regulation of the gut-specific carboxypeptidase: a study using the binary Gal4/UAS system in the mosquito *Aedes aegypti*. Insect Biochem Mol Biol 2014;54:1−10.

[160] Potter CJ, Luo L. Using the Q system in *Drosophila melanogaster*. Nat Protoc 2011;6 (8):1105−20.

[161] Choi PS, Meyerson M. Targeted genomic rearrangements using CRISPR/Cas technology. Nat Commun 2014;5:3728.

[162] Gantz V, Bier E. The mutagenic chain reaction: A method for converting heterozygous to homozygous mutations. Science 2015;348(6233):442−4.

[163] Esvelt KM, Smidler AL, Catteruccia F, Church GM. Concerning RNA-guided gene drives for the alteration of wild populations. ELife 2014;e03401.

Chapter 9

Gene Drive Strategies for Population Replacement

John M. Marshall[1] and Omar S. Akbari[2]

[1]*Division of Biostatistics, School of Public Health, University of California, Berkeley, CA, USA,*
[2]*Department of Entomology, University of California, Riverside, CA, USA*

INTRODUCTION

After 3.8 billion years of research and development, Nature has provided inspiration for a plethora of human design problems. During the Renaissance, Leonardo da Vinci designed a flying machine inspired by the anatomy of birds. Today, Nature's evolutionary solutions are informing the design of solar panels from photosynthesis, and digital displays using the light-refracting properties of butterfly wings. Nature's intricate structures and processes may also help in the fight against mosquito-borne diseases. Gene drive—the process whereby natural mechanisms for spreading genes into populations are used to drive desirable genes into populations (e.g., genes conferring refractoriness to malaria or dengue fever in mosquitoes)—is another example of Nature's processes being applied for the benefit of humanity. Gene drive systems may either spread from low initial frequencies or display threshold properties such that they are likely to spread if released above a certain frequency in the population and are otherwise likely to be eliminated.

Population replacement, in this context, refers to the process whereby a population of disease-transmitting mosquitoes is replaced with a population of disease-refractory ones. Several approaches are being explored to engineer mosquitoes unable to transmit human diseases, and there have been a number of notable successes. For example, Isaacs et al. have engineered *Anopheles stephensi* mosquitoes expressing single-chain antibodies that prevent *Plasmodium falciparum* malaria parasites from developing in the mosquito, thus preventing onward transmission of the parasite [1]. Gene drive systems are expected to be instrumental in spreading disease-refractory genes into wild mosquito populations, given the wide geographical areas that these species inhabit and the expectation that refractory genes will be associated with

Genetic Control of Malaria and Dengue.
© 2016 Elsevier Inc. All rights reserved.

at least modest fitness costs [2]. Gene drive systems are also being considered to implement population suppression strategies whereby genes conferring a fitness load or gender bias are instead driven into the vector population, thereby reducing disease transmission.

Early Inspiration

Initial suggestions for spreading desirable genes into insect pest populations date back to the early 1940s and involved the proposition of translocations [3,4] and transposable elements (TEs) [5], inspired from natural systems. Translocations are rearrangements of parts between nonhomologous chromosomes. If insects homozygous for a translocation are introduced into a population at high frequency, they are predicted to spread to fixation [6], and if the translocation is linked to a disease-refractory gene, it is predicted to consequently be driven into the population as well. Initial field trials with translocations were unsuccessful in demonstrating spread [7]; but this is likely a result of those translocations being generated using X-rays, which often induce high fitness costs.

The suggestion of using TEs to drive disease-refractory genes into mosquito populations was largely inspired by the observation that a TE known as the *P* element spread through most of the global *Drosophila melanogaster* population within the span of a few decades following natural acquisition from *Drosophila willistoni* [8]. TEs are able to spread through a population due to mechanisms that enable them to increase their copy number within a host genome and hence to be inherited more frequently in subsequent generations. As a result, they are able to spread into a population from very low initial frequencies even if they incur a fitness cost to their host [9]. It was hoped that the *P* element invasion of *Drosophila* could be repeated in disease-transmitting mosquito species using a TE attached to a disease-refractory gene; however, early laboratory work on TEs in mosquito vector species has failed to identify elements with high remobilization rates following integration into mosquito lines [10].

Promising New Systems

Two of the most promising gene drive systems at present also involve technologies inspired by Nature—the use of homing endonuclease genes (HEGs) observed to spread in fungi, plants, and bacteria [11], and a selfish genetic element known as *Medea* observed to spread in *Tribolium* beetles [12,13]. A synthetic *Medea* element has been developed in *Drosophila* that works by the hypothesis that *Medea* encodes both a maternally expressed toxin and a zygotically expressed antidote [14]. This combination results in the death of wild-type offspring of *Medea*-bearing mothers, thus favoring the *Medea* allele in subsequent generations and mimicking the behavior of the natural element in *Tribolium. Medea* was the first synthetic gene drive system to be developed and has a number of desirable design features; however, significant work is still ongoing to develop a *Medea* element in a mosquito disease vector.

Recently, there has been much excitement around HEGs as, while *Medea* was first engineered in *Drosophila*, a naturally occurring HEG has been shown to spread in a laboratory population of *Anopheles gambiae*, the main African malaria vector, containing an engineered target sequence for the HEG [15]. HEGs spread by expressing an endonuclease that creates a double-stranded break at specific target sequences lacking the HEG. Homologous DNA repair then copies the HEG to the cut chromosome, increasing its representation in subsequent generations. Similar to the aforementioned gene drive systems, HEGs are being considered to drive disease-refractory genes into mosquito populations; however, a number of additional strategies for their application are also being considered, which aim to suppress rather than replace mosquito populations [11], and progress has been made toward these ends as well [16].

Design Criteria

As the technology for developing gene drive systems for population replacement develops on a number of fronts, it is useful to consider design criteria for assessing the safety and efficacy of the various approaches. An excellent review by Braig and Yan [17] proposes several biological properties that an ideal gene drive system should or must have:

1. The gene drive system must be effective. That is, it must be strong enough to compensate for any loss in host fitness due to the presence of both itself and its transgenic load (manifest as a reduction in host fertility, life span, or competitiveness). It must be able to spread to very high frequency in a population on a timescale relevant to disease control (i.e., a few years) and must be unimpeded by wild-type vectors immigrating into the target area.
2. The gene drive system must be able to carry with it several large genes and associated regulatory elements. At the very least, a disease-refractory and marker gene will be needed along with regulatory elements; but multiple disease-refractory genes are preferable in order to slow the rate at which the pathogen evolves resistance to each of them.
3. Features should be included to minimize the rate at which linkage is lost between the drive system and disease-refractory genes, as even rare recombination events could be significant for wide-scale spread over a long time period.
4. It should be possible to use the gene drive system to introduce waves of refractory genes over time to counteract the effects of evolution

of pathogen resistance, mutational inactivation of the refractory gene, or loss of linkage between the refractory gene and drive system.

5. The gene drive system should be easily adapted to multiple vector species. Human malaria, for instance, is transmitted by approximately 50 species of mosquitoes belonging to the genus *Anopheles*. In sub-Saharan Africa, the most important transmitters are *An. gambiae*, *Anopheles coluzzii*, *Anopheles arabiensis*, and *Anopheles funestus*, ideally all of which should be rendered refractory in a population replacement strategy.

Additional features of an ideal gene drive system were proposed by James to address ecological, epidemiological, and social issues, including safety [2]. Safety is a broad criterion that should be assessed through risk assessment in which potential hazards are identified along with their corresponding magnitudes and likelihoods. This provides a framework for managing the most significant risks and for the overall safety of the system to be scored. However, prior to a comprehensive risk assessment, a few general safety criteria for gene drive systems can be imagined.

6. The behavior of the gene drive system in the target species should be stable and predictable, thus minimizing the likelihood of unpredictable side effects in target species.

7. A mechanism should be available to prevent horizontal transfer of the gene drive system and/or refractory gene to nontarget species, thus minimizing the wider ecological impact of the release.

8. The gene drive system and refractory gene should not cause undesirable effects for human health, for instance, by selecting for increased virulence in the pathogen population. The gene drive system should also include a mechanism for removing the refractory gene from the population in the event of any adverse effect.

9. The gene drive system must be consistent with the social and regulatory requirements of the affected communities. For instance, public attitude surveys in Mali [18] highlight the importance of confined field trials prior to a wide-scale release, which could be achieved through the initial use of gene drive systems with high release thresholds followed by subsequent releases with more invasive systems.

10. The gene drive system should be cost-effective, as budgets for disease control are limited and a number of alternative interventions are available. The initial development of gene drive systems is expensive; but ongoing investment can be minimized by designing systems that are resilient to evolutionary degradation.

Cost-effectiveness is an important consideration, as it is not only relevant to the choice of gene drive system, but to whether gene drive should be used at all. In a recent modeling study, Okamoto et al. demonstrated the economic feasibility of releasing large numbers of insects carrying a dengue-refractory gene without a gene drive system in order to reduce the dengue transmission potential of *Aedes aegypti* mosquitoes in Iquitos, Peru [19]. Wide-scale control of *Anopheles* malaria vectors in sub-Saharan Africa is less likely amenable to the mass release strategy; however, it is essential to assess this in terms of efficacy, safety, and cost-effectiveness prior to implementation.

In this chapter, we review a range of gene drive systems being considered to drive disease-refractory genes into mosquito vector populations. We divide gene drive systems into two broad categories: (i) those that spread by causing a double-stranded break at a specific target sequence and insert themselves at this location through DNA repair (e.g., HEGs) and (ii) those that use combinations of toxins and antidotes, active at different life stages, to favor their own inheritance (e.g., *Medea*). We also review modern approaches to developing translocations as form of gene drive, which do not fit into either category. Systems using symbiotic or commensal microorganisms to mediate gene drive are covered in another chapter (e.g., *Wolbachia*). For each system, we review the biological mechanisms involved, the system's current stage of development, and its alignment with the abovementioned design criteria.

GENE DRIVE SYSTEMS THAT SPREAD VIA TARGET SITE CLEAVAGE AND REPAIR

We begin by reviewing gene drive systems that manipulate inheritance in their favor by causing a double-stranded break at one or more specific target sites in the host's genome and utilize the host's homologous DNA repair mechanism to increase their genomic copy number. Gene drive systems of this type include TEs, HEGs, and a number of recently proposed HEG analogs, such as zinc-finger nucleases (ZFNs), transcription-activator-like effector nucleases (TALENs), and clustered, regularly interspaced, short palindromic repeats (CRISPRs).

Transposable Elements

TEs are genomic components capable of changing their position and sometimes replicating within a genome. Consequently, they show widespread prevalence throughout the genomes of many taxa, with various families of TEs accounting for ∼90% of the Salamander genome, 50% of the *Ae. aegypti* genome, and 45% of the human genome. There are various classes of TEs, and those being considered for population replacement in mosquitoes belong to class 2. Class 2 elements contain both repeat sequences that mark their boundaries and their own transposase gene that catalyzes transposition. They move via a cut-and-paste mechanism [20], whereby transposition results in excision of the TE via two double-stranded breaks, leaving behind a gap where they have been excised. In some cases, this gap is filled by homologous gap repair from a chromatid also having the TE. The excised TE is then inserted at another genomic location, resulting in their genomic copy number being increased by one. In a second replication mechanism, some TEs transpose during the S phase of the cell cycle. If a recently replicated element transposes to an unreplicated region of the genome, it will be replicated a second time, resulting in a net gain of one element in the genome.

Current Status. The widespread distribution of TEs in Nature together with observations of the rapid spread of the *P* element in *Drosophila* [8] inspired initial hopes that class 2 TEs could be inserted, along with disease-refractory and marker genes, into transgenic lines of *Ae. aegypti* (the main vector of dengue fever) and *Anopheles* vectors of malaria. Class 2 TEs lacking their transposase gene are often used as vectors for introducing novel genes into mosquitoes; hence, integration into mosquito lines is relatively straightforward. More problematic, however, has been the remobilization of TEs containing their own transposase gene once they have been integrated. An excellent review by O'Brochta et al. describes results from experiments in which four class 2 TEs—*Hermes*, *Mos1*, *Minos*, and *piggyback*—were used to create transgenic lines of *Ae. aegypti* [10]. In all cases, remobilization was shown to be highly inefficient. More recently, attempts were made to improve the post-integration mobility of *Hermes* in *Ae. aegypti* using an additional construct to express a transposase gene under the control of a testis-specific promoter [21]; however, remobilization was still only observed in less than 1% of the transgenic lines.

Design Criteria. The observed remobilization of natural TEs suggests that remobilization of introduced elements should also be possible; however, the regulation of TE mobility is complex, and it may require much experimentation to find TEs compatible with mosquito vectors. This work is likely not cost-effective, as TEs fail to satisfy most of the design criteria outlined earlier, and have been superseded by more recently proposed systems like HEGs and *Medea*. Of particular note, it is unlikely that TEs will be able to carry large inserts containing disease-refractory genes as transposition events are known to be imprecise and prone to DNA loss. Furthermore, a study on the *Himar1 mariner* element suggests that transposition rates decline substantially with increasing insert size [22], suggesting that elements which have lost their transgenic load will outspread those which have not [23]. Finally, the large numbers of target sites that TEs have undermine their predictability and stability in target species, and their wide species host range highlights the risk of horizontal gene transfer and spread in nontarget species.

Homing Endonuclease Genes

HEGs are highly efficient selfish genetic elements that spread by expressing an endonuclease that recognizes and cleaves a highly specific target sequence of 14−40 base pairs usually only present at a single site in the host genome [24]. As the HEG is positioned directly opposite its target site, actually within its own recognition sequence, it induces a double-stranded break only in chromosomes lacking the HEG. The HEG is effectively copied to the target site, in a process referred to as "homing," when the cell's repair machinery uses the HEG-bearing chromosome as a template for homology-directed repair. When homing occurs in the germ line of the host organism, a HEG can be transmitted to progeny at a higher than Mendelian inheritance ratios, enabling its spread through a population (Figure 9.1A).

On the basis of observations of homing activity in a number of nonmetazoan organisms including yeast, fungi, algae, and plants, Burt proposed that HEGs could be used as a gene drive system for population replacement in mosquito disease vectors; however, he also proposed and favored their use as a population suppression system [11]. Burt proposed a suite of HEG-based

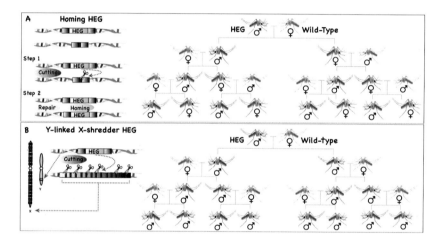

FIGURE 9.1   Preferential inheritance of homing-based gene drive systems. (A) Left panel: A homing HEG (green) encodes an endonuclease that recognizes and cleaves a specific target sequence (red) on a wild-type chromosome (step 1). Once the target site is cleaved, the cell repairs the chromosomal break through homologous recombination using the HEG-bearing chromosome as a template (step 2). This two-step process results in the HEG effectively being copied to the wild-type chromosome in a process referred to as "homing," thereby converting a HEG heterozygote into a HEG homozygote. Right panel: When a HEG-bearing male (green mosquito) is released into the wild and mates with a wild-type female (gray mosquito), the majority of their progeny inherit the HEG, and over time the HEG invades entire populations. (B) Left panel: For HEG-based population suppression, an X-shredder HEG is positioned on the Y chromosome (Y-linked X-shredder HEG). This HEG encodes an endonuclease that recognizes and cleaves chromosomal sequences that are repeated exclusively on the X-chromosome of the mosquito. When expressed during spermatogenesis, X-bearing spermatids are disrupted by the HEG, resulting in the majority of functional sperm being Y-bearing and containing the HEG. Right panel: When a Y-linked X-shredder HEG-bearing male (green mosquito) is released into the wild and mates with a wild-type female (gray mosquito), all resulting progenies are HEG-bearing males. Over time, this is predicted to induce an all-male population crash and potentially eventual extinction of the vector species.

strategies for genetic control of mosquito vectors—two involving population replacement and three involving population suppression:

1. First, the HEG could be linked to a disease-refractory gene and engineered to target a gene-sparse region of a chromosome (so as to reduce impacts on mosquito host fitness), thus carrying the disease-refractory gene with it as it spreads into the population.

2. In a related population replacement approach, the HEG could be engineered to target an endogenous gene involved in the development or transmission of the pathogen, thus reducing vector competence as it spreads [25]. This approach has the benefit that it does not involve an effector gene and hence is more resilient to evolutionary degradation; however, it does require a gene to be identified, the disruption of which would block pathogen transmission, and for a HEG to be engineered to target this, which is quite arduous.

3. In terms of population suppression, a HEG could be engineered that targets a native mosquito gene required in at least one copy for either mosquito survival or fertility. If a HEG of this type is active in the mosquito germ line, then it will increase in frequency in the population, inducing a genetic fitness load on the population as it spreads. This could lead to either population suppression or an eventual population crash.

4. An alternative to the homing-based applications of HEGs is to rely entirely on their target site cleavage activity. In the first of these approaches, known as the "autosomal X-shredder" strategy, a HEG can be designed to specifically cleave the X chromosome at multiple locations, effectively destroying it. If an X-shredder HEG is expressed during male meiosis, it will result in destruction of X-bearing male sperm. If females mate with males having the X-shredder, most viable sperm will be Y-bearing and hence most of the progeny will be male. This strategy will reduce the reproductive potential of the population; but it requires regular releases since the X-shredding gene is associated with a fitness cost and will only persist in the population for a few generations.

5. Finally, Burt proposed a "Y-linked X-shredder" strategy whereby, if the X-shredder HEG is located on the Y chromosome, then it will be driven into the population along with the transgenic Y chromosome as it induces an increasingly male gender bias. This approach would mimic naturally existing meiotic drive systems that bias sex ratios, although it could potentially induce a much larger gender bias than those observed in Nature [26−28], causing a cascade of male-only population crashes that could potentially lead to species extinction (Figure 9.1B).

Current Status. An encouraging result for homing-based HEG strategies has been the engineering of a naturally occurring HEG, I-*Sce*I, which has been shown to cleave in *Ae. aegypti* [29] and spread in laboratory populations of both *D. melanogaster* and *An. gambiae* containing an engineered target sequence for the HEG [15,30,31]. These results are encouraging because they show that, although HEGs have not been discovered in any metazoan species to date, there is nothing intrinsic about metazoan biology that prevents HEGs from homing. Furthermore, the fact that this was achieved in *An. gambiae*, the most important African malaria vector, is hopeful for its application to disease control. For the population replacement strategy to work in the wild, a HEG must be engineered or identified which has a target sequence in the wild mosquito genome. Engineering HEGs to recognize and cleave new target sequences has proven difficult thus far [32−34], and future research should focus on the development of novel approaches to circumvent these difficulties.

Population suppression strategies that rely solely on the target site cleave activity of HEGs have shown remarkable progress in recent years. A HEG originally discovered in the slime mold *Physarum polycephalum*, I-*Ppo*1 [35], was integrated into the *An. gambiae* genome and shown to recognize and cleave a conserved DNA sequence, repeated hundreds of times and located exclusively on the X chromosome cluster of ribosomal DNA genes in *An. gambiae* [36]. This cleavage activity is highly applicable to both the autosomal and Y-linked X-shredder strategies of HEG-driven population suppression and has also provided a novel genetic approach to the sterile insect technique for *An. gambiae*. The expression of I-*Ppo*1 during spermatogenesis in *An. gambiae* resulted in cleavage of the paternal X chromosome in differentiating spermatozoa, which was expected to result in a male bias among progeny. However, it turned out that the I-*Ppo*1 from mature sperm cells was carried over into the zygote, thus shredding the zygotic X chromosomes as well and rendering the transgenic males completely sterile [37]. It was later shown that transgenic mosquitoes engineered with *I-Ppo*1 could induce high levels on sterility in large cage populations, confirming the suitability of this technology for use in sterile insect population suppression programs [38]. This could be a useful first application of HEG technology in the wild given the self-limiting nature of sterile insect releases.

For X-shredder strategies to work, I-*Ppo*1 would need to be destabilized in order to minimize its carryover into the zygote by mature sperm. To this end, recent work by Galizi et al. has succeeded in expressing destabilized autosomal versions of I-*Ppo*1, which result in efficient shredding of the paternal X chromosome and are restricted to male meiosis [16]. Consequently, males carrying this construct are fully fertile and some insertions produce >95% male offspring bias. Males inheriting the autosomal I-*Ppo*1 gene also produce a male bias in their progeny, showing that the gender-biasing effect of autosomal X-shredders will remain in the population for several generations; however, continued releases would be required, as the X-shredder gene is not favored through inheritance when located on an autosome and is expected to be eliminated due to fitness costs. Nevertheless, for repeated releases, population suppression is expected, which would be more efficient than the previously mentioned sterile male releases and would also be self-limiting, albeit over a longer period. Autosomal X-shredders could therefore be an appropriate second application of HEG technology.

The only remaining steps in order to realize the Y-linked X-shredder strategy are to dock the destabilized I-*Ppo*1 HEG onto the *An. gambiae* Y chromosome and ensure that it is expressed during spermatogenesis. To this end, recent progress has been made in developing a Y chromosome docking line in *An. gambiae* [39]. Future work will focus on docking the HEG onto the Y chromosome and ensuring it can be expressed and function as anticipated.

Design Criteria. HEG-based strategies for genetic control of vector-borne diseases are extremely promising given the remarkable progress made recently, most notably in the malaria vector *An. gambiae*. HEGs are highly effective as a gene drive system, capable of spreading for low initial frequencies to high frequency on a short timescale. They are also relatively short sequences targeting very precise regions of the genome, suggesting both stability and a low rate of corruption due to evolutionary degradation. Species-specific regulatory sequences can be included to limit their horizontal transfer to nontarget species, and furthermore, a strategy has been proposed to reverse the spread of a deleterious HEG through the release of HEG-resistant alleles in the event of unforeseen consequences [11]. Additionally, a wide range of HEG strategies are available displaying different levels of confineability, allowing them to be used at all stages of a phased release and to be tailored to the social and regulatory requirements of affected communities.

Target site cleavage strategies show more promise than those reliant on homing activity as they sidestep many of the abovementioned design criteria and are independent of disease-refractory genes. Target site mutagenesis and gap repair through nonhomologous end joining can both result in disruption of the HEG cleavage site, rendering certain individuals immune to the HEG and preventing the HEG from spreading through an entire population. For strategies in which a HEG disrupts a gene required for mosquito survival or fertility, HEG-resistant mutants will be favored in a population once they emerge. Furthermore, there is a possibility of losing the disease-refractory gene either through mutagenesis or during homology-directed repair—a concern that becomes more serious for larger inserts, and would render a population replacement strategy futile. The Y-linked X-shredder strategy is less vulnerable to target site mutagenesis as it targets so many loci on the X chromosome at once. It is, however, dependent on germ line gene expression on the *An. gambiae* Y chromosome although this could potentially be achieved through the use of insulator sequences.

TALENs and ZFNs

TALENs and ZFNs have been proposed as alternative platforms for engineering homing-based gene drive systems [40]—that is, systems that spread by cleaving a specific target sequence and then using the cell's repair machinery to copy themselves to the target site. The benefit of TALENs and ZFNs over HEGs is that they can be easily engineered to target desired DNA sequences due to the modular nature of their DNA-binding domains. TALENs are derived from naturally occurring proteins that are secreted by the pathogenic bacteria *Xanthomonas* spp. to alter gene expression in host plant cells [41,42]. These proteins contain arrays of highly conserved, repetitive DNA-binding domains, each recognizing only a single base pair, with specificity being determined by repeat-variable di-residues [43,44]. The relationship between these repeats and DNA recognition can be exploited to design TALENs that target virtually any desired DNA sequence. For ZFNs, DNA-binding specificity can be similarly manipulated, being determined by an array of finger modules that can be generated either by selection using large combinatorial libraries, or by rational design [45].

For both TALENs and ZFNs, DNA-binding modules can be combined with several types of domains, including transcriptional activators, nucleases, and recombinases, allowing for a comprehensive range of genetic modifications [46]. In terms of cleavage activity, a wide range of tailored recognition sequences can be cleaved efficiently as TALENs and ZFNs are fusion proteins consisting of a nonspecific fok1 nuclease linked to a DNA-binding motif [47,48]. The TALEN or ZFN may then be copied to the cleaved target side by homology-directed repair, and hence used as a gene drive system for driving disease-refractory genes into mosquito populations.

Current Status and Design Criteria. Both TALENs and ZFNs rely upon homing activity and thus, for the purposes of population replacement and control, are functionally similar to HEGs. Given this similarity, the range of replacement and suppression strategies outlined earlier is also applicable to these systems and many of the design issues are similar too. For example, TALENs and ZFNs are also expected to spread from low initial frequencies, species-specificity can be incorporated through the addition of regulatory elements, and a deleterious TALEN or ZFN can be removed from a population through the release of TALEN- or ZFN-resistant alleles. However, there are some important differences. In terms of cost-efficiency, both TALENs and ZFNs are easier to engineer to target specific DNA sequences, and consequently, they could be straightforwardly adapted to multiple vector species, which is particularly important for malaria control. However, concerns arise regarding their stability, as their repetitive nature makes them more prone to mutation and evolutionary degradation. Recent progress toward developing both TALEN- and ZFN-based gene drive systems in *D. melanogaster* have successfully demonstrated DNA-binding specificity, cleavage, and homing through homology-directed DNA repair; however, mutational inactivation led to a decline in effectiveness over just a short period of time [40]. Thus, if TALENs or ZFNs are to be useful as gene drive systems in the future, their stability issues must first be overcome.

Clustered, Regularly Interspaced, Short Palindromic Repeats

CRISPR is another promising system proposed, although not yet demonstrated, as an alternative platform for homing-based gene drive. The system is based on an adaptive immune process in bacteria whereby sequences derived from invading bacteriophages or plasmids are integrated into the bacterial CRISPR locus. This essentially provides bacterial cells with the ability to "remember" and protect themselves against previously encountered viral genomes and invasive, mobile genetic elements [49]. To perform nuclease activities, CRISPR systems use an array of CRISPR RNAs (crRNAs) derived from exogenous DNA targets (e.g., viral genomes), noncoding transactivating RNAs, and a cluster of CRISPR-associated (Cas) genes. Three types of CRISPR systems have been discovered, with type II CRISPR systems being best characterized. These consist of a Cas9 nuclease and a crRNA array encoding guide RNAs and auxiliary transactivating crRNAs to mediate target site cleavage [50]. As for the homing-based systems described earlier, if the double-stranded break is repaired by homology-directed repair, the CRISPR system may be copied to the cleaved target site and hence used as a gene drive system for population replacement similar to HEGs. If the target site cleavage activity is directed toward the X chromosome, then the population suppression strategies initially described for HEGs could also be realized.

Current Status. Recent encouragement for CRISPR-based gene drive has been provided by proof-of-principle studies showing that the type II CRISPR system from *Streptococcus pyogenes* can be modified to target endogenous genes in bacteria [51] and human cell lines [52,53]. It has subsequently been shown that CRISPR can be used to alter genes in a range of other species including insects such as *D. melanogaster* [54,55] and mosquitoes. Straightforwardly, utilizing this system in other organisms requires only two components—the Cas9 nuclease and guide RNAs [52,56]. DNA-binding specificity is determined by the first 20 nucleotides of the guide RNA as these designate the DNA target side that Cas9 will be guided to according to Watson−Crick DNA−RNA base pairing rules. The only restriction for the target site selection is that it must lie directly upstream of a protospacer adjacent motif sequence that matches the canonical form 5′-NGG. Aside from that, it is possible that the CRISPR system can be engineered to target and cleave essentially any genomic location, with subsequent homing and gene drive occurring via homology-directed repair, however this remains to be demonstrated.

Design Criteria. CRISPR-based gene drive has yet to be implemented; however, its mechanisms imply that the approach is achievable. In terms of design criteria, the system is very similar to TALENs and ZFNs—it is expected to spread from low initial frequencies, species-specificity can be incorporated through regulatory elements, and a deleterious CRISPR can be removed through release of CRISPR-resistant alleles. The system is active in a range of species and target sites are even easier to engineer than for TALENs, suggesting the system would be easily adapted to multiple vector species. Another advantage of the CRISPR system is that it can be used to target multiple sequences in a single experiment [57], increasing its potential efficacy and decreasing the rate at which target site mutagenesis could slow its spread. A major concern, however, is that the CRISPR system itself may be degraded. The CRISPR system is quite large, consisting of promoters, the Cas9 gene, guide RNAs and, depending on the strategy being implemented, multiple disease-refractory genes and associated regulatory elements. A system this size is prone to mutation and errors introduced during homing, including potential loss of function of disease-refractory genes. These considerations may lead to population suppression strategies being favored for CRISPR-based drive systems; however, this would place selection pressure on mutant CRISPR alleles having lost their function and so the evolutionary stability of the CRISPR system will need to be explored and optimized if it is to provide a cost-effective alternative to the relatively stable yet difficult-to-engineer X-shredding HEGs.

TOXIN−ANTIDOTE GENE DRIVE SYSTEMS

We now move on to gene drive systems that use combinations of toxins and antidotes, active at different life stages, to favor their own inheritance [58]. Gene drive systems of this type include *Medea*, engineered forms of underdominance such as UD^{MEL}, self-limiting systems such as killer-rescue, and other toxin−antidote possibilities such as *Semele*, *Medusa*, and inverse *Medea*.

Medea

The story of *Medea* has origins in both Greek mythology and beetle biology. In Greek mythology, Medea was the wife of the hero Jason, to whom she had two children. Her marriage to Jason was hard-earned, transpiring only after she enabled him to plough a field with fire-breathing oxen, among other achievements; but despite this, he left her when the king of Corinth offered him his daughter. As a form of revenge, Medea killed their two children. From a biological perspective, such infanticide would make Medea an unfit mother; but if the trait is genetic and children that inherit it also have the ability to defend themselves, then mathematical models show that it actually has a selective advantage and, if present at modest levels in a population, is expected to become present among all individuals within a matter of generations [59,60]. This is simply because children who are able to defend themselves against a murderous parent are more fit than those who cannot.

The Greek analogy sounds bizarre; but genes displaying these properties do actually exist in Nature and have been discovered and characterized in various regions of the world [12,61,62]. The first such element to be identified was in the flour beetle *Tribolium castaneum* [12] and was given the name *Medea* after both the character from Greek mythology, and as an acronym for "maternal-effect dominant embryonic arrest." By crossing individuals from geographically isolated locations, it was found that *Medea*-bearing males gave rise to both wild-type and *Medea*-bearing offspring; but that *Medea*-bearing females only gave rise to *Medea*-bearing offspring. It appeared that *Medea*-bearing mothers were selectively killing non-*Medea*-bearing offspring; or alternatively that they were trying to kill all offspring and the *Medea*-bearing offspring were able to defend themselves.

The genetic factors involved in this behavior remain obscure; but the dynamics suggest a model in which *Medea* consists of two tightly linked genes—a maternally expressed toxin gene, the product of which causes all eggs to become unviable and a zygotically expressed antidote gene, the product of which rescues *Medea*-bearing eggs from the effects of the toxin [12,63]. In *Tribolium*, *Medea* dynamics are attributed to an insertion of a composite Tc1 transposon inserted between two genes both having maternal and zygotic components [13]. Remarkably, this system was reverse-engineered using entirely synthetic components in laboratory populations of *D. melanogaster* and was shown to rapidly drive population replacement [14,64]. These synthetic elements were constructed using two unique, tightly linked components—a maternal toxin consisting of maternally deposited microRNA designed to target an essential embryonic gene; and a zygotic antidote consisting of a tightly linked, zygotically expressed, microRNA-resistant version of the embryonic essential gene. The combination of these components results in the death of wild-type offspring of *Medea*-bearing mothers, thus favoring the *Medea* allele in subsequent generations and mimicking the behavior of the natural element in *Tribolium* (Figure 9.2A).

Current Status. *Medea* was the first synthetic gene drive system to be developed, in this case in *D. melanogaster* [14]. Given that the synthetic *Medea* elements were constructed using rationally designed synthetic components and well-understood, conserved molecular and genetic mechanisms, it should be possible to engineer *Medea* elements in a range of other insects including mosquitoes. The *Medea* drive strategy is particularly well-suited to driving disease-refractory genes into mosquito populations, and hence the development of several efficient refractory genes for each disease of interest is encouraged.

Design Criteria. In many ways, *Medea* is the ideal system for replacement of wild mosquito populations with disease-refractory varieties. Solutions are available for all of the design criteria outlined earlier, and *Medea* has an advantage over homing-based strategies for population replacement since it is stably integrated into the host chromosome, thus not affected by the substantial risk of loss during homology-directed repair. If introduced at modest population frequencies, *Medea* can spread and rapidly

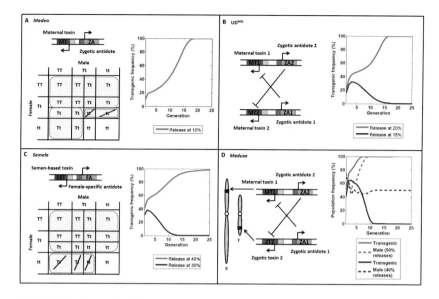

FIGURE 9.2 Dynamics of toxin–antidote-based gene drive systems. (A) *Medea* elements distort the offspring ratio in their favor through the action of a maternally expressed toxin (MT) and a zygotically expressed antidote (ZA). This results in the death of wild-type offspring of heterozygous mothers and enables the *Medea* element to spread into a population from very low initial frequencies. Dynamics here are shown for a *Medea* element with no fitness cost, released at 10% in the population. Transgenic frequency refers to any individual carrying at least one copy of the element. (B) UD^{MEL} (maternal-effect lethal underdominance) is a toxin–antidote-based underdominant system consisting of two constructs, each of which possesses a maternally expressed toxin (MT1 and MT2) whose activity is manifest during progeny embryogenesis and a zygotic antidote (ZA1 and ZA2) capable of neutralizing the maternal toxin expressed by the opposite construct. This results in heterozygous females being sterile if mated to wild-type individuals, thus leading to the characteristic bistable dynamics of underdominant systems. Dynamics here are shown for UD^{MEL} constructs at independently assorting loci having no fitness costs. If released at a population frequency of 20%, the system spreads to fixation in the population; but if released at 15%, the system is eliminated. (C) *Semele* elements distort the offspring ratio in their favor through the action of a semen-based toxin (SBT) and a female-specific antidote (FA). This results in unviable crosses between transgenic males and wild-type females and favors transgenic individuals provided the *Semele* element is present at population frequencies exceeding ∼36% (above this frequency, the selective advantage of the antidote exceeds the selective disadvantage of the toxin). Dynamics here are shown for a *Semele* element with no fitness cost. If released at a population frequency of 40%, the element spreads to fixation in the population; but if released at 30%, the system is eliminated. (D) *Medusa* is a two-construct, sex chromosome-linked drive system capable of inducing confineable and reversible population suppression. The system consists of four components—a maternally expressed, X-linked toxin (MT1) causes suppression of the female population and selects for the transgene-bearing Y since only transgenic male offspring have the corresponding Y-linked zygotically expressed antidote (ZA1). A zygotically expressed, Y-linked toxin (ZT2) and a zygotically expressed, X-linked antidote (ZA2) then selects for the transgene-bearing X when the transgene-bearing Y is present, creating a balanced lethal system. When present above a certain threshold frequency, *Medusa* spreads while creating a strong male gender bias leading to population suppression. Dynamics here are shown for *Medusa* constructs having no fitness costs. For two consecutive male-only releases at a population frequency of 50%, the population becomes entirely male as the system spreads to fixation in the population; but for two consecutive male-only releases at a population frequency of 40%, the system is eliminated.

replace a population, even in the presence of modest fitness costs [60]; how-ever, *Medea* is unlikely to spread following a small-scale accidental release because its driving ability is low at low population frequencies [18].

Tight linkage between the toxin, antidote, and refractory genes by placing the toxin and refractory genes within an intron of the antidote gene can improve system stability and reduce the rate of loss of the refractory gene through recombination. However, in the event that the *Medea* element or refractory gene become unlinked, mutated, or rendered ineffective through parasite evolution, second-generation *Medea* elements can be generated that utilize toxin−antidote combinations distinct from those of the first-generation elements [14], making it possible to carry out multiple cycles of population replacement. This strategy can also be used to remove refractory genes from populations in the event of adverse effects. As the functional components of *Medea* are developed in mosquito species, it will become more cost-efficient to develop these elements and to adapt them to multiple vector species.

Toxin−Antidote-Based Underdominance

Underdominant systems display the property that heterozygotes, or their progeny, have lower fitness than either homozygote [65]. In the simplest case of a single biallelic locus for which matings between opposite homozygotes are sterile, whichever allele is more frequent in the population will tend to spread to fixation. Underdominant systems therefore display features similar to that of a bistable switch at the population level—if the system is present above a critical threshold frequency, it will tend to spread to fixation, while if it is present below the threshold, it will tend to be eliminated in favor of the alternative allele or chromosome. A variety of toxin−antidote systems have been proposed to achieve these underdominant dynamics and the critical threshold frequency depends on the system and fitness cost.

A range of underdominant systems is available in Nature, including chromosomal alternations such as inversions, translocations, and compound chromosomes [3,4]. We will return to translocations in the Translocation section; but will concentrate here on novel forms of underdominance that are in principle straightforward to engineer using combinations of toxins and antidotes. Toxin−antidote approaches to underdominance were originally proposed by Davis et al., who suggested an elegant system having two transgenic constructs, each of which possesses a gene whose expression induces lethality and a gene that suppresses the expression or activity of the gene inducing lethality carried by the other construct [66]. The constructs can either be inserted at the same locus on a pair of homologous chromosomes or at different loci on nonhomologous chromosomes. These systems display underdominant properties because individuals carrying neither or both constructs are viable; but a proportion of their offspring—those carrying just one of the constructs—are unviable. The critical threshold for the two-locus system is $\sim 27\%$, above which it is predicted to spread to fixation, and for the single-locus system is $\sim 67\%$ [66].

Current Status. Attempts to engineer the underdominance system proposed by Davis et al. have thus far been unsuccessful [66]; however, a related novel underdominant system known as maternal-effect lethal underdominance (UD^{MEL}) has recently been engineered in *D. melanogaster* and demonstrated to replace wild-type laboratory populations in a threshold-dependent manner [67,68]. In the UD^{MEL} system, there are two transgenic constructs, each of which possesses a maternally expressed toxin gene whose activity is manifest during progeny embryogenesis and a zygotic antidote gene capable of neutralizing the maternal toxin expressed by the opposite construct. From the crosses produced by this system (Figure S1 of Akbari et al. [67]), it can be seen that heterozygous females are sterile if mated to wild-type individuals, while populations of transgenic homozygotes are fully viable, as are wild-type populations. This leads to the characteristic bistable dynamics of underdominant systems. As per the system proposed by Davis et al., the UD^{MEL} constructs can be inserted at the same locus or on a pair of homologous chromosomes [66]. The critical threshold for the two-locus system is $\sim 19\%$ and for the single-locus system is $\sim 64\%$, assuming no fitness costs [67], and threshold-dependent drive has been demonstrated in the laboratory for both cases (Figure 9.2B).

Design Criteria. Toxin–antidote-based underdominant systems such as UD^{MEL} are an excellent option during the testing phase of population replacement, or whenever a confined release is desired. The threshold nature of these systems has three advantages in these scenarios. First, they are unlikely to spread following an accidental released because escapees will inevitably be present at subthreshold levels and be eliminated from the environment [18]. Second, they are expected to be confineable to isolated release sites because transgenic insects released at superthreshold frequencies are expected to spread transgenes locally while they remain at subthreshold levels at nearby locations. And third, releases are reversible as transgenes can be eliminated by diluting them to subthreshold frequencies through a sustained release of wild-type insects.

It should be noted that the confineability of these systems, although likely, is not guaranteed.

In theory, chance events could lead to underdominant systems gaining a foothold and spreading in structured populations, presumably beginning from a single individual; however, this is more likely to occur on an evolutionary timescale than on a human timescale. Underdominant systems may be better-suited to *An. gambiae* because it disperses quickly over the range of a single village [69,70], reducing the chance of its spread being confined to smaller subpopulations. The small-scale population structure of *Ae. aegypti*, however, may prevent its village-wide spread in natural populations of these vectors.

Otherwise, similarly to *Medea*, solutions are available for all of the design criteria outlined earlier. As the functional components are developed and identified in mosquitoes—microRNAs, maternal and early-zygote-specific promoters and essential genes—these systems will be highly useful for confined population replacement of vector species such as *An. gambiae*.

Killer-Rescue

Killer-rescue is an intriguingly simple two-locus gene drive system proposed by Gould et al. for both its ease of engineering and its ability to spread into a population in a time-limited way [71]. Both these qualities are desirable in the early stages of a population replacement program. The system consists of two alleles at unlinked loci—one that encodes a toxin (a killer allele) and another that confers immunity to the toxin (a rescue allele), which could be tightly linked to a gene for disease refractoriness. A release of individuals homozygous for both alleles results in temporary drive as the alleles segregate and the presence of the killer allele in the population confers a benefit to those also carrying the rescue allele. In an alternative configuration, a second killer allele can be included at an independently assorting locus to enhance the selective benefit of the rescue allele. However, regardless of the conformation, the killer allele soon declines in frequency due to its inherent fitness cost and, as it does, the selective benefit of the rescue allele is lost. As this happens, if the rescue allele or disease-refractory gene confers a fitness cost to the host, then it will gradually be eliminated from the population as well over a timeframe determined by the magnitude of its fitness cost—a higher fitness cost leading to it being eliminated more quickly.

Design Criteria. As mentioned earlier, the killer-rescue system is intriguing for its ability to spread in a time-limited manner, thus reducing risks, as appropriate during field trials of transgenic mosquitoes carrying disease-refractory genes. The system is also spatially limited, as it only has a window of time in which to disperse to neighboring populations, and will spread to much lower levels in these populations than at the population of release [72]. Similar to underdominant systems, it will not persist following an accidental release, and its elimination from a population can be accelerated through large-scale releases of wild-type insects. Also, similar to other toxin−antidote systems, solutions are available for all of the design criteria outlined earlier.

Some consideration should go into the fitness cost of the rescue allele and refractory gene, as a high fitness cost will lead to rapid elimination, but the maximum frequency of the disease-refractory allele in the population will be compromised; while small fitness costs will allow the system to spread to very high maximum frequencies, but it may take several years for the system to be eliminated from the population entirely. Further complicating this, fitness costs are exceedingly difficult to quantify in the field. The bistable nature of underdominant systems therefore makes them more controllable in terms of confinement and reversibility; however, the major benefit of the killer-rescue system is its ease of engineering. Molecular tools are already available to engineer the system in a variety of mosquito species, allowing the system to be implemented with relative ease in a range of disease vectors.

Other Confineable Toxin–Antidote Systems

As *Medea*, killer-rescue, and the various forms of engineered underdominance highlight, there are many ways in which toxins and antidotes can be used to favor the inheritance of one allele over another. For example, even if we limit ourselves to single-locus systems like *Medea*, either the toxin or antidote gene could be placed under the control of a paternal, maternal, or zygote-specific promoter, function through a recessive or dominant mechanism, and be located on a sex chromosome or autosome [73]. The possibilities multiply if we also consider multilocus systems. A few additional toxin–antidote systems displaying unique population dynamics are *Semele*, inverse *Medea*, and *Medusa*, all of which are also confineable to partially isolated populations.

Semele. Semele is a single-locus system consisting of a toxin gene expressed in the semen of transgenic males that either kills or renders infertile wild-type females and an antidote gene expressed in females that protects them against the effects of the toxin [74]. The name is an acronym for "semen-based lethality" and, like *Medea*, also has Greek origins. In Greek mythology, Semele was a mortal female who attracted the attention of Zeus while slaughtering a bull at his altar (Zeus, at this point, was flying overhead disguised as an eagle). Zeus became infatuated with Semele and impregnated her, but Semele died after witnessing his godliness because she was not herself a god. The story parallels the biology of the *Semele* construct, in which wild-type females die (or become infertile) upon mating with transgenic males.

Semele has several interesting population dynamic properties. If only males carrying the *Semele* allele are released into a wild population, they are expected to suppress the population size when released in large numbers. This happens because all of the wild females that mate with the *Semele* males are susceptible to their toxic semen. If both males and females carrying the *Semele* allele are released, the system displays bistable dynamics with a threshold frequency of $\sim 36\%$ in the absence of fitness costs [74]. Above the threshold, the selective advantage of the female antidote outweighs the reproductive disadvantage conferred by the toxic semen and the system spreads into the population. In combination, this means that an initial release of *Semele* males could be used to suppress a population preceding a superthreshold release of males and females, thus reducing the release size required to exceed the critical population frequency (Figure 9.2C).

Inverse Medea. Inverse *Medea* is another single-locus system capable of achieving confined population replacement [75]. The system consists of a zygotic toxin and maternal antidote—essentially the same components as the *Medea* system with the promoters switched. This has the effect of rendering heterozygous offspring of wild-type mothers unviable and leads to bistable dynamics in which the system spreads when it represents a majority of the alleles in a population, and is otherwise eliminated. While similar dynamic properties are displayed by other underdominant toxin–antidote systems, the benefit of inverse *Medea* is its ease of engineering once the components to generate *Medea* elements in mosquito vectors have been identified. Several approaches to engineering these elements are available—for example, the toxin could be a microRNA that silences expression of a gene whose activity is required for early embryo development, and the antidote could be a maternally expressed RNA that restores the necessary activity to the zygote and is resistant to silencing.

Medusa. *Medusa* is a two-construct, sex chromosome-linked drive system capable of inducing confineable and reversible population suppression [76]. The system consists of four components—two at a locus on the X chromosome and two at a locus on the Y chromosome. The combination of a maternally expressed, X-linked toxin and a zygotically expressed, Y-linked antidote causes suppression of the female population and selects for the transgene-bearing Y since only transgenic male offspring of *Medusa*-bearing females are protected from the effects of the toxin. At the same time, the combination of a zygotically expressed, Y-linked toxin and a zygotically expressed, X-linked antidote selects for the transgene-bearing X when the transgene-bearing Y is present. Together, this creates a balanced lethal system that, when present above a certain threshold frequency, spreads while creating a strong male gender bias, hence causing population suppression (Figure 9.2D). Characteristic of all drive systems with thresholds, releases of *Medusa* mosquitoes are confineable and reversible, making the system an ideal tool for confined population suppression. This could be particularly useful in the lead-up to releases of invasive population suppression systems such as Y-linked X-shredder HEGs [76].

The name *Medusa* is an acronym for "sex chromosome-associated *Medea* underdominance," as its components are identical to those of *Medea* and engineered underdominance. The name also has origins in Greek mythology, where *Medusa* is a beautiful yet terrifying woman who caused onlookers to be turned to stone (toxin) but was ultimately beheaded by Perseus who distracted himself with Athena's mirrored shield (antidote). Simple population dynamic models show that an all-male release of *Medusa* males, carried out over six generations, is expected to induce a population crash within 12 generations for modest release sizes [76]. Reinvasion of wild-type insects can result in a population rebound; however, this can be prevented through regular releases of modest numbers of *Medusa* males.

Design Criteria. The vast range of possible toxin—antidote combinations highlights the versatility of this approach to engineering gene drive systems. *Semele* is an excellent option for confined population replacement due to its ability to suppress a vector population prior to replacement, inverse *Medea* is an excellent underdominant system that is easy to engineer once the components of the *Medea* system have been identified in mosquito vectors, and *Medusa* is an ideal system for confined population suppression in preparation for invasive X-shredder strategies [76]. Other toxin—antidote systems are imaginable and may be favored depending on the components first identified in molecular work on vector species [73]. As toxin—antidote systems, the design criteria outlined earlier are generally satisfied, and as largely confineable systems, the systems highlighted here are excellent options during the testing phase of population replacement, or whenever a confined release is desired.

TRANSLOCATIONS

As the first gene drive system to be proposed [4], translocations have since undergone a lull in interest following the observation that radiation-generated translocations failed to spread in the field, likely due to high fitness costs induced by X-rays [7]. However, recent developments in molecular biology permit the creation of translocations without relying upon radiation suggesting that, after several decades of inactivity, the application of this gene drive system could be revisited. Translocations result from the mutual exchange of chromosomal segments between nonhomologous chromosomes. Translocation heterozygotes are usually partially sterile, while translocation homozygotes are usually fully fertile. This effect is manifest during meiosis when nearly half of the gametes from a translocation heterozygote have a duplication of one chromosomal segment and a deficiency of another. The haploid gametes are functional, but when they fuse with native gametes following fertilization, the resulting zygotes are inviable. This produces the bistable dynamics described for other underdominant systems.

Current Status. Curtis proposed that if a translocation strain was developed that had a disease-refractory gene tightly linked to the translocation break point, disease-resistance would spread into that population as the translocation fixes [4]. To test this hypothesis, mosquito strains with chromosomal translocations were developed using X-ray mutagenesis; however, the low fitness associated with these strains and the difficulty of bringing disease-refractory genotypes into appropriate genetic backgrounds inhibited these approaches from further development. It is now possible to generate translocations at almost any genomic location without irradiation as a result of progress in genome sequencing and synthetic biology [77,78]. This will reduce the fitness costs associated with translocations and will allow disease-refractory genes to be more easily linked to translocation break points, making them a feasible, future gene drive system for confined population replacement.

Design Criteria. As an underdominant system displaying bistable dynamics, translocations provide another option for confined population replacement. As modern molecular techniques are yet to be applied to the development of this system, its agreement with several of the design criteria mentioned earlier are yet to be determined, and its attractiveness as a local gene drive system will depend on its ease of engineering and satisfaction of these criteria in comparison to toxin−antidote-based underdominant systems. Toxin−antidote-based systems may be preferable for phased releases as their components are more similar to invasive *Medea* elements that could be used for subsequent wide-scale population replacement. That said there is a theoretical expectation that translocations are an effective gene drive system for local population replacement [4] and that the loss of disease-refractory genes will be minimized by inserting them at translocation break points. As a bistable system, translocations could be eliminated from a population through mass release of wild-type insects and would satisfy social and regulatory requirements when confinement is desired.

CONCLUSION

In 2006, Sinkins and Gould published an excellent review of gene drive systems for insect disease vectors which today provides a testament to how quickly the field has progressed in less than a decade [79]. As the authors state, "ultimately, the drive system that becomes most widely used might be one that is entirely novel and not described here." Interestingly, the majority of the drive systems described in this chapter—TALENs, CRISPRs, UD^{MEL}, killer-rescue, *Semele*, inverse *Medea*, and *Medusa*—were not mentioned in the Sinkins and Gould review as they are have only been recently published.

Of the systems that were mentioned by Sinkins and Gould, progress has been rapid. In mentioning *Medea*, for instance, the authors stated that "a molecular understanding of its function could lead to the development of artificial *Medea*-like constructs"—something that was achieved the following year [14] and is now one of the most promising approaches for population replacement. Regarding HEGs, the authors stated that, "Unfortunately, HEGs have only been reported in fungi, plants, bacteria and bacteriophages ... the potential for developing an HEG-based functional system in insects is unknown." The following year, a HEG isolated from a species of slime mold demonstrated cleavage activity in *An. gambiae* [36], and a few years later, another naturally occurring HEG was shown to spread in laboratory populations of both *D. melanogaster* and *An. gambiae* [15,30,31]. HEGs are now one of the most promising gene drive systems for inducing population suppression (Table 9.1). Research on using *Wolbachia* to control vector-borne diseases has been even more rapid, with large-scale field trials already having taken place in several countries including Australia and Vietnam [80]. This prompts the question of what the gene drive field will look like a decade from now?

Gene Drive for Any Situation

Sinkins and Gould also pointed out that "the various types of drive mechanisms should not be viewed as competing systems," adding that, "Different characteristics will be needed in different situations." Gene drive systems can lead to a number of outcomes in terms of population dynamics, and the optimal system in each case will depend upon the desired outcome. For driving disease-refractory genes into mosquito populations over a wide area, *Medea* seems to be a very promising system, as it is capable of spreading from low initial frequencies and is also stably integrated into the host chromosome. When population replacement is desired over a wide geographic area, stability in the face of evolutionary degradation is an important consideration, and *Medea* may be preferable to homing-based strategies incorporating disease-refractory genes because these are susceptible to DNA loss during homology-directed repair, which is expected to become increasingly significant over large spatial and temporal scales.

Systems with release thresholds are preferable when a confined release is desired because these systems are likely to be confineable to their population of release and to be reversible through releases of wild-type insects [72]. Toxin−antidote-based underdominant systems would be an obvious choice if the goal were to test the concept of population replacement prior to a release of toxin−antidote-based *Medea* elements. The bistable nature of these systems makes them particularly amenable to confinement; however, killer-rescue systems and a mass release of transgenic insects with disease-refractory genes [19,71] should also be considered, as these are significantly easier to engineer in a wide range of vector species and the spreading a disease-refractory gene into an isolated population will not always require gene drive.

For population suppression, Y-linked X-shredder HEGs are an ideal system, assuming the X-shredding HEG can be docked onto the Y chromosome and expressed during spermatogenesis. The major benefits of the X-shredder HEG are the generally small size of HEGs, making them less susceptible to evolutionary degradation, and the large number of loci cleaved on the X chromosome, making the strategy less susceptible to target site mutagenesis [11]. Autosomal X-shredders, as a self-limiting population suppression system acting through the same molecular

TABLE 9.1 Alignment of Potential Gene Drive Systems with Design Criteria Outlined in the Introduction[a]

Design Criteria	Target Site Cleavage-Based Gene Drive Systems				Toxin–Antidote-Based Gene Drive Systems		Engineered Translocations
	Engineered TEs	Engineered HEGs	ZFNs, TALENs	CRISPRs	Medea, UD^{MEL}	Killer-rescue, *Semele*, *Medusa*	
Effectiveness of spread	Maybe [10] (not yet effective in vector species)	Yes [11] (very effective, first drive system shown to spread in a malaria vector)	Probably [40] (homing demonstrated, currently compromised by mutational inactivation)	Probably [51–57] (components identified, can target multiple sequences at once)	Probably [14,64,67] (observed to spread in laboratory *Drosophila* populations)	Theoretically [71,74,76] (not yet engineered, models predict spread)	Theoretically [4,77,78] (models predict spread, not yet engineered using modern components)
Ability to carry large effector genes	No [22] (transposition rate declines with increasing insert size)	Possibly [11] (could be lost during homology-directed repair)	Possibly [40] (could be lost during homology-directed repair)	Possibly (could be lost during homology-directed repair)	Yes [14,67] (stably integrated into host chromosome)	Yes [71,74,76] (stably integrated into host chromosome)	Yes [4] (stably integrated into host chromosome)
Tight linkage with effector genes	No [23] (transposition events prone to DNA loss)	No [11] (homology-directed repair prone to DNA loss)	No [40] (homology-directed repair prone to DNA loss)	No (homology-directed repair prone to DNA loss)	Yes [14,67] (place toxin and effector genes within intron of antidote gene)	Yes [71,74,76] (place effector gene within intron of antidote gene)	Yes [4] (very tight if effector gene linked to a translocation break point)
Waves of introductions	Maybe [10] (difficult to engineer multiple TEs)	Maybe [11] (difficult to engineer)	Yes [40] (easier to engineer than HEGs)	Yes (easier to engineer than HEGs)	Yes [14,67] (use distinct toxin–antidote combinations)	Yes [71,74,76] (use distinct toxin–antidote combinations)	Yes [4] (use threshold properties)

Easily adapted to other species	No [10] (difficult to find TEs compatible with vector species)	No (difficult to engineer target site)	Maybe [40] (once components identified in species)	Maybe (once components identified in species)	Maybe [14,67] (once components identified in species)	Maybe [71,74,76] (once components identified in species)	Maybe [77,78] (once components identified in species)
Stability in target species	No [8] (large number of target sites undermine predictability)	Yes [11] (short sequences targeting precise genomic regions)	Moderate to low [40] (prone to mutation due to repetitive nature)	Moderate to low (more prone to mutation due to repetitive nature and large size)	Yes [14,67] (stably integrated into host chromosome)	Yes [71,74,76] (stably integrated into host chromosomes)	Yes [77,78] (very stable)
Minimal horizontal gene transfer	No [8] (wide species host range)	Yes [11] (include species-specific regulatory sequences)	Yes [11,40] (include species-specific regulatory sequences)	Yes [11] (include species-specific regulatory sequences)	Yes [14,67] (include species-specific regulatory sequences)	Yes [71,74,76] (include species-specific regulatory sequences)	Yes
Mechanism for removal	No	Yes [11] (design second HEG to target first HEG)	Yes [11,40] (design second ZFN or TALEN to target first ZFN or TALEN)	Yes [11] (design second CRISPR to target first CRISPR)	Yes [14,67] (use threshold properties or second-generation element to remove refractory gene)	Yes [71,74,76] (use threshold properties or dilution with wild-types)	Yes [4] (use threshold properties)
Social and regulatory requirements	No [79] (not confineable or reversible)	Yes [11] (wide range of strategies with different levels of confineability)	Yes [11,40] (wide range of strategies with different levels of confineability)	Yes [11] (wide range of strategies with different levels of confineability)	Yes [18,72] (confineable or unlikely to spread following small accidental release)	Yes [72] (confineable to partially isolated populations)	Yes [72] (confineable to partially isolated populations)

[a] In many cases, data supporting satisfaction of design criteria are preliminary. TEs, transposable elements; HEGs, homing endonuclease genes; ZFNs, zinc-finger nucleases; TALENs, transcription-activator-like effector nucleases; CRISPRs, clustered, regularly interspaced, short palindromic repeats; UD^{MEL}, maternal-effect lethal underdominance.

mechanism, are an obvious choice for testing this drive system prior to a wide-scale release. Similar approaches using ZFNs, TALENs, and CRISPRs should also be considered, especially considering their relative ease of engineering. However, the repetitive nature of ZFNs and TALENs and the large size of CRISPRs generally will make them more susceptible to mutation and evolutionary degradation (Figure 9.3).

Outstanding Issues and Future Outlook

In 1899, US patent officer Charles Duell famously stated that, "Everything that can be invented already has been invented." It would be just as foolish to say that all imaginable gene drive systems have already been imagined. The coming decades are bound to witness the emergence of a plethora of novel mechanisms for spreading desirable genes into insect populations, and it will be fascinating to see how these systems align with the design criteria mentioned earlier. Furthermore, of the systems for which development has already begun, it will be fascinating to see how their laboratory and field studies progress. Progress on toxin−antidote-based systems will be greatly facilitated by the development of their functional components—toxins, antidotes and regulatory elements—in mosquito vectors. It will also be interesting to see how modern approaches to translocations perform against toxin−antidote-based approaches to underdominance. Regarding homing-based systems, critical developments will be the engineering of HEGs for other vector species, the insertion and expression of X-shredders on the Y chromosome, and determining the resilience of alternative homing-based systems to evolutionary degradation.

As a technology capable of engineering or eliminating entire species, the development of gene drive systems carries with it both great promise and great responsibility. Issues are heightened by the ability of invasive systems to spread into neighboring communities and countries without their consent [81]. Comprehensive risk assessments that address ecological, epidemiological, and social issues are therefore essential, and such technology should only be used in the absence of significant risks. On the flip side, gene drive technology has the potential to make a profound impact on relieving the global vector-borne disease burden [2]. Considering malaria as an example, traditional interventions such as bed nets and antimalarial drugs require human compliance, which never truly exceeds $\sim 80\%$ coverage, meaning that there is always a residual human population capable of sustaining transmission [82]. Replacement of disease-transmitting mosquitoes with disease-refractory ones has the unique benefit that it does not require human compliance, and can spread into areas where interventions are difficult to apply. This makes it one of the most promising components of future integrated strategies for the elimination of vector-borne diseases.

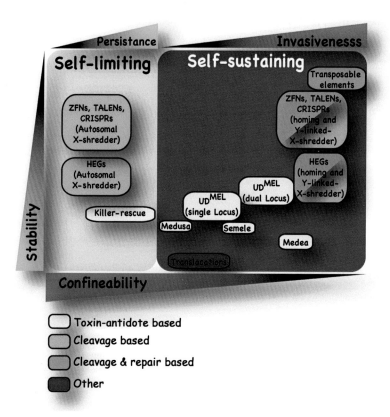

FIGURE 9.3 Confineability and stability of potential gene drive systems. The potential gene drive systems described in this chapter differ in multiple ways, including their confineability (the ability to limit their spatial spread following a release) and their stability (resilience against evolutionary degradation, predictable behavior in the host organism and infrequent spread into nontarget species). Here, we depict the potential gene drive systems in a two-dimensional graph according to these properties. Self-limiting systems eliminate themselves from a population as a result of their own dynamics and hence are highly confineable, although some persist in a population longer than others. Self-sustaining systems are capable of maintaining a high population frequency but are relatively confineable if they display threshold properties in terms of release frequency. Self-sustaining systems not displaying threshold dynamics can be highly invasive. Toxin–antidote-based systems (yellow) are relatively stable but have differing levels of confineability. Cleavage-based population replacement systems (purple) are relatively invasive whether they carry disease-refractory genes or induce a population fitness load. The process of homing also causes them to be relatively unstable due to errors introduced during gap repair. Cleavage-based population suppression systems (salmon) can be either invasive if located on the Y chromosome or self-limiting if located on an autosome. ZFNs, zinc-finger nucleases; TALENS, transcription-activator-like effector nucleases; CRISPRs, clustered, regularly interspaced, short palindromic repeats; HEGs, homing endonuclease genes; UD^{MEL}, maternal-effect lethal underdominance.

ACKNOWLEDGMENTS

JMM acknowledges support from a fellowship from the Medical Research Council/ Department for International Development, UK.

REFERENCES

[1] Isaacs AT, Jasinskiene N, Tretiakov M, et al. Transgenic *Anopheles stephensi* coexpressing single-chain antibodies resist *Plasmodium falciparum* development. Proc Natl Acad Sci USA 2012;109(28):E1922−30.

[2] James AA. Gene drive systems in mosquitoes: rules of the road. Trends Parasitol 2005;21 (2):64−7.

[3] Serebrovskii AS. On the possibility of a new method for the control of insect pests. Zool Zh 1940;19:618−30.

[4] Curtis CF. Possible use of translocations to fix desirable genes in insect pest populations. Nature 1968;218(5139):368−9.

[5] Kidwell MG, Ribeiro JM. Can transposable elements be used to drive disease refractoriness genes into vector populations? Parasitol Today 1992;8(10):325−9.

[6] Curtis CF, Robinson AS. Computer simulation of the use of double translocations for pest control. Genetics 1971;69(1):97−113.

[7] Lorimer N, Hallinan E, Rai KS. Translocation homozygotes in the yellow fever mosquito, *Aedes aegypti*. J Hered 1972;63(4):158−66.

[8] Preston CR, Engels WR. Spread of P transposable elements in inbred lines of *Drosophila melanogaster*. Prog Nucleic Acid Res Mol Biol 1989;36:71−85.

[9] Charlesworth B, Sniegowski P, Stephan W. The evolutionary dynamics of repetitive DNA in eukaryotes. Nature 1994;371(6494):215−20.

[10] O'Brochta DA, Sethuraman N, Wilson R, et al. Gene vector and transposable element behavior in mosquitoes. J Exp Biol 2003;206(Pt. 21):3823−34.

[11] Burt A. Site-specific selfish genes as tools for the control and genetic engineering of natural populations. Proc Biol Sci 2003;270(1518):921−8.

[12] Beeman RW, Friesen KS, Denell RE. Maternal-effect selfish genes in flour beetles. Science 1992;256(5053):89−92.

[13] Lorenzen MD, Gnirke A, Margolis J, et al. The maternal-effect, selfish genetic element *Medea* is associated with a composite Tc1 transposon. Proc Natl Acad Sci USA 2008;105 (29):10085−9.

[14] Chen CH, Huang H, Ward CM, et al. A synthetic maternal-effect selfish genetic element drives population replacement in *Drosophila*. Science 2007;316(5824):597−600.

[15] Windbichler N, Menichelli M, Papathanos PA, et al. A synthetic homing endonuclease-based gene drive system in the human malaria mosquito. Nature 2011;473(7346): 212−15.

[16] Galizi R, Doyle LA, Menichelli M, et al. A synthetic sex ratio distortion system for the control of the human malaria mosquito. Nat Commun 2014;5:3977.

[17] Braig HR, Yan G. The spread of genetic constructs in natural insect populations. In: Letourneau DK, Burrows BE, editors. Genetically engineered organisms: assessing environmental and human health effects. Boca Raton: CRC Press; 2001. p. 251−314.

[18] Marshall JM. The effect of gene drive on containment of transgenic mosquitoes. J Theor Biol 2009;258(2):250−65.

[19] Okamoto KW, Robert MA, Gould F, Lloyd AL. Feasible introgression of an anti-pathogen transgene into an urban mosquito population without using gene-drive. PLoS Negl Trop Dis 2014;8(7):e2827.

[20] Plasterk RHA, van-Luenen HGAM. The Tc1/Mariner family of transposable elements. In: Craig NL, Craigie R, Geller M, Lambowitz AM, editors. Mobile DNA II. Washington, DC: American Society for Microbiology; 2002. p. 519−32.

[21] Smith RC, Atkinson PW. Mobility properties of the Hermes transposable element in transgenic lines of *Aedes aegypti*. Genetica 2011;139(1):7−22.

[22] Lampe DJ, Grant TE, Robertson HM. Factors affecting transposition of the Himar1 mariner transposon *in vitro*. Genetics 1998;149(1):179−87.

[23] Marshall JM. The impact of dissociation on transposon-mediated disease control strategies. Genetics 2008;178(3):1673−82.

[24] Stoddard BL. Homing endonuclease structure and function. Q Rev Biophys 2005;38 (1):49−95.

[25] Deredec A, Burt A, Godfray HC. The population genetics of using homing endonuclease genes in vector and pest management. Genetics 2008;179(4):2013−26.

[26] Craig Jr. GB, Hickey WA, Vandehey RC. An inherited male-producing factor in *Aedes aegypti*. Science 1960;132(3443):1887−9.

[27] Hamilton WD. Extraordinary sex ratios. A sex-ratio theory for sex linkage and inbreeding has new implications in cytogenetics and entomology. Science 1967;156(3774):477−88.

[28] Wood RJ, Newton ME. Sex-ratio distortion caused by meiotic drive in mosquitos. Am Nat 1991;137(3):379−91.

[29] Traver BE, Anderson MA, Adelman ZN. Homing endonucleases catalyze double-stranded DNA breaks and somatic transgene excision in *Aedes aegypti*. Insect Mol Biol 2009;18 (5):623−33.

[30] Chan YS, Huen DS, Glauert R, Whiteway E, Russell S. Optimising homing endonuclease gene drive performance in a semi-refractory species: the *Drosophila melanogaster* experience. PLoS One 2013;8(1):e54130.

[31] Chan YS, Naujoks DA, Huen DS, Russell S. Insect population control by homing endonuclease-based gene drive: an evaluation in *Drosophila melanogaster*. Genetics 2011;188(1):33−44.

[32] Chan YS, Takeuchi R, Jarjour J, Huen DS, Stoddard BL, Russell S. The design and *in vivo* evaluation of engineered I-OnuI-based enzymes for HEG gene drive. PLoS One 2013;8(9):e74254.

[33] Takeuchi R, Choi M, Stoddard BL. Redesign of extensive protein-DNA interfaces of meganucleases using iterative cycles of *in vitro* compartmentalization. Proc Natl Acad Sci USA 2014;111(11):4061−6.

[34] Thyme SB, Boissel SJS, Quadri SA, et al. Reprogramming homing endonuclease specificity through computational design and directed evolution. Nucleic Acids Res 2014;42 (4):2564−76.

[35] Flick KE, Jurica MS, Monnat Jr. RJ, Stoddard BL. DNA binding and cleavage by the nuclear intron-encoded homing endonuclease I-*Ppo*I. Nature 1998;394(6688):96−101.

[36] Windbichler N, Papathanos PA, Catteruccia F, Ranson H, Burt A, Crisanti A. Homing endonuclease mediated gene targeting in *Anopheles gambiae* cells and embryos. Nucleic Acids Res 2007;35(17):5922−33.

[37] Windbichler N, Papathanos PA, Crisanti A. Targeting the X chromosome during spermatogenesis induces Y chromosome transmission ratio distortion and early dominant embryo lethality in *Anopheles gambiae*. PLoS Genet 2008;4(12):e1000291.

[38]   Klein TA, Windbichler N, Deredec A, Burt A, Benedict MQ. Infertility resulting from transgenic I-*PpoI* male *Anopheles gambiae* in large cage trials. Pathog Glob Health 2012;106(1):20−31.

[39]   Bernardini F, Galizi R, Menichelli M, et al. Site-specific genetic engineering of the *Anopheles gambiae* Y chromosome. Proc Natl Acad Sci USA 2014;111(21):7600−5.

[40]   Simoni A, Siniscalchi C, Chan YS, et al. Development of synthetic selfish elements based on modular nucleases in *Drosophila melanogaster*. Nucleic Acids Res 2014;42 (11):7461−72.

[41]   Boch J, Scholze H, Schornack S, et al. Breaking the code of DNA binding specificity of TAL-type III effectors. Science 2009;326(5959):1509−12.

[42]   Moscou MJ, Bogdanove AJ. A simple cipher governs DNA recognition by TAL effectors. Science 2009;326(5959):1501.

[43]   Mak AN, Bradley P, Cernadas RA, Bogdanove AJ, Stoddard BL. The crystal structure of TAL effector PthXo1 bound to its DNA target. Science 2012;335(6069):716−19.

[44]   Deng D, Yan C, Pan X, et al. Structural basis for sequence-specific recognition of DNA by TAL effectors. Science 2012;335(6069):720−3.

[45]   Beerli RR, Barbas III CF. Engineering polydactyl zinc-finger transcription factors. Nat Biotechnol 2002;20(2):135−41.

[46]   Gaj T, Gersbach CA, Barbas III CF. ZFN, TALEN, and CRISPR/Cas-based methods for genome engineering. Trends Biotechnol 2013;31(7):397−405.

[47]   Christian M, Cermak T, Doyle EL, et al. Targeting DNA double-strand breaks with TAL effector nucleases. Genetics 2010;186(2):757−61.

[48]   Carlson DF, Fahrenkrug SC, Hackett PB, Targeting DNA. With Fingers and TALENs. Mol Ther Nucleic Acids 2012;1:e3.

[49]   Wiedenheft B, Sternberg SH, Doudna JA. RNA-guided genetic silencing systems in bacteria and archaea. Nature 2012;482(7385):331−8.

[50]   Garneau JE, Dupuis ME, Villion M, et al. The CRISPR/Cas bacterial immune system cleaves bacteriophage and plasmid DNA. Nature 2010;468(7320):67−71.

[51]   Jiang WY, Bikard D, Cox D, Zhang F, Marraffini LA. RNA-guided editing of bacterial genomes using CRISPR-Cas systems. Nat Biotechnol 2013;31(3):233−9.

[52]   Mali P, Yang LH, Esvelt KM, et al. RNA-guided human genome engineering via Cas9. Science 2013;339(6121):823−6.

[53]   Cong L, Ran FA, Cox D, et al. Multiplex genome engineering using CRISPR/Cas systems. Science 2013;339(6121):819−23.

[54]   Yu ZS, Ren MD, Wang ZX, et al. Highlyefficient genome modifications mediated by CRISPR/Cas9 in *Drosophila*. Genetics 2013;195(1): 289−91.

[55]   Gratz SJ, Cummings AM, Nguyen JN, et al. Genome engineering of *Drosophila* with the CRISPR RNA-guided Cas9 nuclease. Genetics 2013;194(4): 1029−35.

[56]   Jinek M, Chylinski K, Fonfara I, Hauer M, Doudna JA, Charpentier E. A programmable dual-RNA-guided DNA endonuclease in adaptive bacterial immunity. Science 2012;337 (6096):816−21.

[57]   Wang HY, Yang H, Shivalila CS, et al. One-step generation of mice carrying mutations in multiple genes by CRISPR/Cas-mediated genome engineering. Cell 2013;153(4): 910−18.

[58]   Marshall JM. The toxin and antidote puzzle: new ways to control insect pest populations through manipulating inheritance. Bioeng Bugs 2011;2(5):235−40.

[59] Wade MJ, Beeman RW. The population dynamics of maternal-effect selfish genes. Genetics 1994;138(4):1309−14.

[60] Ward CM, Su JT, Huang Y, Lloyd AL, Gould F, Hay BA. Medea selfish genetic elements as tools for altering traits of wild populations: a theoretical analysis. Evolution 2011;65 (4):1149−62.

[61] Hurst LD. scat + is a selfish gene analogous to Medea of *Tribolium castaneum*. Cell 1993;75(3):407−8.

[62] Weichenhan D, Kunze B, Traut W, Winking H. Restoration of the Mendelian transmission ratio by a deletion in the mouse chromosome 1 HSR. Genet Res 1998;71(2):119−25.

[63] Beeman RW, Friesen KS. Properties and natural occurrence of maternal-effect selfish genes ('Medea' factors) in the red flour beetle, *Tribolium castaneum*. Heredity 1999;82 (Pt. 5):529−34.

[64] Akbari OS, Chen CH, Marshall JM, Huang H, Antoshechkin I, Hay BA. Novel synthetic medea selfish genetic elements drive population replacement in *Drosophila*; a theoretical exploration of medea-dependent population suppression. ACS Synth Biol 2014;3 (12):915−28.

[65] Hartl DL, Clark AG. Principles of population genetics. 3rd ed. Sunderland, MA: Sinauer Associates; 1997.

[66] Davis S, Bax N, Grewe P. Engineered underdominance allows efficient and economical introgression of traits into pest populations. J Theor Biol 2001;212(1):83−98.

[67] Akbari OS, Matzen KD, Marshall JM, Huang H, Ward CM, Hay BA. A synthetic gene drive system for local, reversible modification and suppression of insect populations. Curr Biol 2013;23(8):671−7.

[68] Wimmer EA. Insect biotechnology: controllable replacement of disease vectors. Curr Biol 2013;23(10):R453−6.

[69] Silver JB. Mosquito ecology field sampling methods. 3rd ed. New York, NY: Springer; 2008.

[70] Taylor C, Toure YT, Carnahan J, et al. Gene flow among populations of the malaria vector, *Anopheles gambiae*, in Mali, West Africa. Genetics 2001;157(2):743−50.

[71] Gould F, Huang Y, Legros M, Lloyd AL. A killer-rescue system for self-limiting gene drive of anti-pathogen constructs. Proc Biol Sci 2008;275(1653):2823−9.

[72] Marshall JM, Hay BA. Confinement of gene drive systems to local populations: a comparative analysis. J Theor Biol 2012;294:153−71.

[73] Marshall JM, Hay BA. General principles of single-construct chromosomal gene drive. Evolution 2012;66(7):2150−66.

[74] Marshall JM, Pittman GW, Buchman AB, Hay BA. Semele: a killer-male, rescue-female system for suppression and replacement of insect disease vector populations. Genetics 2011;187(2):535−51.

[75] Marshall JM, Hay BA. Inverse Medea as a novel gene drive system for local population replacement: a theoretical analysis. J Hered 2011;102(3):336−41.

[76] Marshall JM, Hay BA. Medusa: a novel gene drive system for confined suppression of insect populations. PLoS One 2014;9(7):e102694.

[77] Golic KG, Golic MM. Engineering the *Drosophila* genome: chromosome rearrangements by design. Genetics 1996;144(4):1693−711.

[78] Egli D, Hafen E, Schaffner W. An efficient method to generate chromosomal rearrangements by targeted DNA double-strand breaks in *Drosophila melanogaster*. Genome Res 2004;14(7):1382−93.

[79] Sinkins SP, Gould F. Gene drive systems for insect disease vectors. Nat Rev Genet 2006;7(6):427−35.

[80] Hoffmann AA, Montgomery BL, Popovici J, et al. Successful establishment of *Wolbachia* in *Aedes* populations to suppress dengue transmission. Nature 2011;476(7361):454−7.

[81] Marshall JM. The Cartagena Protocol and genetically modified mosquitoes. Nat Biotechnol 2010;28(9):896−7.

[82] Griffin JT, Hollingsworth TD, Okell LC, et al. Reducing *Plasmodium falciparum* malaria transmission in Africa: a model-based evaluation of intervention strategies. PLoS Med 2010;7(8):e1000324.

Chapter 10

Exploring the Sex-Determination Pathway for Control of Mosquito-Borne Infectious Diseases

James K. Biedler[1,2], Brantley A. Hall[2,3], Xiaofang Jiang[2,3] and Zhijian J. Tu[1,2,3]
[1]Department of Biochemistry, Virginia Tech, Blacksburg, VA, USA, [2]Fralin Life Science Institute, Virginia Tech, Blacksburg, VA, USA, [3]Interdisciplinary PhD Program in Genetics, Bioinformatics, and Computational Biology, Virginia Tech, Blacksburg, VA, USA

INTRODUCTION

Mosquitoes are the vectors of many disease-causing agents that continue to make a large impact on human health. For example, *Aedes aegypti* is a major vector for the dengue, yellow fever, and chikungunya viruses. According to the WHO, the global incidence of dengue has increased significantly in recent decades and nearly half of the world's population is now at risk of contracting the disease [1]. There is no specific treatment for dengue and current prevention depends solely on effective vector control [1]. *Anopheles* mosquitoes are the primary vectors of malaria, one of the most deadly and costly diseases in human history [2,3].

Current control measures for mosquitoes are under threat as drug- and insecticide-resistance spreads [4–8], and there is a demand for new control measures, including genetic control. Genetic strategies to control mosquito-borne diseases can be divided into two major categories, namely population suppression and population replacement. Examples of population suppression include the release of sterile male mosquitoes (sterile insect technique, SIT) and the release of insects carrying a dominant lethal (RIDL) gene, both having the goal of reducing the vector population density to achieve disease control. Many advances have been made in this area and several strategies have been designed, some with promising results in the field [9]. Population replacement refers to the replacement or conversion of a pathogen-susceptible vector population into a pathogen-resistant one. Population

Genetic Control of Malaria and Dengue.
© 2016 Elsevier Inc. All rights reserved.

replacement strategies involve the release of insects carrying one or more effector genes that confer resistance to the targeted pathogen to prevent its transmission. This strategy is based on the ability to develop effector genes capable of inhibiting pathogen transmission, the ability to introduce these effector genes into the target species through genetic modification, and the ability to fix the introduced genes in the target population through a gene drive mechanism, for example, Refs [9−11]. Gene drive refers to the ability of a gene to be inherited more frequently than Mendelian genetics would dictate, thus, increasing in frequency until reaching fixation [12]. The largest remaining technical obstacle to successful population replacement strategies is the lack of a safe, efficient, and robust gene drive system, although progress is being made in this arena [13−16].

Only female mosquitoes transmit disease pathogens because only females feed on blood. Thus, when implementing genetic approaches through either population suppression or population replacement, it is critical to release only male mosquitoes. Although a few genetic methods have been developed for sex separation, there are remaining challenges to achieving robust, reliable, and cost-effective mass production and sex separation of mosquitoes [17,18]. Rapid progress in mosquito genomics and functional genomics presents exciting opportunities for improving our understanding of sex-determination pathways, which will provide new targets and strategies for genetic sexing as well as population reduction through the introduction of male-bias.

Excellent recent reviews are available on genetic methods for mosquito control and associated ethical and policy discussions [9,19−22]. Readers are encouraged to consult these resources for detailed information on the aforementioned topics. In this chapter, we will briefly review the existing genetic methods and strategies that are being developed to control mosquito-borne infectious diseases. We will then focus on the current understanding of the sex-determination pathway in mosquitoes, including recent progress from genomics and bioinformatics analyses. We will then compare and contrast mosquitoes of the *Culicinae* and *Anopheles* subfamilies, which have different types of sex-determining chromosomes, and we will end by exploring new strategies that build on the discoveries of the molecular mechanisms of mosquito sex determination.

CURRENT GENETIC METHODS AND STRATEGIES

As mentioned above, recent reviews are available on genetic methods for mosquito control [9,19,23]. We will briefly introduce these methods (Table 10.1) to provide a context to the topic of this chapter [9,19] and we will focus on methods that have been demonstrated in mosquitoes. These strategies can be classified as self-limiting or self-sustaining, depending on the degree of pervasiveness, that is, whether they are expected to diminish in

TABLE 10.1 Summary of Mosquito Control Methods

Method	Mechanism	Sustainability	Affecting Stage	Auto-sexing	Pervasiveness	Status	References	Species
SIT	Irradiated males cause sterility	Self-limiting	Embryonic (zygote)	No	First and single generation	Field trial	[24]	Ae. albopictus
RIDL	Conditional expression of lethal gene	Self-limiting	Late larval stage	No	First and single generation	Field trial	[25,26]	Ae. aegypti
Female-specfic RIDL (fsRIDL)	Female-specific conditional expression of lethal gene	Self-limiting	Adult	Yes	Multiple generations, decreasing effect in each subsequent generation	Lab- and field-cage trial	[27–30]	Ae. aegypti, Ae. albopictus, An. stephensi
IIT (noninvasive)	Cytoplasmic incompatibility by Wolbachia (infected-male release)	Self-limiting	Embryonic (zygote)	No	First and single generation	Field trial	[31]	Ae. polynesiensis

(Continued)

TABLE 10.1 (Continued)

Method	Mechanism	Sustainability	Affecting Stage	Auto-sexing	Pervasiveness	Status	References	Species
IIT (invasive)	Cytoplasmic incompatibility by *Wolbachia* (infected-male and female release)	Self-sustaining	Embryonic (zygote)	No	Multiple generations	Field trial	[32]	*Ae. aegypti*
Sex ratio distorter (autosomal, X-shredder)	Meiotic drive—induction of female-specific chromosome breakage	Self-limiting	Meiosis	Yes (up to 95% males)	Multiple generations, decreasing effect in each subsequent generation	Lab trial	[33]	*An. gambiae*

their efficacy or flourish and increase their prevalence in a target population over time [9]. In order for self-limiting strategies to be effective, it is likely that multiple releases will be performed depending on the target population size, release size, and control mechanism. Self-sustaining population replacement strategies have advantages in this regard although extraordinary care and caution are needed to mitigate risk and environmental concerns [20].

Sterile Insect Technique

Over a half century ago, the SIT was proposed as a method to control insects via population suppression [34]. Because female insects of some species only mate once, if they mate with a sterile male their reproductive capacity is nullified. The expectation is that through successive releases the number of progeny is reduced, the ratio of sterile males to wild males is increased, and the population size is reduced until effectively eliminated. After initial success in control programs targeting the screwworm *Cochliomyia hominivorax* population on the island of Curaçao [35], subsequent programs eradicated the screwworm population in the United States [36]. Effective elimination programs have also been executed to control the Mediterranean fruit fly and the tsetse fly [37,38]. Major benefits of SIT are that it is specific, self-limiting as the released sterile males exist for only one generation, and it can generally be considered to have a low impact on the environment. SIT has been used against different mosquito species over the past 40 years, with varying degrees of success (summarized in Ref. [39]). Some challenges to the SIT are the need for releasing large numbers of sterile male insects, which requires large rearing facilities and is costly. Also, mating competitiveness of irradiated males may be compromised to varying degrees compared to their wild-type competitors. Another concern is that accidentally released females, although sterile, could transmit pathogens in regard to the use of SIT in mosquitoes. Therefore, it is critical to be able to efficiently and effectively remove females prior to release.

Incompatible Insect Technique

Wolbachia-infected mosquitoes may be used as a tool for controlling mosquito populations [40]. As a population suppression method, infected males can be released into an uninfected target population such that mated females will not produce any progeny, resulting in a population decline. This unidirectional cytoplasmic incompatibility (CI) is similar to SIT. The feasibility of CI was demonstrated to control a species of *Culex* mosquitoes [41]. Accidental release of infected females poses a greater risk to the unidirectional incompatible insect technique (IIT) strategy compared to SIT, as they could mate with the released infected males, and, thus, drive a population replacement, contrary to the intended population suppression. Again, the need for removal of females prior to release is underscored. If the intention is to perform population replacement, then both infected males and females could be released into a target population. This method was demonstrated in Australia where *Wolbachia* rapidly infected natural populations of *Ae. aegypti* by release of *Wolbachia*-infected mosquitoes, reaching near fixation within a few months [32]. Population replacement could potentially be done using mosquitoes that carry a transgene(s) that renders the mosquito incapable of pathogen transmission. Bidirectional CI could also facilitate population replacement in a population that is infected with a different strain of *Wolbachia*. Bidirectional CI is where a population of mosquitoes infected with strain A is released into a target population infected with strain B. Mating between males and females infected with different strains of *Wolbachia* will fail to yield any viable progeny. If enough mosquitoes infected with strain A are released into a target population infected with strain B, then population replacement results.

Some characteristics of *Wolbachia* infection in *Ae. aegypti* could be of great benefit for disease control. *Wolbachia* infection can reduce life span and pathogen proliferation, therefore, reducing pathogen transmission potential [42−49]. It has been demonstrated that seeding uninfected *Anopheles stephensi* with *Wolbachia*-infected females can result in the spread of *Wolbachia* infection and confer resistance to *Plasmodium falciparum* [50]. In addition, it has been shown that *Wolbachia* can be introduced into *Ae. aegypti* and *Aedes albopictus* and block Dengue virus transmission [51,52]. However, *Wolbachia* has also been found to increase West Nile virus infection and *Plasmodium* infection in some mosquito species [53−56].

Release of Insects Carrying a Dominant Lethal Gene

A major recent advancement in SIT and IIT using genetic modification was the RIDL system [57]. The first RIDL system engineered in mosquitoes utilized a positive feedback loop based on the Tet-off system where the cassette contains the tetracycline operator tetO and the *Drosophila* hsp70 minimal promoter driving expression of the tetracycline-repressible transcriptional activator tTA. When mosquitoes are maintained in the presence of tetracycline, tTA is inactive and expression of tTA via tetO is suppressed. When tetracycline is removed, the positive feedback loop is initiated, resulting in the buildup of tTA that is toxic to cells, and lethal. This bi-sex lethal system allows rearing of males in the presence of tetracycline until numbers are sufficient for release. In the field, all progeny resulting from mating of these RIDL males with wild females will not survive, assuming 100% lethality. Phuc et al. actually obtained 95−97% lethality in their laboratory trials. However, they concluded that this did not compromise the effectiveness compared to a line with 100% penetrance of lethality. This system is highly self-limiting, similar to SIT, as the effect is limited to the first generation.

A female-specific RIDL (fsRIDL) based on a flightless phenotype was engineered in *Ae. aegypti, Ae. albopictus,* and *An. stephensi* [27−29,58], whereby lethality is implemented indirectly by making females reproductively ineffective. Two major advantages of this system are (i) auto-sexing (elimination of females) and (ii) a more persistent effect because its female-specific lethality allows the inheritance of the female-lethal gene by males in subsequent generations. If homozygous males are released, then ideally 100% of the female progeny are expected to be rendered reproductively nullified. Surviving heterozygous males can mate again, whereby only 50% of the female progeny will survive, and so on. Adding multiple copies of the lethal gene cassette can enhance the persistence of this method. Although fsRIDL is more persistent than bi-sex RIDL, this system is still self-limiting. While fsRIDL held great initial promise, large cage field trials indicated lack of feasibility in the current state due to a high degree of male fitness cost with the strain tested [30].

A Sex Ratio Distorter in *Aedes aegypti* and a Synthetic Sex Ratio Distorter in *Anopheles gambiae*

Meiotic drive is a natural phenomenon that favors the non-Mendelian transmission of an allele to progeny to achieve higher than expected allele frequencies. The segregation distorter (SD) of *Drosophila melanogaster* [59] and the *t*-complex in the mouse [60] are two examples of these systems. Interestingly, a third example describes a sex-linked meiotic driver in *Drosophila simulans* [61] that is an X-linked sex ratio distorter resulting in a majority of female progeny by inducing the loss of Y-bearing sperm. A previously described sex ratio distorter in the mosquito *Ae. aegypti* [62] that results in highly male-biased progeny was recently genetically mapped [63]. This meiotic drive gene called M^D is tightly linked to the M locus, the sex-determination region that is responsible for male development. The M^D gene product exerts its effect *in trans*, by inducing breakage of the female-determining chromosome during meiosis, in individuals carrying the drive-sensitive locus [64]. A male-bias of approximately 85% was achieved with specific crosses involving males from the T37 strain [65]. Sex ratio distorters such as M^D could be used to facilitate a population crash as a result of increasing male:female sex ratios [66].

A synthetic sex ratio distorter called "X-shredder" has been engineered in *Anopheles gambiae* [33], which has an X−Y sex-determination system. The "X-shredder" is a homing endonuclease that is expressed by a sperm-specific promoter and targets a sequence specific to ribosomal genes that are only found on the *An. gambiae* X chromosome, thus eliminating X-bearing sperm and resulting in male-biased progeny. This method results in population suppression as fewer females are produced and is self-limiting as the endonuclease is located on an autosome. However, if the driver were placed on the

Y chromosome, then the sex ratio distorter (X-shredder) will be inherited at 100% probability and be sustained over many generations until it crashes a population due to increasing male:female sex ratio. Transgene expression on the Y chromosome of the African malaria mosquito *An. gambiae* has recently been demonstrated [67], which increases the feasibility of a Y-driving mechanism. The "X-shredder" strategy is expected to only function in a limited number of *Anopheles* species as most *Anopheles* mosquitoes have ribosomal genes on both X and Y chromosomes [68−70].

Sex Separation

One critical component that the majority of the above systems have in common is the requirement that only males are released. There are several reasons for this. Females bite and transmit the disease-causing agents. Also, for systems that employ male-sterility or male-induced lethality in offspring, released females can compete with wild-type females to mate with the released males simply due to their presence and perhaps assortative mating. With systems such as unidirectional CI where *Wolbachia* is maternally inherited and infected males are sterile when mated with uninfected females (similar to SIT), accidentally released infected females could mate with the infected released males and, thus, result in population replacement, negating the intended result of population suppression. Therefore, females must be removed during the rearing process. Several sexing/female elimination strategies have been employed and initially these were generally time consuming, costly, and not 100% effective. Some advances have been made that increase efficiency by utilizing conditional female-specific lethal genes and regulatory systems like Tet-off. The fsRIDL systems engineered in *Ae. aegypti*, *Ae. albopictus*, and *An. stephensi* [27−29,58] allow for selection of males at the adult stage as the females cannot fly. The ideal sexing system may be one that allows selection to eliminate females at the embryonic stage or during meiosis such as was demonstrated with the X-shredder in *An. gambiae* [33].

Self-Sustaining Population Replacement Strategies

While *Wolbachia*-induced CI can skew the inheritance of the *Wolbachia* sp., spreading it through the population, the inability to transform *Wolbachia* makes it difficult to drive refractory genes through mosquito populations. However, as described earlier, *Wolbachia* infection alone may confer resistance to some pathogens. The potential of homing endonucleases, an ancient class of selfish DNA that introduces a double-stranded DNA break at a specific genomic target, to serve as gene drives has been described [71]. A homing endonuclease can copy itself to the homologous chromosome or the sister chromatid that lacks the homing endonuclease, a process termed "homing," which results from homology-dependent repair induced by the double-stranded DNA break. Homing endonucleases have been shown to function in both *Aedes* and *Anopheles* mosquitoes [72−74], and its gene drive ability has been demonstrated in *An. gambiae* [15]. CRISPR/cas9-mediated site-specific DNA breaks have also been proposed as a way to induce "homing" to drive genes into mosquito populations [75,76]. Recently, a CRISPR gene drive has been demonstrated in *Drosophila* [77]. A very different gene drive was engineered by Hay and colleagues in *Drosophila*, inspired by the naturally occurring *Medea* (maternal-effect dominant embryonic arrest) element in the flour beetle *Tribolium castaneum*, which successfully drove population replacement in the laboratory [13]. Progress has been made toward identifying and testing the components in *Aedes* and *Anopheles* mosquitoes [14,78−80]. The meiotic drive described above also has the potential to drive refractory genes into mosquito populations. A threshold-dependent underdominance system that could confer local population replacement has been demonstrated in *Drosophila* [81].

SEX DETERMINATION IN MOSQUITOES

M Locus and the Y Chromosome

Anopheles mosquitoes have well-differentiated heteromorphic Y chromosomes. *Anopheles culicifacies* of the XXY genotype was found to be male [82], suggesting that a dominant male-determining factor (M factor) on the Y chromosome initiates sexual differentiation, similar to humans and several non-*Drosophilid* flies [83]. The Y chromosome has also been shown to control the stenogamy−eurygamy mating behavior [84]. Male development in *Ae. aegypti* is also initiated by a dominant male-determining gene (M factor) on a nonrecombining M locus of the homomorphic sex-determining chromosome [85−88]. *Aedes aegypti* has three pairs of chromosomes and the M locus has been mapped to chromosome 1, band 1q21 [88]. According to the nomenclature proposed [89], the copy of chromosome 1 with the M locus is the M chromosome, and the copy without the M locus is the m chromosome. Previous studies suggest that the M locus of *Ae. aegypti* and *Culex pipiens* are linked to the same markers [90,91]. There are also cytological differences between the M locus and the m locus, consistent with clear differentiation between the loci [89,92]. Despite being homomorphic with the m chromosome, the presence of the M locus makes the M-chromosome function as a "Y" chromosome as well as an autosome. Intriguingly, the M locus has been found in noncanonical locations in a few *Culicinae* mosquitoes [93].

Little is known about the molecular aspects of sex determination in mosquitoes except for a few conserved genes such as *doublesex* (*dsx*), *fruitless*

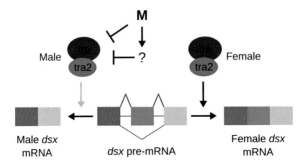

FIGURE 10.1 A simplified model of the sex-determination pathway in species with an M factor. M factor is only present in males, either on the Y chromosome or in an M locus. It may directly inhibit the alternative splicing complex or activate other factor(s) that inhibit the complex. Thus, the presence or absence of the M factor determines the splice isoform of *dsx* and, ultimately, sex. *Fruitless*, which is another sex-determining gene regulated in a similar manner as *dsx*, is not depicted in this figure. No *tra* has been found in mosquitoes, but evidence suggests that there is a TRA-like protein. TRA may have rapidly evolved beyond recognition in mosquitoes or another splicing factor assumed its function [94]. The splicing of the *dsx* pre-mRNA in male and female mosquitoes is much more complex than depicted here and is described in "*Doublesex (dsx), Fruitless (fru)*, and the Elusive *Transformer (tra)*" section.

(*fru*), and *transformer2* (*tra2*). As shown in a simplified model (Figure 10.1), sex-specific splicing of *dsx* pre-mRNA leads to the production of DSX and FRU protein isoforms, which regulate sexual differentiation. This is a much simplified view and detailed discussions are provided in the "*Doublesex* (*dsx*), *Fruitless* (*fru*), and the Elusive *Transformer* (*tra*)" section. The alternative splicing of *dsx* and *fru* pre-mRNA is commonly controlled by a protein complex that includes a fast-evolving *transformer* (TRA) and a conserved TRA2. TRA is the sex-specific protein in this complex. Sex-specific splicing of *dsx* has been described in both *Aedes* and *Anopheles*, suggesting the presence of a TRA-like protein [95,96]. However, a homolog of the fast-evolving TRA has yet to be found in any mosquitoes [94].

As shown in Figure 10.1, the M factor and perhaps additional genes in the Y chromosome or the M locus may directly or indirectly inhibit the TRA/TRA2 splicing machinery to confer male-specific *dsx* splicing [97]. The *Drosophila* Y chromosome contains only a small number of genes, many of which resulted from translocations from other chromosomes [98]. Although sex determination in *D. melanogaster* is not initiated by a dominant Y factor, the *Drosophila* Y is important in spermatogenesis, mating behavior, and even the modulation of chromatin and gene expression [99−101]. The initial signals that trigger sex determination, which are just being understood in a few non-*Drosophila* insects [102,103], are diverse and rapidly evolving. This phenomenon is often referred to as "masters change, slaves remain" [104].

Doublesex (*dsx*), *Fruitless* (*fru*), and the Elusive Transformer (*tra*)

One common thread among the diverse mechanisms that insects use to determine sex is the roles played by two transcription factors, *doublesex* and *fruitless* [94−96,105−107]. In *D. melanogaster*, *doublesex* is alternatively spliced in males and females and the alternative isoforms are responsible for somatic sexual differentiation [107]. Alternative splicing in females results in the inclusion of an extra exon compared to males [107]. This female-specific exon has a noncanonical splice site that can be activated by a protein complex including an upstream gene in the sex-determination pathway, *transformer*, which is only functional in females, resulting in distinct male and female isoforms of *doublesex* [107]. The resulting male and female *doublesex* proteins are both functional but differ in their C-termini [107]. *Fruitless* is a male-specific transcription factor whose expression in the nervous system is crucial for male-specific courtship behaviors [105]. *Fruitless* is also alternatively spliced between males and females resulting in the presence of an extra exon in the female isoform [105]. The female-specific isoform of *fruitless* does not appear to produce a functional protein because the female-specific exon introduces a premature stop codon [105].

Homologs of both *doublesex* and *fruitless* have been identified in the African malaria mosquito *An. gambiae* and the Yellow fever mosquito *Ae. aegypti* [94−96,106]. Mosquito homologs of both genes are generally similar to *Drosophila* and both produce alternative male and female isoforms [94−96,106]. Similar to *D. melanogaster*, the *An. gambiae doublesex* transcript has a single female-specific exon designated exon 5 [95]. However, in contrast to *Drosophila* where a weak 3′ splice site is activated by the *transformer* complex recruited by *cis*-regulatory elements, the 3′ splice site proceeding exon 5 in *An. gambiae* does not appear weak because it adheres to the splice site consensus [95]. Furthermore, *cis*-regulatory elements similar to the elements found in *D. melanogaster* corresponding to *transformer/transformer-2* binding sites were found downstream in the 5′ donor site of intron 5 [95]. Therefore, it is hypothesized that the female-specific isoform of *doublesex* is produced by the activation of the 5′ donor site of intron 5 similar to the regulation of the female-specific exon of *fruitless* in *D. melanogaster* [95]. *Aedes aegypti* has not one but two female-specific exons of *doublesex*, designated as exons 5a and 5b [96]. Correspondingly, two female-specific isoforms of *doublesex* are found in *Ae. aegypti*—the first containing exons 5a and 5b and the second containing only 5b [96]. The 3′ splice site proceeding exon 5a adheres to the splice site consensus but the splice site proceeding exon 5b deviates substantially from the splice site consensus suggesting that it may be a weak splice site [96]. *Cis*-regulatory elements likely corresponding to *transformer/transformer-2* binding sites were identified proceeding and within exon 5b but were more variable in sequence when compared to the motifs in *Drosophila* [96]. Copies of a different *cis*-regulatory sequence were found within exon 5a, indicating that a different splicing factor could repress the exon 5a splice site [96].

Compared to *D. melanogaster fruitless*, *An. gambiae fruitless* has an extra proceeding female-specific exon [106]. However, both female-specific exons contain putative *transformer/transformer-2* binding sites similar to *D. melanogaster* [105,106]. The genomic organization of *fruitless* in *Ae. aegypti* is similar to that in *D. melanogaster* with one female-specific exon [94,105]. However, the *transformer/transformer-2* binding sites are both fewer and more variable in sequence than found in *D. melanogaster* [106].

The well-conserved *cis*-regulatory elements that likely correspond to *transformer/transformer-2* binding sites found in *An. gambiae doublesex* and *fruitless* suggest the presence of a *transformer*-like protein in *Anopheles* mosquitoes [95,106]. In contrast, the lower conservation of *transformer/transformer-2* binding sites in *Ae. aegypti doublesex* and *fruitless* along with the apparently default splicing of exon 5a may suggest the absence of a functional *transformer* in female *Ae. aegypti* [94,96]. The *tra* gene has yet to be found in any mosquitoes.

Dosage Compensation and Sex Determination

In many organisms, the Y chromosome does not contain many genes while the X retains hundreds or thousands of genes. Complete dosage compensation, as shown in *D. melanogaster* and *C. elegans*, is a mechanism to compensate for the loss of one copy of X chromosome by hyperexpressing the entire X chromosome in males [108]. Interestingly, three known master switches of sex determination, *sex-lethal*, *Fem/Masc*, and *xo-lethal 1*, also directly or indirectly regulate dosage compensation in *D. melanogaster*, *Bombyx mori* (a species with ZW sex chromosomes), and *C. elegans*, respectively [103,109,110]. Loss-of-function *sex-lethal* alleles and knockdown of *Masc* cause female embryonic lethality in *D. melanogaster* and *B. mori*,

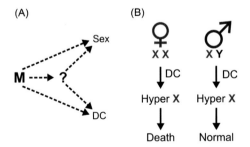

FIGURE 10.2 Relationship of M factors and dosage compensation. (A) The master regulator of sex determination *M* may directly or indirectly control dosage compensation (DC). (B) Misregulation of DC is lethal in XX individuals due to hyperexpression of two X chromosomes.

respectively, most likely due to misregulation of dosage compensation [103,111]. Figure 10.2 depicts a model that explains the connection between sex determination and dosage compensation based on existing experimental evidence. Under this scenario, ectopic expression of the M factor in females will cause misregulation of dosage compensation and, thus, female lethality. Complete dosage compensation in *Anopheles* was implied based on analysis of the *An. gambiae* microarray data [108]. Dosage compensation is not expected for insects with homomorphic sex-determining chromosomes including *Aedes* and *Culex* mosquitoes.

TOWARD THE DISCOVERY OF THE M FACTOR AND *TRA*

Molecular characterizations of Y- or M-linked genes are rare outside of a few model species, mostly due to the repetitive nature of the Y/M chromosome/Locus and the difficulty of cloning repeat-rich heterochromatic sequences. To overcome this bottleneck, a chromosomal quotient (CQ) method was developed to find male-specific genomic sequences by Illumina sequencing of males and females separately [112]. The CQ parameter was designed to allow discovery of unique Y chromosome sequences as well as complex Y sequences that contain repeats or have X or autosomal paralogs. CQ is the normalized ratio of female-to-male alignments to a given reference sequence. For a given sequence S_i, $CQ_{(Si)} = F_{(Si)}/M_{(Si)}$, where $F_{(Si)}$ is the number of female Illumina reads aligned to S_i, and $M_{(Si)}$ is the number of male Illumina reads aligned to S_i. As shown in Figure 10.3, males and females have the same complement of autosomes and, subsequently,

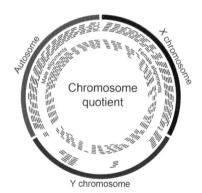

FIGURE 10.3 The chromosome quotient (CQ) method. Y genes have a distinctive pattern of alignments from male and female Illumina sequencing data. Autosomal genes have roughly equal alignments from male and female data. X chromosome genes have twice as many alignments from female data as from male data. Y chromosome genes have alignments from female data only in repetitive regions, and many more alignments from male data distributed between repetitive and nonrepetitive regions. Therefore, Y sequences can be identified by their distinctive CQs.

autosomal sequences have CQ values distributed around one. X chromosomes are twice as numerous in females as in males and have CQ values distributed around two. Y chromosomes are only present in males and, therefore, should have CQs of zero if the sequences are only found on the Y. When similar sequences exist on other chromosomes, Y sequences may have CQ values above zero, depending on the level of similarity and the relative copy numbers of the Y and non-Y sequences. As positive controls, CQ was successfully used to identify human and *Drosophila* Y sequences. The CQ methods and an earlier version of the CQ method were used to identify the first set of Y genes in *Anopheles* mosquitoes [112,113]. These candidate Y genes were verified by experimental means. Included in these Y genes is a novel gene named *Guy1* (*Gene Unique to Y*), which is a single copy Y gene with no autosomal or X paralog [113]. *Guy1* encodes for a small protein with predicted DNA-binding properties [112,113]. *Guy1* is a promising candidate for the *M* factor because *Guy1* transcription initiates at the very onset of embryonic development without the need for any other factor on the Y chromosome [113]. The CQ method was also applied to identify an M-linked gene in *Ae. aegypti* [114].

Manipulations of the *Tra* gene can alter the *dsx* and *fru* splice isoforms, which may lead to sex ratio distortion for vector control. As discussed earlier, *Tra* has not been identified in mosquitoes, but it is believed to exist at least in *Anopheles*. The fast-evolving nature and divergent structure of the TRA protein have made it difficult to uncover via purely homolog sequence searches. A known computational method that successfully identified *tra* in the medfly *Ceratitis capitata* was based on conserved synteny with *Drosophila* [97]. However, this approach cannot be extrapolated to mosquito genomes due to massive genome rearrangements during mosquito evolution. Given the rapidly accumulating RNAseq data from mosquito samples [14,78,80] and the availability of high-throughput methods to sequence full-length cDNA (e.g., IsoSeq) [115], transcriptome analysis may provide the opportunity for *Tra* discovery. Alternatively, biochemical methods may also be used based on TRA−TRA2 interactions. Protein−protein interactions have been successfully used to identify proteins that function in the same complex in mosquitoes [116]. We cannot rule out that a different gene may have acquired the *transformer* function in mosquitoes through convergent evolution.

WAYS AND CONSIDERATIONS TO EXPLORE THE SEX-DETERMINATION PATHWAY FOR CONTROL

The release of males is either required or preferred when genetic methods are implanted to reduce or replace mosquito populations to control mosquito-borne infectious diseases. The fsRIDL methods used female-specific promoters to confer sex-specific lethality [27−29,58]. Manipulation of genes involved in the sex-determination pathway will provide new ways to produce male-only progeny for release. Genetic manipulation of sex ratios could also greatly reduce a local mosquito population [117].

Targets and Sex-Specific Reagents

Doublesex and *fruitless* are both characterized in mosquitoes and could be used as targets to manipulate the sex of mosquitoes. It is possible to ectopically express the male isoforms of *dsx* or *fru*, which may result in abnormality or lethality in females. However, *Doublesex* and *fruitless* are very long and complicated genes that may prove difficult to manipulate. Furthermore, *fruitless* has nonsex-specific isoforms that are crucial to regular development. If full sex conversion is the goal, it will require manipulating both *doublesex* and *fruitless*. Therefore, an approach that uses genes upstream in the sex-determination pathway, either the male-determining factor or a *transformer*-like gene would involve the manipulation of only one gene and would, therefore, be easier to accomplish. Current genomics development also provides exciting opportunities to identify new sex-specific promoters that function at different tissues and developmental stages [14,78–80]. Sex-specific introns may also be identified in genes such as *tra*, which may be easier to manipulate than the long introns in *dsx* and *fru*. These sex-specific promoters and introns will provide useful reagents to design female-lethal strategies as has been demonstrated in mosquitoes and flies [9].

Sex Conversion Versus Female Lethality

Achieving female lethality with an endogenous mosquito gene involved in sex determination provides exciting opportunities to reduce mosquito populations and disease transmission. Such transgenic lines, when made homozygous through conditional expression, will produce all male progeny, providing efficient sex separation. This will be an advantage because current trials using genetically sterile mosquitoes rely on manual separation of males and females [25]. Furthermore, releasing such transgenic males is theoretically much more efficient than classic SITs in achieving population reduction and disease control because of the added benefit of male-bias in subsequent generations. Achieving female-to-male conversion (fertile or sterile) with an endogenous mosquito gene provides further opportunities to reduce mosquito populations. Complete sex conversion could double the productivity of mass-rearing operations. Female-to-male conversion may be easier to accomplish in *Culicinae* mosquitoes because dosage compensation is not expected for organisms with homomorphic sex-determining chromosomes. On the other hand, full female-to-male conversion may only be possible in *Anopheles* when the introduced genes do not affect the dosage compensation in the females.

Timing of the Transgene Effect

It is essential that transgene expression is tightly regulated both temporally and spatially, as expression outside of the necessary time period or in other tissues may create a fitness cost and reduce the effectiveness of the strategy. This requires the identification and testing of different promoters for the desired effect. Early zygotic and maternal-specific promoters have been identified in *Ae. aegypti* and *An. stephensi*, via large-scale transcriptome studies using RNA-Seq [14,78−80]. Early-acting embryonic promoters that confer female-specific lethality are optimal for sex separation during mass production as it reduces the cost per male by 50%. However, promoters that facilitate the induction of embryonic lethality may not be most effective in reducing mosquito populations. For example, it was found that for RIDL, late-acting promoters that were active in the late larval/pupal stage were more effective at population reduction than early-acting embryonic lethality due to density-dependent larval competition for nutrition [57]. For the purpose of sexing, embryonic promoters that drive transgene expression either to overexpress or knockdown a target gene may need to be active at the earliest embryonic stage when the first genes known as early zygotic genes become active. It is important that expression is early enough for transcription, translation, and sufficient accumulation of gene product to induce the desired effect. An alternative strategy may be to target embryonic genes by maternal delivery of transgene products as has been demonstrated in *An. stephensi* [14] and *Drosophila* [13,118]. Some candidate target genes may be expressed later in the embryo [112] and require zygotic promoters with a different expression profile.

Conditional Expression

Conditional gene expression for the purpose of sexing (male-selection) and induced late larval (bi-sex RIDL) and adult female lethality (fsRIDL) has been described above [27−29,57,58]. These methods employ the Tet-off system. Tet-off, as well as a modified Tet-off system, can be utilized for early embryonic sexing via expression of gene(s) involved in sex determination (Figure 10.4). The two systems shown in Figure 10.4A and B achieve the same end result, expression of a male-determining factor (M) in the absence of tetracycline to induce female lethality. In the Tet-off example, the colony can be maintained as homozygotes in the presence of tetracycline, as tetracycline inhibits tTA, expressed by the early zygotic promoter (embryonic promoter P_e), from binding the TRE and activating expression of M, the female-lethal gene. When sufficient numbers are achieved for a release, the removal of tetracycline allows sexing (male-selection) by inducing female lethality. Administration of tetracycline both maternally and by laying eggs on filter paper soaked with tetracycline-containing buffer has been demonstrated

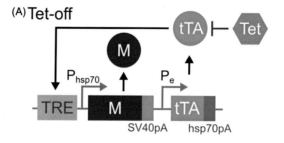

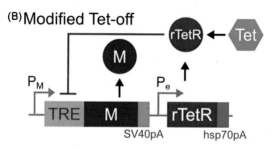

FIGURE 10.4   Tet-off (A) and modified Tet-off (B) methods for generating male-only offspring. TRE, Tetracycline response element; tTA, tetracycline transactivating protein, which is a fusion of the tet repressor (TetR, DNA-binding) and VP16 (transcription activation); Tet, tetracycline or its derivative doxycycline; P_e, an early zygotic promoter; P_m, the promoter of the M-factor gene; hsp70, minimum hsp70; rTetR, a mutant TetR that binds TRE only in the presence of doxycycline. Binding of TRE by tTA will activate transcription, whereas binding of TRE by rTetR will suppress transcription as it does not have VP16. M, tTA, and rTetR proteins are indicated in circles.

(Biedler and Tu, unpublished results). Alternatively, a modified Tet-off system achieves the same effect (Figure 10.4B), whereby the native M promoter drives M expression but is inhibited in the presence of rTetR, which is expressed by P_e. Here, rTetR binds the TRE in the presence of tetracycline and inhibits expression of M because rTetR does not have the VP16 transcription activator. Alternative strategies to achieve conditional expression may be used to activate M expression for sexing at the early embryonic stage. Recent advances in optogenetics make this approach feasible with promoters or transcription factors that are light-inducible [119−121]. Another strategy based on the use of inteins, temperature-sensitive splicing protein sequences, would allow the creation of a temperature-dependent switch when inserted into a gene of interest [122,123].

"Driving Maleness" to Control Mosquito-Borne Infectious Diseases

In addition to the obvious application in sex separation for mass production of males for release, genetic manipulation of sex ratios could also greatly reduce a local mosquito population. Due to the introduction of male-bias in subsequent generations, release of such transgenic males can achieve population suppression and is expected to be more effective than the classic SIT [117]. Higher copy numbers of such transgenes will be sustained longer in the population and are predicted to be more effective than a single copy release [117].

In essence, "maleness" is the ultimate disease-refractory trait because males do not transmit any pathogens. However, due to the high fitness cost of producing only males, such a "refractory" trait will be difficult to spread in natural populations and cannot be sustained in the context of population replacement. On the other hand, coupling a gene that either converts females to males or confers 100% female lethality with a gene drive mechanism such as a homing endonuclease or CRISPR/cas9-mediated homing system (as discussed in "Self-Sustaining Population Replacement Strategies" section) will significantly improve the efficiency of population reduction because more males will be produced at each generation than when a drive is not used. The use of gene drive may even negate the need for conditional expression as homing converts heterozygous transgenes to homozygous transgenes. Another potential benefit of this approach is that the transgene will not be sustained in the environment because the transgene will either be sustained over many generations until it crashes a population due to increasing male: female sex ratio, or it will be lost in the population due to the extremely high fitness cost of only producing males. Population reduction methods discussed in this chapter can also be incorporated with population replacement strategies to achieve sustained disease control. For example, population reduction methods may be employed first to facilitate subsequent population replacement. A recent mathematical modeling effort indicates that releasing insects that have both female-killing and antipathogen genes is potentially effective [124]. Ultimately, all these genetic methods are to be parts of integrated vector management strategies to achieve effective and sustained control of vector-borne infectious diseases.

ACKNOWLEDGMENTS

This work is supported by the Fralin Life Science Institute and the Virginia Experimental Station, and by National Institutes of Health (NIH) grants AI77680, AI105575, and AI113643, and by the National Science Foundation (NSF) Graduate Research Fellowship grant DGE-1519168. We thank Janet Webster for critical reading of the manuscript.

REFERENCES

[1] WHO. Dengue and severe dengue. WHO Fact Sheet, <http://www.who.int/mediacentre/factsheets/fs117/en/>; 2015.

[2] Feachem RG, Phillips AA, Hwang J, et al. Shrinking the malaria map: progress and prospects. Lancet 2010;376(9752):1566–78.

[3] White MT, Conteh L, Cibulskis R, Ghani AC. Costs and cost-effectiveness of malaria control interventions—a systematic review. Malar J 2011;10:337.

[4] Cheeseman IH, Miller BA, Nair S, et al. A major genome region underlying artemisinin resistance in malaria. Science 2012;336(6077):79−82.

[5] Ranson H, N'Guessan R, Lines J, Moiroux N, Nkuni Z, Corbel V. Pyrethroid resistance in African anopheline mosquitoes: what are the implications for malaria control? Trends Parasitol 2011;27(2):91−8.

[6] Catteruccia F. Malaria vector control in the third millennium: progress and perspectives of molecular approaches. Pest Manag Sci 2007;63(7):634−40.

[7] Liu N. Insecticide resistance in mosquitoes: impact, mechanisms, and research directions. Annu Rev Entomol 2015;60:537−59.

[8] Vontas J, Kioulos E, Pavlidi N, Morou E, della Torre A, Ranson H. Insecticide resistance in the major dengue vectors Aedes albopictus and Aedes aegypti. Pestic Biochem Phys 2012;104(2):126−31.

[9] Alphey L. Genetic control of mosquitoes. Annu Rev Entomol 2014;59:205−24.

[10] Meredith SE, James AA. Biotechnology as applied to vectors and vector control. Ann Parasitol Hum Comp 1990;65(Suppl. 1):113−18.

[11] James AA. Engineering mosquito resistance to malaria parasites: the avian malaria model. Insect Biochem Mol Biol 2002;32(10):1317−23.

[12] Sinkins SP, Gould F. Gene drive systems for insect disease vectors. Nat Rev Genet 2006;7(6):427−35.

[13] Chen CH, Huang H, Ward CM, et al. A synthetic maternal-effect selfish genetic element drives population replacement in Drosophila. Science 2007;316(5824):597−600.

[14] Biedler JK, Qi Y, Pledger D, James AA, Tu Z. Maternal germline-specific genes in the Asian malaria mosquito Anopheles stephensi: characterization and application for disease control. G3 2014;5(2):157−66.

[15] Windbichler N, Menichelli M, Papathanos PA, et al. A synthetic homing endonuclease-based gene drive system in the human malaria mosquito. Nature 2011;473(7346): 212−15.

[16] Akbari OS, Chen CH, Marshall JM, Huang H, Antoshechkin I, Hay BA. Novel synthetic Medea selfish genetic elements drive population replacement in Drosophila; a theoretical exploration of Medea-dependent population suppression. ACS Synth Biol 2014;3 (12):915−28.

[17] Papathanos PA, Bossin HC, Benedict MQ, et al. Sex separation strategies: past experience and new approaches. Malar J 2009;8(Suppl. 2):S5.

[18] Gilles JR, Schetelig MF, Scolari F, et al. Towards mosquito sterile insect technique programmes: exploring genetic, molecular, mechanical and behavioural methods of sex separation in mosquitoes. Acta Trop 2014;132(Suppl.):S178−87.

[19] Gabrieli P, Smidler A, Catteruccia F. Engineering the control of mosquito-borne infectious diseases. Genome Biol 2014;15(11):535.

[20] Oye KA, Esvelt K, Appleton E, et al. Biotechnology. Regulating gene drives. Science 2014;345(6197):626−8.

[21] Benedict M, D'Abbs P, Dobson S, et al. Guidance for contained field trials of vector mosquitoes engineered to contain a gene drive system: recommendations of a scientific working group. Vector Borne Zoonotic Dis 2008;8(2):127−66.

[22] Brown DM, Alphey LS, McKemey A, Beech C, James AA. Criteria for identifying and evaluating candidate sites for open-field trials of genetically engineered mosquitoes. Vector Borne Zoonotic Dis 2014;14(4):291−9.

[23] Scott MJ, Benedict MQ. Concept and history of genetic control. Genetic control of malaria and dengue. San Diego, CA: Elsevier; 2015.

[24] Bellini R, Medici A, Puggioli A, Balestrino F, Carrieri M. Pilot field trials with *Aedes albopictus* irradiated sterile males in Italian urban areas. J Med Entomol 2013;50 (2):317−25.

[25] Harris AF, McKemey AR, Nimmo D, et al. Successful suppression of a field mosquito population by sustained release of engineered male mosquitoes. Nat Biotechnol 2012;30 (9):828−30.

[26] Harris AF, Nimmo D, McKemey AR, et al. Field performance of engineered male mosquitoes. Nat Biotechnol 2011;29(11):1034−7.

[27] Fu G, Lees RS, Nimmo D, et al. Female-specific flightless phenotype for mosquito control. Proc Natl Acad Sci USA 2010;107(10):4550−4.

[28] Labbe GM, Scaife S, Morgan SA, Curtis ZH, Alphey L. Female-specific flightless (fsRIDL) phenotype for control of *Aedes albopictus*. PLoS Negl Trop Dis 2012;6(7): e1724.

[29] Marinotti O, Jasinskiene N, Fazekas A, et al. Development of a population suppression strain of the human malaria vector mosquito, *Anopheles stephensi*. Malar J 2013;12:142.

[30] Facchinelli L, Valerio L, Ramsey JM, et al. Field cage studies and progressive evaluation of genetically-engineered mosquitoes. PLoS Negl Trop Dis 2013;7(1):e2001.

[31] O'Connor L, Plichart C, Sang AC, Brelsfoard CL, Bossin HC, Dobson SL. Open release of male mosquitoes infected with a *Wolbachia* biopesticide: field performance and infection containment. PLoS Negl Trop Dis 2012;6(11):e1797.

[32] Hoffmann AA, Montgomery BL, Popovici J, et al. Successful establishment of *Wolbachia* in *Aedes* populations to suppress dengue transmission. Nature 2011;476(7361):454−7.

[33] Galizi R, Doyle LA, Menichelli M, et al. A synthetic sex ratio distortion system for the control of the human malaria mosquito. Nat Commun 2014;5:3977.

[34] Knipling E. Possibilities of insect control or eradication through the use of sexually sterile males. J Econ Entomol 1955;48:459−62.

[35] Bushland RC, Lindquist AW, Knipling EF. Eradication of screw-worms through release of sterilized males. Science 1955;122(3163):287−8.

[36] Krafsur ES, Whitten CJ, Novy JE. Screwworm eradication in North and Central America. Parasitol Today 1987;3(5):131−7.

[37] Hendrichs J, Franz G, Rendon P. Increased effectiveness and applicability of the sterile insect technique through male-only releases for control of Mediterranean fruit-flies during fruiting seasons. J Appl Entomol 1995;119(5):371−7.

[38] Vreysen MJ, Saleh KM, Ali MY, et al. Glossina austeni (Diptera: Glossinidae) eradicated on the island of Unguja, Zanzibar, using the sterile insect technique. J Econ Entomol 2000;93(1):123−35.

[39] Benedict MQ, Robinson AS. The first releases of transgenic mosquitoes: an argument for the sterile insect technique. Trends Parasitol 2003;19(8):349−55.

[40] Xi Z, Joshi D. Genetic control of malaria and dengue using *Wolbachia*. Genetic control of malaria and dengue. San Diego, CA: Elsevier; 2015.

[41] Laven H. Eradication of *Culex pipiens fatigans* through cytoplasmic incompatibility. Nature 1967;216(5113):383−4.

[42] Moreira LA, Iturbe-Ormaetxe I, Jeffery JA, et al. A *Wolbachia* symbiont in *Aedes aegypti* limits infection with dengue, chikungunya, and *Plasmodium*. Cell 2009;139(7):1268−78.

[43] McMeniman CJ, Lane RV, Cass BN, et al. Stable introduction of a life-shortening *Wolbachia* infection into the mosquito *Aedes aegypti*. Science 2009;323(5910):141−4.

[44] Kambris Z, Cook PE, Phuc HK, Sinkins SP. Immune activation by life-shortening *Wolbachia* and reduced filarial competence in mosquitoes. Science 2009;326(5949):134−6.

[45] Bian G, Xu Y, Lu P, Xie Y, Xi Z. The endosymbiotic bacterium *Wolbachia* induces resistance to dengue virus in *Aedes aegypti*. PLoS Pathog 2010;6(4):e1000833.

[46] Frentiu FD, Zakir T, Walker T, et al. Limited dengue virus replication in field-collected *Aedes aegypti* mosquitoes infected with *Wolbachia*. PLoS Negl Trop Dis 2014;8(2): e2688.

[47] Kambris Z, Blagborough AM, Pinto SB, et al. *Wolbachia* stimulates immune gene expression and inhibits plasmodium development in *Anopheles gambiae*. PLoS Pathog 2010;6 (10):e1001143.

[48] Hughes GL, Koga R, Xue P, Fukatsu T, Rasgon JL. Wolbachia infections are virulent and inhibit the human malaria parasite *Plasmodium falciparum* in *Anopheles gambiae*. PLoS Pathog 2011;7(5):e1002043.

[49] Bian G, Zhou G, Lu P, Xi Z. Replacing a native *Wolbachia* with a novel strain results in an increase in endosymbiont load and resistance to dengue virus in a mosquito vector. PLoS Negl Trop Dis 2013;7(6):e2250.

[50] Bian G, Joshi D, Dong Y, et al. *Wolbachia* invades *Anopheles stephensi* populations and induces refractoriness to *Plasmodium* infection. Science 2013;340(6133):748−51.

[51] Walker T, Johnson PH, Moreira LA, et al. The wMel *Wolbachia* strain blocks dengue and invades caged *Aedes aegypti* populations. Nature 2011;476(7361):450−3.

[52] Blagrove MS, Arias-Goeta C, Di Genua C, Failloux AB, Sinkins SP. A *Wolbachia* wMel transinfection in *Aedes albopictus* is not detrimental to host fitness and inhibits Chikungunya virus. PLoS Negl Trop Dis 2013;7(3):e2152.

[53] Dodson BL, Hughes GL, Paul O, Matacchiero AC, Kramer LD, Rasgon JL. *Wolbachia* enhances West Nile virus (WNV) infection in the mosquito *Culex tarsalis*. PLoS Negl Trop Dis 2014;8(7):e2965.

[54] Hughes GL, Rivero A, Rasgon JL. *Wolbachia* can enhance *Plasmodium* infection in mosquitoes: implications for malaria control?. PLoS Pathog 2014;10(9):e1004182.

[55] Hussain M, Lu G, Torres S, et al. Effect of *Wolbachia* on replication of West Nile virus in a mosquito cell line and adult mosquitoes. J Virol 2013;87(2):851−8.

[56] Hughes GL, Vega-Rodriguez J, Xue P, Rasgon JL. *Wolbachia* strain wAlbB enhances infection by the rodent malaria parasite *Plasmodium berghei* in *Anopheles gambiae* mosquitoes. Appl Environ Microbiol 2012;78(5):1491−5.

[57] Phuc HK, Andreasen MH, Burton RS, et al. Late-acting dominant lethal genetic systems and mosquito control. BMC Biol 2007;5:11.

[58] Wise de Valdez MR, Nimmo D, Betz J, et al. Genetic elimination of dengue vector mosquitoes. Proc Natl Acad Sci USA 2011;108(12):4772−5.

[59] Palopoli MF, Wu CI. Rapid evolution of a coadapted gene complex: evidence from the Segregation Distorter (SD) system of meiotic drive in *Drosophila melanogaster*. Genetics 1996;143(4):1675−88.

[60] Hammer MF, Silver LM. Phylogenetic analysis of the alpha-globin pseudogene-4 (Hba-Ps4) locus in the house mouse-species complex reveals a stepwise evolution of T-haplotypes (vol 10, pg 971, 1993). Mol Biol Evol 1994;11(1):158.

[61] Derome N, Metayer K, Montchamp-Moreau C, Veuille M. Signature of selective sweep associated with the evolution of sex-ratio drive in *Drosophila simulans*. Genetics 2004;166(3):1357−66.

[62] Hickey WA, Craig Jr. GB. Genetic distortion of sex ratio in a mosquito, *Aedes aegypti*. Genetics 1966;53(6):1177−96.

[63] Shin D, Mori A, Severson DW. Genetic mapping a meiotic driver that causes sex ratio distortion in the mosquito *Aedes aegypti*. J Hered 2012;103(2):303−7.

[64] Newton ME, Wood RJ, Southern DI. Cytogenetic analysis of meiotic drive in mosquito, *Aedes aegypti* (L). Genetica 1976;46(3):297−318.

[65] Mori A, Chadee DD, Graham DH, Severson DW. Reinvestigation of an endogenous meiotic drive system in the mosquito, *Aedes aegypti* (Diptera: Culicidae). J Med Entomol 2004;41(6):1027−33.

[66] Cha SJ, Mori A, Chadee DD, Severson DW. Cage trials using an endogenous meiotic drive gene in the mosquito *Aedes aegypti* to promote population replacement. Am J Trop Med Hyg 2006;74(1):62−8.

[67] Bernardini F, Galizi R, Menichelli M, et al. Site-specific genetic engineering of the *Anopheles gambiae* Y chromosome. Proc Natl Acad Sci USA 2014;111(21):7600−5.

[68] Collins FH, Mendez MA, Rasmussen MO, Mehaffey PC, Besansky NJ, Finnerty V. A ribosomal RNA gene probe differentiates member species of the *Anopheles gambiae* complex. Am J Trop Med Hyg 1987;37(1):37−41.

[69] Paskewitz SM, Wesson DM, Collins FH. The internal transcribed spacers of ribosomal DNA in five members of the *Anopheles gambiae* species complex. Insect Mol Biol 1993;2(4):247−57.

[70] Rafael MS, Tadei WP, Recco-Pimentel SM. Location of ribosomal genes in the chromosomes of *Anopheles darlingi* and *Anopheles nuneztovari* (Diptera, Culicidae) from the Brazilian Amazon. Mem Inst Oswaldo Cruz 2003;98(5):629−35.

[71] Burt A. Site-specific selfish genes as tools for the control and genetic engineering of natural populations. Proc Biol Sci/R Soc 2003;270(1518):921−8.

[72] Windbichler N, Papathanos PA, Catteruccia F, Ranson H, Burt A, Crisanti A. Homing endonuclease mediated gene targeting in *Anopheles gambiae* cells and embryos. Nucleic Acids Res 2007;35(17):5922−33.

[73] Windbichler N, Papathanos PA, Crisanti A. Targeting the X chromosome during spermatogenesis induces Y chromosome transmission ratio distortion and early dominant embryo lethality in *Anopheles gambiae*. PLoS Genet 2008;4(12):e1000291.

[74] Traver BE, Anderson MA, Adelman ZN. Homing endonucleases catalyze double-stranded DNA breaks and somatic transgene excision in *Aedes aegypti*. Insect Mol Biol 2009;18 (5):623−33.

[75] Esvelt KM, Smidler AL, Catteruccia F, Church GM. Concerning RNA-guided gene drives for the alteration of wild populations. eLife 2014;e03401.

[76] Gantz VM, Bier E. The mutagenic chain reaction: A method for converting heterozygous to homozygous mutations. Genome editing. Science 2015;348(6233):442−4.

[77] Port F, Muschalik N, Bullock S. CRISPR with independent transgenes is a safe and robust alternative to autonomous gene drives in basic research. BioRxiv 2015.

[78] Akbari OS, Antoshechkin I, Amrhein H, et al. The developmental transcriptome of the mosquito *Aedes aegypti*, an invasive species and major arbovirus vector. G3 2013;3 (9):1493−509.

[79] Akbari OS, Papathanos PA, Sandler JE, Kennedy K, Hay BA. Identification of germline transcriptional regulatory elements in *Aedes aegypti*. Sci Rep 2014;4:3954.

[80] Biedler JK, Hu W, Tae H, Tu Z. Identification of early zygotic genes in the yellow fever mosquito *Aedes aegypti* and discovery of a motif involved in early zygotic genome activation. PLoS One 2012;7(3):e33933.

[81] Akbari OS, Matzen KD, Marshall JM, Huang H, Ward CM, Hay BA. A synthetic gene drive system for local, reversible modification and suppression of insect populations. Curr Biol 2013;23(8):671−7.

[82] Baker RH, Sakai RK. Triploids and male determination in the mosquito, *Anopheles culicifacies*. J Hered 1979;70(5):345−6.

[83] Marin I, Baker BS. The evolutionary dynamics of sex determination. Science 1998;281 (5385):1990−4.

[84] Fraccaro M, Tiepolo L, Laudani U, Marchi A, Jayakar SD. Y chromosome controls mating behaviour on *Anopheles* mosquitoes. Nature 1977;265(5592):326−8.

[85] Newton ME, Southern DI, Wood RJ. X and Y chromosomes of *Aedes aegypti* (L.) distinguished by Giemsa C-banding. Chromosoma 1974;49(1):41−9.

[86] Severson DW, Meece JK, Lovin DD, Saha G, Morlais I. Linkage map organization of expressed sequence tags and sequence tagged sites in the mosquito, *Aedes aegypti*. Insect Mol Biol 2002;11(4):371−8.

[87] Toups MA, Hahn MW. Retrogenes reveal the direction of sex-chromosome evolution in mosquitoes. Genetics 2010;186(2):763−6.

[88] McClelland GAH. Sex-linkage in *Aedes aegypti*. Trans Roy Soc Tropical Med Hyg 1962;56(4).

[89] Motara M, Rai K. Giemsa C-banding patterns in *Aedes* (Stegomia) mosquitoes. Chromosoma 1978;70:51−8.

[90] Mori A, Severson DW, Christensen BM. Comparative linkage maps for the mosquitoes (*Culex pipiens* and *Aedes aegypti*) based on common RFLP loci. J Hered 1999;90 (1):160−4.

[91] Malcolm CA, Bourguet D, Ascolillo A, et al. A sex-linked *Ace* gene, not linked to insensitive acetylcholinesterase-mediated insecticide resistance in *Culex pipiens*. Insect Mol Biol 1998;7(2):107−20.

[92] Motara M, Rai K. Chromosomal differentiation in two species of *Aedes* and their hybrids revealed by Giemsa C-banding. Chromosoma 1977;64:125−32.

[93] Venkatesan M, Broman KW, Sellers M, Rasgon JL. An initial linkage map of the West Nile Virus vector *Culex tarsalis*. Insect Mol Biol 2009;18(4):453−63.

[94] Salvemini M, D'Amato R, Petrella V, et al. The orthologue of the fruitfly sex behaviour gene fruitless in the mosquito *Aedes aegypti*: evolution of genomic organisation and alternative splicing. PLoS One 2013;8(2):e48554.

[95] Scali C, Catteruccia F, Li Q, Crisanti A. Identification of sex-specific transcripts of the *Anopheles gambiae* doublesex gene. J Exp Biol 2005;208(Pt. 19):3701−9.

[96] Salvemini M, Mauro U, Lombardo F, et al. Genomic organization and splicing evolution of the doublesex gene, a *Drosophila* regulator of sexual differentiation, in the dengue and yellow fever mosquito *Aedes aegypti*. BMC Evol Biol 2011;11:41.

[97] Pane A, Salvemini M, Delli Bovi P, Polito C, Saccone G. The transformer gene in *Ceratitis capitata* provides a genetic basis for selecting and remembering the sexual fate. Development 2002;129(15):3715−25.

[98] Carvalho AB. Origin and evolution of the *Drosophila* Y chromosome. Curr Opin Genet Dev 2002;12(6):664−8.

[99] Carvalho AB, Lazzaro BP, Clark AG. Y chromosomal fertility factors kl-2 and kl-3 of *Drosophila melanogaster* encode dynein heavy chain polypeptides. Proc Natl Acad Sci USA 2000;97(24):13239−44.

[100] Lemos B, Araripe LO, Hartl DL. Polymorphic Y chromosomes harbor cryptic variation with manifold functional consequences. Science 2008;319(5859):91−3.

[101] Zhou Q, Bachtrog D. Sex-specific adaptation drives early sex chromosome evolution in *Drosophila*. Science 2012;337(6092):341−5.

[102] Hasselmann M, Gempe T, Schiott M, Nunes-Silva CG, Otte M, Beye M. Evidence for the evolutionary nascence of a novel sex determination pathway in honeybees. Nature 2008;454(7203):519−22.

[103] Kiuchi T, Koga H, Kawamoto M, et al. A single female-specific piRNA is the primary determiner of sex in the silkworm. Nature 2014;509(7502):633−6.

[104] Graham P, Penn JK, Schedl P. Masters change, slaves remain. Bioessays 2003;25 (1):1−4.

[105] Demir E, Dickson BJ. fruitless splicing specifies male courtship behavior in Drosophila. Cell 2005;121(5):785−94.

[106] Gailey DA, Billeter JC, Liu JH, Bauzon F, Allendorfer JB, Goodwin SF. Functional conservation of the fruitless male sex-determination gene across 250 Myr of insect evolution. Mol Biol Evol 2006;23(3):633−43.

[107] Burtis KC, Baker BS. *Drosophila* doublesex gene controls somatic sexual differentiation by producing alternatively spliced mRNAs encoding related sex-specific polypeptides. Cell 1989;56(6):997−1010.

[108] Mank JE. Sex chromosome dosage compensation: definitely not for everyone. Trends Genet 2013;29(12):677−83.

[109] Schutt C, Nothiger R. Structure, function and evolution of sex-determining systems in Dipteran insects. Development 2000;127(4):667−77.

[110] Thomas CG, Woodruff GC, Haag ES. Causes and consequences of the evolution of reproductive mode in *Caenorhabditis* nematodes. Trends Genet 2012;28(5):213−20.

[111] Cline TW. Two closely linked mutations in *Drosophila melanogaster* that are lethal to opposite sexes and interact with daughterless. Genetics 1978;90(4):683−98.

[112] Hall AB, Qi Y, Timoshevskiy V, Sharakhova MV, Sharakhov IV, Tu Z. Six novel Y chromosome genes in *Anopheles* mosquitoes discovered by independently sequencing males and females. BMC Genomics 2013;14:273.

[113] Criscione F, Qi Y, Saunders R, Hall B, Tu Z. A unique Y gene in the Asian malaria mosquito *Anopheles stephensi* encodes a small lysine-rich protein and is transcribed at the onset of embryonic development. Insect Mol Biol 2013;22(4):433−41.

[114] Hall AB, Timoshevskiy VA, Sharakhova MV, et al. Insights into the preservation of the homomorphic sex-determining chromosome of *Aedes aegypti* from the discovery of a male-biased gene tightly linked to the M-locus. Genome Biol Evol 2014;6(1):179−91.

[115] Sharon D, Tilgner H, Grubert F, Snyder M. A single-molecule long-read survey of the human transcriptome. Nat Biotechnol 2013;31(11):1009−14.

[116] Li M, Mead EA, Zhu J. Heterodimer of two bHLH-PAS proteins mediates juvenile hormone-induced gene expression. Proc Natl Acad Sci USA 2011;108(2):638−43.

[117] Schliekelman P, Ellner S, Gould F. Pest control by genetic manipulation of sex ratio. J Econ Entomol 2005;98(1):18−34.

[118] Staller MV, Yan D, Randklev S, et al. Depleting gene activities in early *Drosophila* embryos with the "maternal-Gal4-shRNA" system. Genetics 2013;193(1):51−61.

[119] Miao G, Hayashi S. Manipulation of gene expression by infrared laser heat shock and its application to the study of tracheal development in *Drosophila*. Dev Dyn 2015;244 (3):479−87.

[120] Motta-Mena LB, Reade A, Mallory MJ, et al. An optogenetic gene expression system with rapid activation and deactivation kinetics. Nat Chem Biol 2014;10(3):196–202.

[121] Polstein LR, Gersbach CA. A light-inducible CRISPR-Cas9 system for control of endogenous gene activation. Nat Chem Biol 2015;11(3):198–200.

[122] Zeidler MP, Tan C, Bellaiche Y, et al. Temperature-sensitive control of protein activity by conditionally splicing inteins. Nat Biotechnol 2004;22(7):871–6.

[123] Tan G, Chen M, Foote C, Tan C. Temperature-sensitive mutations made easy: generating conditional mutations by using temperature-sensitive inteins that function within different temperature ranges. Genetics 2009;183(1):13–22.

[124] Robert MA, Okamoto K, Lloyd AL, Gould F. A reduce and replace strategy for suppressing vector-borne diseases: insights from a deterministic model. PLoS One 2013;8(9): e73233.

Chapter 11

Disruption of Mosquito Olfaction

Conor J. McMeniman
Department of Molecular Microbiology and Immunology, Malaria Research Institute, Johns Hopkins Bloomberg School of Public Health, Baltimore, MD, USA

INTRODUCTION

Multiple arthropods, including several species of mosquitoes, seek out vertebrate hosts to blood-feed and reproduce. As a direct consequence of this collective suite of behaviors, many blood-sucking arthropods also transmit a range of bacterial, protozoal, and viral pathogens to humans. Some vector species display an innate preference to blood-feed on certain vertebrate hosts, and also vary in their tendency to blood-feed at multiple times during their lifecycle—with important implications for epidemiological risk [1,2]. For instance, of the hundreds of species of mosquitoes that blood-feed on vertebrate hosts, the most dangerous species for human health are those that feed preferentially and frequently on humans [3–5]. The yellow fever mosquito *Aedes aegypti* feeds almost exclusively on the blood of humans, and often takes multiple blood meals during a single gonotrophic cycle [6–9], making this species an extremely efficient vector of dengue, yellow fever, and chikungunya viruses. Similarly, the African malaria mosquito *Anopheles gambiae* s.s. is strongly anthropophilic [10–13], on average obtaining approximately 90% of its blood meals from humans [14]. Such a preference for humans therefore also drives the transmission of human malaria parasites.

To successfully find a human host, anthropophilic mosquitoes are thought to detect and integrate a variety of sensory cues emanating from humans including body odor, CO_2, moisture, heat, and visual contrast [15]. Some of these stimuli, such as thermal convection currents emitted from the human skin, most likely rapidly dissipate to reach equilibrium with the ambient environment within a couple of meters from the host [16], and thus likely contribute toward close range attraction of mosquitoes to humans [17]. Chemical cues, such as CO_2 and other volatile odorants derived from the skin and breath however, most likely dissipate further downwind of a host in

Genetic Control of Malaria and Dengue.
© 2016 Elsevier Inc. All rights reserved.

227

odor plumes where they mediate mosquito attraction toward humans over a range of distances [18,19]. To follow an odor plume toward its source, mosquitoes and other insects most likely rely on their sense of smell coupled with visual feedback from their environmental surrounds. This odor orientation strategy termed odor-guided optomotor-mediated, positive anemotaxis (see Ref. [20]) allows mosquitoes to efficiently calibrate their spatial position upwind along a trail of scent—a phenomenon first described in insects by Kennedy [21] during the 1940s with wind-tunnel studies involving *Ae. aegypti*.

A number of chemical compounds, some of which are commonly found in human odor, that elicit both electrophysiological and behavioral responses in *An. gambiae* and *Ae. aegypti* including ammonia [22−24] as well as several alcohols [25]; aldehydes [25−27]; aliphatic, unsaturated and aromatic carboxylic acids [23−25,27−38]; oxocarboxylic acids [35,39]; ketones [24,25,27,40,41]; and sulfides [24,29,40−42] have been identified (see Refs [43,44]). Identifying the active volatile components of human odor that combine with CO_2 to drive host-seeking behavior, along with their associated receptors in the mosquito olfactory system will help to define critical peripheral sensory pathways needed for mosquitoes to find humans. In this chapter, I will review the molecular basis of olfaction in *Drosophila* and mosquitoes, genetic methods that facilitate identification of mosquito genes that detect human odorants and CO_2, and our current understanding of their relative contribution to mosquito host-seeking behavior. These peripheral sensors represent alluring targets for chemical and genetic strategies that aim to stop mosquitoes from finding humans.

ANATOMY OF THE MOSQUITO OLFACTORY SYSTEM

Human odor is a complex mixture of hundreds of volatile organic compounds [45−47], which blend together to attract mosquitoes and other disease vectors toward humans. To sense these volatile host emissions, mosquitoes use a finely tuned olfactory system that is composed of three major sensory appendages: the antennae, the maxillary palps, and the proboscis. Each of these olfactory organs is populated by a large number of multiporous sensory hairs called sensilla that house olfactory sensory neurons (OSNs) expressing chemosensory receptors tuned to specific odorants (discussed later in "Chemosensory Receptors Detecting Odorants and CO_2 in Drosophila and Mosquitoes" section; Figure 11.1A for anatomy). Within the internal lumen of each sensillum, OSN dendrites are bathed in sensillum lymph that contains water-soluble odorant-binding proteins (OBPs) [48]. OBPs are putatively responsible for binding and transporting volatile organic compounds that enter the sensillum lymph. Although it is evident that varied mosquito species contain large families of OBP genes [49−53], many of which are expressed in sensilla [54,55], their relative functional contributions to odorant detection presently remain unclear.

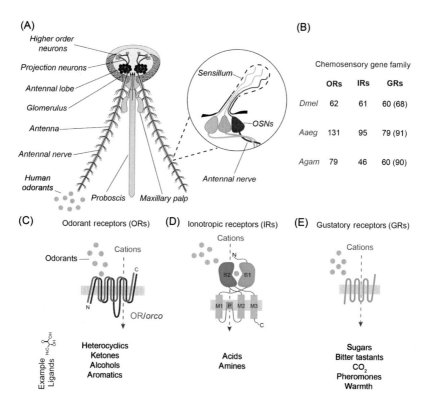

FIGURE 11.1 Overview of the mosquito olfactory system: (A) OSNs on the mosquito antennae, proboscis, and maxillary palps detect varied chemical stimuli, including human volatiles and CO_2. The antennal nerve—comprising of OSN axons—relays olfactory information from the periphery to the antennal lobe of the mosquito brain. Olfactory inputs from the maxillary palps and proboscis to the mosquito brain are schematized with arrows. (B) Total gene numbers of ORs, IRs, and GRs from *D. melanogaster* (*Dmel*), *Ae. aegypti* (*Aaeg*), and *An. gambiae* (*Agam*) genomes. Parentheses indicate the number of predicted GR splice variants. (C–E) Schematic models of ORs, IRs, and GRs. Examples of some ligands for each of these chemoreceptor families are listed.

Diverse sensilla types are found on the antennae, palps, and proboscis of male and female mosquitoes. Mosquito antennae are populated by five morphological types of sensilla [56,57]—of which two: trichodea (hairs) and basiconica (grooved pegs) contain OSNs responsive to odorants. Of these, sensilla trichodea comprise the largest proportion of the antennal sensillum population in *An. gambiae* and *Ae. aegypti* and house between two and three OSNs [56,58], responsive to many alcohols, carboxylic acids, and esters [59,60]. OSNs housed within grooved peg sensilla have also been identified that respond sensitively to the human-derived odorant L-lactic acid [61]. In contrast to antennae, the maxillary palps are populated by two types of sensilla: sensilla chaetica and capitate peg sensilla [62]. In both *An. gambiae* and *Ae. aegypti*, capitate peg sensilla contain three OSNs [63,64], one of which is exquisitely sensitive to the volatile gas CO_2 [65,66]. The labellum on the tip of the mosquito proboscis also contains three types of trichoid sensilla named T1, T2, and T3 [56]. Labellum-associated OSNs housed within T2 sensilla have recently been described, which respond to a small range of human-related odorants including several ketones, isovaleric acid, butylamine, and oxocarboxylic acids—suggesting that this organ may have dual functionality in taste and olfaction [67].

In *Drosophila*, mosquitoes, and other insects, OSNs from the antennae, maxillary palps, and proboscis project centrally to the antennal lobe, the first relay station for olfactory processing in the insect brain. Here, they form defined clusters of synaptic neuropil called glomeruli. In *Drosophila*, OSNs expressing the same odorant receptor (OR) converge onto a single glomerulus [68]. Anterograde dye-filling of peripheral OSNs that respond functionally to the same subset of odors, and subsequent tracing of their glomerular projections have indicated that this pattern is likely to be conserved in mosquitoes [59,69,70]. Antibody labeling and three-dimensional reconstruction using confocal microscopy have revealed that each antennal lobe of female *An. gambiae* and *Ae. aegypti* has 60 and 50 glomerular neuropils, respectively [71,72].

In the insect antennal lobe, odor information is locally processed and coded, via excitatory and inhibitory local neurons that interconnect multiple glomeruli [73−76], before being sent to higher order processing centers of the insect brain, such as the lateral horn and mushroom bodies of the protocerebrum, that orchestrate innate and learned behaviors. How sensory inputs from human odorants and CO_2 are processed in the mosquito antennal lobe, before being relayed onto higher order neurons involved in initiating host-seeking and blood-feeding behavior is an outstanding research question that awaits further characterization [77].

CHEMOSENSORY RECEPTORS DETECTING ODORANTS AND CO_2 IN *DROSOPHILA* AND MOSQUITOES

What chemosensory receptors allow mosquitoes to detect hosts and drive preference for humans? Three large gene families encoding chemosensory receptors that allow insects to sense volatile odorants including CO_2 have been described to date. These include: (i) ORs, (ii) ionotropic receptors (IRs), and (iii) gustatory receptors (GRs). Initially discovered in the vinegar fly *Drosophila melanogaster*, ORs, IRs, and GRs have since been identified in a range of insect species including *An. gambiae* and *Ae. aegypti* using comparative genomics. A description of their discovery, biochemistry, ligand specificity, and progressive characterization in mosquitoes is presented below (Figure 11.1B−E for an overview).

Odorant Receptors

Insect OR genes encode seven-transmembrane domain (7-TMD) receptor proteins that are selectively expressed in subsets of OSNs on olfactory appendages such as the antennae, maxillary palps, and proboscis. Independently described by three research groups in *Drosophila* in 1999 [78–80], this large gene family appears to be evolutionarily divergent from vertebrate ORs [81], which are 7-TMD G-protein-coupled receptors (GPCRs) found in the olfactory tissues of most vertebrate species examined [82–84]. Relative to their vertebrate counterparts, insect OR proteins lack typical DRY and NPXXY amino acid motifs that characterize vertebrate GPCRs [80] and also adopt a different transmembrane topology with their N-terminus localized intracellularly and C-terminus found extracellularly [85–88]. Genome sequencing and expression profiling were first used to describe 62 OR genes present in *D. melanogaster* [78–80]. Subsequent comparative genomic analyses have revealed the presence of this novel gene family in a wide range of insect species including mosquitoes, as first described by Zwiebel and colleagues in *An. gambiae* (79 ORs) [89–91], and *Ae. aegypti* (131 ORs) [92].

Insect ORs most likely function as heteromeric ligand-gated ion channels [87,93,94] that are constituted by a ligand-binding OR tuned to a particular series of odorants, and a broadly expressed and evolutionarily conserved OR coreceptor, termed *orco* (*O*dorant *R*eceptor *Co*-Receptor) [95–97]. Most OSNs in *Drosophila* coexpress *orco* (formerly known as *Or83b* in this species) and one of 61 ligand-specific ORs [98]. This pattern of coexpression appears to be maintained in other insects including mosquitoes [99–101], with variation in their number of ligand-binding ORs. Targeted mutagenesis of *orco* in *D. melanogaster* [98], and more recently in *Ae. aegypti* [102], has revealed that this gene is required for both electrophysiological and behavioral responses to certain odorants. Consistent with this role, biochemical evidence indicates that *orco* traffics ligand-binding ORs from the OSN cell body to the dendritic membrane, where *orco* then functions as a component of a heteromeric *orco*/OR receptor complex [85].

In *D. melanogaster*, electrophysiological recordings from the different morphological classes of sensilla distributed over the fly antenna and palp indicate that OSNs can be classified into distinct functional classes, each with their own unique spectrum of responsiveness to odors [103–106]. For instance, some OSNs respond to broadly to a number of odorants, whereas others appear to be more narrowly tuned to a smaller subset [105]. Toward dissecting the molecular basis of odor coding—that is, how odorant identities are encoded by the complement of olfactory receptors expressed within OSNs—various methods have since been used to quantify responses of ORs to a variety of chemical compounds. One powerful approach for this purpose involves genetically introducing a particular OR of choice into a mutant *Drosophila* antennal OSN (ab3A) that lacks its endogenous ligand-binding

OR, but retains *orco* [107]. This heterologous expression system termed the "empty neuron" typically allows the functional reconstitution of heteromeric OR/*orco* complexes [108], even when the ligand-binding ORs are derived from non-*Drosophild* insects such as mosquitoes [109]. Odor-evoked responses can then recorded from this neuron using single-sensillum electrophysiology [110–112] to assign ligands to the tested receptor and generate a tuning curve for each OR in response to odorant stimulation. Such OR tuning curves quantify both the number of odorants that excite a particular receptor and their associated sensitivity to these compounds.

Analyses using the empty neuron system in *Drosophila* [108,113–115] have revealed three important principles of olfactory coding by the OR repertoire. First, in most instances, individual odorants activate specific subsets of receptors suggestive of a combinatorial model of odor coding. Second, individual receptors vary in their tuning breadth to odors. Third, there appears to be a continuum in tuning breadths across the OR repertoire. Although some olfactory stimuli activate many classes of OSNs and their associated ORs, other odorants, such as the *Drosophila* pheromone cVA [103,116,117] and the microbial product geosmin [118], activate dedicated olfactory circuits known as labeled lines that consist of a single OR, OSN, and cognate antennal lobe glomerulus. This olfactory processing strategy is thought to encode certain odorants of major behavioral significance including some mating pheromones and aversive stimuli.

Large-scale screens that the test odorant responsiveness of a major proportion of the OR repertoires from *D. melanogaster* and *An. gambiae* have been carried out using the empty neuron system [108,119]. In both fly and mosquito, these analyses have revealed that ligand-binding ORs are generally most strongly responsive to heterocyclics, ketones, alcohols, and aromatic compounds. Interestingly, several *An. gambiae* ORs tested using this system appear to be narrowly tuned to several odorants that emanate from humans including indole (a volatile component of human sweat), 1-octen-3-ol (a component of human breath), 2,3-butanedione (a metabolic by-product of human skin microflora), and 2-ethylphenol (found in the urine of many animals) [119]. Such narrow tuning of these ORs may function to improve cue salience, and therefore guide mosquito olfactory responses during their search for humans. Similar large-scale analyses are yet to be undertaken with the *Ae. aegypti* OR repertoire [92], and it will be interesting to compare and contrast the odorant tuning profiles of ORs from these two anthropophilic mosquito species.

Screens using the empty neuron system have also been complemented by other heterologous expression methods, including the expression of insect ORs in *Drosophila* S2 cells [85], lepidopteran cell lines [88,120], human embryonic kidney (HEK293) cells [121–123], HeLa cells [93], and *Xenopus* oocytes [124]. In addition to improving our understanding of the molecular basis of odor coding in mosquitoes and other insects [124–126], these systems have provided key insights into the biochemical stoichiometry, signal transduction, and pharmacology of insect ORs [85,87,88,93,94,123,127]. Indeed, a small molecule screen designed by Zwiebel and colleagues to find modulators of OR function [122] has yielded a valuable class of chemical compounds that broadly modulate the activity of *orco* in a variety of insect species of medical and agricultural importance [128–131]. Although improvements to the volatility of these compounds as well as analyses of their effects on olfactory-guided behaviors in insects are needed, these promising leads may yield novel classes of repellents that are active against a range of insect disease vectors.

Ionotropic Receptors

Of the three types of olfactory sensilla found on the *Drosophila* antennae (basiconic, coeloconic, and trichoid), the vast majority of OSNs housed within coeloconic sensilla do not express either *orco* or other OR members, yet respond to ligands such as acids and ammonia [106]. This observation hinted toward the existence of novel types of insect chemosensory receptors that detect these volatile chemicals. Using leads from a bioinformatic screen for insect-specific genes enriched in OSNs [132], Benton and colleagues recently identified a novel family of chemosensory receptors divergent from ORs, which they named IRs [133]. These genes are expressed in coeloconic OSNs, accumulate at the ciliated endings of sensory dendrites, and are tuned to volatile odorants such as acids and amines [133–136].

In total, 61 predicted IR genes and two IR pseudogenes were identified in the *D. melanogaster* genome [133]. This gene family appears to be closely related to ionotropic glutamate receptors (iGluRs)—which facilitate chemical communication at synapses in invertebrate and vertebrate nervous systems by binding of extracellular glutamate and related ligands [137]. Like iGluRs, IRs have a predicted modular protein structure consisting of an extracellular N-terminus, a bipartite ligand-binding domain gating an ion channel, and a short cytoplasmic C-terminus; but they appear to have lost characteristic amino acid residues in their ligand-binding domain needed for glutamate binding [133]. IRs are expressed in combinations of up to three subunits in an OSN—with each complex consisting of a ligand-binding IR, and one or two broadly expressed coreceptors (*IR8a and IR25a*) [134]. These coreceptors are necessary and sufficient for trafficking IR complexes to ciliated endings of dendrites [134]. OSNs expressing an individual ligand-binding IR converge upon a single antennal lobe glomerulus, mirroring the pattern of glomerular innervation seen OSNs that express ORs [133]. Furthermore, ectopic expression of IRs in nonendogenous OSNs in the *D. melanogaster* olfactory system is sufficient to confer novel odor-sensitivity upon these neurons, clearly implicating these proteins as a novel family of chemosensory receptors [133,134].

Given that compounds such as ammonia, lactic acid, and other carboxylic acids emanating from humans have been shown to blend together in behavioral assays to attract mosquitoes [23], and that *Drosophila* IRs appear to be tuned to acids and amines [135,136], IR genes are prime candidates for encoding receptors that guide mosquitoes toward humans. Genomic analysis using amino acid sequences of *Drosophila* IRs as a guide has facilitated identification of 46 IRs in the *An. gambiae* and 95 IRs in the *Ae. aegypti* genomes [137,138]. Expression profiling has indicated that IRs are expressed in both larval and adult olfactory tissues of these mosquito species [138–142]. Although large-scale heterologous screens that link mosquito IRs to human odor ligands have yet to be undertaken, such analyses coupled with characterization of the complete mosquito OR repertoire promise to yield great insights into the chemosensory receptors that mosquitoes use to detect humans.

Gustatory Receptors

GR genes encode an evolutionarily divergent family of 7-TMD receptor proteins that were identified shortly after the discovery of *Drosophila* ORs [143,144]. GRs appear to be distantly related to insect ORs [145] and were named the GR family as many GR genes are expressed in taste organs such as the labellum of the proboscis, the pharyngeal organs, and the tarsi. Like ORs, GRs lack apparent GPCR amino acid motifs needed for canonical signal transduction through G-proteins, and potentially function as ligand-gated cation channels [146].

Bioinformatics and gene expression profiling initially led to the identification of 68 GRs encoded by 60 genes in the *Drosophila* genome through alternative splicing [143–145,147]. Comparative genomic analyses have since indicated that mosquito species encode similarly sized repertoires of GRs. Excluding pseudogenes, the *An. gambiae* genome putatively expresses 90 functional GRs encoded by 60 GR genes, while the *An. aegypti* genome is predicted to express 91 functional GRs encoded by 79 GR genes [148]. Supporting these computational predictions, many of these GRs appear to be expressed in the maxillary palp, labellum of the proboscis, and tarsi of *Ae. aegypti* as determined using RNAseq transcriptional profiling [139,142].

In *Drosophila*, GRs have been shown to function in wide diversity of sensory modalities. Several *Drosophila* GRs are involved in the detection of sugars and bitter substrates: *Gr5a*, *Gr43a*, *Gr64a*, and *Gr64f* genes function as sugar sensors [149–152], while *Gr33a*, *Gr66a*, and *Gr93a* mediate responses to caffeine and other bitter compounds [153–155]. Consistent with the role of *Gr43a* as an internal sensor for fructose, its *An. gambiae* ortholog *AgamGr25* also responds robustly to this sugar and other sweet tastants [156]. Other *Drosophila* GRs have been implicated in pheromone detection during courtship behavior (e.g., *Gr32a*, *Gr33a*, *Gr39a*, and *Gr68a*)

[155,157−159] and the detection of light and warmth (e.g., *Gr28b*) [160,161]. Whether orthologous genes are responsible for detecting these stimuli in mosquitoes remains to be determined.

When the GR gene family was first described in *Drosophila*, a handful of GRs (*Gr10a*, *Gr21a*, and *Gr63a*) were found to be expressed in OSNs on the fly antenna [144]. This suggested that members of the GR gene family possibly detect volatile odorants. Subsequently, it was demonstrated that two of these genes, *Gr21a* and *Gr63a*, function together to mediate detection of the volatile gas CO_2 [162,163]. Multiple pieces of evidence indicate that these two GRs together form a molecular sensor for CO_2. First, *Gr21a* and *Gr63a* are coexpressed in ab1C neurons on the fly antennae that detect CO_2 [162,163]. Secondly, if *Gr63a* is deleted from the *Drosophila* genome via homologous recombination, *Gr63a* mutants are rendered both electrophysiologically and behaviorally insensitive to CO_2 [162]. *Gr21a* therefore cannot detect CO_2 alone. Finally, ectopic expression of both *Gr21a* and *Gr63a*, but neither receptor by itself, confers partial CO_2 sensitivity on the ab3A empty neuron that does not otherwise sense this volatile gas [162,163]. Additional genetic analyses have revealed signaling by this CO_2 receptor complex that is also modulated by G-proteins: $G\alpha_q$ and $G\gamma30A$—though neither the mechanistic basis for this requirement nor the signal transduction mechanism has presently been characterized [164]. It is further not known whether one of these GR subunits functions as a nonligand-selective coreceptor, whereas the other interacts with the ligand CO_2, as is the case with insect ORs [85] and IRs [134].

Direct orthologs of the *Gr21a* and *Gr63a* from *D. melanogaster* have since been identified in a wide range of insect species including *Ae. aegypti* and *An. gambiae*, where they are named *Gr1* and *Gr3*, respectively [165]. In most non-*Drosophila* insect species examined, a *Gr1* paralog named *Gr2* of unknown function is also present. Consistent with their role as candidate CO_2 receptors, fluorescent RNA *in situ* hybridization has revealed that these three genes are coexpressed within the same OSNs on the mosquito maxillary palps [162,166]—the anatomical site of CO_2 detection in multiple mosquito species [65,66]. Functional studies that reconstitute different combinations of *An. gambiae* orthologs of these three genes (denoted *AgamGr22*, *AgamGr23*, and *AgamGr24* in this species) using heterologous expression in *Drosophila* [166]; and transient knockdown of *AaegGr1*, *AaegGr2*, and *AaegGr3* in *Ae. aegypti* [167] further have suggested a conserved role for these genes as candidate CO_2 receptors.

USE OF GENOME-EDITING TECHNOLOGY TO DECODE MOSQUITO ATTRACTION TO HUMANS

The availability of genome sequences from the vinegar fly *D. melanogaster* [168] and three major mosquito disease vectors species—*An. gambiae* [169],

Ae. aegypti [170], and *Culex quinquefasciatus* [171]—has helped to reveal the molecular identity of large families of chemosensory genes that potentially play a role in mosquito host-seeking and blood-feeding behavior. Recent advances in the development of genome-editing tools [172], such as zinc-finger nucleases (ZFNs), TALE-effector nucleases (TALENs), and CRISPR/Cas9 RNA-guided nucleases (discussed in Chapter 8), which facilitate targeted mutagenesis at any chromosomal site of interest, now present researchers with an exciting opportunity to test the functional contribution of these chemosensory genes to mosquito behavior. The speed and cost-effective nature at which mutations in mosquito genes can now be generated using these reagents [77,102,173–175], particularly with regard to the CRISPR/Cas9 system, suggests that large-scale reverse genetic screens that decode the molecular basis of mosquito attraction to humans may now be possible.

Targeted Mutagenesis of OR-Mediated Odorant Reception

Among the first to employ genome-editing technology in mosquitoes, DeGennaro and colleagues [102] utilized ZFNs targeting the *orco* gene of *Ae. aegypti* to query the role of the OR gene family in mosquito host-seeking behavior. Given that *orco* is required for the efficient trafficking of ligand-binding ORs to OSN dendrites, as well as heteromeric receptor formation, mutations to this gene were predicted to disrupt the trafficking and associated function of 130 other ligand-binding ORs present in the *Ae. aegypti* genome. Consistent with this predicted outcome, *orco*-mutant mosquitoes were shown to be electrophysiologically insensitive to a range OR ligands when these chemicals were applied to OSNs from numerous olfactory sensilla distributed across the mosquito antennae and maxillary palps. In line with previous observations from *Drosophila* [176,177], *orco*-mutant mosquitoes were also not repelled by volatile phase of the insect repellent *N,N*-Diethyl-*meta*-toluamide (DEET) [102], indicating that the OR pathway is a direct molecular target for this behaviorally aversive compound [178] (discussed further in "Next-Generation Chemical Strategies Targeting Mosquito Olfaction" section).

Surprisingly, despite a complete loss of OR signaling, *orco*-mutant female mosquitoes were unimpaired in their ability to find a live human arm when this stimulus was presented with, or without human breath in a dual port olfactometer behavioral assay. Similarly, *orco* mutants were as attracted as wild-type mosquitoes to a nylon sleeve laced with both human scent and CO_2. Intriguingly however, if CO_2 was excluded from this particular assay, *orco* mutants were strongly impaired in their ability to find the human odor source. Wild-type mosquitoes with an intact OR pathway, however, retained their capacity to be attracted to this human-scented nylon sleeve, albeit at intermediate levels relative to assays performed in conjunction with CO_2.

These results suggest that in certain behavioral contexts where mosquitoes are close to a live human, or encounter an odor plume supplemented with CO_2, the presence of multiple redundant sensory cues such as CO_2, heat, moisture, and potentially other human odor ligands not detected by the OR pathway (such as those detected by the IRs [133]) may synergize together to drive host attraction in the absence of OR activity. However, the OR pathway is seemingly critical for mosquito attraction to host odor in contexts where host odor plumes are presented without CO_2—strongly supporting a role for the OR ligands in mosquito attraction to humans.

During subsequent olfactometer trials involving side-by-side comparisons of mosquito attraction to humans or animal hosts such as guinea pigs, it was discovered that *orco*-mutant mosquitoes also lose a strong preference for human odor normally seen in wild-type *Ae. aegypti* [102]. As such, the OR pathway may contribute toward mosquito host preference for humans, consistent with a role for this gene family in driving anthropophily in *Ae. aegypti*.

Excitingly, recent progress has been made toward characterizing the ligand-binding ORs that underlie this phenotype in *Ae. aegypti*. Behavioral assays coupled with transcriptional profiling of antennal tissues from anthrophophilic "domestic" and zoophilic "forest" forms of *Ae. aegypti* collected from a coastal region of Kenya have helped to reveal two candidate ORs (*AaegOr4* and *AaegOr103*) whose expression level and ligand sensitivity are strongly correlated with human host preference in this mosquito species [179]. This study revealed that *AeagOr4* alleles found in human-preferring "domestic" mosquitoes have enhanced ligand sensitivity for sulcatone—a chemical putatively found at higher levels in human odor headspace relative to other animals. Furthermore, these "domestic" *AaegOr4* alleles are expressed at higher levels relative those found in animal-preferring "forest" mosquitoes [179]. Interestingly, *AaegOr103* is also expressed at higher levels in antennal tissue from human-preferring "domestic" mosquitoes, although its ligand specificity is currently unknown [179]. While *AaegOr4* and *AaegOr103* still have to be functionally tested using targeted mutagenesis to truly implicate them as causal drivers of host preference, this research is highly suggestive that ORs played an important role in shaping the host preference of *Ae. aegypti* toward biting humans.

Targeted Mutagenesis of CO_2 Reception

To assess the relative importance of CO_2 reception to mosquito host-seeking behavior, ZFN-targeted mutagenesis and homologous recombination were subsequently applied to mutate the *AaegGr3* subunit of mosquito CO_2 receptor complex [77]. As observed in *DmelGr63a*-mutant flies [162], *AaegGr3*-mutant mosquitoes lack both electrophysiological and behavioral responses to CO_2. Interestingly, while wild-type mosquitoes are attracted to heat or the monomolecular attractant lactic acid when presented in conjunction with CO_2, these mutants show no response, suggesting that CO_2 detection gates responses to these particular sensory cues. Despite these striking behavioral deficits in the laboratory, analogous to *orco* mutants, the CO_2 receptor mutants were only slightly impaired when challenged to find a live human subject in a large semi-field cage in tropical North Queensland, Australia.

These results suggested that the full complement of host cues from a live human are indeed able to compensate for the lack of CO_2 detection in *Gr3-*mutant mosquitoes. Using membrane feeding assays where individual or combinations of sensory cues including heat, CO_2, and body odor were applied to each experiment, it was subsequently demonstrated that binary synergism between heat and body odor can drive mosquito attraction and blood-feeding behavior in the absence of CO_2 detection [77]. When CO_2, heat, or body odor are presented alone to mosquitoes in a range of experimental assays, these single cues do not elicit appreciable host-seeking or blood-feeding behavior. However, presenting any two of the three cues during experimental assays is sufficient to trigger these behaviors.

Taken together, results from the behavioral characterization of *orco-* and *Gr3*-mutant mosquitoes [77,102] reveal that multisensory integration of CO_2, heat, and body odor cues drives host-seeking and blood-feeding behavior in *Ae. aegypti*. Such a multimodal processing strategy may have evolved to allow mosquitoes to efficiently home in on a live host and to blood-feed—a behavior inextricably linked to mosquito reproduction. Interestingly, loss of a single pathway such as CO_2 or odorant reception did not yield an appreciable loss in attraction to live hosts—at least at the spatial scales tested in these two studies. Rather it appears that redundant sensory pathways in the mosquito nervous system must exist, which mediate successful localization and acquisition of a blood meal.

Given that similar synergies between heat and CO_2, and lactic acid and CO_2 have been documented to drive mosquito attraction in *An. gambiae* and *Anopheles stephensi* [23,180−183], multimodal integration of host sensory cues may represent a general processing strategy used by various mosquitoes to guide host-seeking and blood-feeding behavior. In the future, dissecting the role of the IR pathway in detection of human odorants, the role of visual perception in host-seeking behavior, and the identification of mosquito receptors that detect other cues such as moisture and heat are important research priorities that may help to reveal the full suite of molecular sensors used by these disease vectors to locate humans.

PROPOSED GENETIC STRATEGIES TARGETING MOSQUITO OLFACTION

Blood-feeding behavior—itself the product of mosquito host preference and feeding frequency—represents an critical entomological factor influencing mosquito vectorial capacity [1], which along with the density of the vector population, influences the probability of contact between mosquito, pathogen, and host. Given the importance of olfaction in guiding mosquito disease vectors toward their vertebrate hosts, chemical and genetic strategies that disrupt this sensory process could theoretically diminish the probability of dengue and malaria transmission. With this goal in mind, the observation that multiple redundant sensory pathways likely guide mosquitoes to humans [77] has important implications for the design of such strategies. Because integration between any pairing of CO_2, heat, and host odor is sufficient to drive mosquito host-seeking and blood-feeding behavior, simply antagonizing a single sensory modality such as CO_2 or odorant reception, via either the use of chemicals or genetic mutations, may be less effective in reducing the probability of contact between vector and host than targeting multiple sensory pathways. This may be particularly relevant in contexts where mosquitoes live in and around human habitation, or are located proximally to human settlements within the effective range of redundant cues emanating from nearby hosts. With this in mind, approaches simultaneously disrupting a number of different mosquito sensory pathways may ultimately yield the most effective strategy to reduce host-seeking and blood-feeding behavior in these disease vectors.

Reasoning that host-seeking and blood-feeding behaviors are not only critical for obtaining nutrients necessary for egg development but also facilitating pathogen transmission, researchers have recently proposed that chemosensory genes may represent suitable targets for engineering homing endonuclease genes (HEGs) to orchestrate population-wide gene knockouts [184,185]. By selectively mutating genes mediating host-seeking and blood-feeding behaviors, such HEGs would spread to fixation in mosquito populations via non-Mendelian inheritance [186], while synergistically acting to diminish mosquito densities and the probability of contact between vector and host. Together these effects would decrease the vectorial capacity of the targeted mosquito population.

A number of technical advances need to be made with regard to: (i) the ability to reengineer existing HEGs to target endogenous mosquito genes [187] and (ii) the development of mosquito strains that carry multiple HEGs targeting different host-seeking genes to account for multimodal integration, for HEG-based strategies targeting mosquito olfaction to be even realized. However if these barriers can be overcome, such an approach may have applied utility to reduce pathogen transmission. Encouragingly, several recent studies have indicated yeast-derived HEGs can exhibit germline activity in both *Ae. aegypti* [188] and *An. gambiae* [189] and initiate population replacement events in caged laboratory trials with *An. gambiae* [190,191] while targeting genes critical for mosquito fitness—culminating in population crashes.

It has additionally been proposed that the mosquito genes governing host preference could be utilized for applied purposes—with the aim of reverting human-preferring mosquitoes to biting animals [192]. Initially, it was thought that it may be possible to cross populations of *An. gambiae* s.s., which are highly anthropophilic, with its zoophilic sibling species *Anopheles quadriannulatus* to yield zoophilic hybrids with altered host preference and reduced vectorial capacity. However, subsequently it was discovered that anthropophilic behavior in *An. gambiae* s.s. is a dominant or partially dominant trait [192]—complicating strategies to modify host preference via inundated releases of *An. quadriannulatus* into *An. gambiae* s.s. populations. Alternatively, if the genes mediating host preference for humans or animals in mosquitoes can be identified, these may represent suitable candidates for HEG-based strategies that would aim to modify host preference through loss-of-function mutations to these genes. Encouragingly recent progress has been made toward identifying candidate ORs (*AaegOr4* and *AaegOr103*) [179] that may contribute toward human host preference in *Ae. aegypti*.

While genetic strategies targeting mosquito olfaction for disease control are, at this point in time, completely hypothetical, careful consideration of the selective pressures that these approaches will place on both vector and pathogen is paramount. This may be particularly pertinent for mosquito olfaction, as the drive to find humans to blood-feed and reproduce is likely subject to strong natural selection—as exemplified by the presence of redundant mechanisms that underpin mosquito host-seeking and blood-feeding behavior [77].

NEXT-GENERATION CHEMICAL STRATEGIES TARGETING MOSQUITO OLFACTION

While targeted mutagenesis experiments have indicated that antagonizing a single sensory pathway such as CO_2 or odorant reception is unlikely to diminish mosquito attraction to humans [77,102], the combinatorial use of novel chemicals constitutively agonizing chemoreceptor function [122,128–130,193,194] or odorant blends which are innately repulsive to mosquitoes [26,27,32] may provide an alternative strategy to drive human repulsion in mosquitoes. Given the accessibility of chemoreceptors to volatile chemicals in the periphery and their role in mediating olfactory behaviors, ORs, IRs, and GRs represent attractive targets for the development of new mosquito repellents. Indeed, it has recently been established that the vapor phases of the synthetic insect repellent DEET, and the botanical compound citronellal exert their mode of action by targeting the OR and transient receptor potential ion channel families [102,176,177,195]—although how these compounds exert their specific modes of action within these pathways are still not well understood [178]. Toward finding novel candidate repellents, Logan and colleagues [27,196] recently identified a number of human-derived volatile chemicals that interfere with attraction of a range of mosquito species to human hosts. In the future, high-throughput behavioral assays [197] purposed to identify chemicals from synthetic or natural sources innately avoided by these insects, and cheminformatic screens to find structurally related compounds from these leads [198,199], may help identify of a range of pleasant smelling and nontoxic chemicals to formulate next-generation personal and spatial insect repellents that effectively ward mosquitoes away from humans.

Chemical strategies have also been proposed that aim to turn anthropophilic mosquitoes' sense of smell against them, luring these disease vectors away from humans into odor-baited traps. For instance, pioneering research over the past 25 years into the sensory physiology and chemical ecology of *Ae. aegypti* and *An. gambiae* by several research groups has vastly accelerated the development of novel chemical blends mimicking human scent (see Ref. [44]). Using integrative chemical ecology approaches including behavior and electrophysiology, these researchers have successfully identified chemical blends consisting of CO_2, ammonia, lactic acid, and other varying constituents that are highly attractive to these mosquito species in both laboratory and semi-field analyses [22,24,32,34,36,40−42,200−205]. Promisingly, pilot field studies using odor-baited traps containing some of these chemical blends have suggested that such an approach may have applied utility as a strategy to reduce mosquito population densities [201,202,206]. Further optimization in the attractiveness of such chemical blends—as guided by functional genetic analyses of mosquito chemosensory behavior—as well as inclusion of additional attractive cues such as heat in these traps [17], stands to generate a potentially powerful method for control of malaria and dengue.

CONCLUSIONS

Anopheles gambiae and *Ae. aegypti* are devastating vectors for malaria and dengue as a consequence of a strong innate drive within these mosquito species to find and blood-feed on humans. The advent of genomics has rapidly accelerated our understanding of large families of chemosensory genes that potentially mediate olfactory responses in a range of insect species including these medically important disease vectors. Advances in heterologous expression and genome-editing systems now provide researchers with the opportunity to dissect the molecular basis of mosquito olfaction at an unprecedented resolution. These methods will facilitate the identification of chemosensory receptors for human odorants and CO_2 that drive mosquito attraction and preference for humans. Such peripheral sensors represent alluring targets for genetic control strategies; as well as the design of next-generation spatial insect repellents, attractants, and behavioral modulators that aim to stop mosquitoes from finding humans. As mosquitoes are likely to perceive humans and other hosts via multisensory integration, genetic and chemical strategies that target mosquito olfaction should take this point into consideration to yield effective methods for disease control.

ACKNOWLEDGMENTS

I would like to thank Ben Matthews, Meg Younger, and the two anonymous reviewers of this chapter for helpful comments.

REFERENCES

[1] Garrett-Jones C. Prognosis for interruption of malaria transmission through assessment of the mosquito's vectorial capacity. Nature 1964;204:1173−5.

[2] MacDonald G. The epidemiology and control of malaria. London: Oxford University Press; 1957.

[3] Besansky NJ, Hill CA, Costantini C. No accounting for taste: host preference in malaria vectors. Trends Parasitol 2004;20(6):249−51.

[4] Scott TW, Takken W. Feeding strategies of anthropophilic mosquitoes result in increased risk of pathogen transmission. Trends Parasitol 2012;28(3):114−21.

[5] Takken W, Verhulst NO. Host preferences of blood-feeding mosquitoes. Annu Rev Entomol 2013;58:433−53.

[6] Harrington LC, Fleisher A, Ruiz-Moreno D, et al. Heterogeneous feeding patterns of the dengue vector, *Aedes aegypti*, on individual human hosts in rural Thailand. PLoS Negl Trop Dis 2014;8(8):e3048.

[7] Ponlawat A, Harrington LC. Blood feeding patterns of *Aedes aegypti* and *Aedes albopictus* in Thailand. J Med Entomol 2005;42(5):844−9.

[8] Scott TW, Amerasinghe PH, Morrison AC, et al. Longitudinal studies of *Aedes aegypti* (Diptera: Culicidae) in Thailand and Puerto Rico: blood feeding frequency. J Med Entomol 2000;37(1):89−101.

[9] Scott TW, Chow E, Strickman D, et al. Blood-feeding patterns of *Aedes aegypti* (Diptera: Culicidae) collected in a rural Thai village. J Med Entomol 1993;30(5):922−7.

[10] Costantini C, Sagnon N, della Torre A, Coluzzi M. Mosquito behavioural aspects of vector-human interactions in the *Anopheles gambiae* complex. Parassitologia 1999;41 (1−3):209−17.

[11] Dekker T, Takken W, Braks MA. Innate preference for host-odor blends modulates degree of anthropophagy of *Anopheles gambiae* sensu lato (Diptera: Culicidae). J Med Entomol 2001;38(6):868−71.

[12] Mwangangi JM, Mbogo CM, Nzovu JG, Githure JI, Yan G, Beier JC. Blood-meal analysis for anopheline mosquitoes sampled along the Kenyan coast. J Am Mosq Control Assoc 2003;19(4):371−5.

[13] Pates HV, Takken W, Stuke K, Curtis CF. Differential behaviour of *Anopheles gambiae* sensu stricto (Diptera: Culicidae) to human and cow odours in the laboratory. Bull Entomol Res 2001;91(4):289−96.

[14] Garrett-Jones C, Boreham PFL, Pant CP. Feeding habits of anophelines (Diptera: Culicidae) in 1971−78, with reference to the human blood index: a review. Bull Entomol Res 1980;70:165−85.

[15] Gibson G, Torr SJ. Visual and olfactory responses of haematophagous Diptera to host stimuli. Med Vet Entomol 1999;13(1):2−23.

[16] Lewis HE, Foster AR, Mullan BJ, Cox RN, Clark RP. Aerodynamics of the human microenvironment. Lancet 1969;1(7609):1273−7.

[17] Spitzen J, Spoor CW, Grieco F, et al. A 3D analysis of flight behavior of *Anopheles gambiae* sensu stricto malaria mosquitoes in response to human odor and heat. PLoS One 2013;8(5):e62995.

[18] Dekker T, Carde RT. Moment-to-moment flight manoeuvres of the female yellow fever mosquito (*Aedes aegypti* L.) in response to plumes of carbon dioxide and human skin odour. J Exp Biol 2011;214(Pt. 20):3480−94.

[19] Dekker T, Geier M, Carde RT. Carbon dioxide instantly sensitizes female yellow fever mosquitoes to human skin odours. J Exp Biol 2005;208(Pt. 15):2963–72.

[20] Cardé RT, Gibson G. Host finding by female mosquitoes: mechanisms of orientation to host odors and other cues. In: Takken W, Knols BGJ, editors. Olfaction in vector-host interactions. Wageningen, The Netherlands: Wageningen Academic Publishers; 2010. p. 115–41.

[21] Kennedy JS. The visual responses of flying mosquitoes. Proc Zool Soc London 1940; A109:221–42.

[22] Geier M, Bosch OJ, Boeckh J. Ammonia as an attractive component of host odour for the yellow fever mosquito, Aedes aegypti. Chem Senses 1999;24(6):647–53.

[23] Smallegange RC, Qiu YT, van Loon JJ, Takken W. Synergism between ammonia, lactic acid and carboxylic acids as kairomones in the host-seeking behaviour of the malaria mosquito Anopheles gambiae sensu stricto (Diptera: Culicidae). Chem Senses 2005;30 (2):145–52.

[24] Williams CR, Bergbauer R, Geier M, et al. Laboratory and field assessment of some kairomone blends for host-seeking Aedes aegypti. J Am Mosq Control Assoc 2006;22 (4):641–7.

[25] Bernier UR, Kline DL, Schreck CE, Yost RA, Barnard DR. Chemical analysis of human skin emanations: comparison of volatiles from humans that differ in attraction of Aedes aegypti (Diptera: Culicidae). J Am Mosq Control Assoc 2002;18(3):186–95.

[26] Douglas III HD, Co JE, Jones TH, Conner WE, Day JF. Chemical odorant of colonial seabird repels mosquitoes. J Med Entomol 2005;42(4):647–51.

[27] Logan JG, Birkett MA, Clark SJ, et al. Identification of human-derived volatile chemicals that interfere with attraction of Aedes aegypti mosquitoes. J Chem Ecol 2008;34 (3):308–22.

[28] Acree Jr F, Turner RB, Gouck HK, Beroza M, Smith N. L-Lactic acid: a mosquito attractant isolated from humans. Science 1968;161(3848):1346–7.

[29] Allan SA, Bernier UR, Kline DL. Attraction of mosquitoes to volatiles associated with blood. J Vector Ecol 2006;31(1):71–8.

[30] Allan SA, Bernier UR, Kline DL. Laboratory evaluation of avian odors for mosquito (Diptera: Culicidae) attraction. J Med Entomol 2006;43(2):225–31.

[31] Bosch OJ, Geier M, Boeckh J. Contribution of fatty acids to olfactory host finding of female Aedes aegypti. Chem Senses 2000;25(3):323–30.

[32] Costantini C, Birkett MA, Gibson G, et al. Electroantennogram and behavioural responses of the malaria vector Anopheles gambiae to human-specific sweat components. Med Vet Entomol 2001;15(3):259–66.

[33] Erias AE, Jepson PC. Host location by Aedes aegypti (Diptera: Culicidae): a wind tunnel study of chemical cues. Bull Entomol Res 1991;81:151–60.

[34] Geier M, Sass H, Boeckh J. A search for components in human body odour that attract females of Aedes aegypti. Ciba Found Symp 1996;200:132–44 discussion 44–8, 78–83.

[35] Healy TP, Copland MJ. Human sweat and 2-oxopentanoic acid elicit a landing response from Anopheles gambiae. Med Vet Entomol 2000;14(2):195–200.

[36] Smallegange RC, Qiu YT, Bukovinszkine-Kiss G, Van Loon JJ, Takken W. The effect of aliphatic carboxylic acids on olfaction-based host-seeking of the malaria mosquito Anopheles gambiae sensu stricto. J Chem Ecol 2009;35(8):933–43.

[37] Smith CN, Smith N, Gouck HK, et al. L-lactic acid as a factor in the attraction of Aedes aegypti (Diptera: Culicidae) to human hosts. Ann Entomol Soc Am 1970;63(3):760–70.

[38] Steib BM, Geier M, Boeckh J. The effect of lactic acid on odour-related host preference of yellow fever mosquitoes. Chem Senses 2001;26(5):523−8.

[39] Healy TP, Copland MJ, Cork A, Przyborowska A, Halket JM. Landing responses of *Anopheles gambiae* elicited by oxocarboxylic acids. Med Vet Entomol 2002;16 (2):126−32.

[40] Bernier UR, Kline DL, Allan SA, Barnard DR. Laboratory comparison of *Aedes aegypti* attraction to human odors and to synthetic human odor compounds and blends. J Am Mosq Control Assoc 2007;23(3):288−93.

[41] Bernier UR, Kline DL, Posey KH, Booth MM, Yost RA, Barnard DR. Synergistic attraction of *Aedes aegypti* (L.) to binary blends of L-lactic acid and acetone, dichloromethane, or dimethyl disulfide. J Med Entomol 2003;40(5):653−6.

[42] Silva IM, Eiras AE, Kline DL, Bernier UR. Laboratory evaluation of mosquito traps baited with a synthetic human odor blend to capture *Aedes aegypti*. J Am Mosq Control Assoc 2005;21(2):229−33.

[43] Qui YT, van Loon JJA. Olfactory physiology of blood-feeding vector mosquitoes. In: Takken W, Knols BGJ, editors. Olfaction in vector-host interactions. Wageningen, The Netherlands: Wageningen Academic Publishers; 2010. p. 39−61.

[44] Smallegange RC, Takken W. Host-seeking behavior of mosquitoes: responses to olfactory stimuli in the laboratory. In: Takken W, Knols BGJ, editors. Olfaction in vector-host interactions. Wageningen, The Netherlands: Wageningen Academic Publishers; 2010. p. 143−80.

[45] Bernier UR, Booth MM, Yost RA. Analysis of human skin emanations by gas chromatography/mass spectrometry. 1. Thermal desorption of attractants for the yellow fever mosquito (*Aedes aegypti*) from handled glass beads. Anal Chem 1999;71(1):1−7.

[46] Bernier UR, Kline DL, Barnard DR, Schreck CE, Yost RA. Analysis of human skin emanations by gas chromatography/mass spectrometry. 2. Identification of volatile compounds that are candidate attractants for the yellow fever mosquito (*Aedes aegypti*). Anal Chem 2000;72(4):747−56.

[47] Dormont L, Bessiere JM, Cohuet A. Human skin volatiles: a review. J Chem Ecol 2013;39(5):569−78.

[48] Vogt RG, Riddiford LM. Pheromone binding and inactivation by moth antennae. Nature 1981;293(5828):161−3.

[49] Li ZX, Pickett JA, Field LM, Zhou JJ. Identification and expression of odorant-binding proteins of the malaria-carrying mosquitoes *Anopheles gambiae* and *Anopheles arabiensis*. Arch Insect Biochem Physiol 2005;58(3):175−89.

[50] Manoharan M, Ng Fuk Chong M, Vaitinadapoule A, Frumence E, Sowdhamini R, Offmann B. Comparative genomics of odorant binding proteins in *Anopheles gambiae, Aedes aegypti*, and *Culex quinquefasciatus*. Genome Biol Evol 2013;5 (1):163−80.

[51] Pelletier J, Leal WS. Genome analysis and expression patterns of odorant-binding proteins from the Southern House mosquito *Culex pipiens quinquefasciatus*. PLoS One 2009;4(7): e6237.

[52] Pelletier J, Leal WS. Characterization of olfactory genes in the antennae of the Southern house mosquito, *Culex quinquefasciatus*. J Insect Physiol 2011;57(7):915−29.

[53] Xu W, Cornel AJ, Leal WS. Odorant-binding proteins of the malaria mosquito *Anopheles funestus* sensu stricto. PLoS One 2010;5(10):e15403.

[54] Schultze A, Pregitzer P, Walter MF, et al. The co-expression pattern of odorant binding proteins and olfactory receptors identify distinct trichoid sensilla on the antenna of the malaria mosquito *Anopheles gambiae*. PLoS One 2013;8(7):e69412.

[55] Schultze A, Schymura D, Forstner M, Krieger J. Expression pattern of a 'Plus-C' class odorant binding protein in the antenna of the malaria vector *Anopheles gambiae*. Insect Mol Biol 2012;21(2):187–95.

[56] McIver SB. Sensilla of mosquitoes (Diptera: Culicidae). J Med Entomol 1982;19 (5):489–535.

[57] Pitts RJ, Zwiebel LJ. Antennal sensilla of two female anopheline sibling species with differing host ranges. Malar J 2006;5:26.

[58] McIver SB. Structure of sensilla trichodea of female *Aedes aegypti* (L.) with comments on innervation of antennal sensilla. J Insect Physiol 1978;24:383–90.

[59] Ghaninia M, Ignell R, Hansson BS. Functional classification and central nervous projections of olfactory receptor neurons housed in antennal trichoid sensilla of female yellow fever mosquitoes, *Aedes aegypti*. Eur J Neurosci 2007;26(6):1611–23.

[60] Ghaninia M, Larsson M, Hansson BS, Ignell R. Natural odor ligands for olfactory receptor neurons of the female mosquito *Aedes aegypti*: use of gas chromatography-linked single sensillum recordings. J Exp Biol 2008;211(Pt. 18):3020–7.

[61] Davis E, Sokolove P. Lactic acid-sensitive receptors on the antennae of the mosquito, *Aedes aegypti*. J Comp Physiol 1976;105(1):43–54.

[62] McIver S, Charlton C. Studies on the sense organs on the palps of selected culicine mosquitoes. Can J Zool 1970;48(2):293–5.

[63] McIver SB. Fine structure of pegs on the palps of female culicine mosquitoes. Can J Zool 1972;50(5):571–6.

[64] McIver S, Siemicki R. Palpal sensilla of selected anopheline mosquitoes. J Parasitol 1975;61:535–8.

[65] Grant AJ, Wigton BE, Aghajanian JG, O'Connell RJ. Electrophysiological responses of receptor neurons in mosquito maxillary palp sensilla to carbon dioxide. J Comp Physiol [A] 1995;177(4):389–96.

[66] Kellogg FE. Water vapour and carbon dioxide receptors in *Aedes aegypti*. J Insect Physiol 1970;16(1):99–108.

[67] Kwon HW, Lu T, Rutzler M, Zwiebel LJ. Olfactory responses in a gustatory organ of the malaria vector mosquito *Anopheles gambiae*. Proc Natl Acad Sci USA 2006;103 (36):13526–31.

[68] Vosshall LB, Wong AM, Axel R. An olfactory sensory map in the fly brain. Cell 2000;102(2):147–59.

[69] Anton S, van Loon JJ, Meijerink J, Smid HM, Takken W, Rospars JP. Central projections of olfactory receptor neurons from single antennal and palpal sensilla in mosquitoes. Arthropod Struct Dev 2003;32(4):319–27.

[70] Distler P, Boeckh J. Central projections of the maxillary and antennal nerves in the mosquito *Aedes aegypti*. J Exp Biol 1997;200(Pt. 13):1873–9.

[71] Ghaninia M, Hansson BS, Ignell R. The antennal lobe of the African malaria mosquito, *Anopheles gambiae*—innervation and three-dimensional reconstruction. Arthropod Struct Dev 2007;36(1):23–39.

[72] Ignell R, Dekker T, Ghaninia M, Hansson BS. Neuronal architecture of the mosquito deutocerebrum. J Comp Neurol 2005;493(2):207–40.

[73] Huang J, Zhang W, Qiao W, Hu A, Wang Z. Functional connectivity and selective odor responses of excitatory local interneurons in *Drosophila* antennal lobe. Neuron 2010;67 (6):1021–33.

[74] Ng M, Roorda RD, Lima SQ, Zemelman BV, Morcillo P, Miesenbock G. Transmission of olfactory information between three populations of neurons in the antennal lobe of the fly. Neuron 2002;36(3):463–74.

[75] Shang Y, Claridge-Chang A, Sjulson L, Pypaert M, Miesenbock G. Excitatory local circuits and their implications for olfactory processing in the fly antennal lobe. Cell 2007;128(3):601–12.

[76] Yaksi E, Wilson RI. Electrical coupling between olfactory glomeruli. Neuron 2010;67 (6):1034–47.

[77] McMeniman CJ, Corfas RA, Matthews BJ, Ritchie SA, Vosshall LB. Multimodal integration of carbon dioxide and other sensory cues drives mosquito attraction to humans. Cell 2014;156(5):1060–71.

[78] Clyne PJ, Warr CG, Freeman MR, Lessing D, Kim J, Carlson JR. A novel family of divergent seven-transmembrane proteins: candidate odorant receptors in *Drosophila*. Neuron 1999;22(2):327–38.

[79] Gao Q, Chess A. Identification of candidate *Drosophila* olfactory receptors from genomic DNA sequence. Genomics 1999;60(1):31–9.

[80] Vosshall LB, Amrein H, Morozov PS, Rzhetsky A, Axel R. A spatial map of olfactory receptor expression in the *Drosophila* antenna. Cell 1999;96(5):725–36.

[81] Buck L, Axel R. A novel multigene family may encode odorant receptors: a molecular basis for odor recognition. Cell 1991;65(1):175–87.

[82] Freitag J, Krieger J, Strotmann J, Breer H. Two classes of olfactory receptors in *Xenopus laevis*. Neuron 1995;15(6):1383–92.

[83] Nef S, Allaman I, Fiumelli H, De Castro E, Nef P. Olfaction in birds: differential embryonic expression of nine putative odorant receptor genes in the avian olfactory system. Mech Dev 1996;55(1):65–77.

[84] Selbie LA, Townsend-Nicholson A, Iismaa TP, Shine J. Novel G protein-coupled receptors: a gene family of putative human olfactory receptor sequences. Brain Res Mol Brain Res 1992;13(1–2):159–63.

[85] Benton R, Sachse S, Michnick SW, Vosshall LB. Atypical membrane topology and heteromeric function of *Drosophila* odorant receptors *in vivo*. PLoS Biol 2006;4(2):e20.

[86] Lundin C, Kall L, Kreher SA, et al. Membrane topology of the *Drosophila OR83b* odorant receptor. FEBS Lett 2007;581(29):5601–4.

[87] Smart R, Kiely A, Beale M, et al. *Drosophila* odorant receptors are novel seven transmembrane domain proteins that can signal independently of heterotrimeric G proteins. Insect Biochem Mol Biol 2008;38(8):770–80.

[88] Tsitoura P, Andronopoulou E, Tsikou D, et al. Expression and membrane topology of *Anopheles gambiae* odorant receptors in lepidopteran insect cells. PLoS One 2010;5(11): e15428.

[89] Fox AN, Pitts RJ, Robertson HM, Carlson JR, Zwiebel LJ. Candidate odorant receptors from the malaria vector mosquito *Anopheles gambiae* and evidence of down-regulation in response to blood feeding. Proc Natl Acad Sci USA 2001;98(25):14693–7.

[90] Fox AN, Pitts RJ, Zwiebel LJ. A cluster of candidate odorant receptors from the malaria vector mosquito, *Anopheles gambiae*. Chem Senses 2002;27(5):453–9.

[91] Hill CA, Fox AN, Pitts RJ, et al. G protein-coupled receptors in *Anopheles gambiae*. Science 2002;298(5591):176–8.

[92] Bohbot J, Pitts RJ, Kwon HW, Rutzler M, Robertson HM, Zwiebel LJ. Molecular characterization of the *Aedes aegypti* odorant receptor gene family. Insect Mol Biol 2007;16 (5):525–37.

[93] Sato K, Pellegrino M, Nakagawa T, Nakagawa T, Vosshall LB, Touhara K. Insect olfactory receptors are heteromeric ligand-gated ion channels. Nature 2008;452(7190):1002–6.

[94] Wicher D, Schafer R, Bauernfeind R, et al. *Drosophila* odorant receptors are both ligand-gated and cyclic-nucleotide-activated cation channels. Nature 2008;452 (7190):1007−11.

[95] Jones WD, Nguyen TA, Kloss B, Lee KJ, Vosshall LB. Functional conservation of an insect odorant receptor gene across 250 million years of evolution. Curr Biol 2005;15(4): R119−21.

[96] Krieger J, Klink O, Mohl C, Raming K, Breer H. A candidate olfactory receptor subtype highly conserved across different insect orders. J Comp Physiol A Neuroethol Sens Neural Behav Physiol 2003;189(7):519−26.

[97] Vosshall LB, Hansson BS. A unified nomenclature system for the insect olfactory coreceptor. Chem Senses 2011;36(6):497−8.

[98] Larsson MC, Domingos AI, Jones WD, Chiappe ME, Amrein H, Vosshall LB. *Or83b* encodes a broadly expressed odorant receptor essential for *Drosophila* olfaction. Neuron 2004;43(5):703−14.

[99] Melo AC, Rutzler M, Pitts RJ, Zwiebel LJ. Identification of a chemosensory receptor from the yellow fever mosquito, *Aedes aegypti*, that is highly conserved and expressed in olfactory and gustatory organs. Chem Senses 2004;29(5):403−10.

[100] Pitts RJ, Fox AN, Zwiebel LJ. A highly conserved candidate chemoreceptor expressed in both olfactory and gustatory tissues in the malaria vector *Anopheles gambiae*. Proc Natl Acad Sci USA 2004;101(14):5058−63.

[101] Xia Y, Zwiebel LJ. Identification and characterization of an odorant receptor from the West Nile virus mosquito, *Culex quinquefasciatus*. Insect Biochem Mol Biol 2006;36 (3):169−76.

[102] DeGennaro M, McBride CS, Seeholzer L, et al. *orco* mutant mosquitoes lose strong preference for humans and are not repelled by volatile DEET. Nature 2013;498 (7455):487−91.

[103] Clyne P, Grant A, O'Connell R, Carlson JR. Odorant response of individual sensilla on the *Drosophila* antenna. Invert Neurosci 1997;3(2−3):127−35.

[104] de Bruyne M, Clyne PJ, Carlson JR. Odor coding in a model olfactory organ: the *Drosophila* maxillary palp. J Neurosci 1999;19(11):4520−32.

[105] de Bruyne M, Foster K, Carlson JR. Odor coding in the *Drosophila* antenna. Neuron 2001;30(2):537−52.

[106] Yao CA, Ignell R, Carlson JR. Chemosensory coding by neurons in the coeloconic sensilla of the *Drosophila* antenna. J Neurosci 2005;25(37):8359−67.

[107] Dobritsa AA, van der Goes van Naters W, Warr CG, Steinbrecht RA, Carlson JR. Integrating the molecular and cellular basis of odor coding in the *Drosophila* antenna. Neuron 2003;37(5):827−41.

[108] Hallem EA, Ho MG, Carlson JR. The molecular basis of odor coding in the *Drosophila* antenna. Cell 2004;117(7):965−79.

[109] Hallem EA, Nicole Fox A, Zwiebel LJ, Carlson JR. Olfaction: mosquito receptor for human-sweat odorant. Nature 2004;427(6971):212−13.

[110] Benton R, Dahanukar A. Electrophysiological recording from *Drosophila* olfactory sensilla. Cold Spring Harb Protoc 2011;2011(7):824−38.

[111] Olsson SB, Hansson BS. Electroantennogram and single sensillum recording in insect antennae. Methods Mol Biol 2013;1068:157−77.

[112] Pellegrino M, Nakagawa T, Vosshall LB. Single sensillum recordings in the insects *Drosophila melanogaster* and *Anopheles gambiae*. J Vis Exp 2010;36:1−5.

[113] Hallem EA, Carlson JR. Coding of odors by a receptor repertoire. Cell 2006;125 (1):143−60.

[114] Kreher SA, Kwon JY, Carlson JR. The molecular basis of odor coding in the *Drosophila* larva. Neuron 2005;46(3):445−56.

[115] Kreher SA, Mathew D, Kim J, Carlson JR. Translation of sensory input into behavioral output via an olfactory system. Neuron 2008;59(1):110−24.

[116] Datta SR, Vasconcelos ML, Ruta V, et al. The *Drosophila* pheromone cVA activates a sexually dimorphic neural circuit. Nature 2008;452(7186):473−7.

[117] Schlief ML, Wilson RI. Olfactory processing and behavior downstream from highly selective receptor neurons. Nat Neurosci 2007;10(5):623−30.

[118] Stensmyr MC, Dweck HK, Farhan A, et al. A conserved dedicated olfactory circuit for detecting harmful microbes in *Drosophila*. Cell 2012;151(6):1345−57.

[119] Carey AF, Wang G, Su CY, Zwiebel LJ, Carlson JR. Odorant reception in the malaria mosquito *Anopheles gambiae*. Nature 2010;464(7285):66−71.

[120] Kiely A, Authier A, Kralicek AV, Warr CG, Newcomb RD. Functional analysis of a *Drosophila melanogaster* olfactory receptor expressed in *Sf9* cells. J Neurosci Methods 2007;159(2):189−94.

[121] Corcoran JA, Jordan MD, Carraher C, Newcomb RD. A novel method to study insect olfactory receptor function using HEK293 cells. Insect Biochem Mol Biol 2014.

[122] Jones PL, Pask GM, Rinker DC, Zwiebel LJ. Functional agonism of insect odorant receptor ion channels. Proc Natl Acad Sci USA 2011;108(21):8821−5.

[123] Neuhaus EM, Gisselmann G, Zhang W, Dooley R, Stortkuhl K, Hatt H. Odorant receptor heterodimerization in the olfactory system of *Drosophila melanogaster*. Nat Neurosci 2005;8(1):15−17.

[124] Wetzel CH, Behrendt HJ, Gisselmann G, Stortkuhl KF, Hovemann B, Hatt H. Functional expression and characterization of a *Drosophila* odorant receptor in a heterologous cell system. Proc Natl Acad Sci USA 2001;98(16):9377−80.

[125] Pask GM, Romaine IM, Zwiebel LJ. The molecular receptive range of a lactone receptor in *Anopheles gambiae*. Chem Senses 2013;38(1):19−25.

[126] Wang G, Carey AF, Carlson JR, Zwiebel LJ. Molecular basis of odor coding in the malaria vector mosquito *Anopheles gambiae*. Proc Natl Acad Sci USA 2010;107 (9):4418−23.

[127] Pask GM, Bobkov YV, Corey EA, Ache BW, Zwiebel LJ. Blockade of insect odorant receptor currents by amiloride derivatives. Chem Senses 2013;38(3):221−9.

[128] Chen S, Luetje CW. Identification of new agonists and antagonists of the insect odorant receptor co-receptor subunit. PLoS One 2012;7(5):e36784.

[129] Jones PL, Pask GM, Romaine IM, et al. Allosteric antagonism of insect odorant receptor ion channels. PLoS One 2012;7(1):e30304.

[130] Kumar BN, Taylor RW, Pask GM, Zwiebel LJ, Newcomb RD, Christie DL. A conserved aspartic acid is important for agonist (VUAA1) and odorant/tuning receptor-dependent activation of the insect odorant co-receptor (*Orco*). PLoS One 2013;8(7):e70218.

[131] Taylor RW, Romaine IM, Liu C, et al. Structure-activity relationship of a broad-spectrum insect odorant receptor agonist. ACS Chem Biol 2012;7(10):1647−52.

[132] Benton R, Vannice KS, Vosshall LB. An essential role for a CD36-related receptor in pheromone detection in *Drosophila*. Nature 2007;450(7167):289−93.

[133] Benton R, Vannice KS, Gomez-Diaz C, Vosshall LB. Variant ionotropic glutamate receptors as chemosensory receptors in *Drosophila*. Cell 2009;136(1):149−62.

[134] Abuin L, Bargeton B, Ulbrich MH, Isacoff EY, Kellenberger S, Benton R. Functional architecture of olfactory ionotropic glutamate receptors. Neuron 2011;69(1):44−60.

[135] Ai M, Min S, Grosjean Y, et al. Acid sensing by the *Drosophila* olfactory system. Nature 2010;468(7324):691−5.

[136] Silbering AF, Rytz R, Grosjean Y, et al. Complementary function and integrated wiring of the evolutionarily distinct *Drosophila* olfactory subsystems. J Neurosci 2011;31 (38):13357−75.

[137] Croset V, Rytz R, Cummins SF, et al. Ancient protostome origin of chemosensory ionotropic glutamate receptors and the evolution of insect taste and olfaction. PLoS Genet 2010;6(8):e1001064.

[138] Liu C, Pitts RJ, Bohbot JD, Jones PL, Wang G, Zwiebel LJ. Distinct olfactory signaling mechanisms in the malaria vector mosquito *Anopheles gambiae*. PLoS Biol 2010;8(8).

[139] Bohbot JD, Sparks JT, Dickens JC. The maxillary palp of *Aedes aegypti*, a model of multisensory integration. Insect Biochem Mol Biol 2014;48:29−39.

[140] Pitts RJ, Rinker DC, Jones PL, Rokas A, Zwiebel LJ. Transcriptome profiling of chemosensory appendages in the malaria vector *Anopheles gambiae* reveals tissue- and sex-specific signatures of odor coding. BMC Genomics 2011;12:271.

[141] Rinker DC, Zhou X, Pitts RJ, Rokas A, Zwiebel LJ. Antennal transcriptome profiles of anopheline mosquitoes reveal human host olfactory specialization in *Anopheles gambiae*. BMC Genomics 2013;14:749.

[142] Sparks JT, Bohbot JD, Dickens JC. The genetics of chemoreception in the labella and tarsi of *Aedes aegypti*. Insect Biochem Mol Biol 2014;48:8−16.

[143] Clyne PJ, Warr CG, Carlson JR. Candidate taste receptors in *Drosophila*. Science 2000;287(5459):1830−4.

[144] Scott K, Brady Jr R, Cravchik A, et al. A chemosensory gene family encoding candidate gustatory and olfactory receptors in *Drosophila*. Cell 2001;104(5):661−73.

[145] Robertson HM, Warr CG, Carlson JR. Molecular evolution of the insect chemoreceptor gene superfamily in *Drosophila melanogaster*. Proc Natl Acad Sci USA 2003;100(Suppl. 2):14537−42.

[146] Sato K, Tanaka K, Touhara K. Sugar-regulated cation channel formed by an insect gustatory receptor. Proc Natl Acad Sci USA 2011;108(28):11680−5.

[147] Dunipace L, Meister S, McNealy C, Amrein H. Spatially restricted expression of candidate taste receptors in the *Drosophila* gustatory system. Curr Biol 2001;11(11):822−35.

[148] Kent LB, Walden KK, Robertson HM. The *Gr* family of candidate gustatory and olfactory receptors in the yellow-fever mosquito *Aedes aegypti*. Chem Senses 2008;33 (1):79−93.

[149] Dahanukar A, Lei YT, Kwon JY, Carlson JR. Two *Gr* genes underlie sugar reception in *Drosophila*. Neuron 2007;56(3):503−16.

[150] Jiao Y, Moon SJ, Wang X, Ren Q, Montell C. *Gr64f* is required in combination with other gustatory receptors for sugar detection in *Drosophila*. Curr Biol 2008;18 (22):1797−801.

[151] Miyamoto T, Slone J, Song X, Amrein H. A fructose receptor functions as a nutrient sensor in the *Drosophila* brain. Cell 2012;151(5):1113−25.

[152] Slone J, Daniels J, Amrein H. Sugar receptors in *Drosophila*. Curr Biol 2007;17 (20):1809−16.

[153] Lee Y, Moon SJ, Montell C. Multiple gustatory receptors required for the caffeine response in *Drosophila*. Proc Natl Acad Sci USA 2009;106(11):4495−500.

[154] Moon SJ, Kottgen M, Jiao Y, Xu H, Montell C. A taste receptor required for the caffeine response in vivo. Curr Biol 2006;16(18):1812–17.

[155] Moon SJ, Lee Y, Jiao Y, Montell CA. *Drosophila* gustatory receptor essential for aversive taste and inhibiting male-to-male courtship. Curr Biol 2009;19(19):1623–7.

[156] Freeman EG, Wisotsky Z, Dahanukar A. Detection of sweet tastants by a conserved group of insect gustatory receptors. Proc Natl Acad Sci USA 2014;111(4):1598–603.

[157] Bray S, Amrein H. A putative *Drosophila* pheromone receptor expressed in male-specific taste neurons is required for efficient courtship. Neuron 2003;39(6):1019–29.

[158] Miyamoto T, Amrein H. Suppression of male courtship by a *Drosophila* pheromone receptor. Nat Neurosci 2008;11(8):874–6.

[159] Watanabe K, Toba G, Koganezawa M, Yamamoto D. *Gr39a*, a highly diversified gustatory receptor in *Drosophila*, has a role in sexual behavior. Behav Genet 2011;41 (5):746–53.

[160] Ni L, Bronk P, Chang EC, et al. A gustatory receptor paralogue controls rapid warmth avoidance in *Drosophila*. Nature 2013;500(7464):580–4.

[161] Xiang Y, Yuan Q, Vogt N, Looger LL, Jan LY, Jan YN. Light-avoidance-mediating photoreceptors tile the *Drosophila* larval body wall. Nature 2010;468(7326):921–6.

[162] Jones WD, Cayirlioglu P, Kadow IG, Vosshall LB. Two chemosensory receptors together mediate carbon dioxide detection in *Drosophila*. Nature 2007;445(7123):86–90.

[163] Kwon JY, Dahanukar A, Weiss LA, Carlson JR. The molecular basis of CO_2 reception in *Drosophila*. Proc Natl Acad Sci USA 2007;104(9):3574–8.

[164] Yao CA, Carlson JR. Role of G-proteins in odor-sensing and CO_2-sensing neurons in *Drosophila*. J Neurosci 2010;30(13):4562–72.

[165] Robertson HM, Kent LB. Evolution of the gene lineage encoding the carbon dioxide receptor in insects. J Insect Sci 2009;9:19.

[166] Lu T, Qiu YT, Wang G, et al. Odor coding in the maxillary palp of the malaria vector mosquito *Anopheles gambiae*. Curr Biol 2007;17(18):1533–44.

[167] Erdelyan CN, Mahood TH, Bader TS, Whyard S. Functional validation of the carbon dioxide receptor genes in *Aedes aegypti* mosquitoes using RNA interference. Insect Mol Biol 2012;21(1):119–27.

[168] Adams MD, Celniker SE, Holt RA, et al. The genome sequence of *Drosophila melanogaster*. Science 2000;287(5461):2185–95.

[169] Holt RA, Subramanian GM, Halpern A, et al. The genome sequence of the malaria mosquito *Anopheles gambiae*. Science 2002;298(5591):129–49.

[170] Nene V, Wortman JR, Lawson D, et al. Genome sequence of *Aedes aegypti*, a major arbovirus vector. Science 2007;316(5832):1718–23.

[171] Arensburger P, Megy K, Waterhouse RM, et al. Sequencing of *Culex quinquefasciatus* establishes a platform for mosquito comparative genomics. Science 2010;330(6000):86–8.

[172] Cheng JK, Alper HS. The genome editing toolbox: a spectrum of approaches for targeted modification. Curr Opin Biotechnol 2014;30c:87–94.

[173] Aryan A, Anderson MA, Myles KM, Adelman ZN. TALEN-based gene disruption in the dengue vector *Aedes aegypti*. PLoS One 2013;8(3):e60082.

[174] Aryan A, Myles KM, Adelman ZN. Targeted genome editing in *Aedes aegypti* using TALENs. Methods 2014;69(1):38–45.

[175] Smidler AL, Terenzi O, Soichot J, Levashina EA, Marois E. Targeted mutagenesis in the malaria mosquito using TALE nucleases. PLoS One 2013;8(8):e74511.

[176] Ditzen M, Pellegrino M, Vosshall LB. Insect odorant receptors are molecular targets of the insect repellent DEET. Science 2008;319(5871):1838–42.

[177]   Pellegrino M, Steinbach N, Stensmyr MC, Hansson BS, Vosshall LB. A natural polymorphism alters odour and DEET sensitivity in an insect odorant receptor. Nature 2011;478 (7370):511–14.

[178]   Leal WS. The enigmatic reception of DEET—the gold standard of insect repellents. Curr Opin Insect Sci 2014;6:93–8.

[179]   McBride CS, Baier F, Omondi AB, et al. Evolution of mosquito preference for humans linked to an odorant receptor. Nature 2014;515(7526):222–7.

[180]   Geier M, Bosch OJ, Boeckh J. Influence of odour plume structure on upwind flight of mosquitoes towards hosts. J Exp Biol 1999;202(Pt. 12):1639–48.

[181]   Krober T, Kessler S, Frei J, Bourquin M, Guerin PM. An *in vitro* assay for testing mosquito repellents employing a warm body and carbon dioxide as a behavioral activator. J Am Mosq Control Assoc 2010;26(4):381–6.

[182]   Maekawa E, Aonuma H, Nelson B, et al. The role of proboscis of the malaria vector mosquito *Anopheles stephensi* in host-seeking behavior. Parasit Vectors 2011;4:10.

[183]   Omrani SM, Vatandoost H, Oshaghi MA, Rahimi A. Upwind responses of *Anopheles stephensi* to carbon dioxide and L-lactic acid: an olfactometer study. East Mediterr Health J 2012;18(11):1134–42.

[184]   Burt A. Heritable strategies for controlling insect vectors of disease. Philos Trans R Soc Lond B Biol Sci 2014;369(1645):20130432.

[185]   Deredec A, Godfray HC, Burt A. Requirements for effective malaria control with homing endonuclease genes. Proc Natl Acad Sci USA 2011;108(43):E874–80.

[186]   Goddard MR, Greig D, Burt A. Outcrossed sex allows a selfish gene to invade yeast populations. Proc Biol Sci 2001;268(1485):2537–42.

[187]   Chan YS, Takeuchi R, Jarjour J, Huen DS, Stoddard BL, Russell S. The design and in vivo evaluation of engineered I-*OnuI*-based enzymes for HEG gene drive. PLoS One 2013;8(9):e74254.

[188]   Aryan A, Anderson MA, Myles KM, Adelman ZN. Germline excision of transgenes in *Aedes aegypti* by homing endonucleases. Sci Rep 2013;3:1603.

[189]   Windbichler N, Papathanos PA, Catteruccia F, Ranson H, Burt A, Crisanti A. Homing endonuclease mediated gene targeting in *Anopheles gambiae* cells and embryos. Nucleic Acids Res 2007;35(17):5922–33.

[190]   Galizi R, Doyle LA, Menichelli M, et al. A synthetic sex ratio distortion system for the control of the human malaria mosquito. Nat Commun 2014;5:3977.

[191]   Windbichler N, Menichelli M, Papathanos PA, et al. A synthetic homing endonuclease-based gene drive system in the human malaria mosquito. Nature 2011;473 (7346):212–15.

[192]   Pates HV, Curtis CF, Takken W. Hybridization studies to modify the host preference of *Anopheles gambiae*. Med Vet Entomol 2014;28(Suppl. 1):68–74.

[193]   Tauxe GM, Macwilliam D, Boyle SM, Guda T, Ray A. Targeting a dual detector of skin and CO_2 to modify mosquito host seeking. Cell 2013;155(6):1365–79.

[194]   Turner SL, Li N, Guda T, Githure J, Carde RT, Ray A. Ultra-prolonged activation of CO_2-sensing neurons disorients mosquitoes. Nature 2011;474(7349):87–91.

[195]   Kwon Y, Kim SH, Ronderos DS, et al. *Drosophila TRPA1* channel is required to avoid the naturally occurring insect repellent citronellal. Curr Biol 2010;20(18):1672–8.

[196]   Logan JG, Stanczyk NM, Hassanali A, et al. Arm-in-cage testing of natural human-derived mosquito repellents. Malar J 2010;9:239.

[197]   Steck K, Veit D, Grandy R, et al. A high-throughput behavioral paradigm for *Drosophila* olfaction—The Flywalk. Sci Rep 2012;2:361.

[198] Boyle SM, McInally S, Ray A. Expanding the olfactory code by *in silico* decoding of odor-receptor chemical space. eLife 2013;2:e01120.

[199] Kain P, Boyle SM, Tharadra SK, et al. Odour receptors and neurons for DEET and new insect repellents. Nature 2013;502(7472):507−12.

[200] Jawara M, Smallegange RC, Jeffries D, et al. Optimizing odor-baited trap methods for collecting mosquitoes during the malaria season in The Gambia. PLoS One 2009;4(12): e8167.

[201] Mukabana WR, Mweresa CK, Otieno B, et al. A novel synthetic odorant blend for trapping of malaria and other African mosquito species. J Chem Ecol 2012;38(3):235−44.

[202] Okumu FO, Killeen GF, Ogoma S, et al. Development and field evaluation of a synthetic mosquito lure that is more attractive than humans. PLoS One 2010;5(1):e8951.

[203] Qiu YT, Smallegange RC, Hoppe S, van Loon JJ, Bakker EJ, Takken W. Behavioural and electrophysiological responses of the malaria mosquito *Anopheles gambiae* Giles sensu stricto (Diptera: Culicidae) to human skin emanations. Med Vet Entomol 2004;18 (4):429−38.

[204] Qiu YT, Smallegange RC, Ter BC, et al. Attractiveness of MM-X traps baited with human or synthetic odor to mosquitoes (Diptera: Culicidae) in The Gambia. J Med Entomol 2007;44(6):970−83.

[205] Verhulst NO, Mbadi PA, Kiss GB, et al. Improvement of a synthetic lure for *Anopheles gambiae* using compounds produced by human skin microbiota. Malar J 2011;10(1):28.

[206] Hiscox A, Otieno B, Kibet A, et al. Development and optimization of the Suna trap as a tool for mosquito monitoring and control. Malar J 2014;13:257.

Chapter 12

Disruption of Mosquito Blood Meal Protein Metabolism

Patricia Y. Scaraffia
Department of Tropical Medicine, Vector-Borne Infectious Disease Research Center, School of Public Health and Tropical Medicine, Tulane University, New Orleans, LA, USA

INTRODUCTION: AN OVERVIEW OF BLOOD MEAL PROTEIN METABOLISM IN MOSQUITOES

During the ingestion of the blood meal, female mosquitoes can transmit etiological agents that affect millions of people worldwide [1−6]. To combat this serious health problem, new strategies for controlling mosquito populations need to be implemented [7−17]. Attempts to control mosquito populations using biorational approaches depend on a thorough understanding of mosquito biology and metabolism.

Newly emerged female mosquitoes feed on nectar for several days until they are able to take their first blood meal [18]. Anautogenous females, such as *Aedes aegypti*, require at least one vertebrate blood meal to complete oogenesis. This critical role of host blood in reproduction has meant that most investigations related to blood meal protein metabolism in mosquitoes have focused on mosquito reproductive biology [19−27], but these studies have depended on the quantification of the mass of components at the beginning and end of a gonotrophic cycle with no analysis of the dynamics of the process. It was not until 2004 that Prof. Dr Michael Wells and his research group published the first research articles about dynamic measurements of blood meal protein amino acid metabolism in mosquitoes [28−30].

The host blood is a nutritionally rich meal. Proteins represent 20% of the wet weight of blood. Hemoglobin, serum albumin, and immunoglobulins constitute 80% of the total proteins present in a typical blood meal. Compared to proteins, the principal nutrient, lipids and carbohydrates are present in low amounts [30−32].

After female mosquitoes ingest a blood meal, the blood enters the midgut where it is surrounded by a semipermeable extracellular layer known as peritrophic membrane or peritrophic matrix, which is thought to protect midgut epithelial cells from pathogens, abrasion, heme, and other toxins; and in certain cases also to facilitate the blood digestion [33,34]. Although heme is an essential component of many proteins, it can also be toxic to the cells through several mechanisms [35,36]. Blood feeding induces the expression of numerous protease genes in midgut epithelial cells [37]. A rapid and massive diuresis occurs following blood meal ingestion, where excess ions and water are efficiently removed [38−40]. Blood digestion is initiated by exopeptidases and endopeptidases that are secreted into the lumen by midgut epithelial cells in response to the blood meal. Proteins are efficiently converted into peptides and amino acids by a battery of proteases. Although mosquito midgut proteases have been extensively studied, the mechanisms involved in the regulation of protein digestion in blood-fed mosquitoes are complex and not yet completed understood. Numerous cDNAs from three main families of proteases (serine endoproteases, carboxyexopeptidases, and aminoexopeptidases) have been cloned and sequenced, but only a few have been characterized in the midgut of adult *Ae. aegypti* females [41−50]. Enzymological studies are considered essential for elucidating the catalytic properties of mosquito proteases and determining whether small molecule inhibitors of any of these proteases would be useful as mosquito-selective control agents that disrupt or block blood meal proteolysis.

During the digestive process, female mosquitoes retrieve nutrients and excrete waste products very efficiently [51−56] in order to produce viable eggs and be ready for the next blood meal. The metabolic consequence of amino acid deamination is the production of α-keto acids and ammonia. In this chapter, the term ammonia refers to NH_3, NH_4^+, or a combination of the two. The carbon skeletons derived from blood meal protein amino acid metabolism are mainly used for energy production, synthesis of egg reserves, and acquisition of maternal reserves; whereas the loss of nitrogen from amino acid metabolism produces ammonia, which must be excreted or converted to other nitrogen compounds including uric acid, allantoin, allantoic acid, and urea, as discussed in detail later in this chapter. A simplified scenario of the blood meal amino acid metabolism is shown in Figure 12.1.

This chapter provides a comprehensive picture of blood meal protein amino acid metabolism in *Ae. aegypti* females, vectors of several diseases including chikungunya, dengue, and yellow fever [1,57−61]. In this context, the application of traditional and cutting-edge approaches for studying metabolic pathways in mosquitoes that could be useful for discovering possible metabolic targets to disrupt blood feeding, digestion, and excretion is also discussed.

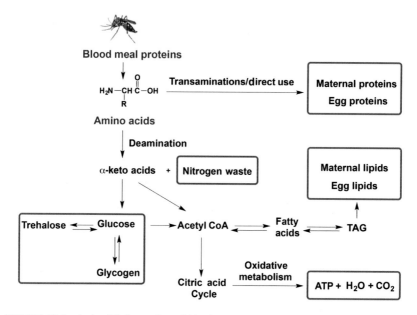

FIGURE 12.1 A simplified overview of blood meal protein amino acid metabolism showing how carbon and nitrogen metabolism are related.

EXPLORING BLOOD MEAL PROTEIN AMINO ACID CARBON SKELETONS USING RADIOACTIVE ISOTOPES

Due to the lack of commercially ready-made ^{14}C-labeled protein products to study the metabolic fate of amino acids derived from dietary protein, a protocol described by Krzywicha and Wagner [62] for synthesizing ^{14}C-amino acid labeled proteins can be used. The algae, *Euglena gracilis*, can synthesize all its amino acids from CO_2 and ammonium salts. If $^{14}CO_2$ is provided to the algae in the light, radioactively-labeled proteins can be made and incorporated into a blood meal, which is then fed to mosquitoes. Thus, by feeding female *Ae. aegypti* with [^{14}C]-labeled *Euglena* proteins, it is possible to follow the labeling patterns of mosquito proteins, amino acids, sugars, lipids, CO_2, and other waste during a gonotrophic cycle, and the endpoint labeling of the female mosquito and egg reserves after the eggs are laid.

Dynamics of Blood Meal Protein Amino Acid Metabolism

Compared to static measurements made only at the beginning and end of a gonotrophic cycle, dynamic measurements made at several time points during the cycle give considerably more information about changing metabolic processes occurring during the cycle. For example, consider the question of whether blood meal protein amino acid carbon is converted to trehalose and glycogen during the gonotrophic cycle. Zhou et al. [28] reported that during the first 36 h following a blood meal there is continual sugar synthesis, mainly trehalose, which is also reflected in transient synthesis of glycogen [28]. If glycogen or trehalose contents were measured only at the beginning and end of the cycle, the activation of gluconeogenesis and glycogenesis would be missed, as well as potential useful control targets. Other conclusions from the studies performed by Zhou et al. [28] are:

- Although the amount of labeled proteins remaining in the mosquito following feeding of a blood meal containing radioactive proteins seems to decline exponentially, there are at least three processes occurring: first and foremost is hydrolysis of the blood meal proteins in the midgut to produce amino acids; second, some of the released amino acids are used to synthesize egg proteins; and third some of the amino acids are used to make proteins that remain in the female after egg laying.
- Gluconeogenesis from amino acids is highly active during the gonotrophic cycle with production of sugars, especially trehalose and glycogen. By the end of the cycle, almost all the labeled sugars have disappeared, presumably through oxidative metabolism or conversion to lipids.
- Lipogenesis from amino acid precursors is also quite active and the lipids produced are deposited in the eggs or retained by the female after egg laying. Thus, it seems that lipids are more important as reserves in the eggs and in the female, than as substrates for energy production.
- During the initial phase of the cycle, a considerable amount of the blood meal protein amino acids are incorporated into midgut proteins — mostly proteases; but after 48 h, when the midgut begins to atrophy [63] these proteins break down and the amino acids are most likely redistributed to proteins in other tissues.

Fate of Blood Meal Amino Acids at the End of the Gonotrophic Cycle and Utilization of Maternal Energy Reserves During a Gonotrophic Cycle

At the end of the gonotrophic cycle only about 10% of the blood meal protein amino acid carbon ends up in the eggs, with about 4% present as proteins and approximately 6% present as lipids; whereas approximately 20% is retained in the female as proteins ($\sim$10%), lipids ($\sim$8%), and sugars ($\sim$2%). Interestingly, the majority of the carbon ($\sim$70%) is oxidized and excreted as waste [28] (Figure 12.2).

In addition to energy produced from the oxidation of blood meal protein amino acids, the female has a certain level of nutrient reserves available for

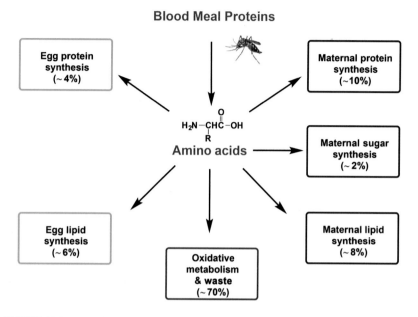

FIGURE 12.2 Fate of blood meal protein amino acid carbon at the end of a gonotrophic cycle in *Ae. aegypti.*

energy production at the beginning of the gonotrophic cycle [63]. These reserves are accumulated during larval development and from nectar feeding after eclosion [18]. In order to compare the utilization of carbohydrate and lipid reserves during a gonotrophic cycle, methods were devised to pre-label either the carbohydrate reserves or lipid reserves in the female mosquito with ^{14}C and follow the fate of the labeled reserves after feeding a meal containing unlabeled proteins [29]. It was reported that more of the carbohydrate reserves are used for energy production than the lipid reserves. This observation coupled with the fact that sugars made from amino acids are also mainly oxidized emphasizes the importance of sugar metabolism as a major source of energy during the gonotrophic cycle [18]. None of the carbohydrate reserves end up in the eggs, but a significant portion of the lipid reserves do so. Thus, it indicates that lipids are an important constituent of eggs, where they may be used for energy production during embryogenesis [29].

The metabolism of free amino acids in a blood meal was also investigated by feeding mosquitoes with ^{14}C-labeled amino acids. In order to obtain a detailed picture about the relative flux of individual amino acids in the blood meal, mosquitoes were fed with ^{14}C-labeled amino acids: leucine (Leu), valine (Val), isoleucine (Ile), phenylalanine (Phe), lysine (Lys), arginine (Arg), histidine (His), and alanine (Ala). These ^{14}C tracer amino acids were either incorporated into an *in vitro* synthesized protein (protein-bound) and then added to the blood meal or individually (free) added to the blood meal. Some interesting differences between the metabolism of those amino acids were observed. In both forms, Leu had the highest incorporation into maternal lipids, whereas His and Arg were the most highly excreted as waste. Most of the Ala was used for energy metabolism. Protein-bound Phe and plasma-free Ile were incorporated at the highest rate into egg proteins. It was also reported that plasma-free Ile incorporated into maternal proteins during the first gonotrophic cycle functions as a Ile reservoir for the second gonotrophic cycle [64]. These data demonstrate that different dietary free amino acids play different roles in building maternal protein and lipid reserves in energy metabolism as well as in building egg reserves. A topic that remains to be extensively explored is how amino acids are transported from the midgut to various mosquito tissues [65–68].

It could also be very enlightening to analyze and compare the dynamics of blood meal protein amino acid metabolism during the gonotrophic cycle between different mosquito species. The data could provide information about the efficiency of each species in converting blood meal protein into eggs, digestive capacity, rate of lipogenesis and gluconeogenesis, etc. and could help to better understand the energetics and reproductive physiology of different mosquito species.

INVESTIGATING BLOOD MEAL PROTEIN AMINO ACID NITROGEN USING TRADITIONAL AND MODERN APPROACHES

The oxidation of blood amino acids for energy production or synthesis of carbohydrates and lipids releases ammonia as a side product as indicated above. However, the ammonia concentration in the mosquito hemolymph remains almost constant during the blood digestion period, suggesting that when the amino acids are deaminated, the ammonia is quickly incorporated into non toxic waste products.

Fate of Nitrogen Derived from Deamination of Amino Acids Using Classical Biochemical and Molecular Techniques

Using high-performance liquid chromatography, previous work has shown that Proline (Pro), glutamine (Gln), and Ala reach a maximum concentration in the mosquito hemolymph at about 18 h after blood feeding [69]. However, Pro is the predominant amino acid in the hemolymph of the adult female *Ae. aegypti* [69,70]. It was proposed that Pro serves as a transient nitrogen sink to store ammonia during blood meal digestion, especially when the rate of ammonia production from deamination of dietary amino acids exceeds the capacity of the mosquito to generate nitrogen waste such as uric acid (Figure 12.3). In addition to its use as an unexpected nitrogen reservoir, Pro can also be utilized

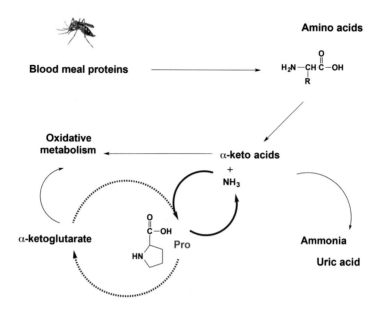

FIGURE 12.3   Pro cycle in *Ae. aegypti* mosquitoes. Pro allows transient storage of ammonia arising from the deamination of blood meal protein amino acids. Ammonia can be recovered from Pro and excreted as nitrogen waste, whereas the carbon skeleton can be used for energy production or for carbohydrate, lipid, and protein synthesis. *Adapted from Goldstrohm et al. [70].*

as a source of energy during flight of *Ae. aegypti* females [71]. It was shown for the first time in mosquitoes that when Pro is oxidized during flight, ammonia is removed through the synthesis of Ala, which shuttles ammonia between flight muscles and fat body by a Pro—Ala cycle (Figure 12.4). Additionally, it was suggested that Gln is involved in an alternative mechanism to shuttle ammonia between flight muscles and the fat body [71].

An essential role of Gln in ammonia detoxification in mosquitoes was confirmed when the effect of feeding a sugar meal containing NH_4Cl was measured. It caused a significant increase in hemolymph Gln, indicating that at least part of the ammonia is rapidly utilized to synthesize this amino acid. In addition, an important role of Pro was also established when mosquitoes were administered a sugar meal containing the glutamine synthetase (GS) inhibitor DL-methionine-DL-sulfoximine [72,73] along with NH_4Cl. The inhibition of GS significantly lowers hemolymph Gln and raises hemolymph Pro, further confirming the role of Pro as a nitrogen sink in *Ae. aegypti* females [69]. Apparently, when GS activity is inhibited, the reductive amination of α-ketoglutarate by glutamate dehydrogenase (GDH) becomes essential. The production of glutamic acid (Glu) via GDH can lead to increased synthesis of Pro via reactions catalyzed by pyrroline-5-carboxylase synthase (P5CS) and pyrroline-5-carboxylase reductase (P5CR). The presence

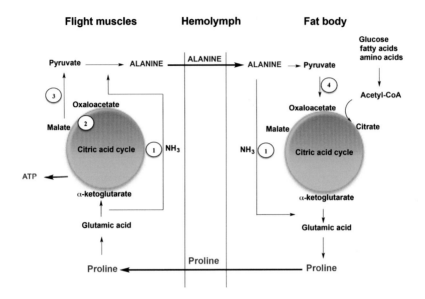

FIGURE 12.4 Pro–Ala cycle in *Ae. aegypti* mosquitoes. Pro serves to shuttle acetyl units from the fat body to flight muscles. Acetyl-CoA is first converted into α-ketogutarate through the citric acid cycle in the fat body. The α-ketogutarate is then converted to Pro in the fat body, and the Pro is transported to the flight muscles. In the flight muscles, Pro is converted back to α-ketogutarate, which then enters the citric acid cycle to be oxidized to ATP. At another step in the citric acid cycle, malate can be decarboxylated to pyruvate, which can serve as an ammonia acceptor to form Ala in flight muscles. Ala is then transported to the fat body to produce Pro. Key enzymes are: (1) alanine aminotransferase, (2) malate dehydrogenase, (3) NAD-linked "malic" enzyme, and (4) pyruvate carboxylase. *Reproduced with permission from Scaraffia and Wells [71].*

of nitrogen waste such as uric acid, urea, and ammonia in the excreta of NH_4Cl-fed mosquitoes suggests that ammonia ingested during feeding is released or metabolized. Thus, ammonia can be rapidly assimilated through the synthesis of Gln and Pro or excreted as nitrogen waste products. Although both Pro and Gln are involved in ammonia homeostasis, Gln seems to play a primary role in the acute response to ammonia, while Pro seems to play a backup role. In 2005, the presence of glutamate synthase (GltS) was discovered in mosquito fat body [69]. GS and GltS form the so-called GS/GltS pathway, which plays a very important role in ammonia fixation and assimilation in plants and bacteria [74–79]. Outside the plant, yeast, and bacterial world, the enzyme activity was only previously reported in *Bombyx mori* [80], *Samia cynthia ricini* [81], and *Spodoptera frugiperda* insect cells [82,83]. The discovery of GltS activity in mosquitoes was unexpected. GltS was cloned and sequenced in *Ae. aegypti* and its expression analyzed in mosquito tissues. The full-length cDNA for AaGltS is 6,252 nucleotides and encodes an open reading frame of 2,084 amino acids with a calculated molecular mass of

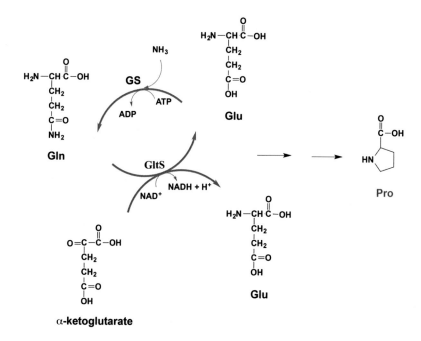

α-ketoglutarate

FIGURE 12.5 Schematic representation of the GS/GltS pathway for the fixation and assimilation of ammonia in mosquitoes. The excess ammonia is fixed by converting Glu to Gln by GS. Gln and α-ketoglutarate are then metabolized by GltS to yield two molecules of Glu, which can be utilized to synthesize Pro. *Adapted from Scaraffia et al. [69].*

229.6 kDa. The amino acid sequence encoding the *Ae. aegypti* NAHDH-dependent GltS constitutes the first complete amino acid sequence of GltS obtained from a metazoan [84]. Inhibition of GltS by azaserine [74] in mosquitoes leads to the inhibition of Glu and Pro synthesis and the accumulation of Gln in hemolymph, as well in mosquito whole body and tissues [69,84,85], demonstrating that GltS enhances the production of Glu for Pro synthesis. Moreover, it was shown that mosquitoes fix and assimilate ammonia efficiently through the GS/GltS pathway. GS can fix ammonia into Glu and produce Gln, whereas GltS can metabolize Gln and produce Glu, which can be converted to Pro or follow other metabolic fates depending on the needs of the cell (Figure 12.5).

Fate of Nitrogen Derived from Deamination of Amino Acids Using Isotopically-Labeled ^{15}N-Compounds, Mass Spectrometry, and RNA Interference Techniques

The monitoring of the fate of ^{15}N from ^{15}NH$_4$Cl in mosquitoes was preceded by the development and optimization of new approaches. The elucidation of the fragmentation mechanisms of Gln by electrospray ionization (ESI) tandem mass spectrometry [86] led to the identification of unlabeled and ^{15}N-labeled Glu and Gln (at different positions). Labeled Gln includes [5-^{15}N]-Gln (^{15}N-amide labeled), [2-^{15}N]-Gln (^{15}N-amine labeled), and [2,5-[15]N$_2$]-Gln ([15]N$_2$-labeled at both amide and amine positions). Based on this knowledge, and on procedures currently used for the detection of amino acids in metabolic disorders in newborn babies [87–89], unlabeled and labeled amino acids and other nitrogen compounds in mosquitoes were later successfully quantified by ESI-selected reaction monitoring in a triple-quadrupole mass spectrometer [84–86,90–94]. The procedure to prepare mosquito ammonia metabolism in whole body, tissues, or excreta for mass spectrometry analysis is illustrated in Figure 12.6.

Selected reaction monitoring is often called multiple reaction monitoring (MRM) when multiple analytes are examined in the same experiment [95]. This is a very powerful mass spectrometry technique, which can detect the compound of interest from a mixture with high sensitivity and selectivity. For example, for the detection of Ala, Glu, and Pro derivatives, a neutral loss of 102 Da ($HCOOC_4H_9$) occurs; for the detection of unlabeled Gln, a neutral loss of 73 Da from Gln derivative is used; whereas for Gln with label(s) at different positions, different neutral losses (73 or 74) are used [84]. Several amino acids and nitrogen waste compounds were analyzed by MRM in mosquito whole body, tissues, and excreta [84–86,90–92,94]. The use of stable isotopically-labeled compounds and tandem mass spectrometry techniques provides several advantages to the study of mosquito metabolism, including: *Sensitivity*: only one mosquito per assay is required; *Selectivity*: amino acids or other nitrogen compounds can be analyzed in a single sample based on mass differences; *Efficiency*: unlabeled and isotopically-labeled compounds can be analyzed; *Temporal resolution*: transient increases or decreases in the level of metabolic intermediates can be monitored over time, which can lead to the discovery of alternate pathways [94].

Fixation and Assimilation of Ammonia

Fixation and assimilation phases of ammonia detoxification are two unique ammonia detoxification processes in female mosquitoes. After feeding *Ae. aegypti* females with a [15]NH$_4$Cl solution for 15 min, mosquitoes are immersed in liquid nitrogen at different times after feeding and prepared for mass spectrometry analysis [84], as briefly described in Figure 12.6.

As shown in Figure 12.7, the dynamics of both ^{14}N and ^{15}N-amino acids can be accurately analyzed in mosquitoes [84]. The rate of incorporation of ^{15}N from [15]NH$_4$Cl into ^{15}N-amino acids of *Ae. aegypti* whole body is rapid. [5-^{15}N]-Gln is the most abundant amino acid quantified at the beginning of the time course, whereas [^{15}N]-Pro is the predominant amino acid detected at the end of the time course. Ammonia is first incorporated into the amine

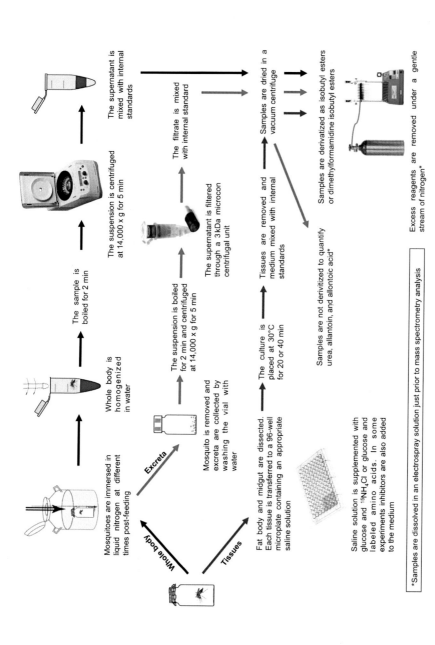

FIGURE 12.6 A simplified illustration of mosquito whole body, tissues, and excreta preparation for mass spectrometry analysis. A detailed description of mosquito sample preparation was previously reported [84,85,90–92,94].

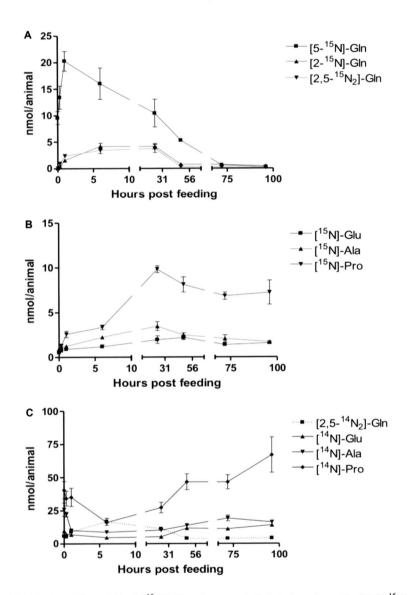

FIGURE 12.7 Effect of 80 mM $^{15}NH_4Cl$ on *Ae. aegypti* whole body amino acids. (A) [5-^{15}N]-Gln, [2-^{15}N]-Gln, and [2, 5-$^{15}N_2$]-Gln. (B) [^{15}N]-Glu, [^{15}N]-Ala, and [^{15}N]-Pro. (C) [$^{14}N_2$]-Gln, [^{14}N]-Glu, [^{14}N]-Ala, and [^{14}N]-Pro. Female mosquitoes were fed on 80 mM $^{15}NH_4Cl$ solution for 15 min and immersed in liquid nitrogen at different times post-feeding (see Figure 12.6 for details). Data are presented as mean ± standard error of three to six independent samples. *$p < 0.05$ (when compared to 0 h by ANOVA). *Reproduced with permission from Scaraffia et al. [84].*

side chain of Gln ([5-^{15}N]-Gln) and then into the amino group of the other amino acids ([^{15}N]-Glu, [^{15}N]-Ala, and [^{15}N]-Pro). At about 1 h post-feeding, [5-^{15}N]-Gln reaches the maximum concentration, whereas for other amino acids the highest concentration is observed at 24 h after feeding. At the end of the time course, the amount of [5-^{15}N]-Gln is significantly reduced, while [^{15}N]-Pro concentration is still high, indicating that Gln can be used as a precursor for Pro synthesis. The changes observed in the concentration of ^{15}N-amino acids are well correlated with the changes observed in ^{14}N-amino acids. Evidence suggests that the detoxification of ammonia in mosquitoes does not follow identical patterns as described in most terrestrial vertebrates (Figure 12.7). To better understand how different mosquito tissues deal with a load of ammonia, an *in vitro* approach was also used [85]. The procedure to analyze ammonia metabolism in fat body or midgut tissues is also illustrated in Figure 12.6. The results obtained from *in vitro* studies indicate that both tissues efficiently detoxify ammonia but use distinct metabolic pathways [85].

Based on the use of specific inhibitors of GS and GltS, the main pathway for ammonia fixation and assimilation in mosquito fat body involves GS/GltS. As shown in Figure 12.8, ^{15}N-labeled ammonia is fixed mainly by converting unlabeled Glu to [5-^{15}N]-Gln. GS seems to be the primary enzyme involved in the first step of ammonia fixation in mosquitoes. [5-^{15}N]-Gln and α-ketoglutarate then react with GltS, which transfers the amide group of labeled Gln to α-ketoglutarate and yields two molecules of Glu: one labeled and one unlabeled. [^{15}N]-Glu can also be produced by the reductive amination of α-ketoglutarate by GDH. In the midgut, GltS is not active and therefore, the GS/GltS is not functional in this tissue. However, GS and GDH are both very active in midgut and fat body [84,85]. In both tissues, [^{15}N]-Glu can be converted to [2, 5-^{15}N]-Gln by GS. Additionally, [^{15}N]-Glu can be used for other metabolic reactions, such as the synthesis of [^{15}N]-Ala by alanine aminotransferase [96], but most of it is used for [^{15}N]-Pro synthesis through P5CS and P5CR. This study demonstrates the role of Gln, Ala, and Pro as transient sinks of nitrogen. Alternately, [^{15}N]-Pro can be converted back into [^{15}N]-Glu by proline dehydrogenase and pyrroline-5-carboxylate dehydrogenase. The ^{15}N from [5-^{15}N]-Gln can also follow another metabolic fate, that is, it can be incorporated into ^{15}N$_2$-uric acid [90] through a series of enzymatic steps. Xanthine dehydrogenase is the enzyme that catalyzes the last two steps of uric acid synthesis in mosquitoes [97,98]. Although not discussed in this chapter, the principle vector of malaria, *Anopheles gambiae*, is related to *Ae. aegypti* and orthologs exist for many of the genes discussed here. Sequence identity of several enzymes involved in ammonia metabolism between *Ae. aegypti*, *Drosophila melanogaster*, and *An. gambiae*, as well as GenBank accession numbers were previously reported [69,90,92,99−104].

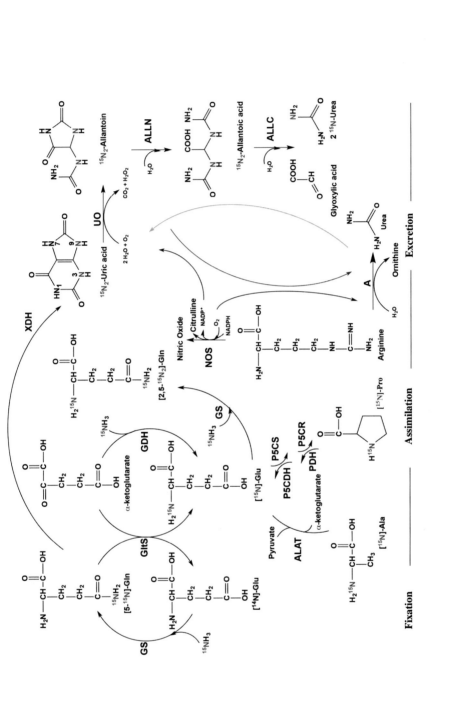

Fixation ——— Assimilation ——— Excretion

Excretion of Nitrogen Waste Products

Diuresis and flux of ions such as Na^+, K^+, and Cl^- have been extensively investigated [40,105,106], and recent studies have added new highlights into this field [107–111]. However, the underlying mechanisms of ion transport, homeostasis, and regulation are not completely understood and little is also known about nitrogen waste transport [51,55].

In *Ae. aegypti* mosquitoes, two ammonia transporters (AaRh50-1 and AaRh50-2) were cloned and sequenced but they have not yet been characterized (GenBank accession numbers: AY926463 and AY926464). Additionally, in the Asian tiger mosquito *Aedes albopictus*, the ammonia transporter AalRh50 (GenBank accession number: GUO13469) was cloned, and its expression analyzed in different mosquito tissues [55,112].

In ureotelic animals, the ammonia is converted to urea using ornithine transcarbamylase as a key enzyme in the synthesis of arginine, the ultimate precursor of urea in the urea cycle. Terrestrial insects are uricotelic and excrete nitrogen mainly as uric acid. Mosquitoes cannot synthesize urea through the conventional arginine–ornithine–urea cycle due to the absence of the gene encoding ornithine transcarbamylase, also called carbamoyltransferase [100,104]. However, mosquitoes do excrete some urea [20,21,69,90,97,98]; it arises from degradation of dietary arginine and turnover of proteins by arginase [98] or through an amphibian-like uricolytic pathway, as shown in Figure 12.8. $^{15}NH_4Cl$-fed *Ae. aegypti* mosquitoes were used to identify uricolytic pathway intermediates by mass spectrometry [90]. It was found that uric acid can be excreted directly or further metabolized to allantoin, allantoic acid, and urea by reactions catalyzed by urate oxidase (UO), allantoinase, and allantoicase, respectively. RNA-mediated knockdown of UO expression in blood-fed mosquitoes confirmed that the uricolytic pathway is functional under normal physiological conditions [90]. Expression patterns of genes involved in ammonia metabolism at different times after blood feeding suggest that fat body and Malpighian tubules are actively involved in the production and excretion of urea through argininolysis and uricolysis. The important role of Malpighian

---

◀ **FIGURE 12.8**   Fixation, assimilation, and excretion phases of ammonia metabolism in *Ae. aegypti* females. The use of isotopically-labeled compounds, mass spectrometry, and RNA interference allowed researchers to discover three phases for ammonia detoxification in *Ae. aegypti* mosquitoes: fixation, assimilation, and excretion. During ammonia detoxification female mosquitoes synthesize specific amino acids, as well as excrete several nitrogen waste products (e.g., uric acid, allantoin, allantoic acid, and urea) through multiple metabolic pathways that are extremely well-coordinated. Abbreviations: Glutamine synthetase (GS), glutamate synthase (GltS), glutamate dehydrogenase (GDH), alanine aminotransferase (ALAT), pyrrolidine-5-carboxylate synthase (P5CS), pyrrolidine-5-carboxylate reductase (P5CR), pyrroline-5-carboxylate dehydrogenase (P5CDH), proline dehydrogenase (PDH), xanthine dehydrogenase (XDH), urate oxidase (UO), arginase (A), nitric oxidase synthase (NOS), allantoinase (ALLN), allantoicase (ALLC). Arrows indicate cross-talk between UO, A, and NOS. *Adapted from Scaraffia et al. [84,85,90] and Isoe and Scaraffia [92].*

tubules in metabolic waste excretion was also recently proposed in *Ae. albopictus* based on transcriptomic studies [113]. The discovery that urea synthesis and excretion is regulated by a unique cross-talk mechanism suggests that metabolic regulation of ammonia in blood-fed *Ae. aegypti* mosquitoes is more complex than was previously thought [92]. The recently developed approach of utilizing low mass MS/MS fragments of protonated amino acids for the distinction of their ^{13}C-isotopomers in metabolic studies [93] is currently being used in our laboratory to distinguish, identify, and quantify ^{13}C-amino acids at various positions, either in the backbone or side chain in mosquito whole body and tissues, with the aim of verifying the hypothesis that partial oxidation of glucose provides some of the α-keto acids necessary for ammonia detoxification. This is a promising approach to identify rate-limiting steps in adult mosquito metabolism that could be potentially exploited as vector control targets.

The identification of mosquito-selective small molecules that inhibit blood meal metabolism could lead to reduced fecundity, and therefore decrease pathogen transmission. In addition to chemical interventions, modern genetic interventions such as transgenesis, paratransgenesis, and delivery of small interfering RNA [11,12,114–118] targeting specific metabolic pathways in mosquitoes should also be considered as an integral part of multiple control strategies.

CONCLUSIONS AND FUTURE DIRECTIONS

The approaches that researchers have used to study blood meal metabolism in mosquitoes have given novel and useful information about the physiology and biochemistry of mosquitoes. Surprisingly, mosquitoes use unpredictable metabolic pathways based on current vertebrate knowledge to metabolize the amino acids derived from protein digestion.

The metabolism of blood meal protein amino acids, as determined by rates of oxidation of ^{14}C-amino acid labeled proteins, shows a complex temporal course during the gonotrophic cycle involving the conversion of amino acid carbon to glucose, fatty acids, and proteins, as well as a substantial conversion to CO_2 as part of the energy-generating reactions which fuel vitellogenesis. In addition, different dietary free amino acids play differential roles such as synthesis of maternal proteins and lipids, egg protein synthesis, and as sources of energy. Although the percentage of amino acids oxidized is high after the digestion of blood meal proteins, producing large quantities of toxic ammonia, female mosquitoes have very efficient mechanisms to avoid lethal accumulation of nitrogen waste in the tissues. By feeding *Ae. aegypti* mosquitoes or incubating mosquito tissues with isotopically-labeled ^{15}N-compounds, it was discovered that mosquitoes can detoxify ammonia by synthesizing specific amino acids and excreting several nitrogen waste products through multiple metabolic pathways including GS/GltS and uricolysis that were not thought to be present in mosquitoes.

As discussed above, blood meal protein metabolism in mosquitoes is a central process for both transmission and reproduction. A better understanding of this process should contribute to generating better tools for the implementation of more efficient chemical and genetic strategies for vector-borne disease control.

The continued use of traditional and state-of-the art approaches such as metabolomics, proteomics, bioinformatics, and RNA interference will increase our understanding of the multiplicity of factors that control the blood meal metabolism in mosquitoes. The development of recent mass spectrometry methodology to quantify ^{13}C-amino acid at different positions to study carbon metabolism will also contribute to improving our knowledge in this field and help to elucidate possible new targets for disrupting blood feeding, digestion, and excretion. A primary focus for the future will also be to develop a complete quantitative mathematical model describing blood meal protein amino acid carbon and nitrogen metabolism. Such a model will allow us to obtain more comprehensive and precise information about various relative changes among different metabolic pathways during a gonotrophic cycle. We need to know if different models are needed for different mosquito species because such differences might provide a direct target for mosquito control. Since each mosquito species occupies a specific ecological niche, the effects of some important environmental factors (e.g., sugar and larval food availability) must be included in the model(s). The effects of physiological status (e.g., mating, parous or non-parous, different gonotrophic cycles, diapause, aging, and feeding behavior) must also be included in the model(s). Finally, field research of blood meal protein metabolism should be extensively performed so that those results can be included in the model.

ACKNOWLEDGMENTS

PYS dedicates this work to the memory of Prof. Dr Michael A. Wells (1938−2006). The author thanks Dr Stacy Mazzalupo for her critical reading of this book chapter and valuable comments. This work was financially supported by NIH Grant R01AI088092 (to PYS).

REFERENCES

[1] Barrett ADT, Higgs S. Yellow fever: a disease that has yet to be conquered. Annu Rev Entomol 2007;52:209−29.

[2] Matthews KR. Controlling and coordinating development in vector-transmitted parasites. Science 2011;331:1149−53.

[3] Bonizzoni M, Gasperi G, Chen X, James AA. The invasive mosquito species *Aedes albopictus*: current knowledge and future perspectives. Trends Parasitol 2013;29:460−8.

[4] Ciota AT, Kramer LD. Vector−virus interactions and transmission dynamics of West Nile virus. Viruses 2013;5:3021−47.

[5] White NJ, Pukrittayakamee S, Hien TT, Faiz MA, Mokuolu OA, Dondorp AM. Malaria. Lancet 2014;383:723−35.

[6] Yauch LE, Shresta S. Dengue virus vaccine development. Adv Virus Res 2014;88: 315−72.

[7] Gatton ML, Chitnis N, Churcher T, et al. The importance of mosquito behavioural adaptations to malaria control in Africa. Evolution 2013;67:1218−30.

[8] Killeen GF, Seyoum A, Sikaala C, et al. Eliminating malaria vectors. Parasit Vectors 2013; 6:172.

[9] Sokhna C, Ndiath MO, Rogier C. The changes in mosquito vector behaviour and the emerging resistance to insecticides will challenge the decline of malaria. Clin Microbiol Infect 2013;19:902−7.

[10] Tabachnick WJ. Nature, nurture and evolution of intra-species variation in mosquito arbovirus transmission competence. Int J Environ Res Public Health 2013;10:249−77.

[11] Wang S, Jacobs-Lorena M. Genetic approaches to interfere with malaria transmission by vector mosquitoes. Trends Biotechnol 2013;31:185−93.

[12] Alphey L, McKemey A, Nimmo D, et al. Genetic control of *Aedes* mosquitoes. Pathog Glob Health 2013;107:170−9.

[13] Dias CN, Moraes DF. Essential oils and their compounds as *Aedes aegypti* L. (Diptera: Culicidae) larvicides: review. Parasitol Res 2014;113:565−92.

[14] Potter CJ. Stop the biting: targeting a mosquito's sense of smell. Cell 2014;156:878−81.

[15] Rainey SM, Shah P, Kohl A, Dietrich I. Understanding the *Wolbachia*-mediated inhibition of arboviruses in mosquitoes: progress and challenges. J Gen Virol 2014;95:517−30.

[16] Slatko BE, Luck AN, Dobson SL, Foster JM. *Wolbachia* endosymbionts and human disease control. Mol Biochem Parasitol 2014;195:88−95.

[17] Dhiman S, Veer V. Culminating anti-malaria efforts at long lasting insecticidal net? J Infect Public Health 2014. Available from: http://dx.doi.org/10.1016/j.jiph.2014.06.002.

[18] Foster WA. Mosquito sugar feeding and reproductive energetics. Annu Rev Entomol 1995; 40:443−74.

[19] France KR, Judson CL. Nitrogen partitioning and blood meal utilization by *Aedes aegypti* (Diptera: Culicidae). J Insect Physiol 1979;25:841−6.

[20] Briegel H. Mosquito reproduction: incomplete utilization of the blood meal protein for oogenesis. J Insect Physiol 1985;31:15−21.

[21] Briegel H. Protein catabolism and nitrogen partitioning during oogenesis in the mosquito *Aedes aegypti*. J Insect Physiol 1986;32:455−62.

[22] Briegel H. Metabolic relationship between female body size, reserves, and fecundity of *Aedes aegypti*. J Insect Physiol 1990;36:165−72.

[23] Briegel H. Fecundity, metabolism, and body size in *Anopheles* (Diptera: Culicidae), vectors of malaria. J Med Entomol 1990;27:839−50.

[24] Briegel H, Horler E. Multiple blood meals as a reproductive strategy in *Anopheles* (Diptera: Culicidae). J Med Entomol 1993;30:975−85.

[25] Briegel H, Timmermann SE. *Aedes albopictus* (Diptera: Culicidae): physiological aspects of development and reproduction. J Med Entomol 2001;38:566−71.

[26] Briegel H, Waltert A, Kuhn AR. Reproductive physiology of *Aedes (Aedimorphus) vexans* (Diptera: Culicidae) in relation to flight potential. J Med Entomol 2001;38:557−65.

[27] Briegel H, Hefti M, DiMarco E. Lipid metabolism during sequential gonotrophic cycles in large and small female *Aedes aegypti*. J Insect Physiol 2002;48:547−54.

[28] Zhou G, Flowers M, Friedrich K, Horton J, Pennington J, Wells MA. Metabolic fate of [^{14}C]-labeled meal protein amino acids in *Aedes aegypti* mosquitoes. J Insect Physiol 2004; 50:337−49.

[29] Zhou G, Pennington JE, Wells MA. Utilization of pre-existing energy stores of female *Aedes aegypti* mosquitoes during the first gonotrophic cycle. Insect Biochem Mol Biol 2004;34:919−25.

[30] Zhou G, Scaraffia PY, Wells MA. Vector nutrition and energy metabolism. In: Marquardt WC, editor. Biology of disease vectors. Burlington, MA: Elsevier Academic Press; 2005. p. 211−5.

[31] Pennington JE, Wells MA. The adult midgut: structure and function. In: Marquardt WC, editor. Biology of disease vectors. Burlington, MA: Elsevier Academic Press; 2005. p. 289−95.

[32] Scaraffia PY, Miesfeld RL. Insect biochemistry/hormones. In: Lennarz WJ, Lane MD, editors. The encyclopedia of biological chemistry. Waltham, MA: Elsevier Academic Press; 2013. p. 590−5.

[33] Hegedus D, Erlandson M, Gillott C, Toprak U. New insights into peritrophic matrix synthesis, architecture, and function. Annu Rev Entomol 2009;54:285−302.

[34] Shao L, Devenport M, Jacobs-Lorena M. The peritrophic matrix of hematophagous insects. Arch Insect Biochem Physiol 2001;47:119−25.

[35] Pascoa V, Oliveira P, Dansa-Petretski M, et al. *Aedes aegypti* peritrophic matrix and its interaction with heme during blood digestion. Insect Biochem Mol Biol 2002;32:517−23.

[36] Graça-Souza A, Maya-Monteiro C, Paiva-Silva G, et al. Adaptations against heme toxicity in blood-feeding arthropods. Insect Biochem Mol Biol 2006;36:322−35.

[37] Sanders HR, Evans AM, Ross LS, Gill SS. Blood meal induces global changes in midgut gene expression in the disease vector, *Aedes aegypti*. Insect Biochem Mol Biol 2003;33: 1105−22.

[38] Beyenbach KW. Transport mechanisms of diuresis in Malpighian tubules of insects. J Exp Biol 2003;206:3845−56.

[39] Beyenbach KW, Piermarini PM. Transcellular and paracellular pathways of transepithelial fluid secretion in Malpighian (renal) tubules of the yellow fever mosquito *Aedes aegypti*. Acta Physiol 2011;202:387−407.

[40] Larsen EH, Deaton LE, Onken H, et al. Osmoregulation and excretion. Compr Physiol 2014;4:405−573.

[41] Barillas-Mury C, Graf R, Hagedorn HH, Wells MA. cDNA and deduced amino acid sequence of the blood meal-induced trypsin from the mosquito, *Aedes aegypti*. Insect Biochem 1991;21:825−31.

[42] Kalhok SE, Tabak LM, Prosser DE, Brook W, Downe AE, White BN. Isolation, sequencing and characterization of two cDNA clones coding for trypsin-like enzymes from the midgut of *Aedes aegypti*. Insect Mol Biol 1993;2:71−9.

[43] Noriega FG, Wang XY, Pennington JE, Barillas-Mury CV, Wells MA. Early trypsin, a female-specific midgut protease in *Aedes aegypti*: isolation, amino terminal sequence determination, and cloning and sequencing of the gene. Insect Biochem Mol Biol 1996; 26:119−26.

[44] Jiang Q, Hall M, Noriega FG, Wells MA. cDNA cloning and pattern of expression of an adult, female-specific chymotrypsin from *Aedes aegypti* midgut. Insect Biochem Mol Biol 1997;27:283−9.

[45] Edwards MJ, Moskalyk LA, Donelly-Doman M, et al. Characterization of a carboxypeptidase A gene from the mosquito, *Aedes aegypti*. Insect Mol Biol 2000;9:33−8.

[46] Bian G, Raikhel AS, Zhu J. Characterization of a juvenile hormone-regulated chymotrypsin-like serine protease gene in *Aedes aegypti* mosquito. Insect Biochem Mol Biol 2008;38: 190−200.

[47] Isoe J, Zamora J, Miesfeld RL. Molecular analysis of the *Aedes aegypti* carboxypeptidase gene family. Insect Biochem Mol Biol 2009;39:68–73.

[48] Isoe J, Rascon Jr AA, Kunz S, Miesfeld RL. Molecular genetic analysis of midgut serine proteases in *Aedes aegypti* mosquitoes. Insect Biochem Mol Biol 2009;39:903–12.

[49] Brackney DE, Isoe J, Black IV WC, et al. Expression profiling and comparative analyses of seven midgut serine proteases from the yellow fever mosquito, *Aedes aegypti*. J Insect Physiol 2010;56:736–44.

[50] Rascón Jr AA, Gearin J, Isoe J, Miesfeld RL. *In vitro* activation and enzyme kinetic analysis of recombinant midgut serine proteases from the Dengue vector mosquito *Aedes aegypti*. BMC Biochem 2011;12:43.

[51] Wright PA. Nitrogen excretion: three end products, many physiological roles. J Exp Biol 1995;198:273–81.

[52] Pereira LO, Oliveira PL, Almeida IC, Paiva-Silva GO. Biglutaminyl-biliverdin IX alpha as a heme degradation product in the dengue fever insect-vector *Aedes aegypti*. Biochemistry 2007;46:6822–9.

[53] Zhou G, Kohlhepp P, Geiser D, Frasquillo MDC, Vazquez-Moreno L, Winzerling JJ. Fate of blood meal iron in mosquitoes. J Insect Physiol 2007;53:1169–78.

[54] O'Donnell MJ. Too much of a good thing: how insects cope with excess ions or toxins in the diet. J Exp Biol 2009;212:363–72.

[55] Weihrauch D, Donini A, O'Donnell MJ. Ammonia transport by terrestrial and aquatic insects. J Insect Physiol 2012;58:473–87.

[56] Hansen IA, Attardo GM, Rodriguez SD, Drake LL. Four-way regulation of mosquito yolk protein precursor genes by juvenile hormone-, ecdysone-, nutrient-, and insulin-like peptide signaling pathways. Front Physiol 2014;5:103.

[57] Rogers DJ, Wilson AJ, Hay SI, Graham AJ. The global distribution of yellow fever and dengue. Adv Parasitol 2006;62:181–220.

[58] Morrison AC, Zielinski-Gutierrez E, Scott TW, Rosenberg R. Defining challenges and proposing solutions for control of the virus vector *Aedes aegypti*. PLoS Med 2008;5:e68.

[59] World Health Organization, Special Programme for Research, Training in Tropical Diseases. Dengue: guidelines for diagnosis, treatment, prevention and control. Geneva, Switzerland: WHO; 2009.

[60] Caglioti C, Lalle E, Castilletti C, Carletti F, Capobianchi MR, Bordi L. Chikungunya virus infection: an overview. New Microbiol 2013;36:211–27.

[61] Powell JR, Tabachnick WJ. Genetic shifting: a novel approach for controlling vector-borne diseases. Trends Parasitol 2014;30:282–8.

[62] Krzywicha AM, Wagner GH. Laboratory culture for ^{14}C-labeling for *Chlorella* and *Oedogonium*. Int J Appl Radiat Isot 1975;26:515–18.

[63] Clements AN. The biology of mosquitoes,. Development, nutrition and reproduction, vol. 1. New York, NY: Chapman & Hall; 1992.

[64] Zhou G, Miesfeld RL. Differential utilization of blood meal amino acids in mosquitoes. Open Access Insect Physiol 2009;1:1–12.

[65] Evans AM, Aimanova KG, Gill SS. Characterization of a blood-meal-responsive proton-dependent amino acid transporter in the disease vector, *Aedes aegypti*. J Exp Biol 2009; 212:3263–71.

[66] Harvey WR, Boudko DY, Rheault MR, Okech BA. NHE$_{VNAT}$: an H$^+$ V-ATPase electrically coupled to a Na$^+$: nutrient amino acid transporter (NAT) forms an Na$^+$/H$^+$ exchanger (NHE). J Exp Biol 2009;212:347–57.

[67] Hansen IA, Boudko DY, Shiao SH, et al. AaCAT1 of the yellow fever mosquito, *Aedes aegypti*: a novel histidine-specific amino acid transporter from the SLC7 family. J Biol Chem 2011;286:10803–13.

[68] Carpenter VK, Drake LL, Aguirre SE, Price DP, Rodriguez SD, Hansen IA. SLC7 amino acid transporters of the yellow fever mosquito *Aedes aegypti* and their role in fat body TOR signaling and reproduction. J Insect Physiol 2012;58:513–22.

[69] Scaraffia PY, Isoe J, Murillo A, Wells MA. Ammonia metabolism in *Aedes aegypti*. Insect Biochem Mol Biol 2005;35:491–503.

[70] Goldstrohm DA, Pennington JE, Wells MA. The role of hemolymph proline as a nitrogen sink during blood meal digestion by the mosquito *Aedes aegypti*. J Insect Physiol 2003; 49:115–21.

[71] Scaraffia PY, Wells MA. Proline can be utilized as an energy substrate during flight of *Aedes aegypti* females. J Insect Physiol 2003;49:591–601.

[72] Eisenberg D, Gill HS, Pfluegl GM, Rotstein SH. Structure–function relationships of glutamine synthetases. Biochim Biophys Acta 2000;1477:122–45.

[73] Meister A. Glutamine-synthetase from mammalian-tissues. Methods Enzymol 1985;113: 185–99.

[74] Lea PJ, Miflin BJ. Glutamate synthase and the synthesis of glutamate in plants. Plant Physiol Biochem 2003;41:555–64.

[75] Raushel FM, Thoden JB, Holden HM. Enzymes with molecular tunnels. Acc Chem Res 2003;36:539–48.

[76] Reitzer L. Nitrogen assimilation and global regulation in *Escherichia coli*. Annu Rev Microbiol 2003;57:155–76.

[77] van den Heuvel RH, Curti B, Vanoni MA, Mattevi A. Glutamate synthase: a fascinating pathway from L-glutamine to L-glutamate. Cell Mol Life Sci 2004;61:669–81.

[78] Vanoni MA, Curti B. Structure–function studies on the iron-sulfur flavoenzyme glutamate synthase: an unexpectedly complex self-regulated enzyme. Arch Biochem Biophys 2005; 433:193–211.

[79] Vanoni MA, Curti B. Structure-function studies of glutamate synthases: a class of self-regulated iron-sulfur flavoenzymes essential for nitrogen assimilation. IUBMB Life 2008; 60:287–300.

[80] Hirayama C, Konno K, Shinbo H. The pathway of ammonia assimilation in the silkworm, *Bombyx mori*. J Insect Physiol 1997;43:959–64.

[81] Osanai M, Okudaira M, Naito J, Demura M, Asakura T. Biosynthesis of L-alanine, a major amino acid of fibroin in *Samia cynthia ricini*. Insect Biochem Mol Biol 2000;30:225–32.

[82] Doverskog M, Jacobsson U, Chapman BE, Kuchel PW, Häggström L. Determination of NADH-dependent glutamate synthase (GOGAT) in *Spodoptera frugiperda* (Sf9) insect cells by a selective ^{1}H/^{15}N NMR *in vitro* assay. J Biotechnol 2000;79:87–97.

[83] Drews M, Doverskog M, Ohman L, et al. Pathways of glutamine metabolism in *Spodoptera frugiperda* (Sf9) insect cells: evidence for the presence of the nitrogen assimilation system, and a metabolic switch by ^{1}H/^{15}N NMR. J Biotechnol 2000;78:23–37.

[84] Scaraffia PY, Zhang Q, Wysocki VH, Isoe J, Wells MA. Analysis of whole body ammonia metabolism in *Aedes aegypti* using [^{15}N]-labeled compounds and mass spectrometry. Insect Biochem Mol Biol 2006;36:614–22.

[85] Scaraffia PY, Zhang Q, Thorson K, Wysocki VH, Miesfeld RL. Differential ammonia metabolism in *Aedes aegypti* fat body and midgut tissues. J Insect Physiol 2010;56: 1040–9.

[86] Zhang Q, Wysocki VH, Scaraffia PY, Wells MA. Fragmentation pathway for glutamine identification: loss of 73 Da from dimethylformamidine glutamine isobutyl ester. J Am Soc Mass Spectrom 2005;16:1192−203.

[87] Chace DH. Mass spectrometry in the clinical laboratory. Chem Rev 2001;101:445−77.

[88] Chace DH, Kalas TA, Naylor EW. The application of tandem mass spectrometry to neonatal screening for inherited disorders of intermediary metabolism. Annu Rev Genomics Hum Genet 2002;3:17−45.

[89] Chace DH, Kalas TA, Naylor EW. Use of tandem mass spectrometry for multianalyte screening of dried blood specimens from newborns. Clin Chem 2003;49:1797−817.

[90] Scaraffia PY, Tan G, Isoe J, Wysocki VH, Wells MA, Miesfeld RL. Discovery of an alternate metabolic pathway for urea synthesis in adult Aedes aegypti mosquitoes. Proc Natl Acad Sci USA 2008;105:518−23.

[91] Bush DR, Wysocki VH, Scaraffia PY. Study of the fragmentation of arginine isobutyl ester applied to arginine quantification in Aedes aegypti mosquito excreta. J Mass Spectrom 2012;47:1364−71.

[92] Isoe J, Scaraffia PY. Urea synthesis and excretion in Aedes aegypti mosquitoes are regulated by a unique cross-talk mechanism. PLoS One 2013;8:e65393.

[93] Ma X, Dagan S, Somogyi A, Wysocki VH, Scaraffia PY. Low mass MS/MS fragments of protonated amino acids used for distinction of their ^{13}C-isotopomers in metabolic studies. J Am Soc Mass Spectrom 2013;24:622−31.

[94] Mazzalupo S, Scaraffia PY. Application of isotopically labeled compounds and tandem mass spectrometry for studying metabolic pathways in mosquitoes. In: Chandrasekar R, Tyagi BK, Gui ZZ, Reeck GR, editors. Short view on insect biochemistry and molecular biology. Manhattan, KS /South India: K-State Union, Kansas State University/International Book Mission; 2014. p. 99−126.

[95] Yost RA, Enke CG. Triple quadrupole mass-spectrometry for direct mixture analysis and structure elucidation. Anal Chem 1979;51:1251−64.

[96] Belloni V, Scaraffia PY. Exposure to L-cycloserine incurs survival costs and behavioral alterations in Aedes aegypti females. Parasit Vectors 2014;7:373.

[97] von Dungern P, Briegel H. Enzymatic analysis of uricotelic protein catabolism in the mosquito Aedes aegypti. J Insect Physiol 2001;47:73−82.

[98] von Dungern P, Briegel H. Protein catabolism in mosquitoes: ureotely and uricotely in larval and imaginal Aedes aegypti. J Insect Physiol 2001;47:131−41.

[99] Holt RA, Subramanian GM, Halpern A, et al. The genome sequence of the malaria mosquito Anopheles gambiae. Science 2002;298:129−49.

[100] Zdobnov EM, von Mering C, Letunic I, et al. Comparative genome and proteome analysis of Anopheles gambiae and Drosophila melanogaster. Science 2002;298:149−59.

[101] Ribeiro J. A catalogue of Anopheles gambiae transcripts significantly more or less expressed following a blood meal. Insect Biochem Mol Biol 2003;33:865−82.

[102] Severson D, DeBruyn B, Lovin D, Brown S, Knudson D, Morlais I. Comparative genome analysis of the yellow fever mosquito Aedes aegypti with Drosophila melanogaster and the malaria vector mosquito Anopheles gambiae. J Hered 2004;95:103−13.

[103] Marinotti O, Calvo E, Nguyen Q, Dissanayake S, Ribeiro J, James A. Genome-wide analysis of gene expression in adult Anopheles gambiae. Insect Mol Biol 2006;15:1−12.

[104] Nene V, Wortman JR, Lawson D, et al. Genome sequence of Aedes aegypti, a major arbovirus vector. Science 2007;316:1718−23.

[105] Pullikuth AK, Filippov V, Gill SS. Phylogeny and cloning of ion transporters in mosquitoes. J Exp Biol 2003;206:3857−68.

[106]   Beyenbach KW, Skaer H, Dow JA. The developmental, molecular, and transport biology of Malpighian tubules. Annu Rev Entomol 2010;55:351−74.

[107]   Kwon H, Pietrantonio PV. Calcitonin receptor 1 (AedaeGPCRCAL1) hindgut expression and direct role in myotropic action in females of the mosquito *Aedes aegypti* (L.). Insect Biochem Mol Biol 2013;43:588−93.

[108]   Pacey EK, O'Donnell MJ. Transport of H^+, Na^+ and K^+ across the posterior midgut of blood-fed mosquitoes (*Aedes aegypti*). J Insect Physiol 2014;61:42−50.

[109]   Benoit JB, Hansen IA, Szuter EM, Drake LL, Burnett DL, Attardo GM. Emerging roles of aquaporins in relation to the physiology of blood-feeding arthropods. J Comp Physiol B 2014;184:811−25.

[110]   Paluzzi JP, Vanderveken M, O'Donnell MJ. The heterodimeric glycoprotein hormone, GPA2/GPB5, regulates ion transport across the hindgut of the adult mosquito, *Aedes aegypti*. PLoS One 2014;9:e86386.

[111]   Hine RM, Rouhier MF, Park ST, Qi Z, Piermarini PM, Beyenbach KW. The excretion of NaCl and KCl loads in mosquitoes: 1. Control data. Am J Physiol Regul Integr Comp Physiol 2014;307:R837−49.

[112]   Wu Y, Zheng X, Zhang M, He A, Li Z, Zhan X. Cloning and functional expression of Rh50-like glycoprotein, a putative ammonia channel, in *Aedes albopictus* mosquitoes. J Insect Physiol 2010;56:1599−610.

[113]   Esquivel CJ, Cassone BJ, Piermarini PM. Transcriptomic evidence for a dramatic functional transition of the Malpighian tubules after a blood meal in the Asian tiger mosquito *Aedes albopictus*. PLoS Negl Trop Dis 2014;8:e2929.

[114]   Burand JP, Hunter WB. RNAi: future in insect management. J Invertebr Pathol 2013;112 (Suppl.):S68−74.

[115]   Katoch R, Sethi A, Thakur N, Murdock L. RNAi for insect control: current perspective and future challenges. Appl Biochem Biotechnol 2013;171:847−73.

[116]   Scott J, Michel K, Bartholomay L, et al. Towards the elements of successful insect RNAi. J Insect Physiol 2013;59:1212−21.

[117]   Blair C, Olson KE. Mosquito immune responses to arbovirus infections. Curr Opin Insect Sci 2014;3:22−9.

[118]   Alphey L. Genetic control of mosquitoes. Annu Rev Entomol 2014;59:205−24.

Chapter 13

Engineering Pathogen Resistance in Mosquitoes

Zach N. Adelman, Sanjay Basu and Kevin M. Myles

Fralin Life Science Institute and Department of Entomology, Virginia Tech, Blacksburg, VA, USA

MALARIA AND DENGUE

Malaria and dengue are the two most important mosquito-borne diseases worldwide. While substantial effort is underway to develop novel vaccines, therapeutic drugs, and other preventative measures, a growing body of literature supports the additional possibility of exploiting the mosquito's own genome to launch new offensives at these deadly scourges. The development of advanced genetic tools for mosquito genome manipulation (reviewed in Ref. [1]), the availability of critical mosquito genome sequences [2,3], and increased knowledge of insect innate immunity and physiology have helped fuel these efforts. Several groups have shown, in a laboratory setting, that mosquitoes can indeed be made resistant to dengue viruses or malaria parasites. This chapter will focus on the specific transgenic strategies that have been used to establish resistance to malaria parasites or dengue viruses in mosquitoes along with the strengths and weaknesses of each strategy. The generation of pathogen resistance, mediated through the microbiome [4] or through *Wolbachia* infection [5], is described elsewhere and will not be discussed here. Technologies that may be used to drive resistance genes into populations also will not be discussed here (see Ref. [6] for a full discussion of these strategies).

The transmission of pathogens by mosquitoes and other arthropod vectors is governed by the vectorial capacity of the insect population (reviewed in Ref. [7]). The most critical component of vectorial capacity is the probability that a mosquito survives from one day to the next; this parameter is heavily targeted by insecticide treatments and source reduction campaigns (these treatments also reduce another important parameter, vector-to-host density). Other interventions such as repellents, baited traps, and physical structures (i.e., screened doors and windows) reduce the biting rate of mosquitoes, which is another important component of vectorial capacity. As these strategies have proven to be insufficient for controlling vector-borne diseases such as malaria and dengue, many research groups have begun targeting the two other parameters of vectorial capacity, vector competence and the length of the extrinsic incubation period (EIP). Vector competence refers to the probability that a vector fed on an infected vertebrate host will eventually be capable of transmitting that pathogen to a new host. The EIP refers to the duration of time that must elapse between ingestion of the pathogen by the vector and the time when the vector is capable of transmitting the pathogen. The EIP and vector competence are heavily influenced by the genetics of both the pathogen and the vector, as well as by interactions with the environment.

Whereas control strategies based on the reduction or elimination of vector populations (either through traditional means or by genetic methods) must be continued indefinitely to prevent reinvasion and reestablishment of competent vector populations, the notion of converting a pathogen-susceptible population to a resistant one is valued for its ecological simplicity. In these strategies, interactions between wild mosquito populations, their hosts, and the environment are preserved. Instead, it is only the pathogen residing within the mosquito that is targeted.

Malaria

A female mosquito bites an individual stricken with malaria. Although she takes only a few microliters of blood before flying off, a small number of gametocytes are ingested with the meal. When the female and male gametocytes are ingested they mature into gametes following exposure to mosquito-specific factors (xanthurenic acid) and environmental cues (increase in pH and drop in temperature of 5°C) [8]. The male gametes undergo exflagellation generating eight microgametes that can fertilize the female macrogamete, forming a diploid zygote, which then develops into a motile ookinete. Ookinetes must traverse the midgut epithelium within 24−48 h after being ingested, where they lodge under the basal lamina and develop into oocysts. The ookinete appears to be the most vulnerable stage of the parasite's development, as the mosquito mounts a significant immune response to the parasite infection leading to comparatively few oocysts (reviewed in Refs [9,10]). Oocysts can be observed 5−7 days after feeding on the infectious blood meal. After approximately 12 days (though this time is strongly influenced by temperature), oocysts mature and rupture releasing thousands of haploid sporozoites into the mosquito hemolymph. The sporozoites are then dispersed throughout the host by its circulatory system. Receptor−ligand interactions between the sporozoite and salivary glands trigger invasion. The invading sporozoites accumulate in bundles within the secretory cavity of the salivary gland. The preferred location is actually the distal portion of the gland within the secretory duct, where the parasites collect in the ducts and wait until egestion with saliva into the vertebrate host.

Dengue

Similar to malaria parasites, dengue virions must infect and escape from the midgut of the mosquito, eventually infecting the distal lateral lobes of the salivary glands before being secreted into the saliva, a process that can take as little as 4 days [11]. Whereas the number of malaria oocysts colonizing the mosquito midgut is directly proportional to the number of sporozoites that can systemically infect the body, for dengue viruses there appears to be little relationship between titers within midgut epithelial cells and the probability of dissemination to the rest of the body [12,13]. That is, in some cases a high level of virus replication in the midgut may still result in poor dissemination to other tissues and organs, and thus low transmission rates. Conversely, limited replication in the midgut may sometimes result in high rates of dissemination and transmission.

For both malaria and dengue, resistance may be evaluated at three critical benchmarks in the life cycle of the pathogen: (i) successful colonization of the mosquito midgut, (ii) escape from the midgut leading to systemic infection in the hemolymph, and (iii) successful infection of the salivary glands with the presence of pathogen in the saliva. In the following sections, we describe the strategies published to date that have attempted, and in some cases succeeded, in generating mosquitoes refractory to either malaria parasites or dengue viruses. We then discuss the choices made in designing and executing such experiments, as a guide to those who may wish to test their own resistance traits.

ENGINEERING RESISTANCE TO PATHOGENS

The initial concept of pathogen-derived resistance was put forward 30 years ago by Sanford and Johnson [14], who used components of the bacteriophage Qβ genome to generate resistance to that phage in *Escherichia coli* bacteria. Sanford and Johnson postulated that this concept would be widely applicable to other host–pathogen systems; subsequent work in a large number of other organisms has proved them correct. In particular, there was an immediate push to use this technology to generate virus-resistant transgenic plants (reviewed in Refs [15–18]). Just a year after the publication of Sanford and Johnson's phage experiments, the first report of successful pathogen-derived resistance in plants appeared [19]; a transgenic tobacco plant carrying a fragment of the tobacco mosaic virus genome rendered the plants resistant to virus infection through an as yet unknown mechanism. Other reports of pathogen-derived resistance rapidly followed, and while a number of potential mechanisms were postulated, it eventually became apparent that most cases were due to a newly discovered mechanism, which later became known as RNA interference (RNAi) (reviewed in Refs [15,16]). A critical arm of the immune system of plants and insects, the RNAi pathway recognizes double-stranded RNA, a specific pattern associated with virus infection. Engineering fragments of a viral genome into a plant genome resulted in the production of virus-specific double-stranded RNA, and thus is a form of genetic vaccination. The first transgenic crop with a pathogen-derived resistance trait, a virus-resistant summer squash, was put into commercial use in 1994. Other crops soon followed such as virus-resistant pepper, tomato, and potato, all of which have since been deregulated and have seen varying degrees of commercial success [15]. Field trials have been completed or are ongoing for many others pathogen-resistant crops including wheat, soybean, sugarcane, and corn [17].

Pathogen-Derived Resistance (RNAi)

Inspired by successes protecting crop plants from RNA viruses, a concerted effort was made to generate similar phenotypes in the dengue vector *Aedes aegypti*, even prior to the development of transgenic technology for mosquitoes. Using a Sindbis virus-based expression system, pathogen-derived resistance was established against a number of arboviruses, including all four serotypes of dengue virus [20−22], yellow fever virus [23], and LaCrosse virus [24,25]. Stable pathogen-derived resistance to dengue virus type 2 was soon established in cultured cells [26] and finally in transgenic mosquitoes [27−29]. As in plants, resistant mosquitoes were generated through an RNAi-dependent mechanism. This RNAi response could be triggered either in the midgut in response to a blood meal [27,29], preventing initial infection of dengue virus type 2, or in the salivary glands [28], preventing virus accumulation in the saliva. While RNAi-based strategies for pathogen-derived resistance appear to be extremely stable in plants [15], resistance phenotypes were lost in mosquitoes after successive (17) generations for reasons that remain unknown [30]. In these mosquitoes, expression of the RNAi trigger appears to have been shut off, with expression of the transgenic reporter gene unaffected. Franz et al. [29] were able to generate a new set of transgenic strains with the same pathogen-derived resistance construct. Again, a strong resistance phenotype was seen against dengue virus type 2. Importantly, resistance was observed against a panel of diverse dengue strains and was maintained for >33 generations, with no loss of resistance or fitness effect even when introgressed into a genetically diverse background. Pathogen-derived resistance has not been reported against malaria parasites and has not yet been extended to other serotypes of dengue virus.

Strengths: Of the various strategies attempted to develop pathogen resistance in mosquitoes, RNAi-based approaches have the strongest track record based on the trail blazed by researchers working on transgenic crops. RNAi-based approaches do not result in the production of any novel protein products, and thus may be likely to have the smallest effects on mosquito fitness. As the double-stranded RNA segments used to trigger the RNAi response are several hundred base pairs in length, it is unlikely that a targeted virus will be able to evolve a sufficient number of mutations to escape the diversity of short-interfering RNAs produced.

Weaknesses: The unexpected loss of silencing phenotypes [30] suggests that strategies to use RNAi-based transgenes in mosquitoes must account for potential transcriptional silencing, an understudied area in mosquitoes. RNAi exhibits extraordinary specificity, since it is based on recognition of the nucleotide sequence of the viral genome itself. While usually a beneficial trait, the extreme specificity of RNAi means that related, but not identical pathogens, may escape interference. Thus, resistance to dengue virus type 2 does not confer resistance to other dengue serotypes [27]; nor may resistance extend to more closely related but still divergent strains. Thus, generating broad resistance may require targeting a highly conserved sequence or stacking multiple sequences from independent viruses. RNAi is also strongly influenced by temperature [31], and it is yet unclear how these observations may affect RNAi-based transgenes.

Immune Activation

Mosquitoes mount a sophisticated and multipronged immune response to infecting parasites or viruses, which has been the subject of much research (reviewed in Refs [32−34]). However, both parasites and viruses may interfere with the recognition, signaling, or effector steps of various immune pathways, reducing their capacity to restrict pathogen replication and hence increasing vector competence. Manipulating the recognition and signaling steps through genetic engineering thus represents a promising approach for efficient targeting of the pathogen to prevent infection and transmission.

The transcription factor Rel2 (homolog of the vertebrate NF-κB) activates a large number of downstream effector genes, many of which have been demonstrated to have anti-*Plasmodium* activity. Transgenic overexpression of the active form of Rel2 in the *Ae. aegypti* fat body following a *Plasmodium gallinaceum*-containing blood meal decreased oocyst intensity [35], while direct overexpression of the effectors defensin A or cecropin in response to a blood meal rendered mosquitoes completely resistant to *P. gallinaceum* and prevented transmission of the parasite to chickens [36]. However, when Kim et al. [37] generated transgenic *Anopheles gambiae* mosquitoes that overexpressed the same parasite-killing cecropin in the gut of the mosquito, decreased parasite intensity was observed but without any corresponding affect on prevalence. Finally, Dong et al. [38] generated transgenic *Anopheles stephensi* that overexpressed the Rel2 transcription factor either in the midgut or the fat body in response to a blood meal. Transgenic, blood meal−induced Rel2-dependent activation of immune effector genes including TEP1, AP1, and

LRRD7 generated strong resistance phenotypes, successfully reducing both the prevalence and intensity of parasite infection.

In additional to Rel2-based signaling, *Anopheles* mosquitoes use reactive nitric oxide (NO) as a critical anti-parasite effector molecule [39]. Suppression of anti-oxidants in transgenic mosquitoes by overexpressing an activated protein kinase in the midgut after blood feeding completely prevented infection of *An. stephensi* with the human malaria parasite in a manner than was dependent on increased reactive oxygen species [40,41]. Global changes in Akt-based signaling also resulted in mitochondrial dysfunction and shortened the lifespan of the transgenic mosquitoes [40,41]. In contrast, overexpression of the phosphatase and tensin homolog (PTEN), which inhibits Akt-based signaling, not only increased the lifespan of transgenic mosquitoes, but also reduced the prevalence and intensity of malaria parasite infection [42].

Strengths: Altering the expression of endogenous transcription factors through transgenic expression bypasses initial pattern recognition steps which may be otherwise suppressed by the pathogen, while overexpression of effectors such as defensin or cecropin bypasses both pattern recognition and signaling steps; both strategies result in much swifter action against the parasite than would be seen in a normal infection. Using an upstream transcription factor has the added benefit of inducing the expression of a large number of downstream genes, both known and unknown, that can act in concert to attack the parasite. This is expected to delay or prevent the development of resistance, as resistance to a large number of effectors would have to evolve simultaneously. Genetic control campaigns based on immune activation may be considered more benign than other approaches, as the overexpressed proteins derive from the vector genome itself, as opposed to being introduced foreign sequences from unrelated organisms.

Weaknesses: Large-scale changes in the immune status of a transgenic mosquito, overexpressing either an upstream transcription factor or downstream effector, are not expected to be specific for the targeted pathogen. Reports of the experiments described above noted that transgenic mosquitoes also became resistant to many types of bacteria that may naturally reside in the midgut [35,36,38]. Greater resistance to bacteria or fungi, along with other metabolic changes that may prolong the life of the mosquito [42], could result in vectors that survive longer than the EIP [43]. While in all studies published to date immune activation was temporally restricted through blood meal−activated promoters, leaky expression from these promoters is common [40,44] and thus may prolong the survival of immature stages and adults. While experimental designs typically evaluate mosquitoes after a single blood meal followed by an incubation period to allow parasite growth, mosquitoes in the wild feed multiple times, which would in a transgenic mosquito result in continual immune activation throughout the adult life [45]. Global changes in the microbiome may also change the susceptibility of mosquitoes to other pathogens [46].

Immune Augmentation

In addition to artificially boosting the native immune response of mosquitoes to human pathogens, substantial effort has been put into identifying and evaluating exogenous effector molecules capable of killing or incapacitating malaria parasites. Rather than boosting the production of endogenous parasite-killing molecules such as defensin or cecropin, such peptides can be identified from a multitude of other insect species where they may also display anti-malarial or anti-dengue properties when introduced into the mosquito. A screen of several naturally occurring and synthetic peptides identified several that reduced the number of viable malaria parasites when incubated in culture [47]. The most promising of these, a 14-amino-acid synthetic peptide called Vida3, was expressed in the midgut of transgenic *An. gambiae* [48]. Expression of Vida3 had no discernible fitness effects on the mosquito [49] and reduced the intensity of *Plasmodium yoelii nigeriensis* infection, but had little effect on prevalence; no effect was seen against *Plasmodium falciparum* [48]. An expanded screen of 33 additional potential anti-malarial peptides identified several additional candidates [50]. The expansion of genomic information from various arthropods should further increase the pool of potential *Plasmodium*-killing peptides; similar screens have not been reported for dengue viruses.

Yoshida et al. [51] generated transgenic *An. stephensi* that overexpressed a Cel-III lectin in the midgut in response to a blood meal. Originally isolated from sea cucumbers, Cel-III, had been shown previously to have cell-killing properties by forming pores in membranes. When challenged with rodent malaria parasites, Cel-III transgenic mosquitoes displayed strong reductions in both the prevalence and the intensity of parasite infection, as well as the transmission of parasites to naive mice. However, when challenged with human malaria parasites, resistance was not nearly as robust, with only moderate reductions in intensity and no change in prevalence [51].

Rather than borrowing from the immune systems of other arthropods, Jasinskiene et al. [52] adapted molecules from the vertebrate adaptive immune response, generating transgenic *Ae. aegypti* strains that overexpressed single-chain antibodies with high affinity to chicken malaria parasites. Much smaller and easier to manipulate than full size antibodies that consist of separate large and heavy chain molecules, single-chain antibodies retain the variable fragment as a single polypeptide. In this case, transient expression of the single-chain fragments in *Ae. aegypti* had previously been shown to reduce parasite numbers [53]. When produced from the mosquito fat body and secreted into the hemolymph following a blood meal, sporozoite invasion of the salivary glands was reduced, but not eliminated; no effect was observed on transmission rates. The authors of this study concluded that the threshold for parasite burden in the mosquito salivary glands should be zero, as mosquitoes with as few as 20 sporozoites were fully capable of infecting naive chickens [52].

Building on these findings, Isaacs et al. [54] generated a series of transgenic *An. stephensi* strains that produced single-chain antibody fragments targeting the human malaria parasite, *P. falciparum*. Expression of single-chain antibodies in gut cells, secreted into the lumen of the gut or secreted into the hemolymph, reduced, but did not eliminate parasite infection of the mosquito [54]. Similarly, expression of single-chain antibodies targeting the sporozoite stage directly in the salivary glands reduced, but did not eliminate transmission to vertebrate hosts [55]. However, combination-based transgenic strategies using single-chain antibodies targeting both the ookinete and the sporozoite completely prevented parasite invasion of the salivary glands [56], validating this experimental approach.

For dengue, few molecules have been tested in a transgenic context for their ability to inactivate the virion or prevent replication. Nawtaisong et al. [57] developed and validated a series of four hammerhead ribozymes capable of cleaving dengue virus type 2 RNA when expressed in transformed mosquito cells; such expression reduced dengue virus type 2 titers by approximately 100 fold. Carter et al. [58] followed this with a description of a trans-splicing effector that targeted the conserved cyclization sequence present in all serotypes of dengue virus, with reductions in dengue virus type 2 titers of approximately 1,000 fold observed in transformed mosquito cells. This was further improved by connecting the trans-splicing event with the expression of a pro-apoptotic factor (Bax) [59]. In transformed cells expressing these trans-splicing effectors, the replication of all four serotypes of dengue could be reduced by five orders of magnitude [59]. However, to date the use of ribozymes to target dengue viruses in the context of a transgenic mosquito has not been reported.

Strengths: Mining the immune responses of organisms other than mosquitoes takes advantage of recent gains in genomics and genome sequencing data to provide a virtually unlimited repertoire of potential anti-malarial or anti-dengue peptides. Transgenic expression of exogenous molecules not previously encountered by these pathogens in nature may prevent or delay the evolution of resistance breaking. Multiplexing different exogenous molecules with different modes of action could further delay resistance breaking and would constitute an entirely novel synthetic arm of the mosquito immune response.

Weaknesses: As with native immune molecules, the exogenous effectors tested as anti-malarial transgenes to date may act against other parasites, bacteria, or fungi that these mosquitoes would normally encounter in nature, and may also act against the mosquito itself. Laboratory fitness experiments may control for the latter; however, experiments in field settings will be required to identify any fitness gains or losses associated with these novel immune effectors and their interactions with native microbiota and natural pathogens. Increasing the specificity of the exogenous effectors (e.g., using single-chain antibodies or ribozymes) should reduce these effects, but increased specificity may increase the likelihood of resistance breaking and pathogen escape.

Host Factor Interference

Various mosquito proteins are required for the successful invasion and replication of parasites and viruses, very few of which have been identified and characterized in detail (reviewed in Ref. [9]). An invading pathogen must bind to one or more proteins on the lumenal surface of the midgut to gain entry to epithelial cells and avoid being digested along with the blood meal. Once inside the cell, various vector components may be necessary for the movement, escape, or replication of the pathogen. Effector molecules that block these pathogen−host interactions therefore represent promising strategies for generating pathogen-resistant mosquitoes.

Ghosh et al. [60] performed a phage library screen for synthetic peptides that bind the surface of the mosquito midgut and salivary glands. A single peptide was found to bind both surfaces and was termed SM1 (salivary and midgut peptide 1). Transgenic expression of SM1 in the midgut of *An. stephensi* mosquitoes substantially reduced the parasite burden of *Plasmodium berghei* and greatly reduced the transmission of this parasite to naive mice. Similar results were suggested when SM1 was expressed in the fat body and secreted into the hemolymph to prevent salivary gland invasion [61]. Subsequent experiments identified the parasite proteins mimicked by the SM1 peptide along with the corresponding mosquito proteins that were essential for invasion of the midgut [62,63] and salivary glands [64]. Interestingly, SM1 peptide protected *An. stephensi* against *P. berghei* but not *P. falciparum*, suggesting that these parasites utilize different host proteins to escape from the midgut [63]. Further investigation revealed a subpopulation of *P. berghei* that could also escape SM1-based resistance, presumably through an independent invasion pathway. A second midgut-binding peptide (MG2) was able to prevent invasion of SM1-resistant *P. berghei* as well as *P. falciparum*, suggesting that these parasites may share an as yet to be discovered receptor. Transgenic expression of MG2 has not yet been reported, and the existence of yet other invasion mechanisms cannot be yet ruled out.

Phospholipase A (PLA) proteins isolated from snake or bee venom were found to inhibit the formation of oocysts when fed along with parasite-infected blood [65]. The protein was found to have no direct affect on the parasite itself, rather it appeared to prevent the binding of *Plasmodium* ookinetes to the mosquito midgut by altering membrane structure [65]. Transgenic expression of the bee venom protein PLA2 in the midgut of *An. stephensi* mosquitoes reduced the prevalence and intensity of rodent malaria parasites (*P. berghei*), and virtually eliminated transmission to new vertebrate hosts [66,67]. However, expression of the PLA2 protein disrupted vesicle trafficking and damaged midgut cells; as a result these mosquitoes produced few progeny [67,68]. These strong negative effects on fitness appear to relate to the enzymatic activity of the PLA2 protein; such activity appears to be dispensable for its anti-*Plasmodium* activity as transgenic

An. stephensi mosquitoes expressing a catalytically inactive version (mPLA2) exhibit similar levels of resistance without the adverse effects [69,70]. In fact, one study found that transgenic mosquitoes expressing mPLA2 had a selective advantage against wild type counterparts when fed continuously on *Plasmodium*-infected blood [70].

Transgenic strategies relying on host factor blocking have not yet been developed for dengue; however, several large-scale projects have implicated a number of mosquito proteins as essential host factors [71−75]. Targeting these potential host factors using synthetic peptides or new gene editing technologies should be a promising line of research for the generation of dengue-resistant mosquitoes.

Strengths: Blocking, altering, deleting, or otherwise preventing pathogen access to one or more critical host factors allows specificity while potentially preserving the antipathogen effect against an entire class of pathogens. Global changes in immune status or metabolic activity are unlikely, thus changes to microbiota or the ability to vector other unrelated organisms are not expected. The slower rate of evolution in the mosquito compared to either malaria parasites or dengue viruses makes it unlikely that the target protein will evolve resistance to the anti-host factor effector. The development of site-specific gene editing technologies introduces the further advantage of potentially deleting critical epitopes or domains from host factor proteins.

Weaknesses: As seen with the SM1 peptide, pathogens with multiple mechanisms of invasion may very rapidly evolve resistance to the host factor−blocking molecule. Thus, it is imperative to characterize all potential alternative pathways or factors that might be redundant to the targeted protein. The most critical host factors may also be responsible for carrying out essential functions in the mosquito, blocking or interfering with these proteins may thus result in unacceptable fitness effects.

EVALUATING PATHOGEN RESISTANCE

New investigations into novel candidate effector molecules, either in the classes described above or those entailing entirely new modes of action, are needed to increase the repertoire of antipathogen molecules. However, investigators who wish to pursue the evaluation of new candidate molecules have numerous choices to make in conducting their experiments, all of which require trade-offs to some extent. The following section highlights some of the possible choices to be confronted in the evaluation of new resistance genes in the context of previous experiments.

Choice of Parasite Species

The rodent malaria parasites *P. berghei* [37,51,55,66,67,76] and *P. yoelii nigeriensis* [48] as well as the bird malaria parasite *P. gallinaceum*

[35,36,52,69] have been commonly used to test engineered resistance phenotypes in transgenic mosquitoes, as these parasites can be manipulated at biosafety level 1, and thus are simpler and safer to handle than human-infective *Plasmodium* spp. which require biosafety level 2 containment practices and procedures. However, with time it has become clear that effector molecules and strategies that are highly active against these model parasites may not be effective against *P. falciparum* [48,51,63]. Thus, while model parasites are still valuable for preliminary experiments to achieve proof-of-principle, and are a good indication of evolutionary conservation of effector function, a demonstration of effectiveness against *P. falciparum* is essential to any new strategy. Indeed, more recent studies have focused exclusively on resistance to *P. falciparum* [38,40,42,56].

Another consideration when choosing a parasite species or strain to use in challenge experiments is the rate of successful infection, both in terms of number of mosquitoes infected (prevalence) and the number of parasites per infected mosquito (intensity). This in turn dictates the size of the experimental group and corresponding statistical power [77]. In studies of field-caught mosquitoes, the numbers of oocysts observed are usually quite low. Billingsley [78] and Taylor [79] found mean numbers of oocysts per mosquito infected with *P. falciparum* were between 1 and 3. Thus, using more natural conditions experimental sample sizes have to be large (e.g., 200 mosquitoes when the control group has a mean infection intensity of 10 oocysts) in order to have sufficient confidence in any antipathogen effect [77]. In contrast, *P. berghei*, *P. gallinaceum*, and even lab-adapted *P. falciparum* parasites can routinely produce 50−100 oocysts per midgut or more. This reduces the number of mosquitoes required per experiment, increasing throughput. Alternatively, manipulating the gametocytemia in the blood meal used to challenge mosquitoes can result it more appropriate oocyst numbers, better reflecting what might occur in a field setting [38,40]. Ultimately though, choice of parasite strains may be limited to those that can reliably produce gametocyte numbers sufficient to infect the target mosquito population. For dengue, lab-adapted strains of dengue virus type 2 such as Jamaica 1409 or New Guinea C are the most commonly used; more recent experiments have added in a more diverse array of dengue genotypes [29]. Once again, reproducibility and consistency in viral titer are likely more important considerations in initial experiments than passage history. Low passage strains can then be used once a specific strain shows promise.

Choice of Vector Species

Most reports of transgene-mediated resistance to malaria have used the Asian malaria vector *An. stephensi*. Alternatively several groups have used *Ae. aegypti*, a competent vector of bird malaria. Just a few reports are based on the modification of the main African malaria vector, *An. gambiae*

[37,48]. This is due to the much higher transformation efficiency of the former two species, though new genetic tools and procedures will hopefully serve to reduce this gap [80]. For dengue, all experiments to date have utilized various strains of *Ae. aegypti*, the primary vector of dengue viruses. To be most relevant, effector genes should be engineered into species/strains where such mosquitoes are likely to be deployed as part of an eventual control strategy [29].

Oral Challenge with Malaria Parasites/Dengue Viruses

While all reports of pathogen resistance to malaria utilize similar methodologies and report similar parameters, each protocol contains unique variations including percent gametocytemia and incubation temperature/time. Standardized protocols would help reduce variation between laboratories and better allow comparisons to other studies. Ouedraogo et al. [81] described a protocol used for a standard membrane feeding assay designed to quantify malaria transmission, but unfortunately the parasitemia and gametocytemia parameters do not form part of that publication and yet heavily influence midgut infection rates. A more rigorous protocol was recently developed that standardizes gametocytemia and exflaggelation rates, achieving consistent *P. falciparum* infections [82].

Performing the Infectious Blood Meal

Water-jacketed glass heated membrane feeders with Parafilm are the simplest apparatus for performing infectious blood feeding although other alternatives exist [83]. Of note, the Hemotech system (Discovery Labs) can be assembled/disassembled safely within a biosafety cabinet, an important consideration when working with dengue viruses which at high titer may represent an aerosol risk to lab workers. Another recent adaptation, the Glytube, has been developed and tested with *Ae. aegypti*, providing a low-tech cost-effective approach to blood feeding [84]. If using rodent or avian-*Plasmodium* infections then the infected host can be anesthetized and both transgenic and non-transgenic mosquitoes can feed from the same infected animal. For dengue, fresh virus must be prepared by passaging frozen viral stocks through cultured cells; directly feeding frozen dengue viruses to mosquitoes results in poor infection rates [85]. Unlike malaria parasites, where the percent gametocytemia can be determined and adjusted by counting parasites prior to feeding, there is no similar method available to measure the amount of dengue virus in the fresh culture. Thus, a portion of the fresh virus suspension should be set aside prior to the feed for later determination of viral titer via plaque assay. In addition, at the completion of the feed, the remaining blood should also be analyzed for the amount of infectious virus remaining to ensure that mosquitoes feeding toward the end of the time limit were exposed to a sufficient dose, as virus titer may decrease over time when incubated at 37°C.

Mosquitoes that fail to imbibe a complete blood meal should always be removed from the experiment. Each mosquito is estimated to imbibe $2-3$ μl of blood; the impact of this variability is often taken into account though the impact of such linear differences is not clear, as body size may not be related to prevalence of infection [86]. Wing size often used as proxy for overall body size, and several groups have used this to verify that mosquitoes in both transgenic and control groups are exposed to a similar sized meal [48,56]. For dengue, there is evidence that both larger [87] and smaller [88] mosquitoes are more likely to be infected, with yet other studies finding no correlation at all [89]. Given these uncertainties, it is best to minimize variation in body size through standardized rearing protocols and sufficient sample sizes for each experiment.

ASSESSING RATE OF INFECTION (MIDGUT)

Many of the *Plasmodium*-resistant transgenic mosquito strains engineered thus far have targeted the parasite at the midgut-invasive stage, as this is a major bottleneck stage in terms of parasite numbers (reviewed in Ref. [9]). Successful prevention of oocyst formation thus negates the need to target hemolymph sporozoites or salivary gland sporozoites, both of which exist in far greater numbers in infected mosquitoes.

To determine the level of parasite resistance at this stage, parasite prevalence and intensity in the midgut can be calculated by counting the number of oocysts about 1 week following delivery of the infectious blood meal, though there is substantial variation in the experimental methods used by various groups, some of which is dictated by the parasite species used (Table 13.1). Prevalence is expressed as percentage of mosquitoes carrying a nonzero number of viable parasites as compared to the total number of mosquitoes that fed on the infected blood. Intensity of infection refers to the mean/median of the number of oocysts present as compared to a control group. Prevalence and intensity have distinct biological interpretations and both values should always be reported. Epidemiologically speaking, the only value of importance is prevalence, as only a single oocyst is required to continue the parasite life cycle and render a mosquito able to transmit the parasite. However, many lab-adapted parasite strains produce abnormally large numbers of oocysts; a strong reduction in intensity using these strains is often interpreted as the functional equivalent to a reduction in prevalence in wild strains that form few oocysts [77].

Oocysts can be observed on the midgut by a variety of means employing bright-field, phase, or interference-contrast microscopy. Once the midgut is dissected from the mosquito, mercurochrome staining ($0.1-2\%$ solution) facilitates counting of the oocysts. Fixing midguts in formalin provides a

TABLE 13.1 Incubation and Analysis Conditions for Plasmodium Parasites After an Infectious Blood Meal

| | | | | Days Post-Infectious Blood Meal | | |
| | | | | Oocysts | Salivary Gland Sporozoites | |
Parasite	Strain[a]	Vector	Temp °C[b]	Oocysts	Salivary Gland Sporozoites	Reference
P. berghei	ANKA 2.34	An. gambiae	19	13	n.r.	[38]
P. berghei	ANKA 2.34	An. gambiae	19	12–14	n.r.	[37]
P. berghei	ANKA 2.34	An. stephensi	21	15	25	[66,76]
P. berghei	PfCSP/Pb	An. stephensi	21	9–10	14 and 21	[55]
P. berghei	ANKA (GFP)	An. stephensi	19	10	n.r.	[70]
P. berghei	ANKA 234	An. stephensi	21	15	21	[51]
P. berghei	ANKA 2.34	An. stephensi	20	15	n.r.	[67]
P. falciparum	NF54	An. gambiae	26	8	n.r.	[48]
P. falciparum	NF54	An. stephensi	25	8	n.r.	[70]
P. falciparum	NF54	An. stephensi	27	10	n.r.	[42]
P. falciparum	3D7	An. stephensi	27	7	n.r.	[38]
P. falciparum	NF54	An. stephensi	28	10	n.r.	[40]
P. falciparum	NF54	An. stephensi	27	9	17–19	[54]
P. gallinaceum	n.r.	Ae. aegypti	27	8–9	13–15	[52]
P. gallinaceum	n.r.	Ae. aegypti	27	7	14	[35]
P. gallinaceum	n.r.	Ae. fluviatilis	27	7	15	[69]
P. yoelii nigeriensis	n.r.	An. gambiae	18	6–7	n.r.	[48]

[a]n.r., not reported.
[b]Holding temperature after infectious blood meal.

longer counting window, particularly when combined with GFP-expressing parasites [90]. *Plasmodium falciparum* oocysts can also be stained with bromo-fluorescein at day 8 for counting [48]. Computer-aided quantification using ImageJ software allows for greater automation and reduces single operator effect [91]. More recently, *P. falciparum* strains expressing luciferase have been used to assess both prevalence and intensity of infection without the need for dissection [92]. PCR-based assays could be used for calculating prevalence but not intensity and may be only useful for high-throughput prevalence index [93].

Though not common, it is also possible to quantify the number of ookinetes prior to invasion of the mosquito gut. This can be accomplished by homogenizing the midgut (with blood meal after 24 h post-blood feed), processing, and Giemsa staining [38]. This therefore gives more accurate data on the number of invasive ookinetes at the time of blood meal compared to the number of oocysts and can be a good demonstration that the resistant mosquitoes did indeed ingest a sufficient number of parasites.

Dengue infection in the midgut peaks at around 7 days post-blood meal [11]; standard plaque assays can be used to determine the amount of infectious virus present, while immunofluorescence assays can help visualize the number of infectious foci per midgut [27,29]. PCR can also be used to identify the number of viral genome equivalents, though this may be somewhat confounded by the presence of incomplete or degraded viral genomes.

ASSESSING RATE OF DISSEMINATION (TO SALIVARY GLAND)

After the maturation of sporozoites within the oocyst, the oocyst ruptures and releases these hair-like forms into the hemolymph of the mosquito's circulatory system. A single ruptured *P. falciparum* (NF54) oocyst will yield approximately 2,100 sporozoites. However, many of these midgut sporozoites do not find the salivary glands successfully due to the action of the mosquito immune response and the nature of the open circulatory system. Thus, for a single ruptured oocyst, only about 1,000 sporozoites may be found in the salivary glands [93].

Sporozoites that have not yet traversed to the salivary glands are not yet infectious to humans; this represents a second opportunity for arresting parasite infection within the mosquito host. The hemolymph also uniquely lends itself to an antiparasitic strategy as the fat body of the mosquito is a major immune organ and its role as a major protein generator can be utilized to produce engineered secreted proteins [36,54,56]. Assessment of any transgene targeted against the pre-salivary gland sporozoite stage ultimately involves counting the number of parasites that reach the salivary glands, a process that again can be broken down into prevalence and intensity. As with midgut infections, prevalence is the more relevant parameter, as just one or a very few parasites in the salivary glands are sufficient to result in parasite transmission and disease.

Sporozoites are most often visualized using microscopy after dissecting the salivary glands and rupturing them using pressure upon the microscope slide. The sporozoites are then counted using a hemocytometer under a microscope. There are a few variations upon this theme with some researchers counting the entire sporozoite load and others counting a representative sample. Yoshida et al. [51] employed a scale approach based upon an earlier study [94]. This scale also accounts for a set low number of 10, that is, the minimum number of *Plasmodium vivax* sporozoites capable of establishing a malaria infection after needle inoculation [95]. A more recent publication also used computer software to analyze stored samples [54]; useful as there can be substantial time pressure when counting large numbers of samples. Briefly, after DAPI staining, the sporozoites were counted using Axiovision and Improvision software programs (parameters detailed in Materials and Methods section of Ref. [54]). Ramakrishnan et al. [96] adopted a more quantitative approach, establishing a transgenic strain of *P. berghei* that expresses luciferase specifically in sporozoites that have reached the mosquito salivary glands. This allowed for the direct processing of mosquitoes and eliminates the need for labor-intensive dissections.

Dengue infection in the salivary glands has been observed as early as 4 days post-infection, but typically peaks between 10 and 14 days after the infectious blood meal. Once again, standard plaque assays can be used to determine the amount of infectious virus present, while immunofluorescence assays can help visualize the number of infectious foci [27]. As noted earlier, the rate of disseminated infections for dengue viruses may or may not relate to the burden of infection observed in the gut [12,13]. Thus, even if strong reductions in virus titer are observed through the action of effector gene expression, it is still essential to determine the effects on virus dissemination throughout the body, including the salivary glands.

ASSESSING RATE OF PATHOGEN TRANSMISSION

Evaluating Malaria Transmission

As it has previously been shown that as few as 10−20 sporozoites can establish an infection in the vertebrate host [52], the presence of any sporozoites in the saliva of the mosquito ultimately represents a failure of the strategy, at least at the level of the individual mosquito. This therefore influences the experimental design as the presence of any sporozoites within the salivary glands may be seen as an end point rather than the transmission of the parasite to another host.

When using rodent or bird malaria parasites, the potential of engineered mosquitoes to transmit sporozoites can be determined by allowing challenged mosquitoes to feed upon an uninfected vertebrate host. For example, sporozoites are present within the salivary glands 17−21 days after an infectious blood meal for the rodent malaria parasite *P. berghei*, and at this point a controlled number of mosquitoes can be allowed to feed upon naïve mice. Those observed to feed are dissected immediately afterward to assess sporozoite prevalence and intensity within the salivary glands. The challenged mice are subsequently monitored by tail vein smears and Geimsa stains to assess their infection status. If no infection is observed after 25−30 days then the mouse is categorized as noninfected.

Plasmodium falciparum transmission studies are obviously inherently more complicated and have not been reported in any of the studies described for engineered resistance in mosquitoes. In other contexts (vaccine evaluation) studies often use an invasive cell assay employing a human liver tissue cell line. The hepatocyte cell line HC-04 allows complete development of *P. falciparum* and *P. vivax* but at low infection efficiencies [97], although sporozoite infectivity *in vitro* involves far less from the parasite than is needed for *in vivo* liver invasion where the sporozoites must first exit the dermal inoculation site and then cross the liver sinusoid whereas *in vitro* the sporozoites are placed directly upon the hepatocyte monolayer. This procedure involves dissecting *An. gambiae* salivary glands between 14 and 17 days after an infectious blood feed. The sporozoites are then added to a monolayer of liver cells that have been grown upon a glass coverslip. The slides are incubated for 3−4 h then fixed and immunohistochemical techniques are employed to visualize the sporozoites that have invaded (*MR4 protocol*—6th edition). The difficulty in working with *P. falciparum* in the context of liver cell invasion may explain why most studies have preferred to use animal model systems rather than *P. falciparum*. The use of fluorescent *P. berghei* [98], *P. yoelii* [99], and *P. falciparum* [100] has, in recent years, allowed more intimate visualization of sporozoite behavior. A recent study generated transgenic *P. berghei* that express *P. falciparum* CSP (circumsporozoite protein) [55], allowing evaluation of an effector active against human malaria, but in the context of a mouse transmission model. This interesting approach thus allows for relevant transmission-blocking quantitative analysis of the salivary glands and transmission assay using naïve mice.

Evaluating Dengue Transmission

The presence of a disseminated infection is not always a true indication of transmission potential [101]. Allowing challenged mosquitoes to feed on vertebrate animals such suckling mice can determine the actual transmission potential through either measurement of direct infection outcomes (morbidity, mortality, presence of infectious virus) or indirect (seroconversion). Alternatively, virus transmission potential can be assessed in the absence of animal models using any number of standard protocols to extract mosquito saliva prior to PCR-based detection or determination of virus titer [101−103]. Briefly, legs and wings can be removed and the proboscis inserted into a capillary tube filled with mineral oil or similar substrate to induce salivation. Virus particles expelled into saliva can be added to mono-layers of cultured cells to detect infectious particles by plaque assay or can be used in PCR to determine the number of viral genome equivalents. Alternatively, groups of mosquitoes can be allowed to probe and salivate into a feeding solution which is then collected and processed [28]. While single individual resolution is lost, this method is much simpler and allows analysis at the population level.

SUMMARY AND OUTLOOK

The general concept of pathogen-derived resistance was proposed and vali-dated 30 years ago, since then various research groups have been attempting to bring this technology, along with other means of engineered resistance to fight malaria and dengue. While earlier experiments largely used model parasites or transient expression systems, within the past 5 years a number of groups have reported generating transgenic mosquitoes resistant to the patho-gens that cause malaria or dengue. However, these experiments have been limited to small-scale studies in the laboratory and have (for practical pur-poses) focused on both lab-adapted mosquitoes and parasites (Table 13.2). To determine the robustness, and thus utility, of each transgenic strategy, larger scale experiments conducted under semi-field conditions are needed. In particular, the durability of resistance phenotypes must be determined across the entire temperature range to be encountered in a potential release location, as both changes in mean temperature as well as diurnal fluctuations can strongly influence the vector competence and hence vectorial capacity of mosquitoes for malaria parasites [106−108] and dengue viruses [109−111] (reviewed in Refs [112,113]). Several reports have started to characterize how the immune responses of both Anopheline and Aedine mosquito vectors vary in response to environmental temperature, in turn affecting the potential transmission of malaria and dengue. For example, changes in tem-perature affect the rate of parasite melanization [107], the expression of effector molecules such as defensin [106,107], the production of NO [107,108], and the effectiveness of RNAi [31]. Since many of the approaches outlined in this chapter attempt to utilize or augment these immune pathways, understanding how each transgenic strategy performs at the full range of conditions to be expected in a natural setting is critical. Likewise, strategies that boost the immune system of the mosquito must be evaluated in the context of environmental microorganisms such as bacteria and fungi present in native breeding locations to determine whether the benefits of pathogen resistance are offset by unwanted gains in mosquito survival

TABLE 13.2 Engineered Resistance to Malaria Parasites

Reference	Vector	Pathogen	Type[a]	Name	Stage Targeted	Midgut Infection Prevalence	Midgut Infection Intensity	Salivary Gland Infection Prevalence	Salivary Gland Infection Intensity	Transmission %Δ	Transmission Rate	Fitness Evaluation Lab	Fitness Evaluation Large Cage	Fitness Evaluation Field	Stability
[68,76, 104]	An. stephensi	P. berghei	HFI	SM1	Ookinete	47%↓	81%↓	67%↓	67%↓	82%↓	10.7%	Advantage			
[66,68]	An. stephensi	P. berghei	HFI	PLA2	Ookinete	78%↓	87%↓	50%↓	50%↓	96%↓	3%	Strong effect			
[37]	An. gambiae	P. berghei	IAct	CecA	Ookinete	No effect	60%↓								
[67]	An. stephensi	P. berghei	HFI	PLA2	Ookinete	33%↓	80%↓					Strong effect			
[52]	Ae. aegypti	P. gallinaceum	IAug	N2	Sporozoite	No effect	No effect	No effect	Strong reduction ↓	33%↓	66%				
[51]	An. stephensi	P. berghei	IAug	CEL-III	Ookinete	84%↓	91%↓	83%↓	Strong reduction ↓	80%↓	20%				
[51]	An. stephensi	P. falciparum	IAug	CEL-III	Ookinete	No effect	69%↓								
[61]	An. stephensi	P. berghei	HFI	SM1	Sporozoite				85%↓ (unpub data)			3/3 lines have strong effect			>25 generations, no effect
[69,105]	Ae. fluviatilis	P. gallinaceum	HFI	mPLA2	Ookinete	No effect	42%↓					No effect, smaller bodies			
[35]	Ae. aegypti	P. gallinaceum	IAct	REL2	Oocyst Sporozoite	95%↓	95%↓	61%↓	61%↓						
[40,41]	An. stephensi	P. falciparum	IAct	Akt	Ookinete	None[b]	None[b]					Strong effect – decrease lifespan			
[36]	Ae. aegypti	P. gallinaceum	IAct	CecA/DefA	Oocyst Sporozoite	None[b]	95%↓	None[b]	None[b]	Complete[b]	None[b]				
[38]	An. stephensi	P. falciparum	Iact	REL2	Ookinete Oocyst Sporozoite	65%↓	83%↓	Strong reduction ↓	Strong reduction ↓			No effect			
[54]	An. stephensi	P. falciparum	IAug	m4B7-cecA m1C3	Ookinete	48%↓	67%↓								>20 generations, no effect

(Continued)

TABLE 13.2 (Continued)

Reference	Vector	Pathogen	Type[a]	Name	Stage Targeted	Midgut Infection Prevalence	Midgut Infection Intensity	Salivary Gland Infection Prevalence	Salivary Gland Infection Intensity	Transmission %Δ	Transmission Rate	Fitness Evaluation Lab	Fitness Evaluation Large Cage	Fitness Evaluation Field	Stability
[54]	An. stephensi	P. falciparum	IAug	m2A10-cecA	Sporozoite	No effect	No effect	Mild reduction	97%↓						
[48]	An. gambiae	P. yoelii	IAug	Vida3	Ookinete	Mild reduction	79%↓					No effect			
[48,49]	An. gambiae	P. falciparum	IAug	Vida3	Ookinete	No effect	No effect					No effect			
[56]	An. stephensi	P. falciparum	IAug	M1C3-m2A10	Ookinete Sporozoite	No effect	No effect	None[b]	None[b]			No effect			
[42]	An. stephensi	P. falciparum	IAct	PTEN	Ookinete	46%↓	53%↓					Increased lifespan			
[70]	An. stephensi	P. falciparum	HFI	mPLA2	Ookinete	No effect	55%↓					Strong effect, but small advantage when fed on infected blood			
[55]	An. stephensi	P. berghei	IAug	m2A10	Sporozoite	No effect	No effect	No effect	No effect	70%↓	25%	No effect			
[27,30]	Ae. aegypti	DENV2	PDR/RNAi	Mnp-IR	Midgut	Strong reduction↓	Strong reduction↓	Strong reduction↓	Strong reduction↓	80%↓					IR-RNA not expressed by G17
[29]	Ae. aegypti	DENV2	PDR/RNAi	Mnp-IR	Midgut	None[b]	None[b]					No effect			IR-RNA stable >33 generations
[28]	Ae. aegypti	DENV2	PDR/RNAi	Mnp-IR	Salivary glands	No effect	No effect	50%↓	90%↓	Complete[b]	None[b]				

[a]IAct, immune activation; IAug, immune augmentation; HFI, host factor interference; PDR/RNAi, pathogen-derived resistance/RNA interference. Gray highlights indicate parameters not investigated/reported in each study.
[b]This indicates where engineered resistance phenotypes were considered complete.

and/or reproductive output, answers that cannot be obtained in the sterile environment of the laboratory.

Each of the strategies discussed in this chapter will have its own range of pathogen specificity; an ideal strategy may be one where all strains of the target organism are all completely blocked by the engineered transgene(s), with no effect (positive or negative) on any other microorganism. For those pathogen-resistant strategies that show promise against one or more lab-adapted pathogen strains, resistance trials are necessary against a wide variety of naturally occurring strains as well as of unrelated microorganisms likely to be encountered in nature. Likewise, the performance of engineered transgenes must be evaluated in the genetic background of the local population where a planned release may occur, to account for unknown genetic interactions between the transgene and the local genotypes that may occur as the introduced transgene is introgressed into the target population.

In addition to continued characterization of existing engineered resistance strains, there is still a pressing need for additional mechanisms that confer parasite or virus resistance in order to better allow transgene stacking, as well as to replace existing technologies in the case of resistance breakdown during trials or after implementation. The recent development of site-specific gene editing and targeting technologies [114,115] brings with it a whole new suite of potential genetic manipulations that may lead to a pathogen-resistant state. Some potential more targeted modifications include new modes of immune activation through directly editing the transcription factor binding sites of critical genes; portions of host factors critical for parasite or viral entry could be selectively deleted or altered, leaving the normal function of the gene product intact; or new effector genes could simply be placed more precisely in a desired chromosomal context to ensure consistent expression and minimal fitness effects. While not likely to replace sterile insect technologies or non-transgenic means of generating pathogen resistance in the near term; transgenic strategies to generate pathogen-resistant phenotypes still show great promise for the potential long-term prevention of both malaria and dengue transmission.

REFERENCES

[1] Adelman ZN, Basu S, Myles KM. Gene insertion and deletion in mosquitoes. In: Adelman ZN, editor. Genetic control of malaria and dengue. San Diego, CA: Elsevier; 2015.

[2] Holt RA, Subramanian GM, Halpern A, et al. The genome sequence of the malaria mosquito *Anopheles gambiae*. Science 2002;298(5591):129−49.

[3] Nene V, Wortman JR, Lawson D, et al. Genome sequence of *Aedes aegypti*, a major arbovirus vector. Science 2007;316(5832):1718−23.

[4] Blumberg BJ, Short SM, Dimopoulos G. Employing the mosquito microflora for disease control. In: Adelman ZN, editor. Genetic control of malaria and dengue. San Diego, CA: Elsevier; 2015.

[5] Xi J, Joshi D. Genetic control of malaria and dengue using Wolbachia. In: Adelman ZN, editor. Genetic control of malaria and dengue. San Diego, CA: Elsevier; 2015.

[6] Marshall JM, Akbari OS. Gene drive strategies for population replacement. In: Adelman ZN, editor. Genetic control of malaria and dengue. San Diego, CA: Elsevier; 2015.

[7] Sinden RE. The challenge of disrupting vectorial capacity. In: Adelman ZN, editor. Genetic control of malaria and dengue. San Diego, CA: Elsevier; 2015.

[8] Billker O, Lindo V, Panico M, et al. Identification of xanthurenic acid as the putative inducer of malaria development in the mosquito. Nature 1998;392(6673):289−92.

[9] Smith RC, Jacobs-Lorena M. Plasmodium−mosquito interactions: a tale of roadblocks and detours. Adv Insect Physiol 2010;39:119−49.

[10] Marois E. The multifaceted mosquito anti-*Plasmodium* response. Curr Opin Microbiol 2011;14(4):429−35.

[11] Salazar MI, Richardson JH, Sanchez-Vargas I, Olson KE, Beaty BJ. Dengue virus type 2: replication and tropisms in orally infected *Aedes aegypti* mosquitoes. BMC Microbiol 2007;7:9.

[12] Bosio CF, Beaty BJ, Black WC. Quantitative genetics of vector competence for dengue-2 virus in *Aedes aegypti*. Am J Trop Med Hyg 1998;59(6):965−70.

[13] Bennett KE, Olson KE, Munoz Mde L, et al. Variation in vector competence for dengue 2 virus among 24 collections of *Aedes aegypti* from Mexico and the United States. Am J Trop Med Hyg 2002;67(1):85−92.

[14] Sanford JC, Johnston SA. The concept of pathogen derived resistance-deriving resistance genes from the parasite's own genome. J Theor Biol 1985;113:395−405.

[15] Gottula J, Fuchs M. Toward a quarter century of pathogen-derived resistance and practical approaches to plant virus disease control. Adv Virus Res 2009;75:161−83.

[16] Simon-Mateo C, Garcia JA. Antiviral strategies in plants based on RNA silencing. Biochim Biophys Acta 2011;1809(11−12):722−31.

[17] Collinge DB, Jorgensen HJ, Lund OS, Lyngkjaer MF. Engineering pathogen resistance in crop plants: current trends and future prospects. Annu Rev Phytopathol 2010;48:269−91.

[18] Thompson JR, Tepfer M. Assessment of the benefits and risks for engineered virus resistance. Adv Virus Res 2010;76:33−56.

[19] Abel PP, Nelson RS, De B, et al. Delay of disease development in transgenic plants that express the tobacco mosaic virus coat protein gene. Science 1986;232(4751):738−43.

[20] Gaines PJ, Olson KE, Higgs S, Powers AM, Beaty BJ, Blair CD. Pathogen-derived resistance to dengue type 2 virus in mosquito cells by expression of the premembrane coding region of the viral genome. J Virol 1996;70(4):2132−7.

[21] Olson KE, Higgs S, Gaines PJ, et al. Genetically engineered resistance to dengue-2 virus transmission in mosquitoes. Science 1996;272(5263):884−6.

[22] Adelman ZN, Blair CD, Carlson JO, Beaty BJ, Olson KE. Sindbis virus-induced silencing of dengue viruses in mosquitoes. Insect Mol Biol 2001;10(3):265−73.

[23] Higgs S, Rayner JO, Olson KE, Davis BS, Beaty BJ, Blair CD. Engineered resistance in *Aedes aegypti* to a West African and a South American strain of yellow fever virus. Am J Trop Med Hyg 1998;58(5):663−70.

[24] Powers AM, Kamrud KI, Olson KE, Higgs S, Carlson JO, Beaty BJ. Molecularly engineered resistance to California serogroup virus replication in mosquito cells and mosquitoes. Proc Natl Acad Sci USA 1996;93(9):4187−91.

[25] Allen-Miura T. Evaluation of expression systems and antiviral genes to inhibit LaCrosse virus replication in mosquito cells. Evaluation of expression systems and antiviral genes to inhibit LaCrosse virus replication in mosquito cells. Ph.D. dissertation, Department of Microbiology, Colorado State University; 2000.

[26] Adelman ZN, Sanchez-Vargas I, Travanty EA, et al. RNA silencing of dengue virus type 2 replication in transformed C6/36 mosquito cells transcribing an inverted-repeat RNA derived from the virus genome. J Virol 2002;76(24):12925–33.

[27] Franz AW, Sanchez-Vargas I, Adelman ZN, et al. Engineering RNA interference-based resistance to dengue virus type 2 in genetically modified *Aedes aegypti*. Proc Natl Acad Sci USA 2006;103(11):4198–203.

[28] Mathur G, Sanchez-Vargas I, Alvarez D, Olson KE, Marinotti O, James AA. Transgene-mediated suppression of dengue viruses in the salivary glands of the yellow fever mosquito, *Aedes aegypti*. Insect Mol Biol 2010;19(6):753–63.

[29] Franz AW, Sanchez-Vargas I, Raban RR, Black WC, James AA, Olson KE. Fitness impact and stability of a transgene conferring resistance to dengue-2 virus following introgression into a genetically diverse *Aedes aegypti* strain. PLoS Negl Trop Dis 2014;8(5): e2833.

[30] Franz AW, Sanchez-Vargas I, Piper J, et al. Stability and loss of a virus resistance phenotype over time in transgenic mosquitoes harbouring an antiviral effector gene. Insect Mol Biol 2009;18(5):661–72.

[31] Adelman ZN, Anderson MA, Wiley MR, et al. Cooler temperatures destabilize RNA interference and increase susceptibility of disease vector mosquitoes to viral infection. PLoS Negl Trop Dis 2013;7(5):e2239.

[32] Clayton AM, Dong Y, Dimopoulos G. The *Anopheles* innate immune system in the defense against malaria infection. J Innate Immun 2014;6(2):169–81.

[33] Merkling SH, van Rij RP. Beyond RNAi: antiviral defense strategies in *Drosophila* and mosquito. J Insect Physiol 2013;59(2):159–70.

[34] Blair CD, Olson KE. Mosquito immune responses to arbovirus infections. Curr Opin Insect Sci 2014;3:22–9.

[35] Antonova Y, Alvarez KS, Kim YJ, Kokoza V, Raikhel AS. The role of NF-kappaB factor REL2 in the *Aedes aegypti* immune response. Insect Biochem Mol Biol 2009;39 (4):303–14.

[36] Kokoza V, Ahmed A, Woon Shin S, Okafor N, Zou Z, Raikhel AS. Blocking of Plasmodium transmission by cooperative action of Cecropin A and Defensin A in transgenic *Aedes aegypti* mosquitoes. Proc Natl Acad Sci USA 2010;107(18):8111–16.

[37] Kim W, Koo H, Richman AM, et al. Ectopic expression of a cecropin transgene in the human malaria vector mosquito *Anopheles gambiae* (Diptera: Culicidae): effects on susceptibility to *Plasmodium*. J Med Entomol 2004;41(3):447–55.

[38] Dong Y, Das S, Cirimotich C, Souza-Neto JA, McLean KJ, Dimopoulos G. Engineered anopheles immunity to *Plasmodium* infection. PLoS Pathog 2011;7(12):e1002458.

[39] Oliveira Gde A, Lieberman J, Barillas-Mury C. Epithelial nitration by a peroxidase/NOX5 system mediates mosquito antiplasmodial immunity. Science 2012;335(6070):856–9.

[40] Corby-Harris V, Drexler A, Watkins de Jong L, et al. Activation of Akt signaling reduces the prevalence and intensity of malaria parasite infection and lifespan in *Anopheles gambiae* mosquitoes. PLoS Pathog 2010;6(7):e1001003.

[41] Luckhart S, Giulivi C, Drexler AL, et al. Sustained activation of Akt elicits mitochondrial dysfunction to block *Plasmodium falciparum* infection in the mosquito host. PLoS Pathog 2013;9(2):e1003180.

[42] Hauck ES, Antonova-Koch Y, Drexler A, et al. Overexpression of phosphatase and tensin homolog improves fitness and decreases *Plasmodium falciparum* development in *Anopheles gambiae*. Microbes Infect 2013;15(12):775–87.

[43] Gendrin M, Rodgers FH, Yerbanga RS, et al. Antibiotics in ingested human blood affect the mosquito microbiota and capacity to transmit malaria. Nat Commun 2015;6:5921.

[44] Moreira LA, Edwards MJ, Adhami F, Jasinskiene N, James AA, Jacobs-Lorena M. Robust gut-specific gene expression in transgenic *Aedes aegypti* mosquitoes. Proc Natl Acad Sci USA 2000;97(20):10895−8.

[45] Chen XG, Marinotti O, Whitman L, Jasinskiene N, James AA, Romans P. The *Anopheles gambiae* vitellogenin gene (VGT2) promoter directs persistent accumulation of a reporter gene product in transgenic *Anopheles gambiae* following multiple bloodmeals. Am J Trop Med Hyg 2007;76(6):1118−24.

[46] Dennison NJ, Jupatanakul N, Dimopoulos G. The mosquito microbiota influences vector competence for human pathogens. Curr Opin Insect Sci 2014;3:6−13.

[47] Arrighi RB, Nakamura C, Miyake J, Hurd H, Burgess JG. Design and activity of antimicrobial peptides against sporogonic-stage parasites causing murine malarias. Antimicrob Agents Chemother 2002;46(7):2104−10.

[48] Meredith JM, Basu S, Nimmo DD, et al. Site-specific integration and expression of an anti-malarial gene in transgenic *Anopheles gambiae* significantly reduces *Plasmodium* infections. PLoS One 2011;6(1):e14587.

[49] McArthur CC, Meredith JM, Eggleston P. Transgenic *Anopheles gambiae* expressing an antimalarial peptide suffer no significant fitness cost. PLoS One 2014;9(2):e88625.

[50] Carter V, Underhill A, Baber I, et al. Killer bee molecules: antimicrobial peptides as effector molecules to target sporogonic stages of *Plasmodium*. PLoS Pathog 2013;9(11): e1003790.

[51] Yoshida S, Shimada Y, Kondoh D, et al. Hemolytic C-type lectin CEL-III from sea cucumber expressed in transgenic mosquitoes impairs malaria parasite development. PLoS Pathog 2007;3(12):e192.

[52] Jasinskiene N, Coleman J, Ashikyan A, Salampessy M, Marinotti O, James AA. Genetic control of malaria parasite transmission: threshold levels for infection in an avian model system. Am J Trop Med Hyg 2007;76(6):1072−8.

[53] de Lara Capurro M, Coleman J, Beerntsen BT, et al. Virus-expressed, recombinant single-chain antibody blocks sporozoite infection of salivary glands in *Plasmodium gallinaceum*-infected *Aedes aegypti*. Am J Trop Med Hyg 2000;62(4):427−33.

[54] Isaacs AT, Li F, Jasinskiene N, et al. Engineered resistance to *Plasmodium falciparum* development in transgenic *Anopheles gambiae*. PLoS Pathog 2011;7(4):e1002017.

[55] Sumitani M, Kasashima K, Yamamoto DS, et al. Reduction of malaria transmission by transgenic mosquitoes expressing an antisporozoite antibody in their salivary glands. Insect Mol Biol 2013;22(1):41−51.

[56] Isaacs AT, Jasinskiene N, Tretiakov M, et al. Transgenic *Anopheles gambiae* coexpressing single-chain antibodies resist *Plasmodium falciparum* development. Proc Natl Acad Sci USA 2012;109(28):E1922−30.

[57] Nawtaisong P, Keith J, Fraser T, et al. Effective suppression of dengue fever virus in mosquito cell cultures using retroviral transduction of hammerhead ribozymes targeting the viral genome. Virol J 2009;6:73.

[58] Carter JR, Keith JH, Barde PV, Fraser TS, Fraser Jr MJ. Targeting of highly conserved dengue virus sequences with anti-dengue virus trans-splicing group I introns. BMC Mol Biol 2010;11:84.

[59] Carter JR, Keith JH, Fraser TS, et al. Effective suppression of dengue virus using a novel group-I intron that induces apoptotic cell death upon infection through conditional expression of the Bax C-terminal domain. Virol J 2014;11:111.

[60] Ghosh AK, Ribolla PE, Jacobs-Lorena M. Targeting Plasmodium ligands on mosquito salivary glands and midgut with a phage display peptide library. Proc Natl Acad Sci USA 2001;98(23):13278−81.

[61] Li C, Marrelli MT, Yan G, Jacobs-Lorena M. Fitness of transgenic *Anopheles gambiae* mosquitoes expressing the SM1 peptide under the control of a vitellogenin promoter. J Hered 2008;99(3):275−82.

[62] Ghosh AK, Coppens I, Gardsvoll H, Ploug M, Jacobs-Lorena M. *Plasmodium* ookinetes coopt mammalian plasminogen to invade the mosquito midgut. Proc Natl Acad Sci USA 2011;108(41):17153−8.

[63] Vega-Rodriguez J, Ghosh AK, Kanzok SM, et al. Multiple pathways for *Plasmodium* ookinete invasion of the mosquito midgut. Proc Natl Acad Sci USA 2014;111(4): E492−500.

[64] Ghosh AK, Devenport M, Jethwaney D, et al. Malaria parasite invasion of the mosquito salivary gland requires interaction between the *Plasmodium* TRAP and the *Anopheles* saglin proteins. PLoS Pathog 2009;5(1):e1000265.

[65] Zieler H, Keister DB, Dvorak JA, Ribeiro JM. A snake venom phospholipase A(2) blocks malaria parasite development in the mosquito midgut by inhibiting ookinete association with the midgut surface. J Exp Biol 2001;204(Pt. 23):4157−67.

[66] Moreira LA, Ito J, Ghosh A, et al. Bee venom phospholipase inhibits malaria parasite development in transgenic mosquitoes. J Biol Chem 2002;277(43):40839−43.

[67] Abraham EG, Donnelly-Doman M, Fujioka H, Ghosh A, Moreira L, Jacobs-Lorena M. Driving midgut-specific expression and secretion of a foreign protein in transgenic mosquitoes with AgAper1 regulatory elements. Insect Mol Biol 2005;14(3):271−9.

[68] Moreira LA, Wang J, Collins FH, Jacobs-Lorena M. Fitness of anopheline mosquitoes expressing transgenes that inhibit *Plasmodium* development. Genetics 2004;166 (3):1337−41.

[69] Rodrigues FG, Santos MN, de Carvalho TX, et al. Expression of a mutated phospholipase A2 in transgenic *Aedes fluviatilis* mosquitoes impacts *Plasmodium gallinaceum* development. Insect Mol Biol 2008;17(2):175−83.

[70] Smith RC, Kizito C, Rasgon JL, Jacobs-Lorena M. Transgenic mosquitoes expressing a phospholipase A(2) gene have a fitness advantage when fed *Plasmodium falciparum*-infected blood. PLoS One 2013;8(10):e76097.

[71] Sessions OM, Barrows NJ, Souza-Neto JA, et al. Discovery of insect and human dengue virus host factors. Nature 2009;458(7241):1047−50.

[72] Jupatanakul N, Sim S, Dimopoulos G. *Aedes aegypti* ML and Niemann-Pick type C family members are agonists of dengue virus infection. Dev Comp Immunol 2014;43(1):1−9.

[73] Kang S, Shields AR, Jupatanakul N, Dimopoulos G. Suppressing dengue-2 infection by chemical inhibition of *Aedes aegypti* host factors. PLoS Negl Trop Dis 2014;8(8):e3084.

[74] Mairiang D, Zhang H, Sodja A, et al. Identification of new protein interactions between dengue fever virus and its hosts, human and mosquito. PLoS One 2013;8(1):e53535.

[75] Colpitts TM, Cox J, Vanlandingham DL, et al. Alterations in the *Aedes aegypti* transcriptome during infection with West Nile, dengue and yellow fever viruses. PLoS Pathog 2011;7(9):e1002189.

[76] Ito J, Ghosh A, Moreira LA, Wimmer EA, Jacobs-Lorena M. Transgenic anopheline mosquitoes impaired in transmission of a malaria parasite. Nature 2002;417(6887):452−5.

[77] Churcher TS, Blagborough AM, Delves M, et al. Measuring the blockade of malaria transmission − an analysis of the Standard Membrane Feeding Assay. Int J Parasitol 2012;42 (11):1037−44.

[78] Billingsley PF. Vector—parasite interactions for vaccine development. Int J Parasitol 1994;24(1):53—8.

[79] Taylor LH. Infection rates in, and the number of *Plasmodium falciparum* genotypes carried by *Anopheles* mosquitoes in Tanzania. Ann Trop Med Parasitol 1999;93 (6):659—62.

[80] Volohonsky G, Terenzi O, Soichot J, et al. Tools for *Anopheles gambiae* Transgenesis. G3 2015;5(6):1151—63.

[81] Ouédraogo AL, Guelbéogo WM, Cohuet A, et al. A protocol for membrane feeding assays to determine the infectiousness of *P. falciparum* naturally infected individuals to *Anopheles gambiae*. MalariaWorld J 2013;4(16).

[82] Li T, Eappen AG, Richman AM, et al. Robust, reproducible, industrialized, standard membrane feeding assay for assessing the transmission blocking activity of vaccines and drugs against *Plasmodium falciparum*. Malar J 2015;14(1):150.

[83] Bhattacharyya MK, Kumar N. Transmission blocking and sporozoite invasion assay. In: Moll K, Kaneko A, Scherf A, Wahlgren M, editors. Methods in malaria research. Glasgow, UK: EVIMalaR; 2013. p. 162—5.

[84] Costa-da-Silva AL, Navarrete FR, Salvador FS, et al. Glytube: a conical tube and parafilm M-based method as a simplified device to artificially blood-feed the dengue vector mosquito, *Aedes aegypti*. PLoS One 2013;8(1):e53816.

[85] Richards SL, Pesko K, Alto BW, Mores CN. Reduced infection in mosquitoes exposed to blood meals containing previously frozen flaviviruses. Virus Res 2007;129(2):224—7.

[86] Lyimo EO, Koella JC. Relationship between body size of adult *Anopheles gambiae* s.l. and infection with the malaria parasite *Plasmodium falciparum*. Parasitology 1992;104(Pt. 2):233—7.

[87] Sumanochitrapon W, Strickman D, Sithiprasasna R, Kittayapong P, Innis BL. Effect of size and geographic origin of *Aedes aegypti* on oral infection with dengue-2 virus. Am J Trop Med Hyg 1998;58(3):283—6.

[88] Alto BW, Reiskind MH, Lounibos LP. Size alters susceptibility of vectors to dengue virus infection and dissemination. Am J Trop Med Hyg 2008;79(5):688—95.

[89] Schneider JR, Mori A, Romero-Severson J, Chadee DD, Severson DW. Investigations of dengue-2 susceptibility and body size among *Aedes aegypti* populations. Med Vet Entomol 2007;21(4):370—6.

[90] Usui M, Fukumoto S, Inoue N, Kawazu S. Improvement of the observational method for *Plasmodium berghei* oocysts in the midgut of mosquitoes. Parasit Vectors 2011;4:118.

[91] Delves MJ, Sinden RE. A semi-automated method for counting fluorescent malaria oocysts increases the throughput of transmission blocking studies. Malar J 2010;9:35.

[92] Stone WJ, Churcher TS, Graumans W, et al. A scalable assessment of *Plasmodium falciparum* transmission in the standard membrane-feeding assay, using transgenic parasites expressing green fluorescent protein-luciferase. J Infect Dis 2014;210(9):1456—63.

[93] Stone WJ, Eldering M, van Gemert GJ, et al. The relevance and applicability of oocyst prevalence as a read-out for mosquito feeding assays. Sci Rep 2013;3:3418.

[94] Collins WE, Warren M, Skinner JC, Richardson BB, Kearse TS. Infectivity of the Santa Lucia (El Salvador) strain of *Plasmodium falciparum* to different anophelines. J Parasitol 1977;63(1):57—61.

[95] Ungureanu E, Killick-Kendrick R, Garnham PC, Branzei P, Romanescu C, Shute PG. Prepatent periods of a tropical strain of *Plasmodium vivax* after inoculations of tenfold dilutions of sporozoites. Trans R Soc Trop Med Hyg 1976;70(5—6):482—3.

[96] Ramakrishnan C, Rademacher A, Soichot J, et al. Salivary gland-specific *P. berghei* reporter lines enable rapid evaluation of tissue-specific sporozoite loads in mosquitoes. PLoS One 2012;7(5):e36376.

[97] Tao D, King JG, Tweedell RE, Jost PJ, Boddey JA, Dinglasan RR. The acute transcriptomic and proteomic response of HC-04 hepatoma cells to hepatocyte growth factor and its implications for *Plasmodium falciparum* sporozoite invasion. Mol Cell Proteomics 2014;13(5):1153–64.

[98] Franke-Fayard B, Trueman H, Ramesar J, et al. A *Plasmodium berghei* reference line that constitutively expresses GFP at a high level throughout the complete life cycle. Mol Biochem Parasitol 2004;137(1):23–33.

[99] Tarun AS, Baer K, Dumpit RF, et al. Quantitative isolation and *in vivo* imaging of malaria parasite liver stages. Int J Parasitol 2006;36(12):1283–93.

[100] Talman AM, Blagborough AM, Sinden REA. *Plasmodium falciparum* strain expressing GFP throughout the parasite's life-cycle. PLoS One 2010;5(2):e9156.

[101] Anderson SL, Richards SL, Smartt CT. A simple method for determining arbovirus transmission in mosquitoes. J Am Mosq Control Assoc 2010;26(1):108–11.

[102] Hurlbut HS. Mosquito salivation and virus transmission. Am J Trop Med Hyg 1966;15 (6):989–93.

[103] Colton L, Biggerstaff BJ, Johnson A, Nasci RS. Quantification of West Nile virus in vector mosquito saliva. J Am Mosq Control Assoc 2005;21(1):49–53.

[104] Marrelli MT, Li C, Rasgon JL, Jacobs-Lorena M. Transgenic malaria-resistant mosquitoes have a fitness advantage when feeding on *Plasmodium*-infected blood. Proc Natl Acad Sci USA 2007;104(13):5580–3.

[105] Santos MN, Nogueira PM, Dias FB, Valle D, Moreira LA. Fitness aspects of transgenic *Aedes fluviatilis* mosquitoes expressing a *Plasmodium*-blocking molecule. Transgenic Res 2010;19(6):1129–35.

[106] Murdock CC, Moller-Jacobs LL, Thomas MB. Complex environmental drivers of immunity and resistance in malaria mosquitoes. Proc Biol Sci/R Soc 2013;280(1770):20132030.

[107] Murdock CC, Paaijmans KP, Bell AS, et al. Complex effects of temperature on mosquito immune function. Proc Biol Sci/R Soc 2012;279(1741):3357–66.

[108] Murdock CC, Blanford S, Luckhart S, Thomas MB. Ambient temperature and dietary supplementation interact to shape mosquito vector competence for malaria. J Insect Physiol 2014;67:37–44.

[109] Carrington LB, Armijos MV, Lambrechts L, Scott TW. Fluctuations at a low mean temperature accelerate dengue virus transmission by *Aedes aegypti*. PLoS Negl Trop Dis 2013;7(4):e2190.

[110] Carrington LB, Seifert SN, Armijos MV, Lambrechts L, Scott TW. Reduction of *Aedes aegypti* vector competence for dengue virus under large temperature fluctuations. Am J Trop Med Hyg 2013;88(4):689–97.

[111] Lambrechts L, Paaijmans KP, Fansiri T, et al. Impact of daily temperature fluctuations on dengue virus transmission by *Aedes aegypti*. Proc Natl Acad Sci USA 2011;108 (18):7460–5.

[112] Lefevre T, Vantaux A, Dabire KR, Mouline K, Cohuet A. Non-genetic determinants of mosquito competence for malaria parasites. PLoS Pathog 2013;9(6):e1003365.

[113] Murdock CC, Paaijmans KP, Cox-Foster D, Read AF, Thomas MB. Rethinking vector immunology: the role of environmental temperature in shaping resistance. Nat Rev Microbiol 2012;10(12):869–76.

[114]   Basu S, Aryan A, Overcash JM, et al. Silencing of end-joining repair for efficient site-specific gene insertion after TALEN/CRISPR mutagenesis in *Aedes aegypti*. Proc Natl Acad Sci USA 2015;112(13):4038−43.

[115]   Kistler KE, Vosshall LB, Matthews BJ. Genome engineering with CRISPR-Cas9 in the mosquito *Aedes aegypti*. Cell Rep 2015;11(1):51−60.

Chapter 14

Genetic Control of Malaria and Dengue Using *Wolbachia*

Zhiyong Xi and Deepak Joshi
Department of Microbiology and Molecular Genetics, Michigan State University, East Lansing, MI, USA

INTRODUCTION

Wolbachia is a gram-negative endosymbiont bacterium that is estimated to infect more than 65% of insect species and a large number of other taxa, including arachnids, freshwater crustaceans, and nematodes [1,2]. It is maternally transmitted from mother to offspring and is well known for exhibiting a phenotype called cytoplasmic incompatibility (CI), in which embryonic death occurs when infected males mate with either uninfected females or infected females carrying different type of *Wolbachia* (Table 14.1) [3]. Based on hypothetical models, *Wolbachia* causes modification of the sperm during spermatogenesis in the male. The uninfected eggs fertilized by the modified sperm stop developing at an early stage. In order to rescue these modifications, a similar *Wolbachia* type has to be present in the eggs, leading to fertilization, zygote formation, and successful embryonic development [4]. In contrast to uninfected eggs, infected eggs can be successfully fertilized by sperm from both uninfected males and males infected with the same type of *Wolbachia*. Thus, the infected female has an additional advantage in reproduction over the uninfected female, allowing *Wolbachia* to invade an uninfected population. In this situation, the incompatibility occurs in one direction, referred to as unidirectional CI. When individuals within a population are infected with different type of *Wolbachia*, crosses between the individuals carrying the different infections are incompatible in both directions, resulting in a so-called bidirectional CI (Table 14.1).

Although CI is the only *Wolbachia*-related phenotype observed in mosquitoes [3], the presence of *Wolbachia* in insects has been reported to cause numerous other phenotypic changes in reproduction, including male killing, parthenogenesis, and feminization [1]. Among all these phenotypes, the production of infected females is enhanced while the infected males are either killed (male killing), switched to females (feminization), made to be unnecessary (parthenogenesis), or sacrificed to sterilize the uninfected females (CI). Because only females can transmit *Wolbachia* to their offspring and the male is a dead end for transmission, all these phenotypes benefit the spread and maintenance of *Wolbachia*. By manipulating host reproduction in these ways, this vertically transmitted bacterium has been able to become one of the most widely spread endosymbiotic bacteria in the world. However, these *Wolbachia*-mediated reproductive alterations may not be the only reason for its wide distribution in nature. In some *Wolbachia*−host associations, an impact of *Wolbachia* on host reproduction is not observed, and the benefit that *Wolbachia* provides to the host may play major role in its spread and its stable infection of a population [5].

Depending on the specific association, the relationship between *Wolbachia* and its hosts can be commensal, parasitic, or mutual. It presents as a commensal bacterium in most insect hosts, including mosquitoes. The uninfected insect can live and reproduce as well as the infected individuals after the removal of *Wolbachia* through antibiotic treatment except in some forms of parthenogenesis. Strikingly, a *Wolbachia* strain called *w*MelPop can reduce the longevity of *Drosophila melanogaster* [6], while *Wolbachia* in *Aedes albopictus* (*w*AlbA and *w*AlbB) can increase the host's lifespan [7]. In *Drosophila*, the same *w*MelPop infection can decrease longevity in one genetic background and increase it in a different background, indicating the importance of host genetics on this phenotype expression [8]. In the parasitoid wasp *Asobara tabida*, female wasps that are cured of their *Wolbachia* infection fail to produce any mature oocytes, making *Wolbachia* an obligate symbiotic partner [9]. In filarial nematodes, *Wolbachia* has evolved mutualism with its hosts, and its removal inhibits larval development and adult fertility [10].

Phylogenetically, *Wolbachia* is closely related to the Rickettsiales order in the family Anaplasmataceae. Based on 16S ribosomal RNA gene, *ftsZ*, *wsp* (*Wolbachia* surface protein gene), and other sequence information, *Wolbachia* spp. have been classified into eight different supergroups (A−H), with supergroups A and B found mainly in arthropods and supergroups C and D commonly found in filarial nematodes [1,11]. The E and H supergroup encompasses *Wolbachia* spp. from the springtails and termites, respectively [12,13]. The supergroup F is of particular interesting in that it includes *Wolbachia* spp. from both arthropod and nematode hosts, implying an independent horizontal transfer of *Wolbachia* between these host phyla [14]. Although nematode-associated *Wolbachia* species show a general concordance between *Wolbachia* phylogeny and their hosts, there is no concordance between the phylogeny of some arthropod *Wolbachia* and their hosts, indicating extensive lateral movement of *Wolbachia* between host species [1]. The latter is supported by the fact that a number of arthropods, including mosquito, have been shown to be infected by both supergroups A and B and recombination in *wsp* gene is widespread between them [15].

DEVELOPMENT OF *WOLBACHIA* FOR VECTOR CONTROL

The ability of *Wolbachia* to induce CI in its host has led to the proposal of two major strategies to control mosquitoes and the diseases they transmit. One strategy, known as population replacement, would use *Wolbachia* to spread desired traits (such as an anti pathogen phenotype) into the population [16,17]. Because of the advantage in reproduction that it confers on infected females as compared to their uninfected counterparts, *Wolbachia* is able to invade a target population, even given the limited level of fitness cost it also confers. This invasion, together with *Wolbachia*-mediated pathogen interference, would eventually modify the population and reduce its ability to transmit diseases to humans. The success of *Wolbachia*-mediated population replacement has been observed in *Drosophila* with a native infection in nature and in mosquitoes with an artificial infection in both laboratory and field studies [18−21]. The second strategy is referred to as population suppression and entails the release of male mosquitoes infected with *Wolbachia*, resulting in sterile matings and a reduction in the mosquito population. This strategy is similar to the sterile insect technique (SIT) and, by analogy, is also called the "incompatible insect technique" (IIT). In a field test of this strategy, release of bidirectionally incompatible males successfully eliminated a *Culex* mosquito vector population from a village in Burma (Myanmar) [22]. The other efforts to develop population suppression include successful cage experiments with the Polynesian tiger mosquito *Aedes polynesiensis* [39], *Culex pipiens quinquefasciatus* [23], and the medfly *Ceratitis capitata*, demonstrating that *Wolbachia*-infected males can well compete with the wild males to mate with females and induce their sterility [24].

Wolbachia transinfection in mosquitoes (or transfer of *Wolbachia* between different mosquito species through embryonic microinjection) to generate a stable novel *Wolbachia* infection in the target vector species is the first step in developing a *Wolbachia*-based strategy for the control of dengue and malaria. Since *Wolbachia* was first introduced into the primary dengue vector *Aedes aegypti* in 2005 [21], extensive efforts have been dedicated to developing *Wolbachia* as a novel genetic tool for controlling dengue, malaria, and the other vector-borne diseases, with a number of stable transinfected lines being available at present (Table 14.2) [18,20,25−28]. To date, *Wolbachia* has not been successfully cultured in any artificial medium. But the methods to generate artificially infected cell lines or somatic infections in insects are well established [29,30]. Evidence shows that a strong barrier, which is determined by the host's genetic background, the *Wolbachia* strains, and host microbiota, exists to prevent *Wolbachia* from passing from somatic to germline tissues [31−33]. To bypass this barrier, embryonic microinjection is still the most effective approach for transferring *Wolbachia* between different host species, probably because early encounters between *Wolbachia* and the host immune system (i.e., earlier than the development of the host immune system) aid in shaping the host's immune system so that *Wolbachia* can be maintained persistently.

There are several advantages to developing a *Wolbachia*-based control strategy. First, *Wolbachia*-mediated population replacement allows for a permanent reduction in the mosquito's vectorial capacity, leading to a break in the linkage between pathogen and human host. Once population replacement is reached, disease transmission will not occur, even if there are patients serving as a source of infection in the region. Second, *Wolbachia*-mediated population suppression could lead to an area-wide reduction or even eradiation of mosquito vectors, within a certain time frame. It can effectively target cryptic and inaccessible mosquito habitats where insecticide treatments have failed to provide control. Because only male mosquitoes are released and males do not bite people, population suppression is more likely to be accepted in those countries with restrictive biosafety regulation policies. The consistency of population suppression with the traditional effort to reduce mosquito density, such as through breeding site removal, can also aid in accepting this approach for disease control. Third, the *Wolbachia*-based control strategies will cause very low impact on environment and thus is not difficult for both the public and governments to accept this approach. *Wolbachia* can neither live outside of the hosts nor infect vertebrate host. Many native mosquitoes already carry *Wolbachia*, and humans have close contact with a variety of *Wolbachia*-infected insects. Both population replacement and population suppression will target only the specific vector species that transmits disease and will not affect the other non target species, thus their ecological impacts are expected to be minimum. Population suppression will lead to a significant reduction, or even temporary elimination of the target species, without negative impact on closely related mosquito species, and thus its impact on the food chain will be significantly less than the chemical control. Furthermore, population replacement allows to keep the same mosquito species and only reduce its ability to transmit disease, and thus its perturbation to environment will be minimum. Lastly, the *Wolbachia*-based control strategies can be combined with traditional control methods such as the use of insecticides, drugs, and vaccines. Because each method has its own limitation, such an integrated strategy is expected to complement each other and reach the optimized effect in disease control. For example, *Wolbachia*-based approach can be used to target for the dominant malaria vectors species in Africa that is not susceptible to the widely used vector control tools such as indoor residual insecticide sprays and long-lasting insecticide-treated nets.

WOLBACHIA AND THE MOSQUITO

Wolbachia was first described by Hertig et al. in 1924 as a "Rickettsia-like organism" in *Cx. pipiens* mosquitoes and was given the name *Wolbachia* in 1936 [34,35]. CI was first attributed to *Wolbachia* in *Culex* mosquitoes in 1971 [36]. Until recently, *Wolbachia* was found to be present in approximately 28% of the surveyed mosquito species [37,38], and all the known *Wolbachia* in mosquito belonged to the supergroups A and B. The CI phenotype was also well characterized in a variety of mosquito species, including *Ae. albopictus* [7,39−43], *Aedes polyensia, Aedes riversi, Aedes sculletaris*, and *Aedes fluviaitilis* [44−46].

An important vector for dengue and other arboviruses, the *Ae. albopictus* population is naturally superinfected at a prevalence of approximately 100% with two *Wolbachia* types (*w*AlbA and *w*AlbB) throughout the majority of its geographical range [39,47], except that single infection with *w*AlbA is observed in mosquitoes originating from the islands of Oceania (i.e., Mauritius and Samui Island) [48]. The CI phenomenon was clearly demonstrated in this mosquito species, with the expected CI patterns. *w*AlbB-singly infected mosquitoes were artificially generated in the laboratory [41], and matings between uninfected females and infected males or between a single-infected female and a superinfected male were found to be incompatible, but the reciprocal cross was compatible (Table 14.1) [41].

TABLE 14.1 The Observed Cytoplasmic Incompatibility in *Ae. albopictus*

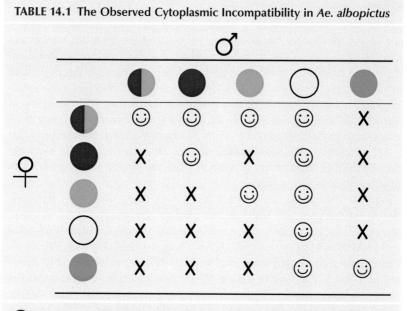

☺, compatible; X, incompatible. Red, green, white, and blue indicate different infection type: *w*AlbA, *w*AlbB, uninfected, and *w*Ri, respectively. *Aedes albopictus* naturally carries *w*AlbA and *w*AlbB. Mosquito strains with a single *w*AlbA, *w*AlbB, and *w*Ri were established in the laboratory. Crosses of mosquitoes with different infections show either unidirectional or bidirectional CI [41,42]. Overall, CI occurs when male carries a *Wolbachia* type which does not present in female.

Crosses between different wAlbA and wAlbB single infections resulted in egg mortality in both directions, indicating that they modified sperms in different way and the modification cannot be rescued each other [42]. Unlike the *Cx. pipiens* complex in which there is the greatest variation in CI crossing patterns of any insect with over 17 distinct CI phenotypes (cytotypes) reported [22,49,50], a diversity of CI in the context of geographical isolates of the mosquito is not observed in the wAlbA and wAlbB strains [51].

Wolbachia transinfection was first successfully developed in *Ae. albopictus* [41]. Thus far, the cured *Ae. albopictus* line has received more artificial infections than any other mosquitoes species (Table 14.2). *Wolbachia* originating from *Drosophila* (wRi, wMclPop, and wMel) and *Culex* spp. (wPip) has been successfully transferred to *Ae. albopictus* with perfect maternal transmission [43,52−55], indicating that this mosquito species is physiologically compatible with a variety of *Wolbachia* strains. Strikingly, each member within the *Cx. pipiens* complex seems to be infected with just a single variant of the wPip strain, and the observed diversity of CI types is caused by intra-strain variation. Unlike CI in other insects, these different cytotypes are not correlated with the phylogeny of *Wolbachia* because all of them belong to wPip strain in the supergroup B [56]. This variation was found to occur within the intergenic region, a stretch of DNA sequences located between genes, of wPip and to be associated with various mobile genetic elements, including the prophage of the *Wolbachia* bacteriophage WO. Although the exact nature and functional significance of these genetic variations are still unknown, identification of the variation in genes can be useful as a marker for discriminating between different cytotypes [57,58]. Because of this uniqueness, it is anticipated that *Cx. pipiens* can serve as a system (i) for the fine-scale mapping of the genetic determinants of CI, (ii) for studying the evolutionary population genetics and phylogeography of wPip in natural (field) populations, and (iii) for developing *Wolbachia*-based control strategies using the available native *Wolbachia* supergroup B variants that are found in geographically distinct populations of *Cx. pipiens* [51].

Although *Wolbachia* has a broad insect host range, infections with this organism do not naturally occur in the primary dengue vector *Ae. aegypti*. However, *Ae. aegypti* can support a *Wolbachia* infection; three *Wolbachia* strains (wAlbB, wMelPop, and wMel) have been successfully transferred into this mosquito species, and infections with all three have been stably maintained [21,27,66]. The lack of natural *Wolbachia* infection in *Ae. aegypti* has made it possible to develop *Wolbachia*-based population replacement without the need for a superinfected system to induce unidirectional CI, as required for *Ae. albopictus*. Since wAlbB was first introduced from *Ae. albopictus* into *Ae. aegypti* and demonstrated to invade the laboratory population, significant progress has been made in developing *Wolbachia* as a biocontrol agent to eliminate dengue. wMel and wMelPop were the second and third *Wolbachia* strains transferred into *Ae. aegypti*. All three strains undergo

TABLE 14.2 Stable Transinfected Mosquito Lines with the Potential to be Used for Dengue and Malaria Control

Transinfected Line	Recipient Embryos	Wolbachia Strain	Donor Embryos	Phenotypes Characterized	References
WB1	Ae. aegypti	wAlbB	Ae. albopictus (Hou strain)	CI and PR	[21]
				PI and longevity	[25,59]
				Fitness	Unpublished
PGYP1 and 2	Ae. aegypti	wMelPop	D. melanogaster	CI and longevity	[27]
				PI	[26,28]
				Mating competitiveness	[60]
				Fecundity and fertility	[61]
				Probing behavior	[62]
				Locomotion and metabolism	[63]
				Adult and immature survival	[64]
				Egg desiccation	[65]
MGYP2	Ae. aegypti	wMel	D. melanogaster	CI, PI, PR, and fitness	[18,66]
HTB	Ae. albopictus, aposymbiotic	wAlbB	Ae. albopictus (Hou strain)	CI	[41]
HTR	Ae. albopictus, aposymbiotic	wRi	D. simulans	CI and mating competitiveness	[42]

(Continued)

TABLE 14.2 (Continued)

Transinfected Line	Recipient Embryos	Wolbachia Strain	Donor Embryos	Phenotypes Characterized	References
ARwP	Ae. albopictus, aposymbiotic	wPip	Cx. pipiens	CI and fitness	[53,67]
HouR	Ae. albopictus	wAlbA, wAlbB, wRi	D. simulans	CI	[43]
HTM	Ae. albopictus, aposymbiotic	wMelPop	D. melanogaster	CI and fitness	[55]
Uju.wMel	Ae. albopictus, aposymbiotic	wMel	D. melanogaster	CI, PI, and fitness	[54,68]
HC	Ae. albopictus	wAlbA, wAlbB, wPip	Cx. pipiens		Unpublished
HM	Ae. albopictus	wAlbA, wAlbB, wMel	MGYP2		Unpublished
LB1	An. stephensi	wAlbB	Ae. albopictus	CI, PI, and PR	[20]
				Fitness	[69]

CI, cytoplasmic incompatibility; PR, population replacement; PI, pathogenic interference.

the expected symbiosis with *Ae. aegypti*, with 100% maternal transmission efficiency [21,27,66]. Thus far, *w*AlbB has been stably maintained in *Ae. aegypti* for over 10 years. *w*Mel is the first *Wolbachia* strain to be released into the field, showing that *Wolbachia* can invade field populations of *Ae. aegypti* [18]. Given the ability of *Wolbachia* to inhibit the replication of dengue virus in mosquitoes, it is anticipated that fixation of *Wolbachia* in field populations would reduce dengue transmission and prevent dengue epidemics in the region.

For a long time it was believed that *Anopheles* spp. did not carry *Wolbachia* in nature. Based on several PCR-based field surveys, none of 38 *Anopheles* species were found to carry natural *Wolbachia* infections, including a number of important malaria vectors such as *Anopheles gambiae*, *Anopheles arabiensis*, *Anopheles funestus*, and *Anopheles stephensi* [37,51]. In contrast, 39.5% of the 147 culicine species were found to be positive for *Wolbachia* [51]. It was thought that the absence of *Wolbachia* might be the result of incompatible physiological environments in the *Anopheles* mosquitoes or a lack of opportunity to acquire *Wolbachia* by horizontal transmission from other taxa, or competitive exclusion by the native microbiota in *Anopheles* spp. [51]. A recent successful transfer of *w*AlbB from *Ae. albopictus* into *An. stephensi*, however, indicates that *Anopheles* spp. can support a *Wolbachia* infection [20]. Interestingly, recent evidence further has supported the claim that *Wolbachia* infection is present in *An. gambiae* in Burkina Faso, West Africa [70]. However, the genetic material identified in *An. gambiae* belongs to a novel *Wolbachia* strain, related to, but distinct from, strains infecting other arthropods, suggesting that there could be more diverse forms of *Wolbachia* that remain largely unidentified by the available molecular diagnostic approaches [70].

FACTORS DETERMINING THE EFFICACY OF *WOLBACHIA*-BASED DISEASE CONTROL

Cytoplasmic Incompatibility

Among the four factors that influence the efficacy of *Wolbachia*-based disease control, CI is the mostly striking feature associated with *Wolbachia*, and it is also the key factor that allows *Wolbachia* to be used for vector-borne disease control. Both unidirectional and bidirectional CI are well observed in *Aedes* and *Culex* mosquitoes [7,22,42,49,50]. A complete or nearly complete CI is commonly observed when *Wolbachia* is used in mosquitoes [23,41,71]. This result differs from that obtained with *Wolbachia* in *Drosophila*, where it induces CI at different levels, ranging from none (e.g., *w*Au) to medium (e.g., *w*Mel) to high (e.g., *w*Ri) levels of CI [72]. The expression of CI in *Drosophila* is also subject to the *Wolbachia* strain used and the host's genetic background, age, and environmental conditions

[73–76]. The CI phenotype includes two essential components: (i) sperm modification (known as the mod function) during spermiogenesis in the male and (ii) rescue of egg viability (the resc function) during embryogenesis in the female. It has been found that modification and rescue can be segregated in a given host, such that the CI can be classified into four categories: mod^+ $resc^+$, mod^+ $resc^-$, mod^- $resc^+$, and mod^- $resc^-$ [4,77,78]. Except for the so-called "suicide" (mod^+ $resc^-$) strain, all the others have been found to occur in *Drosophila* in nature [72]. However, only mod^+ $resc^+$ was observed in mosquitoes. In this aspect, CI in mosquitoes represents a much simpler system than in *Drosophila*. However, if all those different bidirectional CI observed in *Culex* mosquito are considered, mosquito could be viewed as the system with the most complicated CI.

Developmental defects resulting from CI are visible during the first zygotic mitosis. In CI embryos, all events from fertilization to pronuclear juxtaposition appear normal, but nuclear envelope breakdown and chromatin condensation of the paternal pronuclear material are delayed relative to the maternal pronuclear material. This delay causes asynchrony between the male and female pronuclei and a loss of paternal chromosomes at the first mitosis. When *Wolbachia* is present in the egg, synchrony is restored [79]. It seems that the modification of the sperm must take place at an early stage of spermatogenesis, because *Wolbachia* is shed from maturing sperm and eliminated in cytoplasmic "waste-bags" [80].

Although the molecular mechanism of CI has still to be elucidated, three models have been proposed to explain the basis for CI [77]. The "lock and key" model proposes that *Wolbachia* produces a component that can "lock-in" the paternal pronucleus and prevent its normal behavior during the first mitosis, while *Wolbachia* in the egg can produce a "key" that frees the lock. In the "titration–restitution" model, *Wolbachia* removes some proteins normally associated with the paternal nuclear material in the testes, which are given back by *Wolbachia* to restore normal development if they are present in the egg. In the third model, called "slow-motion," modification is a result of the *Wolbachia* producing a factor that first binds to paternal chromosomes and then slows down their movements during the first embryonic mitosis, leading to unsynchronized paternal and maternal sets. A similar modification of maternal chromosomes occurs when *Wolbachia* is present in the egg, restoring a synchronous cycle between paternal and maternal complements. Each of the three models is limited in its ability to explain complicated CI patterns [77]. A better understanding of the CI mechanism could facilitate to develop a novel CI-based approach for vector-borne disease control. For example, an anti pathogen effector could be driven by a CI-determined gene instead of the *Wolbachia* bacterium [16,81], so that the future technology would not be restricted by the host range of *Wolbachia*.

Maternal Transmission

Wolbachia is mainly transmitted vertically from mother to offspring, although evidence also supports the lateral transfer of *Wolbachia* across different host species [1]. To maintain fidelity in maternal transmission, *Wolbachia* usually reaches a high density in the female's reproductive tissues and develops effective mechanisms to infect the progeny during early development. This infection includes a direct colonization of *Wolbachia* into the germ plasm, or a tropism of *Wolbachia* for stem cell niches, from which *Wolbachia* is transmitted to the germ line. *Wolbachia* in the male is at a dead end, and it will not pass to the offspring through mating with a female [82−84]. Because it can be transmitted maternally, *Wolbachia* can develop an intimate relationship with the host without significantly affecting fitness, as do other microbes that rely only on horizontal transmission. In addition, highly effective maternal transmission allows the offspring to carry the same type of *Wolbachia* such that the CI will not occur between them, and selection should favor the development of perfect maternal transmission in infected females if *Wolbachia* induces strong sperm modification in males.

Establishment of a line of maternally transmitted *Wolbachia* is the first hurdle in developing *Wolbachia*-based approaches for controlling dengue and malaria. This goal appears challenging, in that both *Wolbachia* and host need to match each other in order to maintain a balance so that the host does not remove a novel infection through its defense mechanisms, and the *Wolbachia* infection is not so virulent (e.g., develops to a very high titer) that it kills the host. The genetic background of both the *Wolbachia* and the host plays an important role in determining this symbiotic relationship. In addition, the host microbiome can also help determine whether a novel *Wolbachia* can form a stable state of symbiosis and be maternally transmitted [31]. *w*AlbB, originating from *Ae. albopictus*, is the first *Wolbachia* strain successfully introduced into *Ae. aegypti* and *Anopheles* mosquitoes [20,21]. Although more evidence is still needed, *w*AlbB is likely to be invasive and benign enough to make it an ideal candidate for *Wolbachia* transinfection [32].

A very high fidelity in material transmission is typically observed in the infected mosquitoes in both laboratory and field populations. Different strains (*w*AlbB, *w*Mel, and *w*MelPop) of *Wolbachia* can be transmitted at 100% efficacy in the transinfected *Aedes* and *Anopheles* mosquitoes [21,25,27,66]. Over 97% of *Ae. albopictus* has been superinfected by *Wolbachia* in field populations [85]. However, the density of *Wolbachia* in individual mosquitoes collected from different locations can be significantly different [86]. Evidence indicates that age, larval crowding, nutrition, and temperature can influence *Wolbachia* infection density [87,88]. It will be interesting to determine how such a change in density will affect maternal transmission, or whether it will only affect the level of somatic infection.

Pathogenic Interference

One reason that *Wolbachia* has recently become of broad interest for dengue and malaria control is its ability to induce pathogenic interference. This feature has been seen in both novel and native host associations [20,25,26,28]. In transinfected *Ae. aegypti*, all three types of *Wolbachia* (wAlbB, wMelPop, and wMel) show significant inhibition of dengue virus replication and dissemination, resulting in either a complete or partial blockage of viral transmission [25,26,66]. A similar viral inhibition also occurs in *Ae. albopictus* mosquitoes infected with wMel and wPip [54]. After transfer into *An. stephensi*, wAlbB can significantly inhibit infection with *Plasmodium falciparum* and *Plasmodium berghei* at both the oocyst and sporozoite stages in the transinfected lines [20]. Transient somatic infection of *An. gambiae* with wMelPop also reduces the oocyst burden of both *P. falciparum* and *P. berghei* [33,89]. In addition to the transinfected hosts mentioned above, native *Wolbachia* infections in different hosts, including *Ae. albopictus*, *Cx. quinquefasciatus*, and *D. melanogaster*, have also been observed to induce anti viral resistance [90−93].

Wolbachia-mediated interference with pathogens has four important features: (i) The strength of the inhibition depends on the *Wolbachia* density. An *in vitro* cell line has been used to show that there is a strong negative linear correlation between the number of genome copies of *Wolbachia* and of dengue virus, with viral infection being completely removed when the *Wolbachia* density reaches to sufficient high level [94]. Either a very weak anti-dengue resistance or no resistance is observed in an *Ae. albopictus* laboratory strain because of the very low *Wolbachia* titer in its somatic tissues. The ability of *Wolbachia* to develop a significant blockage is essential for a population replacement-based strategy. This would require for *Wolbachia* to grow to a level in mosquito that is high enough such that all the viruses are cleared in mosquito saliva or the number of the viral particles injected into human is too low to cause an infection. One *Wolbachia* strain may grow to a density different from the others with a specific tissue distribution, resulting in a strain-specific variation in the strength of viral inhibition. (ii) Interference with a pathogen can only occur locally, meaning that *Wolbachia* has to be present in the same tissue in which the pathogen resides in order to induce an anti pathogen effect [26]. Because both dengue and *Plasmodium* need to travel through the midgut, hymolymph, and salivary gland before being transmitted to a human through an infectious bite, the presence of *Wolbachia* in these tissues is important for the bacterium to be able to block their transmission. (iii) *Wolbachia*-mediated interference with pathogens is broad-spectrum, with resistance reported against dengue, West Nile, yellow fever, and chikungunya viruses, filarial worms, and *Plasmodium* parasites (*P. falciparum, P. berghei,* and *P. gallinaceum*) [25,26,28,95]. Based on studies of a similar antiviral effect in *Drosophila*, it appears that RNA viruses are especially sensitive to inhibition, whereas DNA viruses do not seem to be affected [93]. (iv) *Wolbachia*-mediated pathogen interference may evolve when the interaction of *Wolbachia* with the host evolves over time. Typically, *Wolbachia* can grow to a high density and is well distributed to the somatic tissues (in addition to the germ line) in recent *Wolbachia*–host associations. It is likely that the density is then reduced, especially in somatic tissues, eventually resulting in the exclusive presence of *Wolbachia* in the reproductive tissues, as observed in a longer association. This change in somatic distribution may lead to an attenuation of protection against pathogens. However, given the strong antiviral inhibition observed even 10 years after *w*AlbB was introduced into *Ae. aegypti*, the process described above may take a very long time and is not expected to influence the value of *Wolbachia* as a tool for dengue and malaria control.

Although the molecular mechanism of pathogen interference is not yet fully elucidated, the existing evidence supports two modes of action: the first is that *Wolbachia* boosts the mosquito's basal immunity to induce pathogen interference, and the second is that *Wolbachia* and the pathogen compete for the same host resources. An activation of innate immunity has clearly been observed in all the transinfected mosquito lines, including activation of Toll pathway genes, anti-dengue effector molecules, and reactive oxygen species (ROS) [25,26,28,59]. In particular, *Wolbachia*-induced ROS production has been observed in both mosquito and cell lines [59,96]. The evidence has repeatedly indicated that elevated ROS makes mosquitoes refractory to *Plasmodium* and dengue virus [59,97−99]. However, *Drosophila* with native *Wolbachia* can induce resistance to injected dengue virus; in this case, upregulation of immune genes was not observed, indicating that, in addition to immunity, other mechanisms may be involved in the resistance to dengue virus [100]. Metabolic competition between *Wolbachia* and a pathogen is supported by the observation that amino acids and cholesterol are utilized by *Wolbachia* during its interactions with its hosts [101,102].

Attention should be paid to recent reports concerning potential *Wolbachia*-mediated pathogen enhancement in mosquito, including the results from a transient somatic infection (*w*AlbB-infected *An. gambiae* and *An. stephensi*) and natural infections (*Cx. pipiens quinquefasciatus*) [103−106]. A transient system represents a very different type of system than that of a stably infected system because *Wolbachia* acts like pathogens to the host and is removed from the next generation in the transient system while it forms symbiosis with host in a stably infection system. It is still uncertain whether a high parasite infection in *Cx. pipiens quinquefasciatus* with a natural *Wolbachia* infection is caused by either its host genetic background or significantly large size [105]. As described earlier, *Wolbachia*-mediated pathogen interference occurs locally [26,94]. In order to link a particular pathogen interference/enhancement to *Wolbachia*, it is important to know the tissue distribution of *Wolbachia*, instead of simply detecting *Wolbachia* in the whole body of a mosquito. However, these reports did bring to the fore the important issue of quality control regarding the mosquitoes to be released into the field. Before a population replacement strategy is implemented, a comprehensive characterization of each *Wolbachia*-infected system is needed to confirm the impact of that specific *Wolbachia* strain on the pathogens that may be transmitted by mosquitoes.

Fitness

The relationship between *Wolbachia* and its host, either a natively infected or an artificially transinfected host, can be symbiotic, pathogenic, or facultative, depending on the specific *Wolbachia*—host association. Evaluating this relationship usually involves measurement of fitness traits associated with *Wolbachia*, including adult longevity, female fecundity, male mating competitiveness, and immature survivorship and development. Overall, the impact of *Wolbachia* on host fitness is a complicated trait that is variable across different *Wolbachia*—host combinations and is subject to change under different environmental conditions. Sometimes it is difficult to determine the net outcome because *Wolbachia* induces a mixture of phenotypes as both a mutualist and a parasite [69].

Based on a mathematical model, the impact of *Wolbachia* on fecundity has the potential to significantly affect *Wolbachia*-based population replacement [107]. *w*MelPop- and *w*Mel-transinfected *Ae. aegypti* females are less fecund than are wild-type uninfected females, and these deleterious fitness effects are similar in both laboratory and field condition [60,108]. *w*MelPop has been shown to affect egg viability (hatching ratios) when fed both on human and non-human blood sources, but the effect is more severe when the mosquitos take non-human blood, such as pig, chicken, or sheep blood [61,109]. A low level of fecundity and egg viability among these lines has been described as a result of a deprivation of essential nutrients, which is thought to be caused by the existing competition between *Wolbachia* and the host for essential nutrients such as amino acids and cholesterol [109]. In addition, low-nutrition diets for males in the larval phase affect the fecundity of *w*Mel-infected females [60]. In the field condition, wMelPop-CLA-infected mosquitoes had 38.3% of the oviposition success of uninfected mosquitoes, and these infected mosquitoes tended to have smaller wings than uninfected mosquitoes during the cooler November in comparison to December [110]. However, *w*AlbB or superinfection with wAlbA can confer a fecundity benefit on *Ae. albopictus* [111], *An. stephensi* (when the transinfected LB1 is compared to the aposymbiotic line LBT) [69], and *Ae. aegypti* (unpublished data). No visible effect of *w*AlbB on egg viability has been observed in native-infected *Ae. albopictus* or the transinfected *Ae. aegypti* WB1 line. However, a low level of hatching in *An. stephensi* LB1 mosquitoes remains the same when either human or non-human blood sources are used [69]. Also, supplementation of the sucrose meal with amino acids and antioxidants does not rescue this phenotype, indicating that a mechanism other than resource competition may be involved in reducing the egg viability in LB1 (unpublished data).

Male mating competitiveness is an important parameter that determines the efficacy of *Wolbachia*-mediated population suppression, and this phenomenon has been well studied in different *Wolbachia*−host associations. The *w*Ri-transinfected *Ae. albopictus* males are less competitive than are superinfected wild-type males, but the *w*Mel-transfected *Ae. albopictus* are reported to be as competitive as wild-type males [41,112]. Neither *w*Mel nor *w*MelPop affects sperm quality or viability or the ability of males to successfully mate with multiple females in *Ae. aegypti* [60]. Only a very minor negative impact on mating competitiveness has been reported in *w*AlbB-infected *An. stephensi* males when the ratio of uninfected to infected males is low (1:1), and this impact disappears when the ratio is increased to 1:2 or 1:4 [69].

Wolbachia can also affect mosquito fitness at immature stages, although the outcome varies across the different systems. A negative impact of *Wolbachia* on immature survivorship and emergence time has been observed in *Ae. albopictus*, but such effects are not seen in the transfected WB1 *Ae. aegytpi* or LB1 *An. stephensi* mosquito lines [60,69,113]. It appears that this type of impact is also influenced by environmental conditions. When intra-specific competition is intense, the survivorship of *Wolbachia*-infected *Ae. albopictus* larvae is compromised when compared to their uninfected counterparts [114].

The impact of *Wolbachia* on host longevity has received considerable attention, mainly concerning the specific *Wolbachia* strain *w*MelPop. *w*MelPop induces pathogenic effects and reduces host longevity in both native (*D. melanogaster*) and transinfected (*Ae. aegypti*) hosts because of its excessive replication within the host [6,27]. This phenomenon has led to a proposal for dengue control involving modifying the mosquitoes' longevity, because typically mosquitoes need to live long enough to transmit disease and old mosquitoes are more dangerous; thus, a reduction in longevity is expected to curtail disease transmission [27]. In comparison to *w*MelPop, the *w*Mel strain is relatively benign, causing only a 10% reduction in longevity [66]. In contrast, *w*AlbB provides a fitness benefit in certain mosquito lines because the infected females survive much better and have a higher fecundity than do *Wolbachia*-free lines [111]. *w*AlbB-infected *An. stephensi* also live significantly longer than do uninfected mosquitoes with sugar feeding [25]. Similarly, *w*Mel-transinfected *Ae. albopictus* males survive longer than do superinfected wild-type males [112]. Taken together, *Wolbachia* may interact with a mechanism that is highly sensitive in control of mosquito life span such that a subtle change in interaction will lead to a variation in host longevity.

In addition to the traits described above, *Wolbachia* has been shown to affect host locomotion, dispersal, immunity, foraging, and mating behavior in a range of insects [63,115−117]. In *Ae. aegypti*, *w*MelPop infections can reduce females' ability to blood-feed and increase the daytime activity of both female and male *Ae. aegypti* adults from 3 to 26 days of age

[62,63,65,118]. In addition, wMelPop has been found to increase the metabolic rate of *Ae. aegypti*, and CO_2 production is increased in both infected females and males relative to uninfected controls [63].

WOLBACHIA-BASED CONTROL STRATEGIES

Population Replacement to Permanently Reduce Mosquitoes' Vectorial Capacity for Dengue and Malaria

The efficacy of *Wolbachia*-mediated population replacement in disease control depends on (i) the effectiveness of *Wolbachia* in invading and maintaining itself in the target population and (ii) its effectiveness in blocking pathogen transmission. In an ideal situation, *Wolbachia* would induce population replacement and be fixed at 100% infection frequency in a population within a few generations or months [18,20,21]. Certain parameters and environmental conditions will influence or determine the process of *Wolbachia* invasion and its blocking effects in the field.

The population dynamic during the spread of *Wolbachia* in an uninfected population was initially described in detail for *Drosophila simulans* as a model system [19]. The mathematical model predicted that three parameters, the strength of the CI, the maternal transmission efficiency, and the fecundity cost associated with *Wolbachia* infection, would determine the invasion process that had two equilibrium thresholds. According to this model, the frequency of infected individuals must first exceed an unstable equilibrium threshold value; then, it will spread to a stable high equilibrium value. Consistent with the model's prediction, a 20% or approximately 40% infection frequency threshold was required to be reached in order for wAlbB or wMel to invade an *Ae. aegypti* population under laboratory or field conditions, respectively [18,21]. Once *Wolbachia* is established in the field, it will reach a stable high equilibrium, with the level mainly determined by the maternal transmission efficiency and the migration of uninfected mosquitoes into the field site. The threshold can become too high to be exceeded, however, if there is a strong fitness cost associated with the particular *Wolbachia* strain, as in the case of wMelPop. Release of wMelPop-infected *Ae. aegypti* mosquitoes at an extremely high frequency did not result in population replacement in either Australia or Vietnam.

Under field conditions, the anti pathogen efficiency is subject to a variety of environmental conditions. Theoretically, any environmental change that affects the *Wolbachia* density in these tissues is expected to influence the blocking effect. Such influences could include the presence of natural antibiotics, residues in environment resulting from the abuse of antibiotics in the cattle industry, or a very high temperature. Furthermore, it is unknown whether that the *Wolbachia* titers in these somatic tissues will become attenuated after the *Wolbachia* adapts to a new host during its evolution and result in a reduction in a blocking effect, even this reduction may be a slow process.

In the first successful field trials to establish *Wolbachia* in the *Ae. aegypti* population in Australia, the *Wolbachia* continued, after more than 2 years under field conditions, to show complete CI, strong antiviral resistance, and persistent deleterious fitness effects, supporting the supposition that host effects had not evolved over time [108]. Similarly, *w*AlbB has been shown to continue to mediate a strong anti-dengue effect even 10 years after having been introduced into *Ae. aegypti* [21,25]. These results point to the stability of *Wolbachia* infections in the transinfected mosquitoes, supporting the potential of a *Wolbachia*-based population replacement strategy for sustainably reducing the mosquitoes' vectorial capacity in disease-endemic areas.

Wolbachia-mediated population replacement has generated great interest because of its potential for dengue control. In addition to Australia, other countries, including China, Brazil, Colombia, Indonesia, and Vietnam, have obtained regulatory approval for field trials [51]. With the exception of China, all of these countries have released *w*Mel-infected *Ae. aegypti*. Models predict that the establishment of *w*Mel-infected *Ae. aegypti*, even with only a partial viral blocking effect, will have a substantial effect on transmission that is sufficient to eliminate dengue in low- or moderate-transmission settings [119].

Results have also supported the feasibility of developing *Wolbachia*-based population replacement for malaria control. In the *w*AlbB-infected *An. stephensi* LB1 strain, *w*AlbB has been associated with both fitness benefit and cost [20,69]. The reduced egg hatching seen in LB1 females has been the major factor that can hinder the rate of population replacement. However, population replacement, assisted by the release of infected males, has proved possible even with an initial female infection frequency as low as 5% [20]. The ability of *w*AlbB to induce malaria resistance and success in small cage experiments in the laboratory have afforded the opportunity to design population replacement in the field to reduce the burden of malaria in countries such as India, in which *An. stephensi* is the major disease vector.

Population Suppression to Reduce Vector Populations

The *Wolbachia*-based population suppression, or IIT, strategy aims to reduce mosquito bites to decrease the mosquito's vectorial capacity. The feasibility of this strategy is supported by the following observations: (i) male mosquitoes neither bite humans nor transmit diseases, and thus their release causes less public concern than does the release of females; (ii) the capacity for mass rearing and sex separation has reached a production level of 1 million *Ae. aegypti* males per week [120]; (iii) release of male mosquitoes to induce CI matings with females in the field has resulted in the successful eradication of a *Cx. pipiens* population in the field [22]; and (iv) SIT, which is similar to IIT, has successfully suppressed or eliminated populations of other insect pests such as the screwworm and Mediterranean fruit fly (medfly) in

Southern America [121], providing a model for addressing the logistics required for *Wolbachia*-based population suppression. The countries currently leading IIT field trial efforts include the United States, China, and Singapore. The Insect Pest Control Laboratory of the International Atomic Energy Agency (IAEA) has also supported and joined this effort because of its potential to complement SIT in mosquito control.

The efficacy of a *Wolbachia*-based population suppression strategy can be influenced by a number of factors, including male mating competitiveness, female mating choice, target mosquito density and distribution, the migration of mosquitoes from the other areas, the release and dispersal range of the mosquitoes, re-mating ability, mating case, and barriers among cryptic species [73]. In most studies, *Wolbachia*-infected males can compete well with wild-type males in mating with females [23,69]. Although a strong fitness cost has been observed in *w*MelPop-infected females, a similar impact was not observed in the infected males [60]. Furthermore, *Wolbachia*-infected males were found to live significantly longer than wild-type males in various mosquito species [68,69]. All these findings support the potential of *Wolbachia*-infected males to induce CI mating in the field. Additional studies are still needed to examine the impact of *Wolbachia* on female mating choice, to determine the minimum release ratio needed to effectively suppress the population in a short time frame, and to measure the impact of migration on persistent suppression.

Probably the most challenging part of *Wolbachia*-based population suppression is the possible population replacement induced by the accidental release of infected females. With continuous inundative release of the infected male mosquitoes every generation, a single release of the infected females to reach a 5% initial female infection frequency has been found to induce population replacement in *An. stephensi* [20]. It is likely that a very low number of females mixed in with the released males will trigger population replacement, particularly if a *Wolbachia* infection induces a unidirectional CI. This issue can be addressed, first, by developing a better operating procedure that gives a sex separation of 100%. One successful example of such a procedure is the development of a temperature-sensitive lethal-based genetic sexing of medfly strains, in which all the females are killed by a simply incubation of the eggs to a temperature of 34°C for 24 h [122]. A similar sexing procedure in mosquitoes would greatly facilitate the implementation of population suppression. A second method involves developing a second transinfected line that is incompatible with the first released line. A rotation in the release of the two different strains will not only prevent population replacement but also enhance the population suppression effect, as predicted by the model [123]. The third alternative is to treat the population replacement, if it happens, as an acceptable outcome. As described above, this population replacement can result in a sustainable reduction in mosquito vectorial capacity, although the population will no longer have been suppressed by the same mosquito strain.

Another challenge is the development and maintenance of the capacity to produce a sufficient number of *Wolbachia*-infected males for release. Because of the limited resources and nutrition in the larval habitat, a density-dependent mortality occurs at the mosquito's larval stage, with more larvae dying from competition when the density is high. Insufficient release of males will reduce this competition, resulting in an increasing survival of immature mosquitoes and the lack of a suppression effect. Thus, the ability to continuously release enough infected males to maintain the suppression is essential for this control strategy. Even when the population is suppressed or eliminated in an area, the release should be maintained, although at a low scale, in order to prevent the population from becoming restored. The current bottleneck in the mass production of males is sex separation. Although genetic modification of mosquitoes would provide a good solution, it would be difficult to obtain regulatory approval for this approach in many countries at the present time. Thus, efforts have been made to separate males from females based on the difference in pupal size through mechanical sorting using an apparatus such as a John W. Hock larvae/pupae sorter. One report has indicated that with 4 sorters and fewer than 10 staff, 1 million *Ae. aegypti* males can be produced per week, a level of production that could satisfy the needs of a field trial in a relatively large open area [120].

Although both unidirectional and bidirectional CI can be used as a population suppression strategy, a design based on bidirectional CI would have several unique advantages. First, it would be difficult to induce population replacement because of an accidental release of a small number of females. In a bidirectional CI system, the released females would suffer from CI imparted by the males in the field with a different infection type, as long as the target population was not eliminated from the field; thus, the threshold to trigger population replacement would be high. Second, sex separation could even be avoided by good management of the CI matings in the field. Two strains of mosquito with different *Wolbachia* types will "battle" each other when co-existing in one region, with the mosquito population being the "victim." Models predict that the population can be continuously suppressed by artificially maintaining this "battle" [123].

Integration of Population Suppression and Population Replacement

In this strategy, the mosquito population will be suppressed to a very low level via the release of infected males, followed by population replacement through the release of infected females to reduce mosquito vector competence. Subsequently, males carrying a different strain of *Wolbachia* are released to further suppress the population of pathogen-resistant mosquitoes. Here, two parameters of vectorial capacity, vector competence and bite rate, are targeted simultaneously, resulting in a rapid interruption of disease transmission. In an extreme case, the target population would first be eradicated in a region through continuous release of infected males, and then a mosquito strain carrying an appropriate *Wolbachia* type (refractory to the pathogens) would be seeded into this region to maintain ecological stability. To a certain extent, this strategy can be viewed as population replacement driven by population suppression.

The rationale for designing this strategy is to resolve the weakness that occurs when either population suppression or population replacement is used alone. A trade-off has been observed between pathogen interference and fitness cost in *Wolbahcia*−host interactions. It is difficult to induce population replacement with a *Wolbachia* strain that causes complete blocking, such as *w*MelPop, given the strong fitness cost associated with this strain. Both *w*AlbB and *w*Mel can induce population replacement, but there will be leakage of a portion of the virus or parasite during the blocking process [20,25]. Before an ideal strain or strain combination is developed to balance the trade-off between blocking effect and fitness cost, we will need to reduce the risk associated with "leaked" pathogens. Such risks will include the possibility that they will evolve to evade the blocking mechanisms, resulting in a failure of disease control. In addition, it is likely that, without population suppression, the release of a sufficient number of *Wolbachia*-mediated females to reach the initial female infection threshold for population replacement may face difficulties in public acceptance because of the nuisance of female biting. On the other hand, population suppression alone requires a long-term and continuous investment in the program in order to maintain the suppression and prevent the return of the population to a high level, an investment that may not be feasible under certain situations or in some developing countries. Concerns have also been raised concerning the negative impact on the ecologic chain if a mosquito species is eradicated in a region. With regard to these challenges, the integrated method is expected to generate a more practical and sustainable effect on vector-borne disease control than is the use of one method alone. Another strategy that could be applied is population suppression driven by population replacement. In this proposed strategy, population replacement would first be used to spread *Wolbachia* into a population, taking advantage of the deleterious fitness effects associated with this *Wolbachia* infection [65]. For example, *w*MelPop is associated with a substantial loss of viability after a few weeks, when eggs enter a quiescent stage of *Wolbachia* infection, which could be even more serious in a dry season than during a wet season [65,124]. Therefore, a successful invasion by *Wolbachia* in the wet season could result in population suppression and potentially extinction in the dry season [125].

Integration of *Wolbachia*-Based Population Suppression with SIT

Although SIT has been successfully used to control both screwworm and medfly [121], its implementation in mosquito control needs to be further explored. One of the hurdles influencing its use in mosquito control is the reduced fitness, including mating competitiveness, of the irradiated male mosquitoes when compared to wild-type males. The other potential concern is that released females, although sterile, will still be able to transmit the disease. These problems can both be resolved by integration of a *Wolbachia*-based approach with SIT [126]. First, the dosage of irradiation used to sterilize the mosquitoes can be reduced to increase mosquito fitness. Male mosquitoes need a higher dosage of irradiation to achieve sterilization than do females [126]. In the integrated approach, a low-dose irradiation would be used to sterilize females, and *Wolbachia*-infected males would be used to sterilize the wild-type females through CI matings. Second, the accidentally released females would be resistant to pathogens because of their *Wolbachia* infection. Integration with SIT would also benefit *Wolbachia*-based population suppression because the accidently released females are sterile and are unable to cause population replacement. This approach would greatly increase the capacity for mass rearing because sex separation is a bottleneck that limits current production. Thus, integration of SIT with *Wolbachia*-based population suppression could provide an effective way to combine two complementary approaches. This strategy is being currently tested for proof-of-concept by the Insect Pest Control Laboratory of the IAEA.

In general, both population replacement and population suppression can be applied independently, or be integrated with each other or with other approaches to attain the maximum effect in disease control. We use SIT as an example in the above because of its successful use in the control of other insect pests. However, such an integration could be broader based and include other methods, such as transgenic mosquito technique, insecticides, vaccines, and/or drugs. For example, a rotated use of a *Wolbachia*-based approach with insecticide treatment may hinder the development of insecticide resistance. Population replacement and vaccines would interrupt the movement of the pathogen at two independent steps of a single transmission cycle, resulting in synergistic effects in disease control.

CONCLUSION, CHALLENGES, AND FUTURE PROSPECTS

In the last decade, significant progress has been made in developing *Wolbachia* as a tool for controlling both dengue and malaria. We can call it the golden time of *Wolbachia* technology development because so many important milestones have been reached. These milestones include success in introducing *Wolbachia* into *Aedes* dengue vectors and the *Anopheles* malaria vector [20,21,27,66], demonstration of the spread of *Wolbachia* into laboratory and field mosquito populations [18,20,21], the discovery of *Wolbachia*-mediated pathogen interference in mosquitoes [20,25,26,28], and accomplishment of regulatory approval for field trials in seven countries, including Australia, the United States, China, and Brazil. We now have a proof of concept at the entomological level for both population replacement and population suppression in the field. This success has led to a number of novel opportunities for taking the next step in this enterprise. Probably the most exciting question is what the impact of these approaches will be on disease transmission. Recent mathematical models predict that, once established in nature, *w*Mel would reduce the basic reproduction number of dengue transmissions by 66−75%, while resulting in complete abatement of dengue transmission [119]. However, this result would depend on whether *Wolbachia* can persistently block dengue virus in the field as effectively as it has done under laboratory conditions, whether environmental conditions, such as temperature, desiccation and nutrition stress, will significantly change the *Wolbachia*−host interactions, and whether *Wolbachia*'s blocking effect can be maintained with the rapid evolution of dengue virus.

Theoretically, if the population can be suppressed to below an epidemiological threshold, disease transmission will stop. On the other hand, questions still exist concerning how long this suppression can be maintained and what strategy is best to maintain the suppression to prevent disease transmission. SIT-based control of the screwworm and medfly provides a great example of how to maintain suppression or eradication after the initial success is achieved [121]. *Wolbachia*-infected males can be released into a buffer zone to prevent the migration of a mosquito species into a vector-free area from the outside, or they can be released periodically to prevent the return of the population. Considering that using *Wolbachia* offers an environmental friendly approach with a relatively low operating cost, *Wolbachia*-based population suppression can serve as an alternative effective way to achieve vector control that can help resolve current concerns regarding the use of chemical insecticides.

Achieving vector control is a challenging task. Previous experience has told us that an integrative strategy will work better than any single method. *Wolbachia*-based population replacement provides a new avenue for controlling vector-borne disease. If it succeeds, it will lead to a sustainable control of disease transmission with minimum ecological impact on the environment. Because of the trade-off between pathogen blocking and fitness cost, the vector population cannot be modified to make it completely resistant to pathogens at the present time. Developing the ability to transform *Wolbachia* may facilitate the design of an ideal strain that can perfectly block the pathogen and effectively invade a population. Before this goal can be accomplished, concerns regarding pathogen leaking from the blocking mechanism can be addressed by an integrative control strategy.

The Wolbachia-based approach has moved to field trials for dengue control. Success has been achieved in stably introducing *Wolbachia* into the *Anopheles* malaria vector. Unlike dengue, malaria is transmitted by multiple and frequently sympatric *Anopheles* species in different parts of the world. Thus, *Wolbachia* would have to be introduced into each target mosquito species. This requirement could be challenging because *Wolbachia* transinfection is a complicated process, and only a very few laboratories have the required capability. Moreover, some of the *Anopheles* mosquito species cannot even be colonized in the laboratory. Thus, good design is critical to developing a *Wolbachia*-based strategy for malaria control. A priority should be given to those mosquito vectors that are the most difficult to control by the currently available methods. For example, *Wolbachia* could be used to target outdoor-biting and -resting species that can evade insecticide-treated nets and residual insecticide sprays.

As a novel vector control approach, the release of *Wolbachia*-infected mosquitoes in the field requires a large effort in community education. Education is essential, based on the experience of countries that have field trials ongoing. Supported by an extensive collection of publications and data, all the biosafety panels in the countries involved have concluded that it is safe to release *Wolbachia*-infected mosquitoes into the field. However, questions from the community still need to be thoroughly addressed. In countries where the population is particularly sensitive to mosquito biting, population suppression via the release of male mosquitoes can start first. As increasingly knowledge is gained concerning *Wolbachia* and mosquitoes and their interrelations, the community is expected to become more supportive of the release of females, and then population replacement can also be deployed for vector control. A similar outcome could be also generated for the other similar genetic control approach. For example, when the proof of concept has been reached through *Wolbachia*-based population replacement and population suppression, there will be less concerns from both government and community on release of genetic modified mosquito for disease control.

REFERENCES

[1] Werren JH, Baldo L, Clark ME. *Wolbachia*: master manipulators of invertebrate biology. Nat Rev Microbiol 2008;6(10):741−51.
[2] Cordaux R, et al. Widespread *Wolbachia* infection in terrestrial isopods and other crustaceans. Zookeys 2012;176:123−31.
[3] Sinkins SP. *Wolbachia* and cytoplasmic incompatibility in mosquitoes. Insect Biochem Mol Biol 2004;34(7):723−9.
[4] Charlat S, Calmet C, Mercot H. On the *mod resc* model and the evolution of *Wolbachia* compatibility types. Genetics 2001;159(4):1415−22.
[5] Kriesner P, Hoffmann AA, Lee SF, Turelli M, Weeks AR. Rapid sequential spread of two Wolbachia variants in *Drosophila simulans*. PLoS Pathog 2013;9(9):e1003607.
[6] Min KT, Benzer S. *Wolbachia*, normally a symbiont of *Drosophila*, can be virulent, causing degeneration and early death. Proc Natl Acad Sci USA 1997;94(20):10792−6.
[7] Dobson SL, Marsland EJ, Rattanadechakul W. *Wolbachia*-induced cytoplasmic incompatibility in single- and superinfected *Aedes albopictus* (Diptera: Culicidae). J Med Entomol 2001;38(3):382−7.
[8] Carrington LB, Hoffmann AA, Weeks AR. Monitoring long-term evolutionary changes following *Wolbachia* introduction into a novel host: the *Wolbachia* popcorn infection in *Drosophila simulans*. Proc Biol Sci 2010;277(1690):2059−68.

[9] Dedeine F, et al. Removing symbiotic *Wolbachia* bacteria specifically inhibits oogenesis in a parasitic wasp. Proc Natl Acad Sci USA 2001;98(11):6247−52.

[10] Taylor MJ, Bandi C, Hoerauf A. *Wolbachia* bacterial endosymbionts of filarial nematodes. Adv Parasitol 2005;60:245−84.

[11] Werren JH, Zhang W, Guo LR. Evolution and phylogeny of *Wolbachia*: reproductive parasites of arthropods. Proc R Soc Lond B Biol Sci 1995;261(1360):55−63.

[12] Lo N, et al. Taxonomic status of the intracellular bacterium *Wolbachia pipientis*. Int J Syst Evol Microbiol 2007;57(Pt. 3):654−7.

[13] Vandekerckhove TT, et al. Phylogenetic analysis of the 16S rDNA of the cytoplasmic bacterium *Wolbachia* from the novel host *Folsomia candida* (Hexapoda, Collembola) and its implications for wolbachial taxonomy. FEMS Microbiol Lett 1999;180(2):279−86.

[14] Casiraghi M, et al. Phylogeny of *Wolbachia pipientis* based on gltA, groEL and ftsZ gene sequences: clustering of arthropod and nematode symbionts in the F supergroup, and evidence for further diversity in the *Wolbachia* tree. Microbiology 2005;151(Pt. 12): 4015−22.

[15] Baldo L, Lo N, Werren JH. Mosaic nature of the *Wolbachia* surface protein. J Bacteriol 2005;187(15):5406−18.

[16] Turelli M, Hoffmann AA. Microbe-induced cytoplasmic incompatibility as a mechanism for introducing transgenes into arthropod populations. Insect Mol Biol 1999;8(2):243−55.

[17] Curtis CF, Sinkins SP. *Wolbachia* as a possible means of driving genes into populations. Parasitology 1998;116 Suppl.:S111−15.

[18] Hoffmann AA, et al. Successful establishment of *Wolbachia* in *Aedes* populations to suppress dengue transmission. Nature 2011;476(7361):454−7.

[19] Turelli M, Hoffmann AA. Cytoplasmic incompatibility in *Drosophila simulans*: dynamics and parameter estimates from natural populations. Genetics 1995;140(4):1319−38.

[20] Bian G, et al. *Wolbachia* invades *Anopheles stephensi* populations and induces refractoriness to *Plasmodium* infection. Science 2013;340(6133):748−51.

[21] Xi Z, Khoo CC, Dobson SL. *Wolbachia* establishment and invasion in an *Aedes aegypti* laboratory population. Science 2005;310(5746):326−8.

[22] Laven H. Eradication of *Culex pipiens fatigans* through cytoplasmic incompatibility. Nature 1967;216:383−4.

[23] Atyame CM, et al. Cytoplasmic incompatibility as a means of controlling *Culex pipiens quinquefasciatus* mosquito in the islands of the south-western Indian Ocean. PLoS Negl Trop Dis 2011;5(12):e1440.

[24] Zabalou S, et al. *Wolbachia*-induced cytoplasmic incompatibility as a means for insect pest population control. Proc Natl Acad Sci USA 2004;101(42):15042−5.

[25] Bian G, Xu Y, Lu P, Xie Y, Xi Z. The endosymbiotic bacterium *Wolbachia* induces resistance to dengue virus in *Aedes aegypti*. PLoS Pathog 2010;6(4):e1000833.

[26] Moreira LA, et al. A *Wolbachia* symbiont in *Aedes aegypti* limits infection with dengue, Chikungunya, and *Plasmodium*. Cell 2009;139(7):1268−78.

[27] McMeniman CJ, et al. Stable introduction of a life-shortening *Wolbachia* infection into the mosquito *Aedes aegypti*. Science 2009;323(5910):141−4.

[28] Kambris Z, Cook PE, Phuc HK, Sinkins SP. Immune activation by life-shortening *Wolbachia* and reduced filarial competence in mosquitoes. Science 2009;326(5949): 134−6.

[29] Dobson SL, Marsland EJ, Veneti Z, Bourtzis K, O'Neill SL. Characterization of *Wolbachia* host cell range via the in vitro establishment of infections. Appl Environ Microbiol 2002;68(2):656−60.

[30] Jin C, Ren X, Rasgon JL. The virulent *Wolbachia* strain wMelPop efficiently establishes somatic infections in the malaria vector *Anopheles gambiae*. Appl Environ Microbiol 2009;75(10):3373–6.

[31] Hughes GL, et al. Native microbiome impedes vertical transmission of *Wolbachia* in *Anopheles* mosquitoes. Proc Natl Acad Sci USA 2014;111(34):12498–503.

[32] Hughes GL, Pike AD, Xue P, Rasgon JL. Invasion of *Wolbachia* into *Anopheles* and other insect germlines in an *ex vivo* organ culture system. PLoS One 2012;7(4):e36277.

[33] Hughes GL, Koga R, Xue P, Fukatsu T, Rasgon JL. *Wolbachia* infections are virulent and inhibit the human malaria parasite *Plasmodium falciparum* in *Anopheles gambiae*. PLoS Pathog 2011;7(5):e1002043.

[34] Hertig M, Wolbach SB. Studies on Rickettsia-like micro-organisms in insects. J Med Res 1924;44(3):329–374.327

[35] Hertig M. The Rickettsia, *Wolbachia pipientis* (gen. et sp. n.) and associated inclusions of the Mosquito, *Culex pipiens*. Parasitology 1936;28(4):453–86.

[36] Yen JH, Barr AR. New hypothesis of the cause of cytoplasmic incompatibility in *Culex pipiens* L. Nature 1971;232(5313):657–8.

[37] Kittayapong P, Baisley KJ, Baimai V, O'Neill SL. Distribution and diversity of *Wolbachia* infections in Southeast Asian mosquitoes (Diptera: Culicidae). J Med Entomol 2000;37(3):340–5.

[38] Ricci I, Cancrini G, Gabrielli S, D'Amelio S, Favi G. Searching for *Wolbachia* (Rickettsiales: Rickettsiaceae) in mosquitoes (Diptera: Culicidae): large polymerase chain reaction survey and new identifications. J Med Entomol 2002;39(4):562–7.

[39] Sinkins SP, Braig HR, O'Neill SL. *Wolbachia pipientis*: bacterial density and unidirectional cytoplasmic incompatibility between infected populations of *Aedes albopictus*. Exp Parasitol 1995;81(3):284–91.

[40] Dobson SL, Marsland EJ, Rattanadechakul W. Mutualistic *Wolbachia* infection in *Aedes albopictus*: accelerating cytoplasmic drive. Genetics 2002;160(3):1087–94.

[41] Xi Z, Dean JL, Khoo C, Dobson SL. Generation of a novel *Wolbachia* infection in *Aedes albopictus* (Asian tiger mosquito) via embryonic microinjection. Insect Biochem Mol Biol 2005;35(8):903–10.

[42] Xi Z, Khoo CC, Dobson SL. Interspecific transfer of *Wolbachia* into the mosquito disease vector *Aedes albopictus*. Proc Biol Sci 2006;273(1592):1317–22.

[43] Fu Y, Gavotte L, Mercer DR, Dobson SL. Artificial triple *Wolbachia* infection in *Aedes albopictus* yields a new pattern of unidirectional cytoplasmic incompatibility. Appl Environ Microbiol 2010;76(17):5887–91.

[44] Dean JL, Dobson SL. Characterization of *Wolbachia* infections and interspecific crosses of *Aedes (Stegomyia) polynesiensis* and *Ae. (Stegomyia) riversi* (Diptera: Culicidae). J Med Entomol 2004;41(5):894–900.

[45] Brelsfoard CL, Sechan Y, Dobson SL. Interspecific hybridization yields strategy for South Pacific filariasis vector elimination. PLoS Negl Trop Dis 2008;2(1):e129.

[46] Baton LA, Pacidônio EC, Gonçalves DS, Moreira LA. wFlu: characterization and evaluation of a native *Wolbachia* from the mosquito *Aedes fluviatilis* as a potential vector control agent. PLoS One 2013;8(3):e59619.

[47] Zhou W, Rousset F, O'Neil S. Phylogeny and PCR-based classification of *Wolbachia* strains using wsp gene sequences. Proc R Soc Lond B Biol Sci 1998;265(1395):509–15.

[48] Sinkins SP, Braig HR, O'Neill SL. *Wolbachia* superinfections and the expression of cytoplasmic incompatibility. Proc R Soc Lond B Biol Sci 1995;261(1362):325–30.

[49] Laven H. Speciation and evolution in *Culex pipiens*. In: Pal JWR, editor. Genetics of insectectors of disease. Amsterdam: Elsevier; 1967. p. 251−75.

[50] Guillemaud T, Pasteur N, Rousset F. Contrasting levels of variability between cytoplasmic genomes and incompatibility types in the mosquito *Culex pipiens*. Proc R Soc Lond B Biol Sci 1997;264(1379):245−51.

[51] Bourtzis K, et al. Harnessing mosquito-*Wolbachia* symbiosis for vector and disease control. Acta Trop 2014;132 Suppl.:S150−63.

[52] Xi Z, Gavotte L, Xie Y, Dobson SL. Genome-wide analysis of the interaction between the endosymbiotic bacterium *Wolbachia* and its *Drosophila* host. BMC Genomics 2008;9:1.

[53] Calvitti M, Moretti R, Lampazzi E, Bellini R, Dobson SL. Characterization of a new *Aedes albopictus* (Diptera: Culicidae)-*Wolbachia pipientis* (Rickettsiales: Rickettsiaceae) symbiotic association generated by artificial transfer of the wPip strain from *Culex pipiens* (Diptera: Culicidae). J Med Entomol 2010;47(2):179−87.

[54] Blagrove MS, Arias-Goeta C, Failloux AB, Sinkins SP. *Wolbachia* strain wMel induces cytoplasmic incompatibility and blocks dengue transmission in *Aedes albopictus*. Proc Natl Acad Sci USA. 2012;109(1):255−60.

[55] Suh E, Mercer DR, Fu Y, Dobson SL. Pathogenicity of life-shortening *Wolbachia* in *Aedes albopictus* after transfer from *Drosophila melanogaster*. Appl Environ Microbiol 2009;75(24):7783−8.

[56] Atyame CM, Delsuc F, Pasteur N, Weill M, Duron O. Diversification of *Wolbachia* endo-symbiont in the *Culex pipiens* mosquito. Mol Biol Evol 2011;28(10):2761−72.

[57] Sanogo YO, Dobson SL. Molecular discrimination of *Wolbachia* in the *Culex pipiens* complex: evidence for variable bacteriophage hyperparasitism. Insect Mol Biol 2004;13 (4):365−9.

[58] Sinkins SP, et al. *Wolbachia* variability and host effects on crossing type in *Culex* mosqui-toes. Nature 2005;436(7048):257−60.

[59] Pan X, et al. *Wolbachia* induces reactive oxygen species (ROS)-dependent activation of the Toll pathway to control dengue virus in the mosquito *Aedes aegypti*. Proc Natl Acad Sci USA 2012;109(1):E23−31.

[60] Turley AP, Zalucki MP, O'Neill SL, McGraw EA. Transinfected *Wolbachia* have minimal effects on male reproductive success in *Aedes aegypti*. Parasit Vectors 2013;6:36.

[61] McMeniman CJ, Hughes GL, O'Neill SL. A *Wolbachia* symbiont in *Aedes aegypti* disrupts mosquito egg development to a greater extent when mosquitoes feed on nonhu-man versus human blood. J Med Entomol 2011;48(1):76−84.

[62] Moreira LA, et al. Human probing behavior of *Aedes aegypti* when infected with a life-shortening strain of *Wolbachia*. PLoS Negl Trop Dis 2009;3(12):e568.

[63] Evans O, et al. Increased locomotor activity and metabolism of *Aedes aegypti* infected with a life-shortening strain of *Wolbachia pipientis*. J Exp Biol 2009;212(Pt. 10):1436−41.

[64] Suh E, Dobson SL. Reduced competitiveness of *Wolbachia* infected *Aedes aegypti* larvae in intra- and inter-specific immature interactions. J Invertebr Pathol 2013;114(2):173−7.

[65] McMeniman CJ, O'Neill SL. A virulent *Wolbachia* infection decreases the viability of the dengue vector *Aedes aegypti* during periods of embryonic quiescence. PLoS Negl Trop Dis 2010;4(7):e748.

[66] Walker T, et al. The wMel *Wolbachia* strain blocks dengue and invades caged *Aedes aegypti* populations. Nature 2011;476(7361):450−3.

[67] Calvitti M, Moretti R, Skidmore AR, Dobson SL. *Wolbachia* strain wPip yields a pattern of cytoplasmic incompatibility enhancing a *Wolbachia*-based suppression strategy against the disease vector *Aedes albopictus*. Parasit Vectors 2012;5:254.

[68]   Blagrove MS, Arias-Goeta C, Di Genua C, Failloux AB, Sinkins SP. A *Wolbachia* wMel transinfection in *Aedes albopictus* is not detrimental to host fitness and inhibits Chikungunya virus. PLoS Negl Trop Dis 2013;7(3):e2152.

[69]   Joshi D, McFadden MJ, Bevins D, Zhang F, Xi Z. *Wolbachia* strain wAlbB confers both fitness costs and benefit on *Anopheles stephensi*. Parasit Vectors 2014;7:336.

[70]   Baldini F, et al. Evidence of natural *Wolbachia* infections in field populations of *Anopheles gambiae*. Nature Commun 2014;5:3985.

[71]   Dobson SL, et al. *Wolbachia* infections are distributed throughout insect somatic and germ line tissues. Insect Biochem Mol Biol 1999;29(2):153−60.

[72]   Zabalou S, et al. Multiple rescue factors within a *Wolbachia* strain. Genetics 2008;178 (4):2145−60.

[73]   Reynolds KT, Hoffmann AA. Male age, host effects and the weak expression or non-expression of cytoplasmic incompatibility in *Drosophila* strains infected by maternally transmitted *Wolbachia*. Genet Res 2002;80(2):79−87.

[74]   Boyle L, O'Neill SL, Robertson HM, Karr TL. Interspecific and intraspecific horizontal transfer of *Wolbachia* in *Drosophila*. Science 1993;260(5115):1796−9.

[75]   Poinsot D, Bourtzis K, Markakis G, Savakis C, Mercot H. *Wolbachia* transfer from *Drosophila melanogaster* into *D. simulans*: host effect and cytoplasmic incompatibility relationships. Genetics 1998;150(1):227−37.

[76]   McGraw EA, Merritt DJ, Droller JN, O'Neill SL. *Wolbachia*-mediated sperm modification is dependent on the host genotype in *Drosophila*. Proc Biol Sci 2001;268(1485): 2565−70.

[77]   Poinsot D, Charlat S, Mercot H. On the mechanism of *Wolbachia*-induced cytoplasmic incompatibility: confronting the models with the facts. Bioessays 2003;25 (3):259−65.

[78]   Bourtzis K, Dobson SL, Braig HR, O'Neill SL. Rescuing *Wolbachia* have been overlooked. Nature 1998;391(6670):852−3.

[79]   Tram U, Sullivan W. Role of delayed nuclear envelope breakdown and mitosis in *Wolbachia*-induced cytoplasmic incompatibility. Science 2002;296(5570):1124−6.

[80]   Bressac C, Rousset F. The reproductive incompatibility system in *Drosophila simulans*: DAPI-staining analysis of the *Wolbachia* symbionts in sperm cysts. J Invertebr Pathol 1993;61(3):226−30.

[81]   Sinkins SP, Gould F. Gene drive systems for insect disease vectors. Nat Rev Genet 2006; 7(6):427−35.

[82]   Toomey ME, Panaram K, Fast EM, Beatty C, Frydman HM. Evolutionarily conserved *Wolbachia*-encoded factors control pattern of stem-cell niche tropism in *Drosophila* ovaries and favor infection. Proc Natl Acad Sci USA 2013;110(26):10788−93.

[83]   Fast EM, et al. *Wolbachia* enhance *Drosophila* stem cell proliferation and target the germline stem cell niche. Science 2011;334(6058):990−2.

[84]   Frydman HM, Li JM, Robson DN, Wieschaus E. Somatic stem cell niche tropism in *Wolbachia*. Nature 2006;441(7092):509−12.

[85]   Kittayapong P, Baisley KJ, Sharpe RG, Baimai V, O'Neill SL. Maternal transmission efficiency of *Wolbachia* superinfections in *Aedes albopictus* populations in Thailand. Am J Trop Med Hyg 2002;66(1):103−7.

[86]   Ahantarig A, Trinachartvanit W, Kittayapong P. Relative *Wolbachia* density of field-collected *Aedes albopictus* mosquitoes in Thailand. J Vector Ecol 2008;33(1):173−7.

[87]   Tortosa P, et al. *Wolbachia* age-sex-specific density in *Aedes albopictus*: a host evolutionary response to cytoplasmic incompatibility?. PLoS One 2010;5(3):e9700.

[88] Wiwatanaratanabutr I, Kittayapong P. Effects of crowding and temperature on *Wolbachia* infection density among life cycle stages of *Aedes albopictus*. J Invertebr Pathol 2009;102(3):220−4.

[89] Kambris Z, et al. *Wolbachia* stimulates immune gene expression and inhibits plasmodium development in *Anopheles gambiae*. PLoS Pathog 2010;6:10.

[90] Mousson L, et al. The native *Wolbachia* symbionts limit transmission of dengue virus in *Aedes albopictus*. PLoS Negl Trop Dis 2012;6(12):e1989.

[91] Glaser RL, Meola MA. The native *Wolbachia* endosymbionts of *Drosophila melanogaster* and *Culex quinquefasciatus* increase host resistance to West Nile virus infection. PLoS One 2010;5(8):e11977.

[92] Hedges LM, Brownlie JC, O'Neill SL, Johnson KN. *Wolbachia* and virus protection in insects. Science 2008;322(5902):702.

[93] Teixeira L, Ferreira A, Ashburner M. The bacterial symbiont *Wolbachia* induces resistance to RNA viral infections in *Drosophila melanogaster*. PLoS Biol 2008;6(12):e2.

[94] Lu P, Bian G, Pan X, Xi Z. *Wolbachia* induces density-dependent inhibition to dengue virus in mosquito cells. PLoS Negl Trop Dis 2012;6(7):e1754.

[95] van den Hurk AF, et al. Impact of *Wolbachia* on infection with chikungunya and yellow fever viruses in the mosquito vector *Aedes aegypti*. PLoS Negl Trop Dis 2012;6(11): e1892.

[96] Brennan LJ, Keddie BA, Braig HR, Harris HL. The endosymbiont *Wolbachia pipientis* induces the expression of host antioxidant proteins in an *Aedes albopictus* cell line. PLoS One 2008;3(5):e2083.

[97] Molina-Cruz A, et al. Reactive oxygen species modulate *Anopheles gambiae* immunity against bacteria and *Plasmodium*. J Biol Chem 2008;283(6):3217−23.

[98] Kumar S, et al. The role of reactive oxygen species on *Plasmodium* melanotic encapsulation in *Anopheles gambiae*. Proc Natl Acad Sci USA 2003;100(24):14139−44.

[99] Cirimotich CM, et al. Natural microbe-mediated refractoriness to *Plasmodium* infection in *Anopheles gambiae*. Science 2011;332(6031):855−8.

[100] Rances E, Ye YH, Woolfit M, McGraw EA, O'Neill SL. The relative importance of innate immune priming in *Wolbachia*-mediated dengue interference. PLoS Pathog 2012;8(2):e1002548.

[101] Caragata EP, Rances E, O'Neill SL, McGraw EA. Competition for amino acids between *Wolbachia* and the mosquito host, *Aedes aegypti*. Microb Ecol 2014;67(1):205−18.

[102] Caragata EP, et al. Dietary cholesterol modulates pathogen blocking by *Wolbachia*. PLoS Pathog 2013;9(6):e1003459.

[103] Dodson BL, et al. *Wolbachia* enhances West Nile virus (WNV) infection in the mosquito *Culex tarsalis*. PLoS Negl Trop Dis 2014;8(7):e2965.

[104] Hughes GL, Vega-Rodriguez J, Xue P, Rasgon JL. *Wolbachia* strain wAlbB enhances infection by the rodent malaria parasite *Plasmodium berghei* in *Anopheles gambiae* mosquitoes. Appl Environ Microbiol 2012;78(5):1491−5.

[105] Zele F, et al. *Wolbachia* increases susceptibility to *Plasmodium* infection in a natural system. Proc Biol Sci 2014;281(1779):20132837.

[106] Graham RI, Grzywacz D, Mushobozi WL, Wilson K. *Wolbachia* in a major African crop pest increases susceptibility to viral disease rather than protects. Ecol Lett 2012;15 (9):993−1000.

[107] Hoffmann AA, Hercus M, Dagher H. Population dynamics of the *Wolbachia* infection causing cytoplasmic incompatibility in *Drosophila melanogaster*. Genetics 1998;148(1): 221−31.

[108] Hoffmann AA, et al. Stability of the wMel *Wolbachia* infection following invasion into *Aedes aegypti* populations. PLoS Negl Trop Dis 2014;8(9):e3115.

[109] Caragata EP, Rancès E, O'Neill SL, McGraw EA. Competition for amino acids between *Wolbachia* and the mosquito host, *Aedes aegypti*. Microb Ecol 2014;67(1):205−18.

[110] Yeap HL, et al. Assessing quality of life-shortening *Wolbachia*-infected *Aedes aegypti* mosquitoes in the field based on capture rates and morphometric assessments. Parasit Vectors 2014;7:58.

[111] Dobson SL, Rattanadechakul W, Marsland EJ. Fitness advantage and cytoplasmic incompatibility in *Wolbachia* single- and superinfected *Aedes albopictus*. Heredity (Edinb) 2004;93(2):135−42.

[112] Blagrove MS, Arias-Goeta C, Failloux AB, Sinkins SP. *Wolbachia* strain wMel induces cytoplasmic incompatibility and blocks dengue transmission in *Aedes albopictus*. Proc Natl Acad Sci USA 2012;109(1):255−60.

[113] Islam MS, Dobson SL. *Wolbachia* effects on *Aedes albopictus* (Diptera: Culicidae) immature survivorship and development. J Med Entomol 2006;43(4):689−95.

[114] Gavotte L, Mercer DR, Stoeckle JJ, Dobson SL. Costs and benefits of *Wolbachia* infection in immature *Aedes albopictus* depend upon sex and competition level. J Invertebr Pathol 2010;105(3):341−6.

[115] Champion de Crespigny FE, Wedell N. *Wolbachia* infection reduces sperm competitive ability in an insect. Proc Biol Sci 2006;273(1593):1455−8.

[116] Peng Y, Nielsen JE, Cunningham JP, McGraw EA. *Wolbachia* infection alters olfactory-cued locomotion in *Drosophila* spp. Appl Environ Microbiol 2008;74(13):3943−8.

[117] de Crespigny FE, Pitt TD, Wedell N. Increased male mating rate in *Drosophila* is associated with *Wolbachia* infection. J Evol Biol 2006;19(6):1964−72.

[118] Turley AP, Moreira LA, O'Neill SL, McGraw EA. *Wolbachia* infection reduces blood-feeding success in the dengue fever mosquito, *Aedes aegypti*. PLoS Negl Trop Dis 2009; 3(9):e516.

[119] Ferguson NM, et al. *Wolbachia*-mediated interference of dengue virus infection in *Aedes aegypti* and its projected epidemiological impact on transmission. Sci Transl Med 2015;7 (279):279ra237.

[120] Carvalho DO, et al. Mass production of genetically modified *Aedes aegypti* for field releases in Brazil. J Vis Exp 2014;83:e3579.

[121] Wyss JH. Screw-worm eradication in the Americas − overview. In: Tan KH, editor. Area-wide control of fruit flies and other insect pests. Penang: Penerbit Universiti Sains Malaysia; 2000. p. 79−86.

[122] Caceres C. Mass rearing of temperature sensitive genetic sexing strains in the Mediterranean fruit fly (*Ceratitis capitata*). Genetica 2002;116(1):107−16.

[123] Dobson SL, Fox CW, Jiggins FM. The effect of *Wolbachia*-induced cytoplasmic incompatibility on host population size in natural and manipulated systems. Proc R Soc Lond B Biol Sci 2002;269(1490):437−45.

[124] Yeap HL, et al. Dynamics of the "popcorn" *Wolbachia* infection in outbred *Aedes aegypti* informs prospects for mosquito vector control. Genetics 2011;187(2):583−95.

[125] Rasic G, Endersby NM, Williams C, Hoffmann AA. Using *Wolbachia*-based release for suppression of *Aedes* mosquitoes: insights from genetic data and population simulations. Ecol Appl 2014;24(5):1226−34.

[126] Brelsford CL, St Clair W, Dobson SL. Integration of irradiation with cytoplasmic incompatibility to facilitate a lymphatic filariasis vector elimination approach. Parasit Vectors 2009;2(1):38.

Chapter 15

Employing the Mosquito Microflora for Disease Control

Benjamin J. Blumberg, Sarah M. Short and George Dimopoulos
W. Harry Feinstone Department of Molecular Microbiology and Immunology and the Johns Hopkins Malaria Research Institute, Bloomberg School of Public Health, Johns Hopkins University, Baltimore, MD, USA

INTRODUCTION

Aedes and *Anopheles* mosquitoes are a continuous public health threat to human beings due to their ability to transmit dangerous helminth, protozoan, and viral pathogens. Among the plethora of pathogens transmitted by these dipterans are *Aedes*-vectored dengue virus and *Anopheles*-vectored *Plasmodium*, which together account for hundreds of millions of disease cases per year [1−3]. A susceptible mosquito acquires human pathogens when it takes a blood meal from an infected host. In order for transmission to occur, the pathogens must cross a series of physical barriers, often replicate, and then disseminate within the mosquito in order to reach the medium of transmission, which in most cases is the mosquito salivary gland. This process can take several days, and during this period the mosquito immune system coordinates antipathogen defenses in various tissues. The first tissue encountered by ingested pathogens is the mosquito midgut, and in the case of the malaria parasite *Plasmodium falciparum*, this tissue is often cited as the most important barrier to pathogen transmission because it represents a bottleneck where the numbers of pathogens are at their lowest count during their reproductive cycle [4]. For this reason, immune defense in the mosquito gut and the nature of interactions between pathogens and the mosquito immune system is an important area of research for vector-borne disease control. It is becoming increasingly clear, however, that these host−pathogen interactions are shaped by microorganisms that reside in the same tissues invaded by pathogens. These microorganisms can dramatically influence the outcome of infection either by altering mosquito immune system activity and/or by directly affecting pathogen biology [5]. The collective genomes of the microflora are defined as the microbiome, and these along with the host's

Genetic Control of Malaria and Dengue.
© 2016 Elsevier Inc. All rights reserved.

genome form the mosquito holobiont, which is a dynamic genetic unit potentially responsive to selective pressure [6]. The importance of nonpathogenic microbes in shaping pathogen infection suggests that it is more appropriate to reframe host—pathogen interactions as tripartite interactions among the mosquito, the pathogen, and the mosquito's resident microorganisms or microflora [7—9]. These reciprocal tripartite relationships can be exploited through manipulation of the mosquito microflora, which has evolved into a promising avenue of research for disrupting mosquito-transmitted human pathogens.

The Mosquito Innate Immune System

The mosquito innate immune system is a key mediator of tripartite interactions among the mosquito, pathogens, and the mosquito microflora. Pattern recognition receptors (PRRs) recognize microbial pathogen-associated molecular patterns, and these PRRs then activate intracellular signaling cascades that result in the translocation of transcription factors to the nucleus [10]. These transcription factors regulate the production of immune gene mRNAs which code for PRRs, components of antipathogen effector mechanisms, and immune activity regulators [10]. Although seemingly simple, the actual mosquito immune response to a pathogen is quite complex. Despite this complexity, molecular approaches have enabled the dissection of the pathways involved in controlling mosquito innate immunity.

The major immune pathways in the mosquito are the Toll, Imd, and Jak/Stat pathways [11]. The Toll pathway is activated by Gram-positive bacteria, fungi, viruses, and the rodent parasite *Plasmodium berghei* [10,12—15]. The immunodeficiency or IMD pathway is primarily activated in response to the presence of Gram-negative bacteria as well as the human malaria parasite *P. falciparum* [16—18]. The Jak/Stat pathway is involved in defense against viruses and fungi [10,14,19]. The RNA interference or RNAi pathway is an endogenous pathway that regulates host mRNA transcript abundance, but it can also function in antiviral defense by targeting viral mRNAs for destruction [20]. Though the respective roles of these pathways have been elucidated through careful molecular approaches, the pathways and their responsiveness to pathogens still requires further dissection. For instance, the JNK pathway is found downstream of the IMD adapter protein and has also been linked to anti-*Plasmodium* defense [21]. The Jak/Stat pathway has also been shown to function in a late-stage anti-*Plasmodium* immune response [22]. In addition to these immune signaling pathways, the mosquito immune defense also includes physical barriers such as the peritrophic matrix (PM). The PM is a thick yet permeable barrier composed of cross-linked chitin and glycoproteins that completely surrounds an ingested blood meal in the midgut [23]. In addition to regulating the movement of molecules to and from the blood meal, the PM is also capable of limiting *Plasmodium* invasion [23]. The mosquito immune system also includes phagocytosis by hemocytes (blood cells), the production of reactive oxygen and nitrogen species, and melanotic encapsulation [10,24–26].

THE MOSQUITO'S BACTERIOME

The bacterial microbiota plays an important role in modulating physiological processes in the mosquito including the outcome of pathogen infection. We will refer to all the bacteria that share or occupy space within or on the body of the mosquito as the microbiota. Much like humans, bacteria reside in the digestive organs of mosquitoes as well as other tissues [27]. The microbes found in the mosquito gut are of particular interest because they share a physical space with invading human pathogens. They therefore have the potential to affect infection success either through direct interactions with pathogenic organisms or by eliciting mosquito immune system signaling.

While our understanding of the factors determining mosquito microbiota composition remains incomplete, important insights are gradually emerging. A meta-analysis of multiple bacterial sampling studies showed that the mosquito microbiota differs to some degree between mosquito species [6]. However, it is also highly variable between individuals of the same species or population, suggesting that environmental or physiological factors are likely to be important in shaping the bacterial gut microbiota [28–30]. Mosquito larvae encounter microorganisms in their aqueous environment and these microbes are inevitably introduced into the mosquito gut when larvae consume microbes for food [6]. Similarly, adult mosquitoes can ingest microbes from larval water as they eclose from the pupal casing or as adults during nectar feeding, as flower nectar contains many bacteria commonly found in adult mosquitoes [31–34]. Differences in the larval or adult environment could therefore influence which microbes are introduced into the mosquito gut. For example, *Aedes* utilize a wide variety of water sources for larval development, whereas *Anophelines* prefer clear water, which may account for differences in the gut microbiota composition between species [6,35]. Collection location has been reported to be a major determinant of gut microbe composition within species, and it has been suggested that this may largely be due to differences in vegetation type or vertebrate host availability between populations [29,30].

Physiological processes, such as molting, sugar and/or blood feeding, and reproduction also have the potential to exert influence on which microbes are introduced and can persist within the mosquito. As larvae develop, bacteria can persist between larval stages and from larvae to pupae, though it is not clear in all cases whether this is due to transstadial transmission or to reingestion of microbes from shared water sources

[31,36]. During metamorphosis from pupae to adult, the number of bacteria in the gut is markedly reduced, and the composition is also dramatically altered [37−42]. Multiple reports indicate shared bacterial species between pupae and adults, suggesting that some bacteria may persist during this developmental transition [39−41]. However, it is unclear whether bacteria that persist into adulthood are transstadially transmitted or simply reacquired after eclosion through adult ingestion of breeding water [31]. As adults, feeding behavior can also dramatically affect the gut microbiota. Sugar feeding has been shown to alter the microbial composition of the gut; in one study, a single sterile sugar meal resulted in a slight reduction in bacterial diversity, with a corresponding increase in bacteria from Flavobacteriaceae and a decrease in bacteria from Enterobacteriaceae [39]. Blood feeding results in dramatic increases in bacterial number in the female gut, though it also causes a corresponding decrease in bacterial diversity [39,43−45]. Reproduction can also act as an avenue for acquisition of gut bacteria. For example, *Asaia* bacteria can be sexually transmitted from males to females and after sexual transmission are found in the female midgut [36,46]. There is also evidence that bacteria can be passed vertically from mother to offspring, either through egg-smearing, transovarial transmission, or maternal inoculation of larval water [36,47,48].

Efforts to catalog the bacteria present in the digestive tracts of field and laboratory mosquitoes from diverse species, populations, and collection locales have provided an important initial characterization of the mosquito gut microbiota. Traditionally, the identification of mosquito-associated bacteria has relied on culture-dependent methods, which has high utility but can fail to isolate some uncultivable specimens [9,11,42,44,45,49−51]. The use of culture-independent pyrosequencing of bacterial 16s rRNA has revealed substantial diversity in mosquito-associated bacteria as well as shifts in bacterial composition in relation to life stage or feeding status [28,30,39,40,52]. Recent studies have identified hundreds of Operational Taxonomic Units (OTUs) as being associated with the mosquito gut, though the vast majority of these OTUs are often found at very low frequency [28,39]. In fact, adult mosquitoes generally harbor low diversity microbial communities dominated by a few taxa [28,30,39]. Overall diversity in the mosquito gut microbiota may arise primarily from interindividual differences, as individual mosquitoes from the same species or population can have limited overlap in their gut microbial composition [9,28]. Despite the considerable diversity, these studies also reveal some interesting trends that appear to be generally consistent across individuals. Many studies have shown that the adult mosquito microbiota is often dominated by the phylum Proteobacteria [6,7,28−30,39,41,52−55]. Bacteria from the phylum Bacteroidetes are also commonly present, though this result seems to be primarily attributable to the presence of two genera: *Chryseobacterium* and *Elizabethkingia*

[28,39,41]. Firmicutes are also found in mosquito guts, though in adult females they represent only a minor component of the gut microbiota [28,30,39]. A few genera of Proteobacteria in particular are commonly reported to reside in the guts of adult mosquitoes, including *Pseudomonas*, *Enterobacter*, *Serratia*, and *Acinetobacter* [28–30,39–44,49,51,53–56]. Interestingly, *Pantoea*, *Acinetobacter*, *Asaia*, and *Enterobacter* tend to be prevalent in both *Aedes* and *Anophelines* [6,28,44,55,57]. This finding may represent constraints imposed upon microbes by the mosquito host environment or perhaps indicate that different mosquito species acquire microbes from similar sources. In some cases, it appears that field-derived mosquitoes harbor a greater diversity of bacteria than laboratory strains [30,41]. However, a recent study in *Anopheles* suggests that dominant bacterial families are found in both lab and field-derived mosquitoes [39]. Further studies are therefore needed to recapitulate the differences in microbiota composition between lab and field strains.

The Mosquito's Bacteriome Modulates Pathogen Infection

The mosquito microbiota has been shown in many studies to be an important determinant of pathogen infection outcome. For example, reduction of the bacteria in the gut through antibiotic treatment results in increased susceptibility to *P. falciparum* and dengue virus infection [9,15], and ingestion of certain bacterial species can result in dramatically reduced infection levels [7,9,11,15,58–60]. Interestingly, in some studies, the presence of certain bacterial species in the gut correlated with increased susceptibility to *P. falciparum* and dengue virus infection, suggesting that the effect of gut bacteria on pathogen infection is complex and may depend on the specific composition of the gut microbiota [30,61]. In order to cause reduced infection levels, bacteria may interact directly and/or indirectly with invading pathogens. One way gut microbes indirectly affect pathogen survival is by modulating mosquito immune system activity. The presence of midgut bacteria results in basal immune stimulation, and reduction of the adult mosquito midgut microbiota by antibiotic treatment results in decreased production of antipathogen immune effector molecules [7,9]. This may in turn cause the increase in pathogen susceptibility mentioned above [9]. Multiple species of bacteria have been shown to cause increased expression of immune system effector genes when present in the midguts of both *Anopheles* and *Aedes* mosquitoes [7,58]. In fact, the initiation of many anti-*Plasmodium* immune defenses correlate with the rapid expansion of midgut bacteria in response to increased nutrients from an ingested blood meal [10,43]. Additional evidence that bacteria are capable of indirectly antagonizing a pathogen by activating the mosquito immune system comes from research on peptidoglycan recognition receptor protein-LC (PGRP-LC), an extracellular PRR upstream of the

IMD pathway [8]. Work by Meister et al. showed that PGRP-LC-mediated signaling results in the production of antimicrobial effector molecules in response to gut bacteria proliferation after blood feeding [8]. Knockdown of PGRP-LC results in higher *Plasmodium* infection intensities but antibiotic treatment eliminates this effect. This suggests that PGRP-LC mediates *Plasmodium* infection in a gut microbe-dependent manner, possibly due to the production of antimicrobial effectors that are active against both bacteria and *Plasmodium* [7,8,10].

A second way the gut microbiota can antagonize mosquito pathogens is through direct interaction with the pathogens themselves. Bacteria are capable of producing a variety of proteins and metabolites that have varying effects on vector competence [62]. Recently, multiple species of bacteria with anti-*Plasmodium* as well as anti-dengue activity have been identified [7,11,58,63]. Importantly, many of these bacteria not only inhibit pathogen infection in the mosquito but also cause pathogen killing *in vitro*, suggesting that the bacteria produce antipathogen molecules that directly inhibit pathogen survival [7,11,58,63]. Cirimotich et al. showed that a strain of *Enterobacter* sp. (*Esp_Z*) isolated from an *Anopheles arabiensis* mosquito in Zambia caused dramatic reductions in *P. falciparum* development when present in *Anopheles gambiae* midguts. This antipathogen activity was also evident *in vitro*, suggesting that the bacteria produce molecule(s) with antipathogen activity. The authors hypothesized that reactive oxygen species may be one such molecule, and indeed addition of an antioxidant to *Plasmodium* culture reduced the antipathogen activity of *Esp_Z*, supporting this hypothesis. Bahia et al. described multiple species of gut bacteria that had anti-*Plasmodium* activity *in vitro*, including a strain of *Serratia marcescens* that caused very strong pathogen mortality. Antipathogen activity of *S. marcescens* was also detectable in the bacterial culture media after removal of the bacteria via filtration, suggesting that *S. marcescens* produces and secretes antipathogen molecule(s). Ramirez and Short et al. reported broad antipathogen activity of a *Chromobacterium* sp. (*Csp_P*) isolated from an *Aedes* mosquito in Panama [7,63]. They showed that exposure of dengue virus to *Csp_P in vitro* almost completely abolished viral infectivity in both mammalian and insect cells. They also showed that filtered (i.e., bacteria-free) *Csp_P* culture had *in vitro* anti-*Plasmodium* activity against asexual, ookinete, and gametocyte stage parasites. Because the anti-*Plasmodium* activity persisted even after *Csp_P* was removed from culture, the authors concluded that pathogen inhibition was likely due to secreted, stable molecule(s) produced by *Csp_P* and released into the media [63]. These investigations serve to demonstrate that microbes naturally associated with mosquitoes can have powerful antipathogen activity. An important next step will involve investigating ways to introduce and successfully maintain these or other naturally protective microbes in mosquito field populations.

Paratransgenesis of Bacteria

Paratransgenesis is the process of genetically transforming an organism's symbionts to confer specific function(s) that decrease vector competence to pathogens. In order for paratransgenesis to be successful, a symbiont must successfully associate with the vector in the field, be genetically tractable, pose no fitness cost to the vector, be able to produce and secrete the anti-pathogen product of an inserted transgene, and to be able to colonize the vector population [64]. This process is undoubtedly made easier if the vector has obligate symbionts that are required for its survival. An example of an obligate symbiont is the tsetse fly-associated bacteria *Wigglesworthia*, which produces certain vitamins required by tsetse flies, and is vertically transmitted from parent to offspring as an obligate mutualist [65]. Unfortunately, *Wigglesworthia* is not an ideal candidate for paratransgenesis because it is extremely difficult to genetically manipulate *in vitro*. An example of an obligate symbiont that is genetically tractable is the bacteria *Rhodococcus rhodnii*. *Rhodococcus rhodnii* has been genetically modified to produce an antimicrobial peptide that antagonizes the etiological agent of Chagas disease, *Trypanasoma cruzi*, in the gut of the triatome vector *Rhodnius prolixus* [66]. *Rhodnius rhodnii* is essential for the survival of *R. prolixus*, which acquires its symbionts as a nymph through consumption of triatome feces. In a proof-of-principle study, a transgene encoding an antibody fragment was inserted into *R. rhodnii*, which in turn was stably integrated into the host microbiota by placing the transformed bacteria into feces, which were then consumed by *R. prolixus* nymphs [67]. The transformed *R. rhodnii* successfully secreted the antibody fragment into the midgut of *R. prolixus* demonstrating this approach is feasible.

Paratransgenic approaches utilizing antipathogen effector molecules to reduce vector competence for pathogens are reliant on a toolbox of effectors that antagonize the pathogen without compromising the host's fitness. Such effectors have been identified through the dissection of the mosquito immune response to pathogens, by high-throughput peptide screens such as phage-display libraries, and even from investigations exploring the components of animal toxins [68−70]. Useful effectors include those that kill the pathogen and molecules that interact with either the pathogen or vector to block entry of the pathogen into specific cells or tissues. A number of anti-*Plasmodium* effector molecules have been identified, and include mosquito innate immune peptides (e.g., defensins, cecropins), lytic peptides derived from other organisms (e.g., scorpine, a component of scorpion toxin), synthetic lytic peptides (e.g., Shiva-1), peptides that bind to the parasite or parasite-produced factors (e.g., enolase−plasminogen interaction peptide or EPIP, antibody single-chain variable fragments or scFvs), and peptides that bind to mosquito receptors blocking uptake of parasites (e.g., salivary gland and midgut peptide 1 or SM1, mutant phospholipase A2 or mPLA2) [68−77]. The effectors scorpine and EPIP have been successfully expressed in the mosquito-associated bacteria *Pantoea agglomerans*, and these molecules strongly inhibit parasite development in the mosquito [72].

In regard to dengue virus, fewer antiviral effectors have been described and tested in proof-of-principle studies. Similar to *Plasmodium*, a number of antimicrobial effectors are produced by the mosquito upon dengue virus infection, and effectors that efficiently antagonize dengue virus could be exploited in a paratransgenic approach [12,14,15]. Recently, a cecropin-like peptide from dengue virus-infected salivary glands was reported to have strong anti-dengue, anti-chickungunya, and antibacteria activity [78]. Dengue subverts C-type lectins (CTLs) to gain entry into mosquito cells and antibodies against these CTLs may block uptake of the virus [79]. The fact that arboviruses are not required to undergo lengthy developmental transitions in the midgut lumen prior to midgut cell infection limits the use of paratransgenic approaches using extracellular bacteria.

The toolbox of antiparasitic and antiviral molecules is growing and some of these effectors are likely to be compatible with a successful paratransgenic disease control strategy. Perhaps, the most important and rate-limiting step in a paratransgenic approach is the identification and characterization of bacterial candidates for genetic modification.

Bacterial Candidates for Paratransgenesis

Unlike *Wigglesworthia* in tsetse flies and *R. rhodnii* in *R. prolixus*, a true obligate symbiont has yet to be identified in *Aedes* or *Anopheles* mosquitoes. The lack of an obligate symbiont in mosquitoes presents a unique challenge to any paratransgenic approach in regard to the natural maintenance and spread of a transgenic microbe. Despite this, paratransgenesis research is advancing and promising bacterial candidates for use in paratransgenesis are emerging. In fact, a number of bacteria have been explored as candidate vectors for transgenes encoding antipathogen effectors. A successful paratransgenic bacterium must be able to harbor the genetic code for an effector molecule, express, display or excrete the molecule, and survive in the mosquito long enough for the molecule to be able to exert its effects against the pathogen [80]. As a proof-of-principle, a transgene encoding either the SM1 peptide or mPLA2 was inserted into *Escherichia coli*, and ingestion of these transformed *E. coli* by *Anopheles stephensi* caused a significant increase in mosquito resistance to *P. berghei* [80]. Although this study successfully demonstrated the concept of interrupting *Plasmodium* development in the mosquito using genetically modified bacteria, the transformed *E. coli* did not survive long enough in the mosquito to be a viable strategy [80]. Furthermore, the effector molecules produced by these genetically modified *E. coli*, while active, were not efficiently secreted and remained in close proximity to the bacteria [80]. In another study, *E. coli* were transformed with green fluorescent protein and observed to be transstadially transmitted and maintained in the adult midgut for up to 13 days posteclosion [81]. The differences in these findings between the two studies could be due to differences in the mosquito strain used, *E. coli* strain used, or perhaps due to other differences not readily apparent [81]. Although *E. coli* may never be a top candidate for blocking *Plasmodium* infection in the field, it can be used in the laboratory to readily test the expression of antipathogen effectors.

Genetic modification of common gut bacteria of *Aedes* and *Anopheles* mosquitoes may be a more successful strategy for paratransgenesis. The gammaproteobacteria *P. agglomerans* is found in *Aedes* and *Anopheles* midguts across the world, and this bacterium has been identified as a candidate for paratransgenesis [56,82]. A strain of *P. agglomerans* adapted for increased persistence in the mosquito was hypothesized to be genetically tractable in regard to the insertion and expression of transgenes [80,83]. Indeed, using secretion signals from closely related bacterial species including *E. coli*, researchers were able to transform *P. agglomerans* with plasmids encoding anti-*Plasmodium* effectors and these modified bacteria survived in the laboratory as well as wild-type bacteria [83]. Building on this work, another research group transformed *P. agglomerans* to produce anti-*Plasmodium* effectors and then introduced the modified bacteria into *Anopheles* mosquitoes via a sugar meal prior to *Plasmodium* infection [72]. These transformed *Pantoea* blocked the development of both *P. berghei* as well as *P. falciparum* and significantly decreased the prevalence of *Plasmodium* in the mosquito [72]. Though a successful experiment, this system requires introduction of the bacteria into the mosquito midgut each generation. An even more attractive candidate for paratransgenesis is a bacterium that can be passed from one mosquito generation to the next.

Although no obligate bacterial symbiont has been identified in mosquitoes, there are a few candidates that appear to be vertically transmitted from mother to offspring. The gammaproteobacteria *Serratia* is found stably associated with both *Aedes* and *Anopheles*, and in other insect systems it has been found to be vertically transmitted [41,61,84]. A strain of *Serratia* was shown to efficiently colonize the midguts of *Anopheles* mosquitoes, block *Plasmodium* development, and also cause insecticidal activity [58]. Genetic modification of *Serratia* may be most preferable for interrupting *Plasmodium* transmission because recently *Serratia* has been shown to actually increase dengue and chikungunya infection in mosquitoes [61,85]. The insecticidal activity of some *Serratia* strains could create strong selective pressure for mosquito resistance. Therefore, more studies are necessary to characterize *Serratia* strains with reduced insecticidal activity, which may be more optimal for paratransgenic applications. Alternatively, pathogenic strains of *Serratia* could be explored for applied vector control strategies.

Also promising are bacteria from the genus *Asaia*. *Asaia* are Gram-negative alphaproteobacteria found throughout the world and are thought to be associated with flowering plants [86]. Similar to *Serratia*, *Asaia* are also found in both *Aedes* and *Anopheles* mosquitoes [36,47,51,87]. In fact, *Asaia* is stably associated with *An. stephensi*, can be sexually transmitted from males to females, and is vertically transmitted from mother to offspring [36,46,50]. These positive qualities are enhanced by the finding that *Asaia* transformed with green fluorescent protein localize to the midgut and salivary glands of adult female *An. stephensi* [36]. *Asaia* has yet to be modified to express antipathogen effector molecules in the context of pathogen infection, but it may be the most promising bacteria for paratransgenesis because it meets most of the requirements for successful paratransgenesis.

Conclusion

Progress in the study of mosquito–microbiota and mosquito–pathogen interactions has allowed paratransgenesis research to advance in earnest. Promising results in proof-of-principle experiments have demonstrated that pathogen transmission can be reduced through genetic manipulation of the mosquito microbiota. Furthermore, bacteria have already been introduced into mosquitoes that were subsequently released into the environment [88,89]. Nevertheless, there may be challenges in utilizing genetically modified bacteria to reduce the prevalence of mosquito-borne disease. For instance, the ability of target pathogens to develop resistance to chosen effector mechanisms should be explored in greater detail. To date, no obligate symbiont of mosquitoes has been identified. An obligate mosquito symbiont is desired by researchers wishing to employ a paratransgenic strategy similar to genetically modified *R. rhodnii* in triatomes [47]. Genetically modified bacteria have to be introduced into the mosquito, and more studies are required to determine the optimal route of introduction. Similarly, more research is necessary to determine if genetically modified bacteria will be competitive against wild-type bacteria. Despite the challenges, bacterial paratransgenesis is a promising approach in the war against mosquito-borne disease and addressing these challenges in the laboratory will refine the methodology behind field deployment of paratransgenic organisms.

THE MOSQUITO FUNGAL MYCOBIOTA

Fungi are eukaryotic organisms with distinctive cell walls composed of chitin that can be found in nearly every ecosystem across the world. Many species are only observed in their asexual form producing nonmotile spores called conidia [90]. Common fungi include the molds and yeasts, such as the medically important genus *Aspergillus* and the commonly used baker's yeast *Saccharomyces* [91,92]. Fungi play diverse roles in the ecosystem spanning from commensal to mutualist to pathogenic organisms [93−95]. Fungi are capable of producing a rich array of proteins and secondary metabolites, which allows them to adapt to changing environmental conditions. Some fungi produce toxic secondary metabolites, called mycotoxins, which are extremely resistant to environmental- and temperature-mediated degradation [96]. Most mycotoxins are immunosuppressive and mycotoxin-producing fungi are capable of modulating pathogen susceptibility in animals [96,97]. The ability to adapt to diverse environments allows some fungi to become opportunistic pathogens. For example, fungi belonging to the genus *Aspergillus* are predominantly soil-dwelling saprotrophs, meaning they feed on decaying organic matter, but in the human lung *Aspergillus* can become pathogenic [98]. Although the term microbiota is primarily used to describe bacteria, fungi are frequently identified in genetic screens of microbiome samples, albeit their numbers tend to be much lower than bacteria [94]. Still, these fungi are capable of modifying the immune response to pathogens as well as the population size of other microbiota [94,99]. Recently, there has been renewed interest in the role of fungi in the mycobiota of humans and other organisms such as mosquitoes [94,100,101]. These endogenous fungi are collectively referred to as the mycobiota.

There are far fewer studies exploring the mosquito mycobiota as opposed to the bacterial microbiota, and the effects of fungi on pathogen susceptibility in mosquitoes are largely unknown. Most of the research addressing mosquito−fungi interactions is descriptive or focuses on the use of pathogenic fungi or toxic fungal products to control mosquito populations [102−105]. Much like bacteria, mosquitoes are exposed to fungi as larvae in the water, and as adults either by ingesting fungi in sugar meals or by external physical contact with conidia [104,106−108]. There is evidence to suggest that *Penicillium* molds and other filamentous fungi dominate the mosquito mycobiota without any apparent detriment to the health of the mosquito [104,106]. However, other filamentous fungi, such as some species of *Aspergillus* and *Penicillium*, are pathogenic [102,109]. A number of true entomopathogenic fungi parasitize mosquitoes, including fungi that belong to the generas *Beauveria* and *Metarhizium* [110]. These pathogens are capable of reducing mosquito life span as well as vector competence for pathogens [111,112]. In addition to filamentous fungi, many yeasts have been isolated from mosquitoes in both the lab and field [51,100,101,113,114]. It is clear that fungi are naturally associated with mosquitoes and this has warranted investigation into their potential for paratransgenic applications.

Entomopathogenic Fungi for Disease Control

Beauveria bassiana and *Metarhizium anisopliae* are naturally occurring pathogens of mosquitoes, and they have historically been used to control insect pests like caterpillars [108,115]. The conidia or spores of these fungi germinate on the surface of their insect host, penetrate through the cuticle, spread systemically in the hemolymph, and then candidate resulting in the death of the host. Fungal infection causes a reduced propensity for blood feeding and a reduction in pathogen numbers that lowers an infected mosquito's ability to transmit disease [111,116,117]. The spores of these fungi can easily be mixed with oil and then applied to surfaces such as the inside of a house similar to long-lasting insecticides [118−122]. This approach exploits the resting behavior of mosquitoes after a blood meal, which can often be found resting on walls indoors [123]. Some insecticides have a strong repelling effect on mosquitoes, whereas mosquitoes may be attracted to entomopathogenic fungi [93,124]. *Beauveria bassiana* and *M. anisopliae* may be safer than pesticides as they are demonstrably nonpathogenic toward humans with extremely few case reports of disease [125,126]. In contrast to some insecticides, entomopathogenic fungi tend to kill the mosquito slowly. Slow killing is advantageous because it causes only moderate reductions in reproductive fitness, resulting in low selective pressure for resistance [103,127]. Slow killing can still effectively prevent pathogen transmission, because the time from when a mosquito takes an infectious blood meal until it can transmit the pathogen through an infectious bite (the extrinsic incubation period, or EIP) can be days or even weeks [128]. The average EIP for *Plasmodium* is 12−14 days, but in nature, this process can take as long as 30 days [103,127]. *Beauveria bassiana* infection results in high mortality of *An. stephensi* at 14 days after a blood meal [127]. Modeling predicts that most mosquitoes infected with *B. bassiana* shortly after a blood meal would be dead prior to being able to transmit the parasite [108,116]. However, this strategy assumes that mosquitoes become infected with fungi immediately following a blood meal [103,127].

Genetic manipulation of entomopathogenic fungi can improve their efficacy in reducing vector competence. As proof-of-principle, a strain of *M. anisopliae* has been genetically modified to express transgenes that inhibit parasite development in *An. gambiae* [103]. *Metarhizium anisopliae* was manipulated to produce SM1, scorpine or an anti-*Plasmodium* antibody under the control of a promoter expressed upon contact with the mosquito hemolymph [103]. Infection with these transgenic fungi dramatically reduces the number of *Plasmodium* sporozoites in the salivary gland, effectively reducing the likelihood of transmission [103]. The advantage of using transgenic fungi over wild-type strains is that it avoids the requirement of the mosquito becoming infected immediately following a blood meal [103]. Instead, the capacity for transmission is reduced even if the mosquito becomes infected with the fungus at a late stage of *Plasmodium* infection [103]. Multiple lines of transgenic *M. anisopliae* can be created to produce different effector molecules to reduce the evolution of resistance [103]. Furthermore, *M. anisopliae* could be engineered to produce antiviral effector molecules to interrupt dengue virus transmission. In summary, genetic manipulation of entomopathogenic fungi is a highly feasible strategy to decrease mosquito vector competence for pathogens.

Manipulation of Mosquito Yeast Symbionts

Yeasts are found in many insect species and benefit insects by accounting for nutritional dietary deficiency [101,129−132]. Yeasts belonging to the genera *Candida*, *Pichia*, and *Wickerhamomyces* have been identified in *Aedes* and *Anopheles* mosquitoes in both the lab and field [51,113,114]. In nature, yeasts are typically found in sugar-rich environments such as plant nectars, which may be the source of acquisition for mosquitoes [100,114]. These yeasts can be found in the mosquito midgut, which places them in close proximity to invading pathogens. Yeasts do not appear to be pathogenic to mosquitoes, but their role in modulating pathogen susceptibility is unknown. In other biological systems, *Wickerhamomyces anomalus* is used as a biocontrol agent to prevent fungal contamination of grains and produce [133,134]. It is thought that some strains of *W. anomalus* can compete with other microbes by producing a substance known as killer toxin (KT) [135,136]. KTs are known to be active against many microorganisms including *Leishmania* parasites, and future studies on a KT-producing strain of *W. anomalus* should determine if KT is active against *Plasmodium* [101,137]. *Wickerhamomyces anomalus* has also been isolated from the gonads of mosquitoes [114]. These are nutritionally rich tissues involved in egg production, and the presence of *W. anomalus* has been postulated to be indicative of a type of mutualistic role in nutrient regulation [101,130]. The discovery of the yeast outside of the midgut may also indicate that yeasts are able to escape the mosquito immune response through an unknown mechanism [114]. It is unknown if *W. anomalus* is vertically transmitted by mosquitoes, but related *Saccharomyces* yeasts can survive passage through the *Drosophila* midgut [138]. It will be important to study the potential for vertical transmission as this could aid in dispersal of a paratransgenic *W. anomalus*. Mosquitoes are attracted to carbon dioxide which is readily produced by yeast. Traps baited with yeast-produced CO_2 could therefore potentially serve as a cost-effective delivery system for genetically modified yeast [139]. Similar to modified bacteria, *W. anomalus* could be genetically modified to produce antipathogen effector molecules. Related *Pichia* yeasts are highly amenable to genetic manipulation and are frequently used for protein expression [140]. Nevertheless, further research exploring the genetic tractability of *W. anomalus* is necessary.

Conclusion

Although our knowledge of the mosquito mycobiota is quite limited compared to the bacterial microbiota, the prospect of fungi for the control of vector-borne disease is high. Entomopathogenic fungi can be used to shorten the life span of mosquitoes and reduce pathogen numbers in the mosquito. Their ability to reduce mosquito vector competence can be enhanced through genetic manipulation. Similarly, the commensal yeasts that reside in close proximity to pathogens in the mosquito midgut may be tractable to genetic manipulation to reduce mosquito vector competence. Still, further investigations into the environmental and evolutionary impacts of introducing paratransgenic fungi are required. The release of modified fungi into the environment could increase selective pressure for mosquito resistance, thereby reducing their efficacy. It is possible that entomopathogenic fungi could have negative effects in nontarget species. Furthermore, the use of antipathogen transgenes that block parasite infection in the mosquito could select for *Plasmodium* escape mutants. Nevertheless, natural and modified fungi have a high likelihood of being used in any future integrative vector and disease management strategy.

THE MOSQUITO VIROME

Many viruses are capable of infecting mosquitoes, and mosquito-associated virus families include baculoviruses, bunyaviruses, flaviviruses, mesoniviruses, parvoviruses, and rhabdoviruses [141−147]. Human pathogens, such as dengue virus, chikungunya virus, West Nile virus, and Rift Valley fever, are specifically vectored by mosquitoes. Upon infection, the Toll, Jak/Stat, and RNAi pathways participate in mosquito antiviral defense [10]. Viruses are not typically considered part of the microbiome, but viral infection results in modification of the composition of the mosquito microbiota, possibly through the induced immune response [7]. Mosquitoes also serve as exclusive hosts for some viruses, which do not appear to be human pathogens [148]. Some mosquito-specific viruses have been used as tools for elucidating molecular processes and explored for their applicability in mosquito control. A number of viruses including mosquito-specific viruses are capable of being vertically transmitted from adult females to offspring [149]. Densoviruses infect mosquitoes and they can efficiently express protein and RNA products of transgenes both *in vitro* and *in vivo* [150−152]. Mosquito densoviruses (MDVs) are candidates for paratransgenesis because of their genetic tractability and exclusive infection of mosquitoes.

Manipulation of Mosquito Densoviruses

MDVs are members of the *Parvoviridae* family of viruses, which are nonsegmented, single-stranded DNA viruses containing a relatively small genome composed of around 4,000 nucleotides on average [153]. The tiny genome of these viruses makes them genetically tractable because the entire genome can fit into an infectious plasmid amenable to standard cloning techniques [143,154]. Densoviruses replicate in the nucleus of their host cell causing nuclear hypertrophy and create dense masses of virions, hence the origin of the common name densovirus [155,156]. A number of MDVs have been identified as persistent contaminants in cell culture and tend to have minimal cytopathic effects [153,157−159]. These viruses may be adapted to cell culture, given that densoviruses isolated from live mosquitoes produce lower amounts of virus *in vitro* [153,154]. MDVs have been isolated from both *Aedes* and *Anopheles* in all life stages [143,160,161]. Some strains of MDV have been transformed to express green fluorescent protein in order to examine tissue tropism. Overall, MDVs can infect cells in many mosquito tissues although specific viruses may vary in their ability to infect the midgut and salivary glands [152,162]. In general, the viral load of MDV seems to increase with mosquito age and a large proportion of adult mosquitoes remain infected with MDV if they were exposed to the virus as larvae [143,154,160,163]. Similarly, a high proportion of adult mosquitoes that survive a challenge with MDV remain infected [164]. MDVs vary in pathogenicity *in vivo* with some strains causing severe mortality and others causing no difference in mortality between infected and uninfected controls [143,163−165].

Genetic manipulation of MDVs is an attractive option for controlling mosquito populations and in turn mosquito-transmitted diseases. MDVs that reduce the life span of mosquitoes could be genetically modified with different transgenes to improve their efficacy [152,166]. For instance, the insertion of transgenes expressing toxins could increase killing efficiency and reduce selective pressure for resistance to a single mechanism [103,166]. Similarly, MDVs could be engineered to express an RNAi-vector targeting an essential mosquito gene in order to increase killing [167]. RNAi-vectors targeting negative regulators of the immune pathways involved in antipathogen defense would synergize with life-shortening effects to decrease vector competence [167]. Strains of MDV with tropism for the mosquito midgut or salivary gland could be engineered to produce antipathogen effector molecules. This strategy would further reduce the selective pressure imposed by the killing efficiency, similar to what has been modeled for entomopathogenic fungi.

Conclusion

Viral paratransgenesis and in particular paratransgenic MDV is a potentially promising strategy for interrupting dengue and *Plasmodium* transmission. There is a great need to study known viruses to optimize their utility in research and translational applications. Advanced bioinformatic tools will aid in the discovery of new viruses, which may provide additional paratransgenic targets. A possible limitation of this approach is that the genetic makeup of viruses effectively limits the size of any introduced transgenes. Viruses are inefficient at packaging large amounts of genetic material; thus, it may be difficult to create viruses expressing multiple transgenes that is important to prevent the evolution of mosquito resistance to a single mechanism. Nevertheless, the specificity of viruses such as MDV may account for the weakness imposed by its small genome size. Viruses can be produced in devices such as bioreactors, which makes them feasible for mass production, thereby minimizing the cost of such a control strategy [168,169]. Viral paratransgenesis may not be the only solution to interrupting mosquito-borne disease transmission, but it certainly has its place as part of an integrated approach utilizing many strategies.

PROSPECTS, CHALLENGES, AND FUTURE DIRECTIONS

As the war against dengue and malaria expands from traditional control measures (drugs, bed nets, and insecticides) to include next-generation approaches (transgenic mosquitoes and natural and paratransgenic microbes), our capacity to control these diseases also expands. Nevertheless, challenges to utilizing the vector microflora for disease control still remain. There are a plethora of antipathogen effectors that could be utilized in paratransgenic applications, but additional research is necessary to identify those with strong, specific antipathogen activity. A number of microbial candidates for paratransgenesis have been described and some are quite promising (Table 15.1). However, barriers remain before any could be reliably utilized in field applications. Future research improving current strategies and also to identify and characterize more mosquito symbionts is warranted. In theory, the most optimal microbe would be able to spread efficiently throughout a mosquito population with little initial input. In other transgenic models, a genetic drive system can be incorporated along with the transgene in order to force the gene into the target population. In regard to mosquito symbionts, we do not have enough data on specific pressures that could be exploited to provide modified microbes an advantage over their wild-type counterparts. Future studies addressing the physiological processes in mosquitoes that enhance prevalence of specific gut microbes could provide the community with tools to drive transgenic organisms into mosquito populations. Genetically modified microbes or microbes with natural antipathogen activity could be delivered to mosquitoes through water sources or by baiting sugar traps with these organisms. However, multiple introductions in different geographic locations of the target vector may be required. Despite the potential challenges, manipulation of the vector microbiota remains a promising approach to interrupt mosquito-borne disease transmission and has a place in a future, integrated vector management strategy.

TABLE 15.1 Promising Candidates for Microbe-Mediated Control of Mosquito-Borne Diseases

	Associated with Anopheles and/or Aedes	Tissue Localization	Mechanistic Effects	Other Characteristics
Bacteria				
Asaia sp.	Both [29,36,87,170]	Midgut, salivary gland, ovary, eggs [170]		Sexual transmission, vertical transmission [47,57]
Enterobacter sp. Zambia (Esp_Z)	Anopheles [11]	Midgut [11]	Anti-Plasmodium [11,59]	Produces reactive oxygen species, in vitro anti-Plasmodium activity [11]
Serratia marcescens and Serratia sp.	Both [41,171]	Midgut, salivary glands [61,172]	Anti-Plasmodium, enhances dengue virus infection [58,84,85]	Mosquitocidal activity [58]
Pantoea agglomerans	Both [42,51]	Midgut, ovary [51]		Genetically tractable, can efficiently secrete effector molecules [72,83]
Chromobacterium sp. Panama (Csp_P)	Aedes [7]	Midgut [7]	Anti-Plasmodium, anti-dengue, mosquitocidal [7,63]	Has in vitro anti-Plasmodium and anti-dengue activity [63]
Fungi				
Beauveria bassiana	Both [112,116]	Penetrates cuticle, systemic infection [116]	Mosquitocidal [117,173]	Indirectly decreases dengue infection by activating Toll/Jak-Stat pathways [112]

(Continued)

TABLE 15.1 (Continued)

	Associated with *Anopheles* and/or *Aedes*	Tissue Localization	Mechanistic Effects	Other Characteristics
Metarhizium anisopliae	Both [111,116]	Penetrates cuticle, systemic infection [116]	Mosquitocidal [121,122]	Genetically tractable, transgenic strains secrete effector molecules [103]
Wickerhamomyces anomalus (*Pichia anomala*)	Both [100]	Midgut, reproductive tract [101,114]		Produces antimicrobial killer toxins [101]
Viruses				
Densovirus	Both [143,160]	Midgut, salivary gland, ovaries [153]	Inhibits dengue virus replication [174]	Genetically tractable [152], negligible effects on adult mosquito survival [175]

REFERENCES

[1] Hay SI, Okiro EA, Gething PW, et al. Estimating the global clinical burden of *Plasmodium falciparum* malaria in 2007. PLoS Med 2010;7:e1000290.

[2] Murray NEA, Quam MB, Wilder-Smith A. Epidemiology of dengue: past, present and future prospects. Clin Epidemiol 2013;5:299–309.

[3] Gulland A. Death toll from malaria is double the WHO estimate, study finds. BMJ 2012;344:e895.

[4] Smith RC, Vega-Rodríguez J, Jacobs-Lorena M. The *Plasmodium* bottleneck: malaria parasite losses in the mosquito vector. Mem Inst Owaldo Cruz 2014;109:644–61.

[5] Cirimotich CM, Ramirez JL, Dimopoulos G. Native microbiota shape insect vector competence for human pathogens. Cell Host Microbe 2011;10:307–10.

[6] Minard G, Mavingui P, Moro CV. Diversity and function of bacterial microbiota in the mosquito holobiont. Parasit Vectors 2013;6:146.

[7] Ramirez JL, Souza-Neto J, Torres Cosme R, et al. Reciprocal tripartite interactions between the *Aedes aegypti* midgut microbiota, innate immune system and dengue virus influences vector competence. PLoS Negl Trop Dis 2012;6:e1561.

[8] Meister S, Agianian B, Turlure F, et al. *Anopheles gambiae* PGRPLC-mediated defense against bacteria modulates infections with malaria parasites. PLoS Pathog 2009;5: e1000542.

[9] Dong Y, Manfredini F, Dimopoulos G. Implication of the mosquito midgut microbiota in the defense against malaria parasites. PLoS Pathog 2009;5:e1000423.

[10] Cirimotich CM, Dong Y, Garver LS, Sim S, Dimopoulos G. Mosquito immune defenses against *Plasmodium* infection. Dev Comp Immunol 2010;34:387–95.

[11] Cirimotich CM, Dong Y, Clayton AM, et al. Natural microbe-mediated refractoriness to *Plasmodium* infection in *Anopheles gambiae*. Science 2011;332:855–8.

[12] Ramirez JL, Dimopoulos G. The Toll immune signaling pathway control conserved anti-dengue defenses across diverse *Ae. aegypti* strains and against multiple dengue virus serotypes. Dev Comp Immunol 2010;34:625–9.

[13] Frolet C, Thoma M, Blandin S, Hoffmann JA, Levashina EA. Boosting NF-kappaB-dependent basal immunity of *Anopheles gambiae* aborts development of *Plasmodium berghei*. Immunity 2006;25:677–85.

[14] Souza-Neto JA, Sim S, Dimopoulos G. An evolutionary conserved function of the JAK-STAT pathway in anti-dengue defense. Proc Natl Acad Sci U S A 2009;106:17841–6.

[15] Xi Z, Ramirez JL, Dimopoulos G. The *Aedes aegypti* toll pathway controls dengue virus infection. PLoS Pathog 2008;4:e1000098.

[16] Meister S, Kanzok SM, Zheng X-L, et al. Immune signaling pathways regulating bacterial and malaria parasite infection of the mosquito *Anopheles gambiae*. Proc Natl Acad Sci USA 2005;102:11420–5.

[17] Garver LS, Bahia AC, Das S, et al. *Anopheles* Imd pathway factors and effectors in infection intensity-dependent anti-*Plasmodium* action. PLoS Pathog 2012;8:e1002737.

[18] Garver LS, Dong Y, Dimopoulos G. Caspar controls resistance to *Plasmodium falciparum* in diverse anopheline species. PLoS Pathog 2009;5:e1000335.

[19] Dostert C, Jouanguy E, Irving P, et al. The Jak-STAT signaling pathway is required but not sufficient for the antiviral response of *Drosophila*. Nat Immunol 2005;6:946–53.

[20] Kakumani PK, Ponia SS, S RK, et al. Role of RNA interference (RNAi) in dengue virus replication and identification of NS4B as an RNAi suppressor. J Virol 2013;87: 8870–83.

[21] Garver LS, de Almeida Oliveira G, Barillas-Mury C. The JNK pathway is a key mediator of *Anopheles gambiae* antiplasmodial immunity. PLoS Pathog 2013;9:e1003622.

[22] Gupta L, Molina-Cruz A, Kumar S, et al. The STAT pathway mediates late-phase immunity against *Plasmodium* in the mosquito *Anopheles gambiae*. Cell Host Microbe 2009;5:498−507.

[23] Dinglasan RR, Devenport M, Florens L, et al. The *Anopheles gambiae* adult midgut peritrophic matrix proteome. Insect Biochem Mol Biol 2009;39:125−34.

[24] Barillas-Mury C. CLIP proteases and *Plasmodium* melanization in *Anopheles gambiae*. Trends Parasitol 2007;23:297−9.

[25] An C, Budd A, Kanost MR, Michel K. Characterization of a regulatory unit that controls melanization and affects longevity of mosquitoes. Cell Mol Life Sci 2011;68:1929−39.

[26] Ligoxygakis P, Pelte N, Ji C, et al. A serpin mutant links Toll activation to melanization in the host defence of *Drosophila*. EMBO J 2002;21:6330−7.

[27] Chao J, Wistreich GA, Moore J. Failure to isolate microorganisms from within mosquito eggs. Ann Entomol Soc Am 1963;56:559−61.

[28] Osei-Poku J, Mbogo CM, Palmer WJ, Jiggins FM. Deep sequencing reveals extensive variation in the gut microbiota of wild mosquitoes from Kenya. Mol Ecol 2012;21: 5138−50.

[29] Zouache K, Raharimalala FN, Raquin V, et al. Bacterial diversity of field-caught mosquitoes, *Aedes albopictus* and *Aedes aegypti*, from different geographic regions of Madagascar. FEMS Microbiol Ecol 2011;75:377−89.

[30] Boissière A, Tchioffo MT, Bachar D, et al. Midgut microbiota of the malaria mosquito vector *Anopheles gambiae* and interactions with *Plasmodium falciparum* infection. PLoS Pathog 2012;8:e1002742.

[31] Lindh JM, Borg-Karlson A-K, Faye I. Transstadial and horizontal transfer of bacteria within a colony of *Anopheles gambiae* (Diptera: Culicidae) and oviposition response to bacteria-containing water. Acta Trop 2008;107:242−50.

[32] Shi Y, Lou K, Li C. Growth and photosynthetic efficiency promotion of sugar beet (*Beta vulgaris* L.) by endophytic bacteria. Photosynth Res 2010;105:5−13.

[33] Alvarez-Pérez S, Herrera CM, de Vega C. Zooming-in on floral nectar: a first exploration of nectar-associated bacteria in wild plant communities. FEMS Microbiol Ecol 2012;80:591−602.

[34] Yamada Y, Yukphan P. Genera and species in acetic acid bacteria. Int J Food Microbiol 2008;125:15−24.

[35] Clements AN. The biology of mosquitoes: sensory reception and behaviour, vol. 2.; 1999. CABI, UK ISBN: 9780851993133.

[36] Favia G, Ricci I, Damiani C, et al. Bacteria of the genus *Asaia* stably associate with *Anopheles stephensi*, an Asian malarial mosquito vector. Proc Natl Acad Sci USA 2007;104:9047−51.

[37] Moll RM, Romoser WS, Modrzakowski MC, Moncayo AC, Lerdthusnee K. Meconial peritrophic membranes and the fate of midgut bacteria during mosquito (Diptera: Culicidae) metamorphosis. J Med Entomol 2001;38:29−32.

[38] Moncayo AC, Lerdthusnee K, Leon R, Robich RM, Romoser WS. Meconial peritrophic matrix structure, formation, and meconial degeneration in mosquito pupae/pharate adults: histological and ultrastructural aspects. J Med Entomol 2005;42:939−44.

[39] Wang Y, Gilbreath TM, Kukutla P, Yan G, Xu J. Dynamic gut microbiome across life history of the malaria mosquito *Anopheles gambiae* in Kenya. PLoS One 2011;6:e24767.

[40] Chavshin AR, Oshaghi MA, Vatandoost H, et al. Identification of bacterial microflora in the midgut of the larvae and adult of wild caught *Anopheles stephensi*: a step toward finding suitable paratransgenesis candidates. Acta Trop 2012;121:129−34.

[41] Rani A, Sharma A, Rajagopal R, Adak T, Bhatnagar RK. Bacterial diversity analysis of larvae and adult midgut microflora using culture-dependent and culture-independent methods in lab-reared and field-collected *Anopheles stephensi*—an Asian malarial vector. BMC Microbiol 2009;9:96.

[42] Dinparast Djadid N, Jazayeri H, Raz A, Favia G, Ricci I, Zakeri S. Identification of the midgut microbiota of *Anopheles stephensi* and *Anopheles maculipennis* for their application as a paratransgenic tool against malaria. PLoS One 2011;6:e28484.

[43] Pumpuni CB, Demaio J, Kent M, Davis JR, Beier JC. Bacterial population dynamics in three anopheline species: the impact on Plasmodium sporogonic development. Am J Trop Med Hyg 1996;54:214−18.

[44] Terenius O, Lindh JM, Eriksson-Gonzales K, et al. Midgut bacterial dynamics in *Aedes aegypti*. FEMS Microbiol Ecol 2012;80:556−65.

[45] Oliveira JHM, Gonçalves RLS, Lara FA, et al. Blood meal-derived heme decreases ROS levels in the midgut of *Aedes aegypti* and allows proliferation of intestinal microbiota. PLoS Pathog 2011;7:e1001320.

[46] Damiani C, Ricci I, Crotti E, et al. Paternal transmission of symbiotic bacteria in malaria vectors. Curr Biol 2008;18:R1087−8.

[47] Mitraka E, Stathopoulos S, Siden-Kiamos I, Christophides GK, Louis C. *Asaia* accelerates larval development of *Anopheles gambiae*. Pathog Glob Health 2013;107:305−11.

[48] Akhouayri IG, Habtewold T, Christophides GK. Melanotic pathology and vertical transmission of the gut commensal *Elizabethkingia meningoseptica* in the major malaria vector *Anopheles gambiae*. PLoS One 2013;8:e77619.

[49] DeMaio J, Pumpuni CB, Kent M, Beier JC. The midgut bacterial flora of wild *Aedes triseriatus*, *Culex pipiens*, and *Psorophora columbiae* mosquitoes. Am J Trop Med Hyg 1996;54:219−23.

[50] Chouaia B, Rossi P, Montagna M, et al. Molecular evidence for multiple infections as revealed by typing of *Asaia* bacterial symbionts of four mosquito species. Appl Environ Microbiol 2010;76:7444−50.

[51] Gusmão DS, Santos AV, Marini DC, Bacci M, Berbert-Molina MA, Lemos FJA. Culture-dependent and culture-independent characterization of microorganisms associated with *Aedes aegypti* (Diptera: Culicidae) (L.) and dynamics of bacterial colonization in the midgut. Acta Trop 2010;115:275−81.

[52] Minard G, Tran F-H, Dubost A, Tran-Van V, Mavingui P, Moro CV. Pyrosequencing 16S rRNA genes of bacteria associated with wild tiger mosquito *Aedes albopictus*: a pilot study. Front Cell Infect Microbiol 2014;4:59.

[53] Pidiyar VJ, Jangid K, Patole MS, Shouche YS. Studies on cultured and uncultured microbiota of wild *Culex quinquefasciatus* mosquito midgut based on 16s ribosomal RNA gene analysis. Am J Trop Med Hyg 2004;70:597−603.

[54] Lindh JM, Terenius O, Faye I. 16S rRNA gene-based identification of midgut bacteria from field-caught *Anopheles gambiae* sensu lato and *A. funestus* mosquitoes reveals new species related to known insect symbionts. Appl Environ Microbiol 2005;71:7217−23.

[55] Valiente Moro C, Tran FH, Raharimalala FN, Ravelonandro P, Mavingui P. Diversity of culturable bacteria including Pantoea in wild mosquito *Aedes albopictus*. BMC Microbiol 2013;13:70.

[56]   Straif SC, Mbogo CN, Toure AM, et al. Midgut bacteria in *Anopheles gambiae* and *An. funestus* (Diptera: Culicidae) from Kenya and Mali. J Med Entomol 1998;35:222−6.

[57]   Crotti E, Rizzi A, Chouaia B, et al. Acetic acid bacteria, newly emerging symbionts of insects. Appl Environ Microbiol 2010;76:6963−70.

[58]   Bahia AC, Dong Y, Blumberg BJ, et al. Exploring Anopheles gut bacteria for *Plasmodium* blocking activity. Environ Microbiol 2014; published online January 15. http://dx.doi.org/10.1111/1462-2920.12381.

[59]   Eappen AG, Smith RC, Jacobs-Lorena M. *Enterobacter*-activated mosquito immune responses to *Plasmodium* involve activation of SRPN6 in *Anopheles stephensi*. PLoS One 2013;8. Available from: http://dx.doi.org/10.1371/journal.pone.0062937.

[60]   Tchioffo MT, Boissiere A, Churcher TS, et al. Modulation of malaria infection in *Anopheles gambiae* mosquitoes exposed to natural midgut bacteria. PLoS One 2013;8. Available from: http://dx.doi.org/10.1371/journal.pone.0081663.

[61]   Apte-Deshpande A, Paingankar M, Gokhale MD, Deobagkar DN. *Serratia odorifera* a midgut inhabitant of *Aedes aegypti* mosquito enhances its susceptibility to dengue-2 virus. PLoS One 2012;7:e40401.

[62]   Azambuja P, Garcia ES, Ratcliffe NA. Gut microbiota and parasite transmission by insect vectors. Trends Parasitol 2005;21:568−72.

[63]   Ramirez JL, Short SM, Bahia AC, et al. Chromobacterium Csp_P reduces malaria and dengue infection in vector mosquitoes and has entomopathogenic and *in vitro* anti-pathogen activities. PLoS Pathog 2014;10:e1004398.

[64]   Hurwitz I, Fieck A, Read A, et al. Paratransgenic control of vector borne diseases. Int J Biol Sci 2011;7:1334−44.

[65]   Akman L, Yamashita A, Watanabe H, et al. Genome sequence of the endocellular obligate symbiont of tsetse flies, *Wigglesworthia glossinidia*. Nat Genet 2002;32:402−7.

[66]   Fieck A, Hurwitz I, Kang AS, Durvasula R. *Trypanosoma cruzi*: synergistic cytotoxicity of multiple amphipathic anti-microbial peptides to *T. cruzi* and potential bacterial hosts. Exp Parasitol 2010;125:342−7.

[67]   Durvasula RV, Gumbs A, Panackal A, et al. Expression of a functional antibody fragment in the gut of *Rhodnius prolixus* via transgenic bacterial symbiont *Rhodococcus rhodnii*. Med Vet Entomol 1999;13:115−19.

[68]   Kokoza V, Ahmed A, Woon Shin S, Okafor N, Zou Z, Raikhel AS. Blocking of *Plasmodium* transmission by cooperative action of Cecropin A and Defensin A in transgenic *Aedes aegypti* mosquitoes. Proc Natl Acad Sci USA 2010;107:8111−16.

[69]   Conde R, Zamudio FZ, Rodríguez MH, Possani LD. Scorpine, an anti-malaria and anti-bacterial agent purified from scorpion venom. FEBS Lett 2000;471:165−8.

[70]   Ghosh AK, Ribolla PE, Jacobs-Lorena M. Targeting *Plasmodium* ligands on mosquito salivary glands and midgut with a phage display peptide library. Proc Natl Acad Sci USA 2001;98:13278−81.

[71]   Moreira LA, Ito J, Ghosh A, et al. Bee venom phospholipase inhibits malaria parasite development in transgenic mosquitoes. J Biol Chem 2002;277:40839−43.

[72]   Wang S, Ghosh AK, Bongio N, Stebbings KA, Lampe DJ, Jacobs-Lorena M. Fighting malaria with engineered symbiotic bacteria from vector mosquitoes. Proc Natl Acad Sci USA 2012;109:12734−9.

[73]   Zieler H, Keister DB, Dvorak JA, Ribeiro JM. A snake venom phospholipase A(2) blocks malaria parasite development in the mosquito midgut by inhibiting ookinete association with the midgut surface. J Exp Biol 2001;204:4157−67.

[74]   Wang S, Jacobs-Lorena M. Genetic approaches to interfere with malaria transmission by vector mosquitoes. Trends Biotechnol 2013;31:185–93.

[75]   Kim W, Koo H, Richman AM, et al. Ectopic expression of a cecropin transgene in the human malaria vector mosquito *Anopheles gambiae* (Diptera: Culicidae): effects on susceptibility to *Plasmodium*. J Med Entomol 2004;41:447–55.

[76]   Possani LD, Zurita M, Delepierre M, Hernández FH, Rodríguez MH. From noxiustoxin to Shiva-3, a peptide toxic to the sporogonic development of *Plasmodium berghei*. Toxicon 1998;36:1683–92.

[77]   Ghosh AK, Coppens I, Gårdsvoll H, Ploug M, Jacobs-Lorena M. Plasmodium ookinetes coopt mammalian plasminogen to invade the mosquito midgut. Proc Natl Acad Sci USA 2011;108:17153–8.

[78]   Luplertlop N, Surasombatpattana P, Patramool S, et al. Induction of a peptide with activity against a broad spectrum of pathogens in the *Aedes aegypti* salivary gland, following Infection with dengue virus. PLoS Pathog 2011;7:e1001252.

[79]   Liu Y, Zhang F, Liu J, et al. Transmission-blocking antibodies against mosquito C-type lectins for dengue prevention. PLoS Pathog 2013;10:e1003931.

[80]   Riehle MA, Moreira CK, Lampe D, Lauzon C, Jacobs-Lorena M. Using bacteria to express and display anti-Plasmodium molecules in the mosquito midgut. Int J Parasitol 2007;37:595–603.

[81]   Chavshin AR, Oshaghi MA, Vatandoost H, Yakhchali B, Raeisi A, Zarenejad F. *Escherichia coli* expressing a green fluorescent protein (GFP) in *Anopheles stephensi*: a preliminary model for paratransgenesis. Symbiosis 2013;60:17–24.

[82]   Joyce JD, Nogueira JR, Bales AA, Pittman KE, Anderson JR. Interactions between La Crosse virus and bacteria isolated from the digestive tract of *Aedes albopictus* (Diptera: Culicidae). J Med Entomol 2011;48:389–94.

[83]   Bisi DC, Lampe DJ. Secretion of anti-Plasmodium effector proteins from a natural *Pantoea agglomerans* isolate by using PelB and HlyA secretion signals. Appl Environ Microbiol 2011;77:4669–75.

[84]   Stathopoulos S, Neafsey DE, Lawniczak MKN, Muskavitch MAT, Christophides GK. Genetic dissection of *Anopheles gambiae* gut epithelial responses to *Serratia marcescens*. PLoS Pathog 2014;10:e1003897.

[85]   Apte-Deshpande AD, Paingankar MS, Gokhale MD, Deobagkar DN. Serratia odorifera mediated enhancement in susceptibility of *Aedes aegypti* for chikungunya virus. Indian J Med Res 2014;139:762–8.

[86]   Moore JE, McCalmont M, Xu J, Millar BC, Heaney N. *Asaia* sp., an unusual spoilage organism of fruit-flavored bottled water. Appl Environ Microbiol 2002;68:4130–1.

[87]   Minard G, Tran FH, Raharimalala FN, et al. Prevalence, genomic and metabolic profiles of *Acinetobacter* and *Asaia* associated with field-caught *Aedes albopictus* from Madagascar. FEMS Microbiol Ecol 2013;83:63–73.

[88]   Frentiu FD, Zakir T, Walker T, et al. Limited dengue virus replication in field-collected *Aedes aegypti* mosquitoes infected with *Wolbachia*. PLoS Negl Trop Dis 2014;8:e2688.

[89]   McNaughton D, Duong TTH. Designing a community engagement framework for a new dengue control method: a case study from central Vietnam. PLoS Negl Trop Dis 2014;8:e2794.

[90]   Osherov N. The molecular mechanisms of conidial germination. FEMS Microbiol Lett 2001;199:153–60.

[91]   Virginio ED, Kubitschek-Barreira PH, Batista MV, et al. Immunoproteome of *Aspergillus fumigatus* using sera of patients with invasive Aspergillosis. Int J Mol Sci 2014;15:14505–30.

[92] Jayaram VB, Cuyvers S, Verstrepen KJ, Delcour JA, Courtin CM. Succinic acid in levels produced by yeast (*Saccharomyces cerevisiae*) during fermentation strongly impacts wheat bread dough properties. Food Chem 2014;151:421−8.

[93] George J, Jenkins NE, Blanford S, Thomas MB, Baker TC. Malaria mosquitoes attracted by fatal fungus. PLoS One 2013;8:e62632.

[94] Underhill DM, Iliev ID. The mycobiota: interactions between commensal fungi and the host immune system. Nat Rev Immunol 2014;14:405−16.

[95] Bressano M, Curetti M, Giachero L, et al. Mycorrhizal fungi symbiosis as a strategy against oxidative stress in soybean plants. J Plant Physiol 2010;167:1622−6.

[96] Marroquín-Cardona AG, Johnson NM, Phillips TD, Hayes AW. Mycotoxins in a changing global environment—a review. Food Chem Toxicol 2014;69:220−30.

[97] Antonissen G, Martel A, Pasmans F, et al. The impact of *Fusarium* mycotoxins on human and animal host susceptibility to infectious diseases. Toxins (Basel) 2014;6: 430−52.

[98] Warris A. The biology of pulmonary *Aspergillus* infections. J Infect 2014; published online August 15. http://dx.doi.org/10.1016/j.jinf.2014.07.011.

[99] Chiapello LS, Baronetti JL, Aoki MP, Gea S, Rubinstein H, Masih DT. Immunosuppression, interleukin-10 synthesis and apoptosis are induced in rats inoculated with *Cryptococcus neoformans* glucuronoxylomannan. Immunology 2004;113:392−400.

[100] Ricci I, Mosca M, Valzano M, et al. Different mosquito species host *Wickerhamomyces anomalus* (*Pichia anomala*): perspectives on vector-borne diseases symbiotic control. Antonie Van Leeuwenhoek 2011;99:43−50.

[101] Cappelli A, Ulissi U, Valzano M, et al. A *Wickerhamomyces anomalus* killer strain in the malaria vector *Anopheles stephensi*. PLoS One 2014;9:e95988.

[102] Maketon M, Amnuaykanjanasin A, Kaysorngup A. A rapid knockdown effect of *Penicillium citrinum* for control of the mosquito *Culex quinquefasciatus* in Thailand. World J Microbiol Biotechnol 2014;30:727−36.

[103] Fang W, Vega-Rodríguez J, Ghosh AK, Jacobs-Lorena M, Kang A, St Leger RJ. Development of transgenic fungi that kill human malaria parasites in mosquitoes. Science 2011;331:1074−7.

[104] Da Costa GL, de Oliveira PC. *Penicillium* species in mosquitoes from two Brazilian regions. J Basic Microbiol 1998;38:343−7.

[105] Geris R, Rodrigues-Fo E, Garcia da Silva HH, Garcia da Silva I. Larvicidal effects of fungal Meroterpenoids in the control of *Aedes aegypti* L., the main vector of dengue and yellow fever. Chem Biodivers 2008;5:341−5.

[106] Da S. Pereira E, de M. Sarquis MI, Ferreira-Keppler RL, Hamada N, et al. Filamentous fungi associated with mosquito larvae (Diptera: Culicidae) in municipalities of the Brazilian Amazon. Neotrop Entomol 2009;38:352−9.

[107] Tajedin L, Hashemi J, Abaei M, Hosseinpour L, Rafei F, Basseri H. Study on fungal flora in the midgut of the larva and adult of the different populations of the malaria vector *Anopheles stephensi*. Iran J Arthropod Borne Dis 2009;3:36−40.

[108] Lynch PA, Grimm U, Thomas MB, Read AF. Prospective malaria control using entomopathogenic fungi: comparative evaluation of impact on transmission and selection for resistance. Malar J 2012;11:383.

[109] Mohanty SS, Prakash S. Comparative efficacy and pathogenicity of keratinophilic soil fungi against *Culex quinquefasciatus* larvae. Indian J Microbiol 2010;50:299−302.

[110] Scholte E-J, Knols BGJ, Samson RA, Takken W. Entomopathogenic fungi for mosquito control: a review. J Insect Sci 2004;4:19.

[111] Garza-Hernández JA, Rodríguez-Pérez MA, Salazar MI, et al. Vectorial capacity of *Aedes aegypti* for dengue virus type 2 is reduced with co-infection of *Metarhizium anisopliae*. PLoS Negl Trop Dis 2013;7:e2013.

[112] Dong Y, Morton JC, Ramirez JL, Souza-Neto JA, Dimopoulos G. The entomopathogenic fungus *Beauveria bassiana* activate toll and JAK-STAT pathway-controlled effector genes and anti-dengue activity in *Aedes aegypti*. Insect Biochem Mol Biol 2012;42:126−32.

[113] Frants TG, Mertvetsova OA. [Yeast associations with mosquitoes of the genus Aedes Mg. (Diptera, Culicidae) in the Tom-Ob river region]. Nauchnye Doki Vyss Shkoly Biol Nauki 1986;94−8.

[114] Ricci I, Damiani C, Scuppa P, et al. The yeast *Wickerhamomyces anomalus* (*Pichia anomala*) inhabits the midgut and reproductive system of the Asian malaria vector *Anopheles stephensi*. Environ Microbiol 2011;13:911−21.

[115] Liu Y-J, Liu J, Ying S-H, Liu S-S, Feng M-G. A fungal insecticide engineered for fast per os killing of caterpillars has high field efficacy and safety in full-season control of cabbage insect pests. Appl Environ Microbiol 2013;79:6452−8.

[116] Blanford S, Chan BHK, Jenkins N, et al. Fungal pathogen reduces potential for malaria transmission. Science 2005;308:1638−41.

[117] Blanford S, Jenkins NE, Read AF, Thomas MB. Evaluating the lethal and pre-lethal effects of a range of fungi against adult *Anopheles stephensi* mosquitoes. Malar J 2012;11:365.

[118] Lefèvre T, Gouagna L-C, Dabiré KR, et al. Beyond nature and nurture: phenotypic plasticity in blood-feeding behavior of *Anopheles gambiae* s.s. when humans are not readily accessible. Am J Trop Med Hyg 2009;81:1023−9.

[119] Chadee DD. Resting behaviour of *Aedes aegypti* in Trinidad: with evidence for the reintroduction of indoor residual spraying (IRS) for dengue control. Parasit Vectors 2013;6:255.

[120] Mnyone LL, Koenraadt CJ, Lyimo IN, Mpingwa MW, Takken W, Russell TL. Anopheline and culicine mosquitoes are not repelled by surfaces treated with the entomopathogenic fungi *Metarhizium anisopliae* and *Beauveria bassiana*. Parasit Vectors 2010;3:80.

[121] Bukhari T, Takken W, Koenraadt CJM. Development of *Metarhizium anisopliae* and *Beauveria bassiana* formulations for control of malaria mosquito larvae. Parasit Vectors 2011;4:23.

[122] Mnyone LL, Lyimo IN, Lwetoijera DW, et al. Exploiting the behaviour of wild malaria vectors to achieve high infection with fungal biocontrol agents. Malar J 2012;11:87.

[123] Animut A, Balkew M, Lindtjørn B. Impact of housing condition on indoor-biting and indoor-resting *Anopheles arabiensis* density in a highland area, central Ethiopia. Malar J 2013;12:393.

[124] Mongkalangoon P, Grieco JP, Achee NL, Suwonkerd W, Chareonviriyaphap T. Irritability and repellency of synthetic pyrethroids on an *Aedes aegypti* population from Thailand. J Vector Ecol 2009;34:217−24.

[125] De García MC, Arboleda ML, Barraquer F, Grose E. Fungal keratitis caused by *Metarhizium anisopliae* var. anisopliae. J Med Vet Mycol 1997;35:361−3.

[126] Figueira L, Pinheiro D, Moreira R, et al. *Beauveria bassiana* keratitis in bullous keratopathy: antifungal sensitivity testing and management. Eur J Ophthalmol 2012;22:814−18.

[127] Thomas MB, Read AF. Can fungal biopesticides control malaria? Nat Rev Microbiol 2007;5:377−83.

[128] Chan M, Johansson MA. The incubation periods of dengue viruses. PLoS One 2012;7: e50972.

[129] Suh S-O, Gibson CM, Blackwell M. *Metschnikowia chrysoperlae* sp. nov., *Candida picachoensis* sp. nov. and *Candida pimensis* sp. nov., isolated from the green lacewings *Chrysoperla comanche* and *Chrysoperla carnea* (Neuroptera: Chrysopidae). Int J Syst Evol Microbiol 2004;54:1883−90.

[130] Gibson CM, Hunter MS. Negative fitness consequences and transmission dynamics of a heritable fungal symbiont of a parasitic wasp. Appl Environ Microbiol 2009;75: 3115−19.

[131] Rosa CA, Lachance MA, Silva JOC, et al. Yeast communities associated with stingless bees. FEMS Yeast Res 2003;4:271−5.

[132] Benda ND. Detection and characterization of *Kodamaea ohmeri* associated with small hive beetle *Aethina tumida* infesting honey bee hives. J Apic Res 2008;47:194−201.

[133] Olstorpe M, Passoth V. *Pichia anomala* in grain biopreservation. Antonie Van Leeuwenhoek 2011;99:57−62.

[134] Lahlali R, Serrhini MN, Jijakli MH. Efficacy assessment of *Candida oleophila* (strain O) and *Pichia anomala* (strain K) against major postharvest diseases of citrus fruits in Morocco. Commun Agric Appl Biol Sci 2004;69:601−9.

[135] Wang X, Chi Z, Yue L, Li J, Li M, Wu L. A marine killer yeast against the pathogenic yeast strain in crab (*Portunus trituberculatus*) and an optimization of the toxin production. Microbiol Res 2007;162:77−85.

[136] Guo F-J, Ma Y, Xu H-M, Wang X-H, Chi Z-M. A novel killer toxin produced by the marine-derived yeast *Wickerhamomyces anomalus* YF07b. Antonie Van Leeuwenhoek 2013;103:737−46.

[137] Savoia D, Scutera S, Raimondo S, Conti S, Magliani W, Polonelli L. Activity of an engineered synthetic killer peptide on *Leishmania major* and *Leishmania infantum* promastigotes. Exp Parasitol 2006;113:186−92.

[138] Coluccio AE, Rodriguez RK, Kernan MJ, Neiman AM. The yeast spore wall enables spores to survive passage through the digestive tract of *Drosophila*. PLoS One 2008;3: e2873.

[139] Smallegange RC, Schmied WH, van Roey KJ, et al. Sugar-fermenting yeast as an organic source of carbon dioxide to attract the malaria mosquito *Anopheles gambiae*. Malar J 2010;9:292.

[140] Mattanovich D, Branduardi P, Dato L, Gasser B, Sauer M, Porro D. Recombinant protein production in yeasts. Methods Mol Biol 2012;824:329−58.

[141] De Araujo Coutinho CJP, da C, Alves R, Sanscrainte ND, et al. Occurrence and phylogenetic characterization of a baculovirus isolated from *Culex quinquefasciatus* in São Paulo State, Brazil. Arch Virol 2012;157:1741−5.

[142] Sangdee K, Pattanakitsakul S-N. Comparison of mosquito densoviruses: two clades of viruses isolated from indigenous mosquitoes. Southeast Asian J Trop Med Public Health 2013;44:586−93.

[143] Ren X, Hoiczyk E, Rasgon JL. Viral paratransgenesis in the malaria vector *Anopheles gambiae*. PLoS Pathog 2008;4:e1000135.

[144] Vasilakis N, Guzman H, Firth C, et al. Mesoniviruses are mosquito-specific viruses with extensive geographic distribution and host range. Virol J 2014;11:97.

[145] Vasilakis N, Castro-Llanos F, Widen SG, et al. Arboretum and Puerto Almendras viruses: two novel rhabdoviruses isolated from mosquitoes in Peru. J Gen Virol 2014;95:787−92.

[146] May LP, Watts SL, Maruniak JE. Molecular survey for mosquito-transmitted viruses: detection of Tensaw virus in north central Florida mosquito populations. J Am Mosq Control Assoc 2014;30:61−4.

[147] Huhtamo E, Cook S, Moureau G, et al. Novel flaviviruses from mosquitoes: mosquito-specific evolutionary lineages within the phylogenetic group of mosquito-borne flaviviruses. Virology 2014;464−465C:320−9.

[148] Ma M, Huang Y, Gong Z, et al. Discovery of DNA viruses in wild-caught mosquitoes using small RNA high throughput sequencing. PLoS One 2011;6:e24758.

[149] Reese SM, Beaty MK, Gabitzsch ES, Blair CD, Beaty BJ. *Aedes triseriatus* females transovarially infected with La Crosse virus mate more efficiently than uninfected mosquitoes. J Med Entomol 2009;46:1152−8.

[150] Boonsuepsakul S, Luepromchai E, Rongnoparut P. Characterization of *Anopheles minimus* CYP6AA3 expressed in a recombinant baculovirus system. Arch Insect Biochem Physiol 2008;69:13−21.

[151] Pham HT, Jousset F-X, Perreault J, et al. Expression strategy of *Aedes albopictus* densovirus. J Virol 2013;87:9928−32.

[152] Suzuki Y, Niu G, Hughes GL, Rasgon JL. A viral over-expression system for the major malaria mosquito *Anopheles gambiae*. Sci Rep 2014;4:5127.

[153] Carlson J, Suchman E, Buchatsky L. Densoviruses for control and genetic manipulation of mosquitoes. Adv Virus Res 2006;68:361−92.

[154] Paterson A, Robinson E, Suchman E, Afanasiev B, Carlson J. Mosquito densonucleosis viruses cause dramatically different infection phenotypes in the C6/36 *Aedes albopictus* cell line. Virology 2005;337:253−61.

[155] Mukha DV, Chumachenko AG, Dykstra MJ, Kurtti TJ, Schal C. Characterization of a new densovirus infecting the German cockroach, *Blattella germanica*. J Gen Virol 2006;87:1567−75.

[156] Vago C, Duthoit JL, Delahaye F. Les lésions nucléaires de la "Virose à noyaux denses" du Lépidoptère Galleria mellonella. Arch für die gesamte Virusforsch 1966;18:344−9.

[157] Boublik Y, Jousset FX, Bergoin M. Complete nucleotide sequence and genomic organization of the *Aedes albopictus* parvovirus (AaPV) pathogenic for *Aedes aegypti* larvae. Virology 1994;200:752−63.

[158] Chen S, Cheng L, Zhang Q, et al. Genetic, biochemical, and structural characterization of a new densovirus isolated from a chronically infected *Aedes albopictus* C6/36 cell line. Virology 2004;318:123−33.

[159] O'Neill SL, Kittayapong P, Braig HR, Andreadis TG, Gonzalez JP, Tesh RB. Insect densoviruses may be widespread in mosquito cell lines. J Gen Virol 1995;76(Pt. 8): 2067−74.

[160] Kittayapong P, Baisley KJ, O'Neill SL. A mosquito densovirus infecting *Aedes aegypti* and *Aedes albopictus* from Thailand. Am J Trop Med Hyg 1999;61:612−17.

[161] Rwegoshora RT, Kittayapong P. Pathogenicity and infectivity of the Thai-strain densovirus (AThDNV) in *Anopheles minimus* S.L. Southeast Asian J Trop Med Public Health 2004;35:630−4.

[162] Afanasiev BN, Ward TW, Beaty BJ, Carlson JO. Transduction of *Aedes aegypti* mosquitoes with vectors derived from Aedes densovirus. Virology 1999;257:62−72.

[163] Barreau C, Jousset FX, Bergoin M. Pathogenicity of the *Aedes albopictus* parvovirus (AaPV), a denso-like virus, for *Aedes aegypti* mosquitoes. J Invertebr Pathol 1996;68: 299−309.

[164] Ledermann JP, Suchman EL, Black WC, Carlson JO. Infection and pathogenicity of the mosquito densoviruses AeDNV, HeDNV, and APeDNV in *Aedes aegypti* mosquitoes (Diptera: Culicidae). J Econ Entomol 2004;97:1828−35.

[165] Hirunkanokpun S, Carlson JO, Kittayapong P. Evaluation of mosquito densoviruses for controlling *Aedes aegypti* (Diptera: Culicidae): variation in efficiency due to virus strain and geographic origin of mosquitoes. Am J Trop Med Hyg 2008;78:784−90.

[166] Gu J-B, Dong Y-Q, Peng H-J, Chen X-G. A recombinant AeDNA containing the insect-specific toxin, BmK IT1, displayed an increasing pathogenicity on *Aedes albopictus*. Am J Trop Med Hyg 2010;83:614−23.

[167] Gu J, Liu M, Deng Y, Peng H, Chen X. Development of an efficient recombinant mosquito densovirus-mediated RNA interference system and its preliminary application in mosquito control. PLoS One 2011;6:e21329.

[168] Rajendran R, Lingala R, Vuppu SK, et al. Assessment of packed bed bioreactor systems in the production of viral vaccines. AMB Express 2014;4:25.

[169] Micheloud GA, Gioria VV, Eberhardt I, Visnovsky G, Claus JD. Production of the *Anticarsia gemmatalis* multiple nucleopolyhedrovirus in serum-free suspension cultures of the saUFL-AG-286 cell line in stirred reactor and airlift reactor. J Virol Methods 2011;178:106−16.

[170] Damiani C, Ricci I, Crotti E, et al. Mosquito-bacteria symbiosis: the case of *Anopheles gambiae* and *Asaia*. Microb Ecol 2010;60:644−54.

[171] Gusmão DS, Santos AV, Marini DC, et al. First isolation of microorganisms from the gut diverticulum of *Aedes aegypti* (Diptera: Culicidae): new perspectives for an insect-bacteria association. Mem Inst Oswaldo Cruz 2007;102:919−24.

[172] Sharma P, Sharma S, Maurya RK, et al. Salivary glands harbor more diverse microbial communities than gut in *Anopheles culicifacies*. Parasit Vectors 2014;7:235.

[173] Darbro JM, Johnson PH, Thomas MB, Ritchie SA, Kay BH, Ryan PA. Effects of *Beauveria bassiana* on survival, blood-feeding success, and fecundity of *Aedes aegypti* in laboratory and semi-field conditions. Am J Trop Med Hyg 2012;86:656−64.

[174] Wei W, Shao D, Huang X, et al. The pathogenicity of mosquito densovirus (C6/36DNV) and its interaction with dengue virus type II in *Aedes albopictus*. Am J Trop Med Hyg 2006;75:1118−26.

[175] Ren X, Hughes GL, Niu G, Suzuki Y, Rasgon JL. *Anopheles gambiae* densovirus (AgDNV) has negligible effects on adult survival and transcriptome of its mosquito host. PeerJ 2014;2:e584.

Chapter 16

Regulation of Transgenic Mosquitoes

Hector Quemada

Biosafety Resource Network, Institute of International Crop Improvement, Donald Danforth Plant Science Center, St. Louis, MI, USA

INTRODUCTION

Genetic modification of animals can be understood to encompass a wide range of methods for altering the genotype of an organism. The term can cover methods of traditional breeding, the use of transgenic techniques, or recent methods of selectively editing genomes via zinc-finger nucleases, TALENs, or CRISPR/Cas9 nucleases. However, this review will cover only transgenesis, for two reasons: (i) traditional breeding is typically not regulated by most regulatory agencies in the world and is usually used for "genetic improvement," an activity not associated mosquitoes and (ii) the regulatory status of genome-editing techniques remains unclear; transgenic techniques are clearly covered under the Cartagena Protocol [1], while it is not yet universally agreed whether genome-editing processes are [2]. For example, one US agency does not consider such methods as falling under their regulatory authority, while the European Union has treated modifications resulting from the use of zinc-finger nucleases as falling within the scope of directives covering genetically modified organisms—although a clear decision on such methods generally has yet to be made [3,4]. Lack of consistency between countries will create a confusing regulatory landscape where such methods are concerned. Furthermore, the use of endosymbionts or parasitic organisms, whether transgenic or not, will also not be covered in this review, although many of the considerations for transgenic mosquitoes will be similar for transgenic organisms used for the so-called paratransgenic applications [5].

FRAMEWORK FOR REGULATION OF TRANSGENIC INSECTS

In most countries, the main regulatory framework under which research and development of transgenic mosquitoes will take place is that which governs

Genetic Control of Malaria and Dengue.
© 2016 Elsevier Inc. All rights reserved.

the handling of transgenic organisms. This is a result of their status as parties to the Cartagena Protocol on Biosafety (CPB; [1]) and its parent treaty, the Convention on Biological Diversity (CBD; [6]). The CPB maintains a Web site (http://bch.cbd.int), including an information node called the Biosafety Clearing House, that is a valuable resource for biosafety information and regulatory decisions made by all signatories. While the CPB serves as the basis for regulation of transgenic organisms for most countries, notable nonsignatories such as the United States, Australia, Canada, and Argentina also have established systems under which transgenic mosquitoes could be regulated, and could set regulatory precedents that CPB countries may follow.

In CPB countries, a central regulatory authority is usually established. This authority is charged with assuring that genetically modified organisms are developed and used in such a way that adverse effects on human health and the environment are prevented. Article 8(g) of the CBD, the parent treaty to the Protocol, requires Parties to:

> [e]stablish or maintain means to regulate, manage or control the risks associated with the use and release of LMOs [Living Modified Organisms, i.e. genetically engineered, or transgenic, organisms] resulting from biotechnology which are likely to have adverse environmental impacts that could affect the conservation and sustainable use of biological diversity, taking also into account the risks to human health.

Furthermore, Article 8(g) requires signatories to the CBD to regulate LMOs via national legislation and regulations. Most parties have met this obligation through the establishment of a national competent authority. The government department or ministry within which these regulatory bodies are housed may vary from country to country. For example, in African countries where malaria is endemic and where transgenic mosquitoes might eventually be deployed, the regulatory authority may reside in the Ministry of Research and Innovation (Burkina Faso), Environment (Mali) or the Ministry of Education, Science and Technology (Kenya, currently moving to the Ministry of Agriculture). Therefore, while transgenic mosquitoes are intended for public health applications, regulatory oversight is through a country's biosafety authority, which represents multiple government sectors, of which the agency, department, or ministry responsible for public health is only a part.

The Cartagena Protocol itself, in Articles 8−10 and Article 12, outlines a procedure by which the first transboundary movement of an LMO should be handled, through an advanced informed agreement procedure. Annex I outlines the information that must accompany the transboundary movement, including a risk assessment consistent with Annex III. These provisions have been incorporated into national legislation and regulations and form the basis of requests for regulatory approval for conducting various phases of research and development of transgenic organisms, beyond transboundary movement. Furthermore, while the Cartagena Protocol exempts transit and contained use from the requirements of advanced informed agreement, Party countries do in fact regulate such activities.

It should also be noted that Article 27 of the Protocol provides for the process of establishing rules and procedures to deal with liability and redress, in case of environmental damage caused by LMOs. In accordance with this article, many countries have incorporated provisions concerning liability and redress into their biosafety legislation. These laws may have provisions for the attribution of responsibility to certain parties, regardless of fault. For example, in Burkina Faso, the inventor of a genetically modified organism may in some cases be held liable (LOI N° 064-2012/AN, Portant Regime De Securite En Matiere De Biotechnologie, Chapter II). In Mali, liability extends to the "provider, supplier, depository or developer" of a genetically modified organism (Law N°08-042/AN-RM of December 1, 2008 relating to Safety in Biotechnology in the Republic of Mali, Article 58). The existence of legal provisions that potentially expose researchers to liability for damage to the environment (as well as socioeconomic damage in the case of some countries) is an important consideration for any researcher who might be contemplating working with transgenic organisms.

Not all countries take the approach described above. One notable exception is the framework developed by the United States, which regulates genetically engineered organisms under their Coordinated Framework for Regulation of Biotechnology [7]. Under this framework, different agencies have the authority to regulate transgenic organisms, depending upon the characteristics of the organism and its intended use. Thus, transgenic pink bollworm, *Pectinophora gossypiella*, intended for controlling populations of this pest is regulated by USDA-APHIS-BRS [8]. Transgenic diamondback moth, *Plutella xylostella*, is also regulated by the same agency [35]. On the other hand, the Food and Drug Agency—Center for Veterinary Medicine is responsible for regulating a proposed release of transgenic *Aedes aegypti* mosquito [9].

Because of the different legal and administrative approaches to regulation in various countries where transgenic mosquito research will be conducted, researchers should establish good working relationships with the relevant regulatory authorities. Proper working relationships will establish mutual trust, will cultivate confidence in a research group's ability to conduct its work safely and responsibly, and will generate support for the work from the regulators. Such relationships will result in a hospitable regulatory environment for research.

Research with transgenic mosquitoes may also be subject to relevant laws and regulations that cover research with animals generally. For example, in many countries or research organizations, laws, regulations or policies governing human and animal subjects may require the establishment and review of research by ethical committees or other bodies responsible for safe and ethical conduct of research. Laws and regulations governing research on specific diseases, development of drugs or other therapies, insect pest control, biodiversity or genetic resources, public consultation, or public access to information would also be applicable. An overview of the broader regulatory context for transgenic mosquito work is described by Benedict et al. [10].

STATUS OF REGULATORY CAPACITY

Transgenic technology applied to mosquitoes is relatively new to regulators, even in countries with mature regulatory systems and extensive experience. It is difficult to determine the exact number of contained use applications for transgenic insects that have been submitted and approved worldwide, but the number of approved cage or field trial applications is very few, and only one application for commercial release has been approved. Field or cage trials of male sterile *An. aegypti* have been conducted in the Cayman Islands, Malaysia, Panama, Brazil, and Mexico [11–17]. In the United States, confined field trials of transgenic pink bollworm have been conducted since 2001 [18], but these insects have not been deployed as a biocontrol strategy. Oxitec's male sterile *An. aegypti* was only recently approved for commercial use in Brazil—the first country in the world to do so [19]. These applications have all involved approaches where the transgenes confer male sterility—a trait that limits its spread in the environment. These constructs are considered to present a low level of environmental concern [20]. Furthermore, these approaches have conceptual precedents in traditional sterile insect control strategies—a connection recognized by USDA-APHIS in their Environmental Impact Statement, *Use of Genetically Engineered Fruit Fly and Pink Bollworm in APHIS Plant Pest Control Programs* [21], and in their decision document, *Use of genetically engineered fruit fly and pink bollworm in APHIS plant pest control programs; record of decision* [22].

However, some authors have proposed that more specific international agreements covering genetically engineered arthropod vectors [23] should be developed. Others have argued that transgenic insects containing constructs that persist in the wild or containing gene-driving mechanisms present greater environmental concerns [20] and pose problems even in contained use [24]. Furthermore, others have recommended that because of the possible ecosystem-wide impact of gene drive systems, broader public discussion about the application of this technology should be an essential component of decisions to deploy it [25]. As such gene drive mechanisms are still in early research stages [26], no regulatory agency has yet determined whether there are indeed risks that are qualitatively different for transgenic arthropods, or for nonpersistent applications, or that these risks cannot be adequately assessed using existing regulatory agency methods.

Most countries where the application of transgenic mosquitoes is most needed have nascent or young regulatory systems that are relatively inexperienced in regulating any genetically engineered organisms. What experience does exist is focused on crop plants. Because the initial experience with crops has shaped most regulatory systems, the prevailing expectation of regulators in the area of genetically engineered organisms is that the spread of transgenes will be limited—even during unrestricted commercial applications—or that the environmental and health impacts of the transgenes will be minimal. For example, approvals (more precisely, "deregulations") of genetically engineered crop plants issued by the US Department of Agriculture-Animal and Plant Health Inspection Service are usually based upon a Finding of Non-Significant Impact, the endpoint of their risk assessment process. Therefore, the nonpersisting strategies that have been tested so far fit into this paradigm. In contrast, transgenic mosquito control strategies may eventually use mosquitoes with gene drive mechanisms to maximize the spread of transgenes, and thereby having at least the potential, if not the intent of significant environmental impact.

Because of this change in the regulatory paradigm, developers of transgenic mosquitoes with gene drive mechanisms should expect that regulatory agencies, especially in those less developed countries where mosquito-borne diseases are prevalent, will want to gain experience and build capacity first in regulating transgenic organisms other than crop plants (specifically insects), then in regulating transgenic mosquitoes designed to have limited persistence, and finally with transgenic mosquitoes containing transgenes that are meant to spread throughout a population.

In parallel, there is a history in crop plants of development that begins in laboratory or contained facility testing, proceeds to confined field trials, and finally to large-scale release. Each phase involves decreasing controls on the dissemination of the transgenic organism: the laboratory/containment facility phase is meant to assure no escape into the environment, while the confined field trial phase is meant to assure that the impact of the transgenic organism on the environment is limited in time and space. Each phase of testing involves experiments not only to determine efficacy but also to answer questions relevant to safety and risk that would be of concern to regulators. Experiments conducted in each phase typically are designed to answer risk questions that would be encountered at the subsequent phase. A similar progression also seems to be advisable in the development of transgenic mosquito strategies, as described in a recent guidance document published by the World Health Organization [10].

In addition to the inexperience of regulators, researchers may not be experienced in working in the type of regulated environment governing research in transgenic organisms, in particular the type of research and development that would produce data for a regulatory approval dossier. Thus, quality standards for data collection, recording and storage, as well as standard operating procedures (SOPs) for laboratory experimentation and compliance may be lacking. Training in adherence to SOPs, regulatory terms and conditions, and regulatory best practices are therefore sorely needed. However, beyond initial training, experience must be gained in order to make established standards, protocols and training fully operational. Therefore, in considering research with transgenic mosquitoes, especially in mosquito-disease endemic countries, experience will be gained first with strategies where the level of risk is minimal (i.e., nonpersistent constructs), and progress toward potentially more risky versions, as the level of competence increases.

CONTAINMENT, CONFINEMENT, AND LARGE-SCALE RELEASE

Fitting the research and development of a self-sustaining mechanism for vector suppression into the existing practice of regulatory approval requires that the product progress through the phases of containment described above. It is important to note that until final large-scale releases are approved, mechanisms must be in place to control the dissemination of transgenic mosquitoes. During the laboratory/contained use, assurance of containment is mostly met by the physical structure itself as well as the proper procedures and protocols for handling the transgenic mosquitoes.

Research involving insects, including mosquitoes, has been conducted safely for many years, even though they present risks should they be inadvertently released into the environment. However, with the advent of recombinant DNA techniques, there has been a focus on assuring the safety of transgenic organisms. As a result, numerous guidance documents and regulations have been established worldwide and through international agreements. In the early stages of research, work is typically conducted in laboratories, under levels of containment that are appropriate for the category of risk presented by the transgenic organism, and in the case of transgenic insects, the combination of transgenic organism and any pathogen that it may harbor. Risk categories for various types of research involving genetic engineering have been established by such bodies such as the United States National Institutes of Health [27]. However, specific Arthropod Containment Guidelines have also been published to address the risks posed by research activities with arthropods, whether transgenic or not [28]. These guidelines provide guidance on the appropriate containment for certain transgenic arthropods, specifying a set of conditions in their second least-stringent level (ACL-2) for "uninfected genetically modified arthropod vectors ... provided the modification has no, or only negative effects on viability, survivorship, host range, or vector capacity." The current nonpersistent strategies would most likely fall under this level of containment, but the appropriate containment for persistent or gene drive constructs is less clear.

In the case of mosquitoes, even in properly constructed containment facilities with strict adherence to biosafety protocols, a low level of escape from the currently designed facilities is possible. This low level of escape from a biosafety facility may be acceptable, even in climates that are conducive to the survival of the mosquito, as the biological containment provided by the constructs would ensure the disappearance of the transgene from the population. However, the consequences of escape of a mosquito containing a gene drive construct from a containment facility would have varying consequences [24]. Regulatory authorities may increase the biosafety level requirements of the facility in order to achieve a higher level of assurance that the level of escapes would be below a certain acceptable threshold. The criteria for establishing that threshold will likely vary among countries, depending upon factors ranging from environmental considerations (e.g., whether the mosquitoes can survive for any length of time in the climate where the containment facility is located) to social issues (e.g., the level of public acceptance of transgenic work in general and transgenic mosquitoes in particular).

In the field trial stage, prior to unrestricted large-scale release, nonpersisting constructs allow for a realistic expectation of containment, and indeed the field trials conducted to date with transgenic insects, including mosquitoes, are evidence that such field trials can achieve regulatory—if not public—approval. On the other hand, the assurance of similar levels of confinement when testing a persistent or gene drive strategy is questionable. Guidance has been provided for considerations when planning contained (caged) field trials of mosquitoes containing a gene drive system [29]. There might be certain genetic constructs, geographical locations, or ecological conditions under which open-field tests could be done, and the research group that reaches this stage with a gene drive construct will have to face the challenge of at least exploring such possibilities. Some researchers have considered conditions under which such tests might be conducted [30]. The extent of risk assessment that should be done prior to the approval of limited field trials is still unclear and will depend upon the ability of an applicant to demonstrate efficacy of confinement. Alternatively, this problem might precipitate a shift in approach from the current crop-established model. Would it be possible to obtain sufficient data from a series of contained use experiments that would provide sufficient assurance of safety that a decision on large-scale release could be reached without the intermediate step of confined field trials? Such a shift would involve the development of a set of safety studies that would provide data robust enough to be applicable to all environments that might be receiving the transgenic mosquito. Typically for previous approval applications (for crops), countries have usually required that some in-country data be obtained to provide assurance that the behavior of the transgenic organism is as predictable as it was in previous environments. However, Annex III of the Cartagena Protocol, addressing risk assessment, only requires that information on the receiving environment be incorporated into the risk assessment [1], not that such data must be locally generated.

The approval of large-scale deployment in Brazil shows that the assessment of risks from transgenic mosquitoes containing nonpersistent constructs can be conducted to support a favorable decision [19]. While the technical decision does not provide details of the data supplied in support of the dossier submitted by the applicant, the issues of concern apparently were the following: (i) the exposure of humans to potential new toxins or allergens in the saliva of the transgenic mosquito, (ii) the possible ability of the transgenic mosquito to transmit diseases more efficiently, (iii) the presence of levels of tetracycline in the environment causing failure of the conditional lethality mechanism, and (iv) impact on the environment, including gene flow to other species and the increase in other species in the absence of the target species. All these issues appear to have been addressed to the satisfaction of the regulatory authority, which concluded that the transgenic mosquito being considered for approval would have no significant human health and environmental impact.

These issues are also likely to arise when considering approval for large-scale release of mosquitoes containing gene drive constructs. While the technical opinion published by the Brazilian regulatory authority does not provide sufficient detail to serve as a guide for the issues to be addressed when considering large-scale release, there are several documents that do provide such guidance [1,10,21,31,32]. The guidance in these documents is varied and requires reconciling to arrive at a consensus. However, it might be useful to consider them in the light of the general areas of concern described by Benedict et al. [10]:

1. Adverse impacts from with gene flow (vertical and horizontal)
2. Adverse impacts of the transgenic organism on the target species
3. Adverse impacts on nontarget species
4. Adverse impacts on management practice to control arthropods vectoring disease
5. Adverse effects on biogeochemical processes
6. Adverse effects on human health

The amount of data that may be required to address these concerns could be a significant barrier to the deployment of transgenic mosquitoes for vector control. The only available indicators of such costs are those expended by the developers of genetically engineered crops, for which similar data are generated. Cost estimates range from approximately US$6,000,000 to US $15,000,000 [33]. These figures are derived from information provided by private sector developers, who have developed the vast majority of products. While such costs may be economically justifiable for commercially deployed products, public sector or charitable donors will most likely fund the interventions contemplated for vector control, especially in developing countries. Whether such sources can sustain that level of input for vector control is still unclear.

The amount of required regulatory data will depend in part on the approach to assessing risks associated with the release of transgenic mosquitoes. If regulators follow a highly precautionary approach, then the data requirements and the resulting costs in terms of funding and time could limit the applications of these technologies. However if an approach that takes into consideration a balance between risks and potential benefits, then prospects for safe deployment of transgenic mosquitoes within reasonable time-frames and expenses are possible. It is important to note that the baseline to be used in considering these applications is not one that is considered safe but one that is already recognized to be hazardous (disease).

PUBLIC ACCEPTANCE

Regulatory decisions do not occur in a vacuum. The work and decisions of regulatory agencies are often influenced by public opinion. From a legal standpoint, National regulatory frameworks in compliance with the Cartagena Protocol should include provisions for consulting the public in decision-making regarding transgenic organisms. This need for consultation will be especially critical in the case of transgenic mosquitoes because research and development will inevitably require interaction with communities, and cannot succeed without the agreement and support of the public. Therefore, developers of transgenic mosquitoes to control vector-borne diseases will inevitably be required to engage the public not only to assure the technical success of their research but also to enable positive decisions by relevant regulatory agencies [15]. This social aspect of work with transgenic mosquitoes provides the informal freedom to operate that is the indispensable counterpart of formal regulatory decisions.

It should be noted that certain organized groups are philosophically opposed the development of transgenic organisms and will be expected to object to any development of transgenic mosquitoes. The reasons for opposition by these groups is not within the scope of this review, but are such that it is unlikely to gain their acceptance of any deployment of transgenic solutions for disease control. Therefore, engagement and consultations with these groups requires a different approach than would be needed to gain acceptance of the general public.

CONCLUSION

The use of transgenic strategies to control mosquito populations brings them under the framework established internationally to regulate such organisms. The national laws and regulations governing research and development of transgenic mosquitoes have just been established in many countries where the benefits will be greatest. Furthermore, the experience of regulators in these countries is quite low, and based on paradigms developed through the regulation of transgenic crops. Working within this system to increase the capacity of regulatory agencies in disease-endemic countries to make appropriate decisions about the deployment of advanced control strategies will require time and appropriate effort and will rely also on development of regulatory policies regarding transgenic mosquitoes in countries with more experienced and better-resourced regulatory agencies. Without a favorable regulatory pathway, researchers will be unable to make transgenic tools available to the areas where they are most needed.

REFERENCES

[1] Cartagena Protocol. Text of the Cartagena Protocol on biosafety [Internet]. Convention on biological diversity. Available from: <http://bch.cbd.int/protocol/text/> [cited November 30, 2014].

[2] Podevin N, Devos Y, Davies HV, et al. Transgenic or not? No simple answer! EMBO Rep 2012;13:1057−61.

[3] Jones HD. Regulatory uncertainty over genome editing. Nat Plants 2015;1:1−3.

[4] Pauwels K, Podevin N, Breyer D, et al. Engineering nucleases for gene targeting: safety and regulatory considerations. New Biotechnol 2014;31:18−27.

[5] Durvasula RV, Gumbs A, Panackal A, et al. Prevention of insect-borne disease: an approach using transgenic symbiotic bacteria. Proc Natl Acad Sci USA 1997;94:3274−8.

[6] Convention on Biological Diversity. Text of the CBD [Internet]. Convention on biological diversity. Available from: <http://www.cbd.int/convention/text/> [cited November 30, 2014].

[7] Office of Science and Technology Policy. Coordinated framework for regulation of biotechnology. Executive Office of the President, Office of Science and Technology Policy; 198651 FR 22302.

[8] Simmons GS, et al. Field performance of a genetically engineered strain of pink bollworm. PLoS One 2011;6:e24110.

[9] Oxitec. Florida keys project [Internet]. Oxitec Ltd. Available from: <http://www.oxitec.com/health/florida-keys-project/> [cited November 30, 2014].

[10] Benedict M, Bonsall M, James AA, James S, Lavery J, Mumford J, et al. Guidance framework for testing of genetically modified mosquitoes. Geneva: World Health Organization; 2014.

[11] Harris A, McKemey AR, Nimmo D, et al. Successful suppression of a field mosquito population by sustained release of engineered male mosquitoes. Nat Biotechnol 2012;30:828−30.

[12] Harris A, Nimmo D, McKemey A, et al. Field performance of engineered male mosquitoes. Nat Biotechnol 2011;29:1034−7.

[13] Lacroix R, McKemey A, Raduan N, et al. Open field release of genetically engineered sterile male *Aedes aegypti* in Malaysia. PLoS One 2012;7:e42771.

[14] Oxitec. Our products. Oxitec Ltd. Available from: <http://www.oxitec.com/health/our-products/> [cited November 30, 2014].

[15] Ramsey J, Bond J, Macotela M, et al. A regulatory structure for working with genetically modified mosquitoes: lessons from Mexico. PLoS Negl Trop Dis 2014;8:1−9. Available from: http://dx.doi.org/10.1371/journal.pntd.0002623.

[16] Carvalho DO, McKemey AR, Garziera L, et al. Suppression of a field population of *Aedes aegypti* in Brazil by sustained release of transgenic male mosquitoes. PLoS Negl Trop Dis 2015;9:1−15.

[17] Carvalho DO, Nimmo D, Naish N, et al. Mass production of genetically modified *Aedes aegypti* for field releases in Brazil. J Vis Exp 2014;83:e3579.

[18] Information Systems for Biotechnology. USDA field tests of GM crops [Internet]. Information Systems for Biotechnology. Available from: <http://www.isb.vt.edu/search-release-data.aspx> [cited September 10, 2014].

[19] National Biosafety Technical Commission. Technical opinion no. 3964/2014; 2014. p. 1−14.

[20] Reeves RG, Denton JA, Santucci F, et al. Scientific standards and the regulation of genetically modified insects. PLoS Negl Trop Dis 2012;6:1−15.

[21] USDA-APHIS. Use of genetically engineered fruit fly and pink bollworm in APHIS plant pest control programs. Final environmental impact statement. Washington, D.C., USA: US Department of Agriculture; 2008.

[22] USDA-APHIS. Use of genetically engineered fruit fly and pink bollworm in APHIS plant pest control programs; record of decision. Docket no. APHIS-2006-0166. Federal Register 2009;74:21314−6.

[23] Ostera GR, Gostin LO. Biosafety concerns involving genetically modified mosquitoes to combat malaria and dengue in developing countries. J Am Med Assoc 2011;305:930−1.

[24] Marshall JM. The effect of gene drive on containment of transgenic mosquitoes. J Theor Biol 2009;258:250−65.

[25] Oye K, Esvelt K, Appleton E, et al. Regulating gene drives. Science Express publication; July 17, 2014 (10.1126/science. 1254287).

[26] Esvelt K, Smidler A, Caterruccia F, et al. Concerning RNA-guided gene drives for the alteration of wild populations. eLife 2014. Available from: http://dx.doi.org/10.7554/eLife.03401.

[27] National Institutes of Health. NIH guidelines for research involving recombinant or synthetic nucleic acid molecules [Internet]. National Institutes of Health. Available from: <http://osp.od.nih.gov/sites/default/files/NIH_Guidelines.html> [updated November 6, 2013; cited November 30, 2014].

[28] Benedict MQ, Tabachnick WJ, Higgs S. Arthropod containment guidelines. Vector-Borne Zoonotic Dis 2003;3:57−98.

[29] Benedict M, D'Abbs P, Dobson S, et al. Guidance for contained field trials of vector mosquitoes engineered to contain a gene drive system: recommendations of a Scientific Working Group. Vector-Borne Zoonotic Dis 2008;5:127−66.

[30] Marshall JM, Hay B. Confinement of gene drive systems to local populations: a comparative analysis. J Theor Biol 2012;294:153−71.

[31] Benedict M, Eckerstorfer M, Franz G, et al. Scientific/technical report submitted to EFSA defining environmental risk assessment criteria for genetically modified insects to be placed on the EU market. Vienna, Austria: Environment Agency Austria; 2010.

[32] Biosafety Clearing House. Guidance on risk assessment of living modified organism: risk assessment of living modified mosquitoes [Internet]. Convention on biological diversity. Available from: <http://bch.cbd.int/onlineconferences/guidancedoc_ra_mosquitoes.shtml> [cited November 30, 2014].

[33] Kalaitzandonakes N, Alston JM, Bradford KJ. Compliance costs for regulatory approval of new biotech crops. Nat Biotechnol 2007;25:509−11.

[34] Marshall JM. The Cartagena Protocol and genetically modified mosquitoes. Nat Biotechnol 2010;28:896−7.

[35] USDA-APHIS. Availability of an Environmental Assessment for the Field Release of Genetically Engineered Diamondback Moths. Docket No. APHIS−2014−0056. Federal Register 2014;79:51299−51300.

Chapter 17

Economic Analysis of Genetically Modified Mosquito Strategies

Eduardo A. Undurraga, Yara A. Halasa and Donald S. Shepard
Schneider Institutes for Health Policy, Heller School, Brandeis University, Waltham, MA, USA

INTRODUCTION

Dengue is currently the most important mosquito-borne viral disease in the world and among the most important emerging infectious diseases. Recent studies have estimated that possibly more than half of the world's population is at risk of dengue virus (DENV) infection in more than 140 countries and territories, resulting in about 100 million symptomatic cases and 10,000–20,000 deaths annually [1,2]. The dengue pandemic has been primarily driven by urban population growth [3], lack of effective mosquito control [3–6], and growing international trade and travel [5,7,8]. The economic and disease burden of dengue is substantial [9–13].

Dengue is mainly transmitted by *Aedes aegypti*, an urban-adapted mosquito widely distributed in tropical and subtropical regions of the world [14,15]. Existing vector control programs have not halted the spread of *Ae. aegypti*, which mirrors the distribution of DENV [16–19]. Although comparatively less important, *Aedes albopictus* also contributes to DENV transmission, and its range has also increased in recent years [20]. Currently, vector control is the only prevention tool available to control DENV transmission, but this approach has failed in most countries [4–6]. Vector control strategies are usually based on outdoor fogging and the use of larvicides. While countries spend billions of dollars per year worldwide on vector control, there are few available economic studies about the strategy to guide policy decisions [21]. Countries that allocate a budget for dengue prevention and control may incur additional expenditures during outbreaks. Careful economic evaluations provide comparable output measures, and thus allow objective resource allocation and strategic planning decisions [22]. In April 2008, the Pediatric Dengue Vaccine Initiative (PDVI) sponsored an expert panel on dengue and health economics to systematically assess existing literature on dengue economics and identify research needs and gaps to guide

Genetic Control of Malaria and Dengue.
© 2016 Elsevier Inc. All rights reserved.

evidence-based policy. Vector control assessments were named among the priorities for dengue research [13].

Ongoing efforts to control dengue include the development of a dengue vaccine [23−26] and antiviral drugs [27−29], but suppressing the vector population is the main strategy used to reduce DENV transmission currently. Several strategies of vector control have been implemented, with low degrees of success in reducing dengue episodes [5,30]. Conventional vector control strategies focus on reducing the mosquito population, reducing the life span of adult mosquitoes, and minimizing mosquito−human contact by eliminating breeding sites, using larvicides, and space spraying of insecticides [31−33]. While the WHO recommends an integrated vector-management strategy, including environmental management (e.g., cleanup campaigns), epidemiological and entomological surveillance, education and community outreach, and chemical and biological control [34,35], several factors have rendered vector control strategies insufficient in most countries [5,36,37]. These factors include limitations in existing technologies, poor insecticide deployment and insecticide resistance, unsustained efforts, insufficient education and dengue awareness, market constraints, privacy and logistic constraints against indoor spraying, limited resources and competing priorities in endemic countries, and, of course, the complexities of the vector system [5,6,38−40].

Conventional vector control is mostly based on the use of larvicides and pesticides. Larvicides need to be applied to breeding sites, which are dispersed and many times difficult to reach. Insecticide fogging is unpleasant to people and may potentially damage natural habitats, as it kills other insects as well. Another major limitation is that mosquitoes develop insecticide resistance [39,41,42], and fogging kills only adult mosquitoes, but not eggs. Studies suggest that insecticide fogging has low, if any, impact on dengue transmission [5]. At best, it would be effective only if fumigation is applied early in the transmission season [43]. There have also been recent efforts to develop new strategies using insecticides with residual activity [44−46], but they have not yet proved effective. The effectiveness of these techniques is partially limited by the difficulty in reaching the whole adult *Ae. aegypti* population and by the existence of oviposition sites that are hidden or too difficult to reach. Biological control strategies using parasites and/or mosquito predators, such as guppy fish, are used in some countries such as Cambodia and the Philippines [47].

Community-based interventions potentially provide a sustainable way of controlling dengue through breeding site removal, but have had ambiguous results to date. Community involvement takes time and requires a sustained long-term effort. Some studies have reported an effective reduction in the vector population using a community-based integrated approach, including an educational component to increase awareness and understanding of best practices, and targeted vector control approaches [37,48−53]. However, a systematic review in 2007 found weak evidence of the effectiveness of community-based dengue control programs [40].

Costs of dengue vector control are substantial [21,54]. For countries with published comprehensive estimates of the economic burden of dengue, routine vector control programs represent about half of the total costs of dengue illness. There have been a few studies that have estimated the cost of dengue vector control and surveillance activities, including the aggregated economic costs, while others have focused on the economic costs of specific vector control strategies. We conducted a systematic literature review and found 18 articles in 15 countries (8 in Latin America and the Caribbean, 6 in Southeast Asia, and 1 in Africa) that assessed the cost of dengue vector control and converted the costs to 2013 US dollars (Table 17.1). Nine of these articles and manuscripts analyzed the comprehensive vector control activities [31,55−62]. The average per capita cost of dengue vector control in these nine countries was $2.14 in 2013 US dollars. The remaining 9 focused on the cost of a specific vector control intervention such as community mobilization or source reduction [31,49,51,63−69]. The development of insecticide resistance [39,41,42] makes conventional vector control strategies even more challenging and may increase the magnitude of epidemics. If half of the population at risk of dengue received comprehensive conventional vector control measures, the global costs would amount to about $6 billion annually [21].

The high costs of current vector control strategies and their seemingly low effectiveness have encouraged the search for new technologies. New biological technologies to control *Ae. aegypti* under development are intended to break the cycle of DENV transmission. These technologies include the insertion of *Wolbachia* bacteria in mosquitoes [71−74] and the use of genetically modified (GM) mosquitoes [75−80]. These biological technologies also have the advantage of addressing the problem of hidden or difficult reach breeding sites, which are inaccessible using conventional technologies [30].

GM mosquitoes may potentially prove to be safer than conventional vector control methods and effective in controlling dengue vectors. The reduction of *Ae. aegypti* is expected to have little or no impact on ecosystems. The technology is species-specific (which reduces the potential of affecting nontargeted species), self-limiting, and may potentially reduce pesticide use [81]. Furthermore, *Ae. aegypti* is an invasive species in most regions, is eaten opportunistically but is not the major food source for other species [30]. However, the development of these technologies has not been without controversy [82−86]. Because the first test release of transgenic mosquitoes was a surprise to the public and to many scientists in the field [84,85], some critics have accused developers of secrecy and insufficient engagement with the community. But most critics have pointed at insufficient regulatory measures and guidelines [86,87]. Specific recommendations for an assessment of GM mosquitoes as a vector control strategy are being developed [77,80,88,89].

TABLE 17.1 Cost of Vector Control Activities Based on Empirical Publications from 2000 through 2014 (2013 US Dollars)

Authors	Setting	Country	Period	Vector Control Activities	Unit Costs
Comprehensive vector control activities					
Baly et al. [31]	Trujillo state	Venezuela	2007	Insecticides and larvicides	$0.58[a]
Undurraga et al. [55]	Mexico	Mexico	2010–2011	Surveillance, insecticide, nebulization, indoor spraying, larvicides, educational and awareness campaigns, and community-based participatory control programs	$0.79[b]
Kongsin et al. [56]	Thailand	Thailand	2005	Education, limited use of larvicides, and insecticide	$1.15[a]
Taliberti et al. [57]	Sao Paulo City	Brazil	2005	Active surveillance, inspection, education, larvicide, and insecticide	$1.31
Baly et al. [31]	Laem Chabang Municipality	Thailand	2007	Insecticides and larvicides	$1.42[c]
Armien et al. [58]	Panama Province	Panama	2005	Surveillance, laboratory, and vector control activities	$1.80[a]
Perez-Guerra et al. [59]	Puerto Rico	Puerto Rico	2002–2007	Surveillance, cleanup campaigns, fumigation, inspection, education and management	$2.31[d]
Packierasami et al. [60]	Malaysia	Malaysia	2010	Inspection, surveillance, fogging, larviciding, and health education	$2.82[b]
Baly et al. [61]	Guantanamo	Cuba	January–July 2006	Surveillance, source reduction, larviciding, insecticide, education, and active screening for fever cases	$3.10[e]

Orellano and Pedroni [62]	Clorinda	Argentina	January–April 2007	Surveillance, source reduction, fogging, larviciding, and education.	$4.32[f]
Baly et al. [61]	Guantanamo	Cuba	August–December 2006	Surveillance, source reduction, use of larvicide and insecticide, education, and active screening for fever cases	$6.79[e]

Specific vector control interventions

Kay et al. [63]	Xuan Phong District	Vietnam	2007	Community-based strategies to control dengue	$0.09
Suaya et al. [51]	Phnom Penh and Kandal Province	Cambodia	2001–2005	Larviciding campaigns against *Ae. aegypti*	$0.24
Kay et al. [63]	Tho Nghiep District	Vietnam	2007	Education, larvicide, insecticide, and community participation	$0.28[g]
Tun-Lin et al. [64]	Vietnam	Vietnam	2004	Mesocyclops in productive containers	$0.32
Tun-Lin et al. [64]	Myanmar	Myanmar	2004	Dragon-fly nymphs, fish	$1.13
Rizzo et al. [65]	Poptun, El Peten	Guatemala	2009–2010	Insecticide-treated curtains	$1.30
Baly et al. [31]	Trujillo State	Venezuela	2007	Long-lasting insecticide-treated curtains	$1.53[h]
Baly et al. [49]	Santiago de Cuba	Cuba	2001–2002	Conventional dengue control plus community participation	$2.22
Tun-Lin et al. [64]	Philippines	Philippines	2004	Tire splitting, drum and dish rack cleaning, waste management	$2.42
Tun-Lin et al. [64]	Mexico	Mexico	2004	Bucket and flower pot management	$2.51[i]
Pepin [66]	Minas Gerais	Brazil	2009–2011	Intelligent dengue monitoring system	$3.10[j]

(Continued)

TABLE 17.1 (Continued)

Authors	Setting	Country	Period	Vector Control Activities	Unit Costs
Tun-Lin et al. [64]	Kenya	Kenya	2004	Temephos in large productive container	$3.24[k]
Baly et al. [31]	Laem Chabang Municipality	Thailand	2007	Long-lasting insecticide-treated curtains	$3.35[l]
Tozan et al. [67]	Plaeng Yao District, Chachoengsao Province	Thailand	2014	Insecticide-treated school uniforms	$5.50[m]
Ditsuwan et al. [68]	Muang District, Songkhla province	Thailand	2009	Standard indoor ultra-low-volume space spraying	$6.03[n]
Lorono-Pino [69]	Merida City	Mexico	2012	Insecticide consumer products	$8.50

[a]Study was done during epidemic year or season.
[b]Study was done during nonepidemic year.
[c]Estimates correspond to the average over 2 years.
[d]Estimates correspond to the average over 5 years.
[e]Results were presented as $1.89 per household January to July 2006, and $2.14 per household in August to December 2006.
[f]January through February are the months with higher DENV transmission in Clorinda, so vector control costs are most probably not representative of costs during the rest of the year.
[g]Authors used a discount rate of 6% and did not include the costs of buildings.
[h]Cost per capita was obtained assuming an household size of 3.7 (http://geo-mexico.com/?p = 3162).
[i]Partnership model with supervision, included delivery, training, personnel, amortization capital cost, supplies and materials, and utilities.
[j]Cost per capita was obtained assuming an average household size of 3.2 [70].
[k]Cost derived using factory proprietary method. Their estimate excluded the costs of international shipment, collection of uniforms, and distribution to households.
[l]Partnership and vertical models. Cost estimates included delivery, training, personnel, amortization capital cost, supplies and materials, and utilities.
[m]Cost estimates included microcredit fund and in-kind contribution of health workers and teachers.
[n]Opportunity cost represented 90% of total costs.
Notes: Per capita vector control costs for specific interventions were not annualized because of variations in dengue season across countries; hence, comparisons between specific interventions or countries must be done with caution.

There have been recent small-scale trials of GM mosquitoes in the Cayman Islands (2009, 2010) [90,91], Malaysia (2010) [92,93], and Brazil (2011) [94,95], and of *Wolbachia*-infected mosquitoes in Queensland, Australia [72,73] with promising results. *Wolbachia* trials are also in preparation for Brazil, Colombia, Indonesia, and Vietnam [96,97]. Developers of the GM mosquito strategy developed a model of the potential impact and cost of that strategy for dengue control assuming a 100% lethal genetic construct and fully competitive males. Their model predicted rapid elimination of the dengue vector at a cost between $2.17 and $33.37 per case averted (adjusted to 2013 US dollars) [98]. Economic analysis of these new technologies under development, including *Wolbachia*-infected mosquitoes and GM mosquito methods, is important to inform policy makers and donors and to guide health policy decisions and investments [99−101].

In general, comparing the costs related to GM mosquitoes with more conventional vector control technologies is challenging for two main reasons. First, while conventional vector control approaches have not been able to prevent DENV transmission and dengue outbreaks, it is difficult to measure their effectiveness because dengue incidence varies widely by location and year. Second, GM mosquitoes are a relatively new technology that is still under development and additional studies on implementation strategies and effectiveness are needed to better understand the technology and accurately assess its potential use and costs of implementation. Despite the paucity of data, our goal in this chapter is to provide a preliminary framework for assessing a GM mosquito strategy to technology to decrease vector density below the DENV transmission threshold. We present an evaluation framework for a cost-benefit analysis of GM mosquito technologies and apply this framework to a hypothetical scenario combining epidemiological data from various dengue-endemic countries. We discuss the results, main assumptions, and approaches to refine these estimates.

METHODS FOR ECONOMIC APPRAISAL OF GM MOSQUITO TECHNOLOGIES

General Framework

Cost-effectiveness assessments compare the costs of an intervention to its impact measured in disability-adjusted life years (DALYs), a person's years of life lost due to premature mortality plus the time a person lives with a disability imposed by the disease [102]. A cost-benefit analysis, while largely based on the same estimates, would compare the costs of the technology and its benefits using monetary values. Guidelines for cost and cost-effectiveness analysis of vector control programs have been published elsewhere [103−105].

Strategies to decrease dengue transmission involving GM mosquitoes vary in persistence and intended effect. Some GM mosquitoes are intended to spread persistently among the wild mosquito population (self-sustaining systems), whereas others need to be maintained through constant releases of GM mosquitoes (self-limiting systems) [79,106]. GM mosquito strategies may also differ in their intended effect: some strategies are aimed at reducing the total number of mosquitoes (population suppression), such as Oxitec's male OX513A *Ae. aegypti* mosquitoes, while others are aimed at reducing the ability of mosquitoes of spreading DENV, such as mosquitoes infected with *Wolbachia*.

Intensive suppression programs: Recent trials of a self-limiting population-suppression system using GM male mosquitoes have found encouraging results, as the male mosquitoes successfully mated with wild females and the target population of DENV vector mosquitoes was effectively reduced [91]. Apparently, the efficiency of GM mosquitoes increases as the mosquito population gets smaller, due to an increase in the ratio of GM males to wild mosquito males. In other words, the technology is intensive in an early stage and requires much smaller amounts of GM male releases to maintain areas below the DENV transmission threshold.

The *Ae. aegypti* OX513A strain, developed by Oxitec, cannot persist in the wild due to a lethal genetic element that is suppressed when reared in a laboratory setting [92]. The males of this strain mate with wild female *Ae. aegypti* mosquitoes, producing offspring that will die in the early developmental stage. There have been a few preliminary trials involving GM mosquitoes, carried out by Oxitec and partners, involving rearing and releasing male OX513A *Ae. aegypti* mosquitoes among populations of several thousand people. GM mosquitoes (OX513A *Ae. aegypti* strain) were first released in the East End, Cayman Islands in 2009, by the Cayman Islands' Mosquito Research and Control Unit, with about an 80% reduction in the wild mosquito population [90,91]. This open release was followed by a suppression study in an area with approximately 200−300 inhabitants in 16 hectares (ha) [81]. In 2010, GM mosquitoes were released into an inhabited forest in Bentong, Malaysia, to assess their survival and dispersal [92,93]. Since 2011, GM mosquitoes have been released for suppression studies at various sites in Bahia, Brazil [94,95], including Itaberaba in 2011 and 2012 (about 1,800 inhabitants and 11 ha), Mandacaru in 2012 and 2013 (about 2,800 inhabitants and 40 ha), Pedra Branca in 2012 and 2013 (about 1,200 inhabitants and 23 ha), and Jacobina beginning 2013 and still ongoing (about 50,000 inhabitants) [81], although no results from these trials have been published so far. Preliminary estimates from Oxitec have found an average reduction of about 92−99% in the relative number of eggs per trap between treated and control areas in the Cayman Islands and Brazil [81]. Finally, there is also an ongoing suppression study in Nuevo Chorrillo, Panama (about 1,000 inhabitants, 10 ha), where Oxitec has reported a reduction in *Ae. aegypti* larvae of about 66−76% after 17 weeks of GM male mosquito releases [81]. These field studies are producing valuable information, including mosquito production, release, scale of implementation, operational challenges, resources involved, but the technology is still under development and its implementation is an ongoing learning process. Ideally, the GM mosquito strategy of vector control should be scalable from small towns, where field trials are currently being carried out, to larger towns and cities.

The available evidence suggests that the implementation of a GM mosquito strategy has at least three main components [79,80,107]. The first, and the most important from an economic perspective, is the technical part, which includes the research and development of the GM technology and implementation definitions. The second component is the regulatory process, which includes biosafety approval for the mosquito releases and involves engaging the political and the scientific community. The third component is related to local community engagement and requires educational campaigns and community outreach to raise awareness of dengue and the technology.

For example, Figure 17.1 shows the summary of the biosafety regulatory process, public engagement, and time frame before the release of GM male *Ae. aegypti* mosquitoes in a limited marked release and recapture field

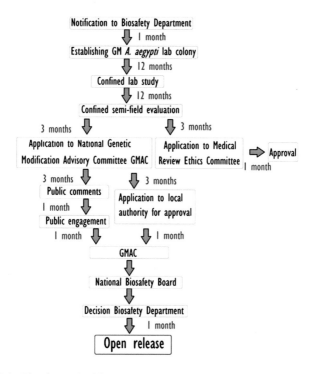

FIGURE 17.1 Biosafety and public engagement process in a limited marked release and recapture field experiment in Malaysia with GM mosquitoes. Summary of the biosafety regulatory process, public engagement, and approximate time frame before the release of GM *Ae. aegypti* male mosquitoes in a limited field experiment in Malaysia. *Source: Ref. [92].*

experiment in Malaysia [92]. The objectives of this field experiment were to evaluate flight distance and longevity of the GM mosquito strain. Dengue control programs need to be understood and accepted by the population, particularly since the vector control strategy involves mosquito releases in the community [85,87,108]. These activities may include community meetings, household visits, distribution of flyers, radio and television broadcasts, cars with loudspeakers, door-to-door visits, and newspaper articles, among others [81].

Economic Costs of GM Mosquitoes

There have been very few economic evaluations of these new technologies. Alphey and others [98] developed a combined model framework to estimate the cost-effectiveness of genetic vector control. By creating a simplified model of the interaction among DENV, mosquito, and human populations, the authors were able to obtain an estimate of the effects of using a dengue control strategy based on GM mosquito vector control. Using several assumptions, including 100% lethality in offspring, fully competitive males, and a simplified model for DENV transmission, their estimates suggest a range of $2—$30 per dengue episode averted (in 2008 US dollars).

Estimating the costs of a GM mosquito strategy to control dengue vectors requires costing all the resources used for vector control activities at the various stages of the project. This section is focused on the potential implementation costs of GM mosquitoes dengue control technology, and the resources involved in the various parts of this process. Figure 17.2 shows the various stages in a potential implementation of a GM mosquito control strategy.

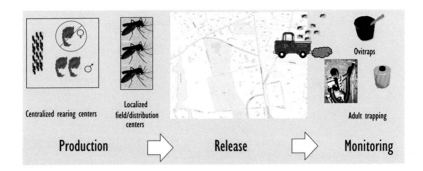

FIGURE 17.2 Example of a potential implementation process of a GM mosquito strategy for vector control. Stages in the field implementation of a potential GM mosquito strategy. (i) Production: a mosquito egg colony is reared in centralized centers were male and female pupae are separated. Male pupae are then sent to localized field/distribution centers, were mosquito pupae are grown to male adult mosquitoes. (ii) Release: mosquitoes are released about three times per week at specific points in an intervention area. A male would seek out females up to 200 m. (iii) Monitoring is based on direct and indirect entomological surveillance using adult trapping (aspiration and BG-sentinel) and ovitraps [81].

Estimating the cost of production is complex and much of the information required for a precise estimate is not yet known. The resources involved in each stage of mosquito production typically consist of capital, personnel, and recurrent costs. Capital costs include all initial investments required for the operations, including vehicles, production facilities, office equipment, field equipment, etc. Recurrent costs may involve, for example, all supplies and materials used and operational costs involved in the suppression and maintenance phases, including the production of eggs and mosquitoes, quality control throughout the production process, releases of adult males, and monitoring and surveillance activities. Supplies and materials include protective clothing, gloves, mosquito cages, office materials, ovitraps, and laboratory equipment, among others. Operational costs may include fuels and lubricants used in the vehicles, maintenance and insurance of vehicles, laboratory equipment, and mosquito production facilities, laboratory supplies, utilities, etc. Personnel costs include all staff involved in the dengue prevention activities. Their costs include base salaries, fringe benefits, and overtime for operational activities and providing and receiving training. Other production costs may include overhead costs or other operating expenses.

The implementation of a GM mosquito strategy is expected to have two major phases:

1. *Phase 1. Suppression*: The objective in this phase is to substantially reduce the total mosquito population by releasing enough GM males to reach a high ratio between GM male mosquitoes and wild male mosquitoes, and reduce the number of surviving larvae. This suppression phase is expected to take 3−6 months of GM male mosquitoes releases.
2. *Phase 2. Maintenance and monitoring*: The objective of this stage is to maintain a low population of *Ae. aegypti* by releasing enough males to maintain a high ratio of GM males to wild males. This low population would maintain the DENV transmission low enough to potentially avoid dengue outbreaks. Once suppression (or near suppression) is achieved, the costs of the GM mosquitoes technology would be much lower and would involve an active program of vector monitoring.

Main Cost Drivers for the GM Mosquito Strategy

Releases of sterile GM mosquitoes need to be tailored to the characteristics of each site. Areas with more abundant mosquitoes will potentially require the release of more GM mosquitoes to outnumber the wild male population. Technological development so far suggests that male GM mosquitoes need to be bred at a dedicated facility. The implementation of this technology requires making several decisions that will affect the operational costs of a GM mosquito vector control strategy:

Production process: An expected key design factor is whether the mosquito eggs should be produced at the site of implementation. Would it be better instead to build a facility in a country that provides comparative productive advantages, such as India, and send only the eggs to target locations? In this case, the only major investment in or near the intervention area would be a facility to grow the adult mosquitoes. Egg production requires both an ongoing colony that generates the eggs and an intensive process of monitoring the quality of GM males. The latter is potentially a substantial part of the costs of the facility and staff time. For example, if the technology were to be implemented in Brazil, would it be better to build a factory in a location like Rio de Janeiro, tackle the mosquito population in that city, and then use the facility to send eggs to other areas in the country? The recipient of the eggs would need a controlled environment to grow the mosquitoes and logistic capacity to release them in strategic locations.

Site implementation: Should an implementation try to tackle a complete city at once? Would a municipality-based strategy work better from a financial point of view? These decisions will probably depend on the size and population of the city, dengue transmission rates, and vector population. In general, using a wedge-implementation strategy may offer the flexibility to divide the city into areas, reduce or suppress mosquito population in each area, and maintain a low-volume release program. A stable control of the mosquito population is expected to be achieved in about 4 years, according to preliminary data from the Oxitec trials, but this estimate may be reduced with ongoing improvements in implementation strategies, technology, and further knowledge about vector ecology.

Maintenance strategy: The intensity of the vector control program at the maintenance phase is another variable that needs to be considered from an economic point of view. For example, if the technology works as expected, one option would be to suppress or almost suppress the mosquito population, which would be more expensive, but most effective in decreasing DENV transmission. Another option would be to release GM mosquitoes less intensively, at a sufficient rate to maintain the mosquito population below the DENV transmission threshold.

Economies of scale and new technologies: The costs of implementation of a GM mosquito strategy for vector control will probably be reduced in time from economies of scale and new technologies. For example, new technologies may not only result in more efficient production costs but also reduce the cost of implementation, such as drones to release male OX513A *Ae. aegypti* mosquitoes to intervention sites.

Time frame: Because the implementation of an effective GM mosquito vector control strategy is expected to take at least 2 years (considering the suppression and maintenance phases), stabilization of the mosquito population would likely require a few more years. Thus, a reasonable time frame for the evaluation of this program would be about 5 years.

Main Sources of Uncertainty in Cost Analysis

Seasonality of dengue: The implementation of GM mosquito technology may vary according to the dengue season. For example, in areas with no clear dengue season, the technology can be implemented throughout the year. In contrast, in areas with high seasonality of dengue, GM mosquito technology may be implemented focusing on the seasonal increase of the *Ae. aegypti* population.

DENV transmission: The impact of this technology largely depends on the dengue transmissibility and its spatiotemporal variation. While there has been some progress in understanding DENV transmission, there is vast uncertainty in the complex relation between transmissibility and mosquito population density thresholds, which is the key variable to assess the impact of a vector control technology [109−112]. Currently, there is no satisfactory answer to the question of the vector density thresholds that would stop transmission or how they vary with other factors, such as herd immunity, nor there is an agreed-upon, reliable entomological measure of risk of DENV infection in a human population [6,110], particularly in hyperendemic areas [113]. Because *Ae. aegypti* is an efficient DENV vector, due to its ecology and behavior [15], the entomological threshold for transmission is particularly low [113,114].

The case of Singapore is particularly illustrative of this complex relation between vector population and DENV transmission. Through the implementation of an intensive vector control system beginning the early 1970s, including entomological surveillance and elimination of breeding sites, public involvement, and law enforcement, Singapore achieved very low vector population densities and dengue incidence for about 15 years [115]. However, despite the maintenance of a reduced vector population, the incidence of dengue has surged since the 1990s, possibly due to lowered herd immunity, increased DENV transmission outside the home, a bigger share of symptomatic dengue, and changes in surveillance [115]. A recent systematic review [111] examined the relation between vector indices and dengue incidence. The results were inconclusive: the authors found little evidence of a correlation between these two variables. One reason may be the complexity introduced by the spatiotemporal heterogeneity of dengue transmission. There are large variations in spatial and time scales in the transmission of dengue [116,117]. A study in Iquitos, Peru, found annual and seasonal variations in transmissibility, in addition to variations by DENV serotype [118].

Figure 17.3 shows the preliminary results from a GM mosquito release trial in Itaberaba, Brazil. The transmission thresholds (in pupae per person) shown on the graph are based on estimates by Focks et al. [119], using simulation models with predefined population seroprevalence, air temperature, and infectious individuals. The *Ae. aegypti* mosquito population was below the 0% transmission threshold in about 9 months following the start of the male OX513A *Ae. aegypti* mosquitoes release program. Preliminary results suggested an 85% reduction in adult mosquito density and a 79% reduction in Ovitrap index relative to untreated areas. While there is still much uncertainty in the relation between vector population and DENV transmission, the preliminary results from the trial are within the theoretical range found by Focks et al. [119] that theoretically would eliminate transmission in most countries.

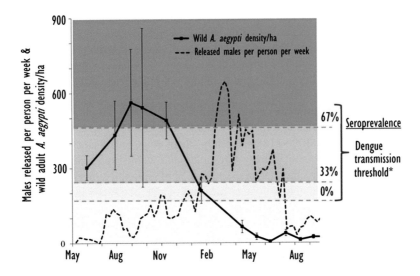

FIGURE 17.3   Itaberaba, Brazil: Oxitec trial preliminary results. RIDL male release rate and wild *Ae. aegypti* density (estimated with mark–release–recapture). Error bars denote 95% confidence intervals. Ha denotes hectares. *Data on mosquito releases were presented by Oxitec [81].* *Dengue transmission thresholds are based on results from a model, not empirical data [119].*

Return on investments: Another source of cost uncertainty relates to the rate of return that would be acceptable for private investors.

Efficacy in implementation: Implementation would probably have varying degrees of success. Will everything work as planned? How does the implementation success vary by seasonality, weather, and dengue epidemiology? To address this variation, we conducted a cost-benefit analysis at various levels of reduction of DENV transmission.

Benefits of Vector Control

Averted Dengue Episodes

The reduction of vector population has various benefits from averted episodes of dengue: (i) reduced direct medical and nonmedical costs, (ii) reduced indirect costs from productivity loss due to symptomatic DENV infection, (iii) reduced impact of other DENV-related costs, such as health system congestion, lost revenue from tourism, complications from comorbidity, and persistent symptoms of dengue [22]. Other benefits from effective *Aedes* vector control strategies include the reduction in transmission of other arboviral diseases, such as chikungunya [120] and yellow fever [121], and of the nuisance of mosquito bites [122].

Considering that health services have a high demand, we can assume that the health care released from averted cases would be put to alternative uses.

We used the following epidemiological indicators: nonfatal hospitalized and ambulatory episodes of dengue averted, fatal episodes of dengue averted, and DALYs averted from a reduction in symptomatic DENV infections.

Main Sources of Uncertainty in Benefits

Dengue incidence: Similar to cost estimates, a substantial source of uncertainty arises from the unclear and complex relationship between DENV transmissibility and mosquito population [109−112]. Because the benefits from vector control technologies depend on the total averted episodes of dengue, any vector control-technologies—including GM mosquitoes—will be more economically attractive in areas with higher dengue incidence. As the cost of vector control technologies decline with improvements in technology and implementation, these technologies will probably become cost-effective in areas with lower incidence of dengue, enabling a wider scale implementation of the technology.

Availability of other technologies to control dengue: Several complementary technologies to control dengue are currently being developed, including vaccines [23−26], antivirals [27−29], and other vector control technologies, such as indoor residual (long-lasting) spray treatments. As these technologies become available, cost-effectiveness evaluations will change substantially as countries will be able to combine some or all of these strategies to improve cost-effectiveness of dengue control. The best strategy will most likely need to be tailored to country-specific dengue epidemiology and public health budgets, and several systematic reviews of epidemiological studies that will assess this decision-making process are already taking place [123−128].

Herd immunity: Will a reduction in herd immunity increase the magnitude of future dengue outbreaks? [115] How will this vary considering several vaccine candidates being currently developed?

Variation by country: How would the cost-benefit analysis vary by GDP per capita and prevailing wages among countries? Our example is based on estimates for mostly middle-income countries, including Brazil, Mexico, Panama, Puerto Rico, and Thailand. How does this technology adapt to a low-income setting, such as Cambodia, or a high-income country, such as Singapore or Australia?

Framework for Cost-Benefit Analysis

Example of Implementation: Dengue Nation

Our goal was to estimate the cost-effectiveness and benefit-cost ratio of a hypothetical GM mosquito strategy to control dengue using available data. The trials so far have been conducted in areas with relatively small populations, and most likely the implementation of this technology will be at a local scale (e.g., municipal or county level) where vector surveillance and monitoring is most effective. Because we were unaware of any publicly available data on dengue incidence, dengue illness costs, and vector control costs at a municipal or county level, our analysis was based on a hypothetical place called Dengue Nation.

We have estimated the potential costs and benefits of a GM mosquito vector control strategy in seven "provinces," each named after the country (or area) where most actual empirical data were obtained. While this theoretical exercise has various limitations, which we discuss later, it has the advantage of providing real data for our estimates. We obtained the approximate costs of conventional vector control strategies for each country. The costs of vector control and the costs per dengue episodes for the "provinces" of Mexico [55], Panama [58], Puerto Rico [129], and Thailand [56,130] were obtained from previous comprehensive dengue cost studies. The costs for the "provinces" of Brazil, Sao Paulo, and Sao José do Rio Preto were obtained by combining various sources of data. The total episodes in Brazil and the costs per dengue episode were obtained from a study of economic costs of dengue in the Americas [9]. We estimated the costs of vector control for Brazil based on the 2002 budget for the Programa Nacional de Controle da Dengue [131,132], assuming that real (inflation-adjusted) expenditures have remained constant in the past decade, and for Sao Paulo and São José do Rio Preto from previous studies [57,133]. Dengue episodes in São José do Rio Preto correspond to the officially reported annual average (2006–2013) [134], unadjusted for underreporting. Our dengue episodes for Brazil were adjusted for underreporting based on a previous study [4], but because of differences in healthcare access and quality, we expect that the municipality of São José do Rio Preto would have substantially better reporting rates than the average for Brazil, and thus opted for a more conservative approach to estimate dengue episodes. We assumed that the ratio between population and reported dengue episodes in Brazil and Sao Paulo was constant. Table 17.2 shows a summary of the cost data for these "provinces" in Dengue Nation.

Model for Cost-Benefit Analysis

A comprehensive cost-benefit analysis of a vector control strategy to decrease or eliminate DENV transmission based on GM mosquitoes is complex to do, as it would need to combine at least a model for mosquito population dynamics, an epidemiological model of disease dynamics—including DENV serotypes, herd immunity, heterogeneities in DENV transmission, apparent and inapparent infections, age distribution of disease, seasonal fluctuations, a model for human demographics and mobility, and cost data for dengue episodes and for surveillance and vector control costs (Figure 17.4) [98,117,136–138]. The model would become even more complex if new control technologies currently in the pipeline, such as vaccines and antivirals, were included.

TABLE 17.2 Data Used for Dengue Nation Estimates (Costs in 2013 US Dollars)

"Province" Estimates (Country or Region)	Population 2010	GDP Per Capita	Dengue (Per 1,000 Pop)	Dengue Illness Cost Per Case	Dengue Illness Cost Per Capita	Vector Control Cost Per Capita
Brazil	195,210,154	11,311	11.11	466.07	5.18	5.27
Sao Paulo	10,953,457	11,311	1.30	466.07	0.61	1.33
São José do Rio Preto	434,000	11,311	34.29	466.07	15.98	37.81
Mexico	117,886,404	10,630	1.18	634.11	0.75	0.77
Panama	3,678,128	10,839	8.96	413.82	3.71	2.17
Puerto Rico	3,721,208	24,009	2.78	3,946.01	10.98	1.82
Thailand	66,402,316	5,674	9.91	319.39	3.17	1.17

Notes: Because we did not have specific costs of illness or GDP per capita estimates for São José do Rio Preto and Sao Paulo, we assumed they were the same as for Brazil.
Sources: Refs. [9,55–58,129,130,134,135].

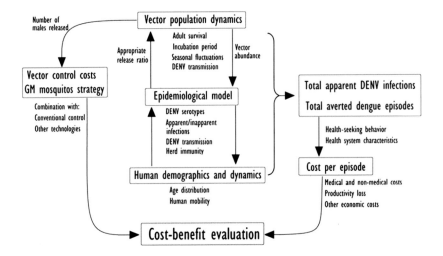

FIGURE 17.4 Overview of an ideal cost-benefit model of a vector control technology. A model for a cost-benefit evaluation would need to combine various models, including at least a model for mosquito population dynamics, DENV epidemiology, human demographics and mobility, and cost data for dengue episodes and for surveillance and vector control. *The model shown is our own, based on a combination of various sources: Refs. [98,117,136–138].*

Here, we have very simple model for a cost-benefit analysis. It includes (i) costs associated with dengue fever in a typical year, (ii) costs of conventional vector control activities, (iii) costs of GM mosquito vector control strategy, and (iv) the effectiveness in reducing transmission of dengue at each stage. The economic costs are determined by Eq. (1):

$$\text{Costs}_i = \sum_{i=1}^{n} \left((\text{episodes}_i \cdot \text{cost\_per\_episode}_i) \cdot (1 - \text{Eff}_i) + \text{CVC}_i + \text{GM\_VC}_i \right)$$

$$(1)$$

where i is the year of intervention, n is the total years of evaluation, and, in year i, episodes$_i$ are the total episodes of dengue fever, cost_per_episode$_i$ are the direct and indirect costs per dengue episode, Eff$_i$ is the effectiveness of the GM mosquito strategy in reducing dengue episodes, CVC$_i$ are the economic costs of conventional vector control technologies as currently practiced, and GM_VC$_i$ are the economic costs of GM mosquito vector control.

The benefits are determined by Eq. (2):

$$\text{Benefits}_i = \sum_{i=1}^{n} \left((\textit{episodes}_i \cdot \text{cost\_per\_episode}_i) \cdot (\text{Eff}_i) + \text{Av\_CVC}_i \right) \quad (2)$$

where Av_CVC$_i$ are averted costs of conventional vector control (if any) in year i.

RESULTS

In this section, we derive preliminary estimates for a cost-benefit analysis. We used the following assumption for our estimates: first, we assumed *Ae. aegypti* population would be maintained below the DENV transmission threshold, which as we discussed above, has not been determined and the relation between vector density and DENV transmission not fully understood. Second, based on preliminary trial results, a substantial reduction of *Ae. aegypti* (approximately 90% reduction in the population) could be expected over 6 months during the suppression phase. Because trials with these results were conducted in small areas with less complex logistics, we assumed instead that a substantial reduction in the mosquito population is achieved at the end of the first year. We also assumed that during the suppression phase, conventional vector control expenditures are maintained and GM mosquitoes vector control technologies are added. We assumed that the effectiveness of the vector control strategy on the suppression phase $Eff_{i=1}$ is different from the maintenance phase $Eff_{i=2-5}$. Third, current costs of conventional vector control techniques are completely displaced during the maintenance phase. Fourth, we assumed that a decrease of the *Ae. aegypti* mosquito population would decrease the risk of DENV transmission, with a resultant decrease in dengue episodes (nonfatal ambulatory and hospitalized, and fatal episodes). The efficiency in the reduction of dengue fever episodes is represented by the variable Eff_i. And fifth, we assumed that a reasonable time frame for evaluating this technology would be a 5-year period, which may also be a sales model: 12-month suppression phase followed by four subsequent 12-month maintenance phases.

Rough cost estimates from Oxitec suggested a preliminary overall cost estimate of the implementation of a GM mosquito vector control strategy of $25−$75 per person in the population in the suppression phase and $10−$20 in the maintenance phase. Despite these considerable preliminary cost estimates, a published study of the costs of genetics-based sterile insect methods for dengue control have estimated much lower costs per person in 2013 of $0.56−$0.73 [98]. Because implementation and development of this technology is at an early stage, with results from field trials helping to optimize implementation strategies, there is substantial variation and uncertainty in cost estimates. When the technology finally becomes available, costs may be lower due to production economies of scale, improved production technologies, optimization of program design, and implementation of GM mosquito technology with other complementary technologies (e.g., vaccines) as part of an integrated dengue control strategy.

Table 17.3 shows the annual costs of the GM mosquito vector control implementation and benefits to avert dengue (in 2013 US dollars) per 1,000 population. We assumed an effectiveness $Eff_{i=1} = 50\%$ in the reduction of dengue episodes during the suppression phase and $Eff_{i=2-5} = 100\%$ during the maintenance phase of the GM mosquitoes vector control strategy, and we used the prices in the upper bound ($75 suppression phase; $20 maintenance phase). For simplicity, we will assume that $Eff_{i=1}/Eff_{i=2-5} = 0.5$ for all models.

TABLE 17.3 Estimated Annual Costs and Benefits of GM Mosquito Vector Control Strategy (2013 US Dollars; Per 1,000 Population), Assuming 100% Effectiveness in Reduction of Dengue Episodes and Upper-Bound Cost Estimates

Dengue Nation		Implementation Phase		Aggregate
Province	Baseline	Suppression (year 1)	Maintenance[a] (years 2–5)	(5 years)
Costs	Dengue + CVC	Dengue[b] + CVC + GMs	Dengue[b] + GMm	
Brazil	−10,443	−82,856	−20,000	−162,856
Sao Paulo City	−1,939	−76,635	−20,000	−156,635
São José do Rio Preto	−53,791	−120,800	−20,000	−200,800
Mexico	−1,518	−76,144	−20,000	−156,144
Panama	−5,876	−79,023	−20,000	−159,023
Puerto Rico	−12,798	−82,306	−20,000	−162,306
Thailand	−4,331	−77,749	−20,000	−157,749
Total costs	−90,696	−595,513	−140,000	−1,155,513
Benefits	None	Dengue[b]	Dengue[b] + Av_CVC	
Brazil	0	2,588	10,443	44,362
Sao Paulo City	0	303	1,939	8,058

São José do Rio Preto	0	7,991	53,791	223,156
Mexico	0	374	1,518	6,445
Panama	0	1,853	5,876	25,356
Puerto Rico	0	5,491	12,798	56,682
Thailand	0	1,583	4,331	18,908
Total benefits	0	20,183	90,696	382,968

[a] Under the various assumptions of the model, years 2–5 have the same benefits and costs.

[b] We assumed an effectiveness Eff = 50% in the reduction of dengue episodes during the suppression phase and Eff = 100% during the maintenance phase of the GM mosquitoes vector control strategy.

Notes: Dengue denotes costs of dengue episodes, CVC denotes conventional vector control strategies, GMs denotes genetically modified mosquito vector control, maintenance phase, GMm denotes genetically modified mosquito vector control, suppression phase.

The results from Table 17.3 suggest that under the assumptions of the models, the only province where the estimated benefits from the implementation of a GM mosquito strategy would be higher than the costs is São José do Rio Preto. Combining the highest cost and most favorable effectiveness in dengue control (50% suppression and 100% maintenance), we obtained a benefit/cost ratio of 1.1 in an area with high DENV transmission, ongoing intensive vector control programs, and relatively high income.

Figure 17.5 shows the 5-year B/C ratios by province at various unit prices of GM mosquito vector control costs, assuming 25−100% of efficacy of the technology in controlling DENV transmission. The results suggest that under the assumptions considered, the program would not be cost-saving in most places we considered. However in places with high DENV transmission and ongoing intensive vector control programs, such as São José do Rio Preto, the GM program would be cost-saving.

It is important to note that because most economic benefits derived from controlling dengue come from averted direct medical and nonmedical expenses and productivity loss, benefits in areas with higher DENV transmission rates, higher population density, and in higher income settings would be considerably more than those in areas with low DENV transmission rates, low population density, and lower income settings. Given the current stage of the technology and the assumptions in our model, the GM strategy appears cost-saving in high-income countries such as Australia and Singapore, or in high DENV transmission areas such as São José do Rio Preto. Had other benefits from a reduction in dengue episodes been considered, such as avoided losses from tourism revenues, complications from comorbidities, or a reduction in transmission of other arboviruses from effective vector control, even more countries might have favorable B/C ratios (i.e., higher than 1.0).

The cost-effectiveness of a health intervention may also be assessed following World Health Organization (WHO) guidelines [139]. WHO suggests that health intervention programs where a year of life gained costs less than the per capita GPD are highly cost-effective, between one and three times the per capita GDP are cost-effective, and more than three times the per capita GDP are not cost-effective. Previous studies have found that the disease burden of dengue is approximately 65 disability-adjusted life-years (DALYs) per million population in Mexico [55] and approximately 136 DALYs per million population in Brazil [9]. Because in this case, we are interested in comparing the cost-effectiveness of two interventions (conventional and GM mosquitoes strategies for vector control), the cost-effectiveness ratio is defined by Eq. (3). Equation (3) assumes that resources devoted to the relatively ineffective conventional vector control could be redirected into GM mosquito technologies.

$$CE = \frac{\text{Costs GM mosquito VC} - \text{Cost conventional VC}}{\text{Incremental health benefits}} \qquad (3)$$

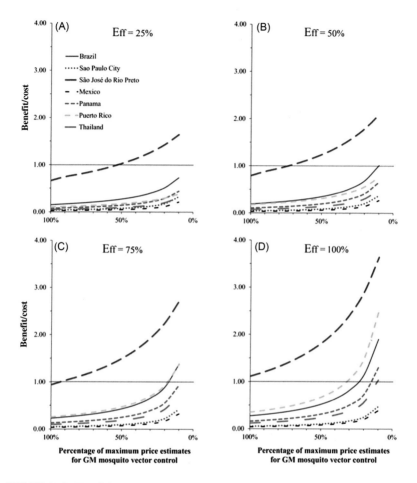

FIGURE 17.5   Benefit/cost (*B/C*) ratio by location under alternative unit prices of GM mosquito vector control costs under effectiveness (Eff) assumptions in years 2−5 of 25−100%. Maximum price estimates for GM mosquito vector control strategies were $75.00 per person in the suppression phase and $20.00 per person in the maintenance phase, based on preliminary estimates by Oxitec. It is important to note that the technology is at an early stage of development and prices are only rough estimates. These estimates are preliminary; a more thorough estimate would be possible when more detailed implementation costs and strategy became available.

For an Eff$_i$ = 100% for years 2−5 and using the upper bound of the price, the cost-effectiveness ratio is $163,051 per DALY gained for Brazil and $457,093 in Mexico. Figure 17.6 shows the scenarios for which the potential intervention would be considered cost-effective (cost per DALY gained equal or below three times the per capita GDP) in these two "provinces," assuming that the DALYs gained are proportional to Eff$_i$ in years 2−5. The "*y*" axis shows the share of the maximum annual price estimate for the GM mosquito technology

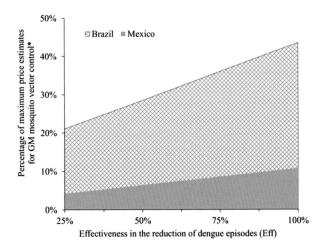

FIGURE 17.6 Maximum cost thresholds for GM to remain cost-effective at alternative levels of effectiveness. Scenarios for which the intervention would be considered cost-effective (cost per DALY gained under three times the per capita GDP) in these two "provinces" of Dengue Nation, assuming that the DALYs gained are proportional to the effectiveness of the intervention in reducing dengue episodes ($Eff_{i=2-5}$). *Maximum price estimates for GM mosquito vector control strategies were $75.00 per person in the suppression phase and $20.00 per person in the maintenance phase, based on preliminary estimates by Oxitec. It is important to note that the technology is at an early stage of development and prices are only rough estimates. These estimates are preliminary; a more thorough estimate would be possible when more detailed implementation costs and strategy became available.

($75 per person in the suppression phase and $20 in the maintenance phase), and the "x" axis shows the potential effectiveness of the technology in reducing DENV transmission (from 25% to 100% effectiveness).

Under the assumptions discussed above, if the GM technology were to be applied in areas with dengue incidence, vector control programs, and income similar to Brazil, the cost-effectiveness of the program would range from $30,000 to $163,000 per disability adjusted life year (DALY). In a lower incidence area such as Mexico, the cost-effectiveness would range from $180,000 to $457,000 per DALY. In the "province" of Brazil in our Dengue Nation, the cost per person per year of the technology would need to be reduced to 21.1% of its current price estimate to be cost-effective at 25% effectiveness ($15.83 suppression; $4.22 maintenance) and to 43.5% of its current price estimate at 100% effectiveness ($36.63 suppression; $8.70 maintenance). These estimates are preliminary; a more thorough estimate would be possible when more detailed implementation costs and strategy became available.

DISCUSSION

The precision of cost-effectiveness evaluations of any technology depends on the depth of available data on effectiveness and costs. The results so far have been based on a rough estimate of the implementation costs of a GM mosquito vector control strategy that is still under development. A more thorough evaluation would be possible if more detailed implementation costs and strategy became available. Despite data limitations and the various assumptions we made, these preliminary results are important because they help to better understand a promising technology for the control of dengue vectors, currently the only available approach to limit DENV transmission. However, a few considerations are important to keep in mind in the implementation of this technology.

The production of mosquitoes could potentially be done in two stages. Eggs and pupae would be generated in centralized rearing centers, which could then send the eggs or pupae to localized field/distribution centers where male adult mosquitoes could then be hatched and released. Rearing centers could be in locations with good linkages to ease transport, and not necessarily in all places where the technology is implemented. Production of mosquitoes could, for example, be grown on a 6-day cycle each week, with a seventh day of rest. Trucks would need to be driven around the implementation area to release the GM male mosquitoes. Release of mosquitoes may need to be done at least three times per week to realize a high percentage of mating, as evidence suggests that males survive for only about 2−3 days. It might also be possible to reduce releases to about 2 per week, with adequate mosquito monitoring, but more data from field trials are needed to understand the process more fully.

A key prerequisite to the use of GM mosquito strategies to control DENV transmission is the ability of researchers and public officials to engage with the community and transparently communicate the potential risks [140] and benefits from the strategy. While there have been important regulatory advances in the field, there are still many important challenges [80,82,83,86,141]. Dengue control programs need to be understood and accepted by the public. Interactions with the community are an important part of a program that releases mosquitoes into the community.

Critics of using GM mosquitoes to control dengue have pointed to perceived secrecy, insufficient engagement with the communities, and underscored potential effects on the environment, mostly from the creation of an "empty niche" and the effects on organisms that feed on mosquitoes [84,85,142]. Transparency and community engagement are essential for the acceptance of this technologies and should be encouraged. While ecological effects might be important and should be investigated with attention, preliminary evidence suggests that GM mosquitoes are safe and might have potential benefits in economic and disease burden, particularly in areas with high DENV transmission.

CONCLUSIONS

Dengue represents a substantial economic and disease burden for most countries in tropical and subtropical regions of the world. Vector control is currently the only prevention tool available to control DENV transmission, but ongoing strategies are not working as evidenced by the ongoing increase in geographical range and intensity of dengue. GM and other new promising technologies to control dengue vectors are being developed, and economic analysis is needed to guide public health decisions. As GM technology is still at a development stage, considerable uncertainty exists around its cost and effectiveness; furthermore, the relation between vector population and DENV transmission is not well understood [6,113]. However, the objective of this chapter was to illustrate the economic analysis of GM mosquito strategies to control dengue. Under our most likely assumptions about effectiveness and our preliminary cost estimates, our analysis suggests that without major cost reductions, GM vector control strategies are unlikely to be cost-effective in most of the places we considered. However, in places with high DENV transmission, ongoing intensive vector programs, and higher income, the program would be highly cost-effective and perhaps cost-saving. A more favorable evaluation in middle-income settings and lower DENV transmission rates may also be the case, if one includes other benefits from a reduction in dengue episodes, such as the costs avoided from losses in tourism revenues or complications from comorbidities that may develop as a consequence of DENV infection.

There is substantial uncertainty in our estimates, mainly due to a paucity of data and insufficient understanding of the relationship between mosquito abundance and DENV transmission. The high estimated costs of a GM strategy for vector control suggest that further refinements in the GM strategy, better data about critical levels of GM mosquitoes and their relation with DENV transmission, appropriate choice of implementation setting, and comprehensive evaluation data will be needed for this technology to be cost-effective on a wide scale. Possible refinements in the GM strategy may be less expensive ways of producing and releasing the GM mosquitoes, such as distributors with backpacks, motorcycles, or bicycles, or community-based distributors, identification of critical levels of GM mosquitoes, and focusing on where the cost-effectiveness would be greatest. These would be areas with high endemicity of dengue, high income, frequent introduction of new serotypes, and regular publicly funded fogging programs. Very likely, the most cost-effective way of controlling dengue transmission will involve applying integrated vector-management programs, that is, using a combination of available technologies including vaccines, antiviral strategies, and both chemical and biological vector control technologies. Effective *Aedes* vector control might also reduce transmission of other arboviral diseases such as chikungunya and yellow fever. If effective, new and improved technologies of vector control could save billions of dollars annually in averted medical costs, productivity losses, and premature deaths caused by DENV.

ACKNOWLEDGMENTS

This work was supported in part by the health theme of the 7th Framework Programme of the European Commission, Grant Agreement No. 282589. The authors are indebted to Kevin Gorman and Luke Alphey of Oxitec for their assistance in supplying information and Clare L. Hurley of Brandeis University for editorial assistance. We also thank the editor and two anonymous reviewers for thoughtful and constructive comments. Responsibility for the data and analysis in this chapter rests entirely with the authors.

REFERENCES

[1] Bhatt S, Gething PW, Brady OJ, et al. The global distribution and burden of dengue. Nature 2013;496(7446):504–7.

[2] Brady OJ, Gething PW, Bhatt S, et al. Refining the global spatial limits of dengue virus transmission by evidence-based consensus. PLoS Negl Trop Dis 2012;6(8):e1760.

[3] Gubler DJ. Dengue viruses: their evolution, history, and emergence as a global public health problem. In: Gubler D, Ooi EE, Vasudevan S, Farrar J, editors. Dengue and dengue hemmorhagic fever. 2nd ed. Wallingford, UK: CAB International; 2014. p. 1–29.

[4] Simmons CP, Farrar JJ, Nguyen VVC, Wills B. Current concepts: dengue. N Engl J Med 2012;366(15):1423–32.

[5] Gubler DJ. Prevention and control of *Aedes aegypti*-borne diseases: lesson learned from past successes and failures. Asia Pac J Mol Biol Biotechnol 2011;19(3):111–14.

[6] Reiter P. Surveillance and control of urban dengue vectors. In: Gubler D, Ooi EE, Vasudevan SG, Farrar J, editors. Dengue and dengue hemorrhagic fever. Wallingford, UK: CAB International; 2014. p. 481–518.

[7] Wilder-Smith A. Dengue infections in travelers. In: Gubler D, Ooi EE, Vasudevan S, Farrar J, editors. Dengue and dengue hemorrhagic fever. 2nd ed. Wallingford, UK: CAB International; 2014. p. 90–8.

[8] Wilder-Smith A, Schwartz E. Current concepts: dengue in travelers. N Engl J Med 2005;353(9):924–32.

[9] Shepard DS, Coudeville L, Halasa YA, Zambrano B, Dayan GH. Economic impact of dengue illness in the Americas. Am J Trop Med Hyg 2011;84(2):200–7.

[10] Shepard DS, Undurraga EA, Halasa YA. Economic and disease burden of dengue in Southeast Asia. PLoS Negl Trop Dis 2013;7(2):e2055.

[11] Shepard DS, Halasa YA, Undurraga EA. Global economic cost of dengue cases treated in the medical system. Am J Trop Med Hyg 2014;91(5S):60.

[12] Stahl H-C, Butenschoen VM, Tran HT, et al. Cost of dengue outbreaks: literature review and country case studies. BMC Public Health 2013;13(1):1048.

[13] Beatty ME, Beutels P, Meltzer MI, et al. Health economics of dengue: a systematic litera- ture review and expert panel's assessment. Am J Trop Med Hyg 2011;84(3):473–88.

[14] Gubler DJ, Ooi EE, Vasudevan SG, Farrar J. Dengue and dengue hemorrhagic fever. Wallingford, UK: CAB International; 2014.

[15] Ritchie SA. Dengue vector bionomics: why *Aedes aegypti* is such a good vector. In: Gubler D, Ooi EE, Vasudevan S, Farrar J, editors. Dengue and dengue hemorrhagic fever. 2nd ed. Wallingford, UK: CAB International; 2014. p. 455–80.

[16] Shepard DS, Halstead SB. Dengue (with notes on yellow fever and Japanese encephalitis). In: Jamison DT, Mosley WH, Measham AR, Bobadilla JL, editors. Disease control

priorities in developing countries. New York, NY: Oxford University Press: The World Bank; 1993. p. 303−20.

[17]   Gubler DJ. Dengue and dengue hemorrhagic fever. Clin Microbiol Rev 1998;11(3): 480−96.

[18]   Gubler DJ. Epidemic dengue/dengue hemorrhagic fever as a public health, social and economic problem in the 21st century. Trends Microbiol 2002;10(2):100−3.

[19]   Gubler DJ, Clark GG. Dengue/dengue hemorrhagic fever: the emergence of a global health problem. Emerg Infect Dis 1995;1(2):55−7.

[20]   Lambrechts L, Scott TW, Gubler DJ. Consequences of the expanding global distribution of *Aedes albopictus* for dengue virus transmission. PLoS Negl Trop Dis 2010;4(5):e646.

[21]   Shepard DS, Halasa YA, Undurraga EA. Cost of dengue vector control: a systematic literature review. Am J Trop Med Hyg 2014;91(5S):253.

[22]   Shepard DS, Undurraga EA, Betancourt-Cravioto M, et al. Approaches to refining estimates of global burden and economics of dengue. PLoS Negl Trop Dis 2014; 8(11):e3306.

[23]   Wilder-Smith A. Dengue vaccines: dawning at last? Lancet 2014;384(9951):1327−9.

[24]   Capeding MR, Tran NH, Hadinegoro SRS, et al. Clinical efficacy and safety of a novel tetravalent dengue vaccine in healthy children in Asia: a phase 3, randomised, observer-masked, placebo-controlled trial. Lancet 2014;384(9951):1358−65.

[25]   Villar L, Dayan GH, Arredondo-García JL, et al. Efficacy of a tetravalent dengue vaccine in children in Latin America. N Engl J Med 2015;372(2):113−23.

[26]   Halstead SB. Dengue vaccines. In: Gubler D, Ooi EE, Vasudevan S, Farrar J, editors. Dengue and dengue hemorrhagic fever. 2nd ed. Wallingford, UK: CAB International; 2014. p. 548−76.

[27]   Barrett AD. Short-course oral corticosteroid therapy is not effective in early dengue infection. Clin Infect Dis 2012;55(9):1225−6.

[28]   Halstead SB. Dengue vascular permeability syndrome: what, no T cells? Clin Infect Dis 2013;56(6):900−1.

[29]   Dow G, Mora E. The maximum potential market for dengue drugs V 1.0. Antiviral Res 2012;96(2):203−12.

[30]   Brown DM, James AA. Dengue vector control: new approaches. In: Gubler D, Ooi EE, Vasudevan S, Farrar J, editors. Dengue and dengue hemorrhagic fever. Wallingford, UK: CAB International; 2014. p. 519−36.

[31]   Baly A, Flessa S, Cote M, et al. The cost of routine *Aedes aegypti* control and of insecticide-treated curtain implementation. Am J Trop Med Hyg 2011;84(5):747−52.

[32]   Massad E, Coutinho FA. The cost of dengue control. Lancet 2011;377:1630−1.

[33]   Erlanger TE, Keiser J, Utzinger J. Effect of dengue vector control interventions on entomological parameters in developing countries: a systematic review and meta-analysis. Med Vet Entomol 2008;22(3):203−21.

[34]   World Health Organization. Dengue and dengue haemorrhagic fever in the Americas: guidelines for prevention and control. Washington, DC: World Health Organization, Pan American Health Organization; 1994.

[35]   World Health Organization. Position statement on integrated vector management, <http://whqlibdoc.who.int/hq/2008/WHO_HTM_NTD_VEM_2008.2_eng.pdf>; 2008 [accessed 30.10.12].

[36]   Tapia-Conyer R, Betancourt-Cravioto M, Mendez-Galvan J. Dengue: an escalating public health problem in Latin America. Paediatr Int Child Health 2012;32(Suppl. 1):14−17.

[37] Ooi E, Gubler D, Nam V. Dengue research needs related to surveillance and emergency response. In: Report of the scientific working group meeting on dengue, Geneva, October 1–5, 2006; 2007, p. 124–33.

[38] Hemingway J, Beaty BJ, Rowland M, Scott TW, Sharp BL. The innovative vector control consortium: improved control of mosquito-borne diseases. Trends Parasitol 2006;22(7):308–12.

[39] Zaim M, Guillet P. Alternative insecticides: an urgent need. Trends Parasitol 2002;18 (4):161–3.

[40] Heintze C, Velasco Garrido M, Kroeger A. What do community-based dengue control programmes achieve? A systematic review of published evaluations. Trans R Soc Trop Med Hyg 2007;101(4):317–25.

[41] Luz PM, Vanni T, Medlock J, Paltiel AD, Galvani AP. Dengue vector control strategies in an urban setting: an economic modelling assessment. Lancet 2011;377(9778):1673–80.

[42] Nauen R. Insecticide resistance in disease vectors of public health importance. Pest Manag Sci 2007;63(7):628–33.

[43] Stoddard ST, Wearing HJ, Reiner Jr. RC, et al. Long-term and seasonal dynamics of dengue in Iquitos, Peru. PLoS Negl Trop Dis 2014;8(7):e3003.

[44] Lenhart A, Trongtokit Y, Alexander N, et al. A cluster-randomized trial of insecticide-treated curtains for dengue vector control in Thailand. Am J Trop Med Hyg 2013;88 (2):254–9.

[45] Vanlerberghe V, Trongtokit Y, Jirarojwatana S, et al. Coverage-dependent effect of insecticide-treated curtains for dengue control in Thailand. Am J Trop Med Hyg 2013;89 (1):93–8.

[46] Wilder-Smith A, Byass P, Olanratmanee P, et al. The impact of insecticide-treated school uniforms on dengue infections in school-aged children: study protocol for a randomised controlled trial in Thailand. Trials 2012;13(1):212.

[47] Chang MS, Christophel EM, Gopinath D, Abdur RM. Challenges and future perspective for dengue vector control in the Western Pacific region. West Pac Surveill Response J: WPSAR 2011;2(2):9–16.

[48] Nam VS, Yen NT, Duc HM, et al. Community-based control of *Aedes aegypti* by using *Mesocyclops* in Southern Vietnam. Am J Trop Med Hyg 2012;86(5):850–9.

[49] Baly A, Toledo ME, Boelaert M, et al. Cost effectiveness of *Aedes aegypti* control programmes: participatory versus vertical. Trans R Soc Trop Med Hyg 2007;101(6): 578–86.

[50] Tapia-Conyer R, Mendez-Galvan J, Burciaga-Zuniga P. Community participation in the prevention and control of dengue: the patio limpio strategy in Mexico. Paediatr Int Child Health 2012;32(Suppl. 1):10–13.

[51] Suaya JA, Shepard DS, Chang MS, et al. Cost-effectiveness of annual targeted larviciding campaigns in Cambodia against the dengue vector *Aedes aegypti*. Trop Med Int Health 2007;12(9):1026–36.

[52] Orellano PW, Pedroni E. Cost-benefit analysis of vector control in areas of potential dengue transmission. Rev Panam Salud Publica 2008;24(2):113–19.

[53] Harris E. Cluster randomized controlled trial to reduce dengue risk in Nicaragua and Mexico though evidence-based community mobilization. In: 61st annual meeting American society of tropical medicine and hygiene, November 11–15, 2012, Atlanta, GA; 2012.

[54] Shepard DS, Halasa YA, Undurraga EA. Economic and disease burden of dengue. In: Gubler DJ, Ooi EE, Vasudevan SG, Farrar J, editors. Dengue and dengue hemorrhagic fever. 2nd ed. Wallingford, UK: CAB International; 2014. p. 50–77.

[55] Undurraga EA, Betancourt-Cravioto M, Ramos-Castañeda J, et al. Economic and disease burden of dengue in Mexico. PLoS Negl Trop Dis 2015;9(3):e0003547.

[56] Kongsin S, Jiamton S, Suaya J, Vasanawathana S, Sirisuvan P, Shepard D. Cost of dengue in Thailand. Dengue Bull 2010;34:77−88.

[57] Taliberti H, Zucchi P. Direct costs of the dengue fever control and prevention program in 2005 in the city of São Paulo. Rev Panam Salud Publica 2010;27(3):175−80.

[58] Armien B, Suaya JA, Quiroz E, et al. Clinical characteristics and national economic cost of the 2005 dengue epidemic in Panama. Am J Trop Med Hyg 2008;79(3):364−71.

[59] Perez-Guerra C, Halasa Y, Rivera R, et al. Economic cost of dengue public prevention activities in Puerto Rico. Dengue Bull 2010;34(13−23).

[60] Packierisamy PR, Ng C, Dahlui M, et al. Cost of dengue vector control activities in Malaysia. Brandeis University, unpublished manuscript.

[61] Baly A, Toledo ME, Rodriguez K, et al. Costs of dengue prevention and incremental cost of dengue outbreak control in Guantanamo, Cuba. Trop Med Int Health 2012;17 (1):123−32.

[62] Orellano P, Pedroni E. Análisis costo-beneficio del control de vectores en la transmisión potencial de dengue. Rev Panam Salud Publica 2008;24(2):113.

[63] Kay BH, Tuyet Hanh TT, Le NH, et al. Sustainability and cost of a community-based strategy against *Aedes aegypti* in northern and central Vietnam. Am J Trop Med Hyg 2010;82(5):822−30.

[64] Tun-Lin W, Lenhart A, Nam VS, et al. Reducing costs and operational constraints of dengue vector control by targeting productive breeding places: a multi-country non-inferiority cluster randomized trial. Trop Med Int Health 2009;14(9):1143−53.

[65] Rizzo N, Gramajo R, Escobar MC, et al. Dengue vector management using insecticide treated materials and targeted interventions on productive breeding-sites in Guatemala. BMC Public Health 2012;12.

[66] Pepin KM, Marques-Toledo C, Scherer L, Morais MM, Ellis B, Eiras AE. Cost-effectiveness of novel system of mosquito surveillance and control, Brazil. Emerg Infect Dis 2013;19(4):542−50.

[67] Tozan Y, Ratanawong P, Louis VR, Kittayapong P, Wilder-Smith A. Use of insecticide-treated school uniforms for prevention of dengue in schoolchildren: a cost-effectiveness analysis. PLoS One 2014;9(9):e108017.

[68] Ditsuwan T, Liabsuetrakul T, Ditsuwan V, Thammapalo S. Cost of standard indoor ultra-low-volume space spraying as a method to control adult dengue vectors. Trop Med Int Health 2012;17(6):767−74.

[69] Loroño-Pino MA, Chan-Dzul YN, Zapata-Gil R, et al. Household use of insecticide consumer products in a dengue-endemic area in México. Trop Med Int Health 2014;19 (10):1267−75.

[70] National Statistical Office. Executive summary. The 2010 population and housing census; 2010 [accessed 15.11.14].

[71] Walker T, Johnson PH, Moreira LA, et al. The wMel *Wolbachia* strain blocks dengue and invades caged *Aedes aegypti* populations. Nature 2011;476(7361):450−3.

[72] Frentiu FD, Zakir T, Walker T, et al. Limited dengue virus replication in field-collected *Aedes aegypti* mosquitoes infected with *Wolbachia*. PLoS Negl Trop Dis 2014;8(2): e2688.

[73] Hoffmann AA, Montgomery BL, Popovici J, et al. Successful establishment of *Wolbachia* in *Aedes* populations to suppress dengue transmission. Nature 2011;476(7361):454−7.

[74] Frentiu FD, Walker T, O'Neill SL. Biological control of dengue and *Wolbachia*-based strategies. In: Gubler D, Ooi EE, Vasudevan S, Farrar J, editors. Dengue and dengue hemorrhagic fever. Wallingford, UK: CAB International; 2014. p. 537−47.

[75] Fu G, Lees RS, Nimmo D, et al. Female-specific flightless phenotype for mosquito control. Proc Natl Acad Sci USA 2010;107(10):4550−4.

[76] Atkinson MP, Su Z, Alphey N, Alphey LS, Coleman PG, Wein LM. Analyzing the control of mosquito-borne diseases by a dominant lethal genetic system. Proc Natl Acad Sci USA 2007;104(22):9540−5.

[77] Scott TW, Takken W, Knols BG, Boëte C. The ecology of genetically modified mosquitoes. Science 2002;298(5591):117−19.

[78] Wise de Valdez MR, Nimmo D, Betz J, et al. Genetic elimination of dengue vector mosquitoes. Proc Natl Acad Sci USA 2011;108(12):4772−5.

[79] Alphey L. Genetic control of mosquitoes. Annu Rev Entomol 2014;59(1):205−24.

[80] Benedict M, D'Abbs P, Dobson S, et al. Guidance for contained field trials of vector mosquitoes engineered to contain a gene drive system: recommendations of a scientific working group. Vector Borne Zoonotic Dis 2008;8(2):127−66.

[81] McKemey A. Genetically engineered mosquitoes for control of dengue vector—*Aedes aegypti*. In: Pan-American Dengue research network meeting, October 19−22, 2014, Belém, Brazil; 2014.

[82] Alphey L, Beech C. Appropriate regulation of GM insects. PLoS Negl Trop Dis 2012;6 (1):e1496.

[83] Beech CJ, Vasan S, Quinlan MM, et al. Deployment of innovative genetic vector control strategies: progress on regulatory and biosafety aspects, capacity building and development of best-practice guidance. Asia Pac J Mol Biol Biotechnol 2009;17:75−85.

[84] Enserink M. GM mosquito trial alarms opponents, strains ties in Gates-funded project. Science 2010;330:1030−1.

[85] Letting the bugs out of the bag. Nature 2011; 470(7333):139.

[86] Reeves RG, Denton JA, Santucci F, Bryk J, Reed FA. Scientific standards and the regulation of genetically modified insects. PLoS Negl Trop Dis 2012;6(1):e1502.

[87] Lavery JV, Harrington LC, Scott TW. Ethical, social, and cultural considerations for site selection for research with genetically modified mosquitoes. Am J Trop Med Hyg 2008; 79(3):312−18.

[88] Beech CJ, Nagaraju J, Vasan S, et al. Risk analysis of a hypothetical open field release of a self-limiting transgenic *Aedes aegypti* mosquito strain to combat dengue. Asia Pac J Mol Biol Biotechnol 2009;17(3):99−111.

[89] Mumford J, Quinlan MM, Beech C, et al. MosqGuide: a project to develop best practice guidance for the deployment of innovative genetic vector control strategies for malaria and dengue. Asia Pac J Mol Biol Biotechnol 2009;17:93−5.

[90] Harris AF, Nimmo D, McKemey AR, et al. Field performance of engineered male mosquitoes. Nat Biotechnol 2011;29(11):1034−7.

[91] Harris AF, McKemey AR, Nimmo D, et al. Successful suppression of a field mosquito population by sustained release of engineered male mosquitoes. Nat Biotechnol 2012;30 (9):828−30.

[92] Subramaniam T, Lee HL, Ahmad NW, Murad S. Genetically modified mosquito: the Malaysian public engagement experience. Biotechnol J 2012;7(11):1323−7.

[93] Lacroix R, McKemey AR, Raduan N, et al. Open field release of genetically engineered sterile male *Aedes aegypti* in Malaysia. PLoS One 2012;7(8):e42771.

[94] Da Silveira E. Genetic solutions: transgenic mosquitoes are set free in Bahia to fight dengue fever. Pesquisa FAPESP 2011;34−7.

[95] de Lima-Oliveira S, Oliveira-Carvalho D, Lara-Capurro M. Mosquito transgênico: do paper para a realidade. Rev Biol 2011;6:38−43.

[96] McNaughton D, Duong TTH. Designing a community engagement framework for a new dengue control method: a case study from Central Vietnam. PLoS Negl Trop Dis 2014;8 (5):e2794.

[97] Eliminate Dengue. Eliminate Dengue program. Project sites around the world. <http://www.eliminatedengue.com/project>; 2015 [accessed 05.03.15].

[98] Alphey N, Alphey L, Bonsall MB. A model framework to estimate impact and cost of genetics-based sterile insect methods for dengue vector control. PLoS One 2011;6(10): e25384.

[99] Vazquez-Prokopec GM. Dengue control: the challenge ahead. Future Microbiol 2011;6 (3):251−3.

[100] DeRoeck D, Deen J, Clemens JD. Policymakers' views on dengue fever/dengue haemorrhagic fever and the need for dengue vaccines in four southeast Asian countries. Vaccine 2003;22(1):121−9.

[101] James S, Simmons CP, James AA. Mosquito trials. Science 2011;334(6057):771−2.

[102] Conteh L, Engels T, Molyneux DH. Socioeconomic aspects of neglected tropical diseases. Lancet 2010;375(9710):239−47.

[103] Walker D. Cost and cost-effectiveness guidelines: which ones to use?. Health Policy Plan 2001;16(1):113−21.

[104] Phillips M, Mills A, Dye C. Guidelines for cost-effectiveness analysis of vector control, <http://www.who.int/docstore/water_sanitation_health/Documents/PEEM3/english/peem3 toc.htm>; 1993 [accessed 10.10.14].

[105] Meltzer MI. Introduction to health economics for physicians. Lancet 2001;358(9286): 993−8.

[106] Alphey L, Alphey N. Five things to know about genetically modified (GM) insects for vector control. PLoS Pathog 2014;10(3):e1003909.

[107] Ramsey JM, Bond JG, Macotela ME, et al. A regulatory structure for working with genetically modified mosquitoes: lessons from Mexico. PLoS Negl Trop Dis 2014;8(3): e2623.

[108] Lavery JV, Tinadana PO, Scott TW, et al. Towards a framework for community engagement in global health research. Trends Parasitol 2010;26(6):279−83.

[109] Wolbers M, Kleinschmidt I, Simmons CP, Donnelly CA. Considerations in the design of clinical trials to test novel entomological approaches to dengue control. PLoS Negl Trop Dis 2012;6(11):e1937.

[110] Scott TW, Morrison AC. *Aedes aegypti* density and the risk of dengue virus transmission. In: Takken W, Scott TW, editors. Ecological aspects for application of genetically modified mosquitoes. Dordrecht, The Netherlands: Kluwer Academic Publishers; 2003. p. 187−206.

[111] Bowman LR, Runge-Ranzinger S, McCall PJ. Assessing the relationship between vector indices and dengue transmission: a systematic review of the evidence. PLoS Negl Trop Dis 2014;8(5):e2848.

[112] Ferguson NM, Cummings DAT. How season and serotype determine dengue transmissibility. Proc Natl Acad Sci USA 2014;111(26):9370−1.

[113] Scott TW, Morrison AC. Vector dynamics and transmission of dengue virus: implications for dengue surveillance and prevention strategies. In: Rothman AL, editor. Dengue virus. Berlin: Springer; 2010. p. 115−28.

[114] Halstead SB. Dengue virus–mosquito interactions. Ann Rev Entomol 2008;56:273–91.

[115] Ooi EE, Goh KT, Gubler DJ. Denque prevention and 35 years of vector control in Singapore. Emerg Infect Dis 2006;12(6):887–93.

[116] Cummings DAT, Irizarry RA, Huang NE, et al. Travelling waves in the occurrence of dengue haemorrhagic fever in Thailand. Nature 2004;427(6972):344–7.

[117] Liebman KA, Stoddard ST, Morrison AC, et al. Spatial dimensions of dengue virus transmission across interepidemic and epidemic periods in Iquitos, Peru (1999–2003). PLoS Negl Trop Dis 2012;6(2):e1472.

[118] Reiner RC, Stoddard ST, Forshey BM, et al. Time-varying, serotype-specific force of infection of dengue virus. Proc Natl Acad Sci USA 2014;111(26):E2694–702.

[119] Focks DA, Brenner RJ, Hayes J, Daniels E. Transmission thresholds for dengue in terms of *Aedes aegypti* pupae per person with discussion of their utility in source reduction efforts. Am J Trop Med Hyg 2000;62(1):11–18.

[120] Weaver SC, Lecuit M. Chikungunya virus and the global spread of a mosquito-borne disease. N Engl J Med 2015;372(13):1231–9.

[121] Monath TP. Yellow fever: an update. Lancet Infect Dis 2001;1(1):11–20.

[122] Shepard DS, Halasa YA, Fonseca DM, et al. Economic evaluation of an area-wide integrated pest management program to control the Asian tiger mosquito in New Jersey. PLoS One 2014;9(10):e111014.

[123] Mohd-Zaki AH, Brett J, Ismail E, L'Azou M. Epidemiology of dengue disease in Malaysia (2000–2012): a systematic literature review. PLoS Negl Trop Dis 2014;8(11): e3159.

[124] Teixeira MG, Siqueira Jr. JB, Ferreira GLC, Bricks L, Joint G. Epidemiological trends of dengue disease in Brazil (2000–2010): a systematic literature search and analysis. PLoS Negl Trop Dis 2013;7(12):e2520.

[125] Limkittikul K, Brett J, L'Azou M. Epidemiological trends of dengue disease in Thailand (2000–2011): a systematic literature review. PLoS Negl Trop Dis 2014;8(11):e3241.

[126] Bravo L, Roque VG, Brett J, Dizon R, L'Azou M. Epidemiology of dengue disease in the Philippines (2000–2011): a systematic literature review. PLoS Negl Trop Dis 2014;8 (11):e3027.

[127] L'Azou M, Taurel A-F, Flamand C, Quénel P. Recent epidemiological trends of dengue in the French Territories of the Americas (2000–2012): a systematic literature review. PLoS Negl Trop Dis 2014;8(11):e3235.

[128] Gómez-Dantés H, Farfán-Ale JA, Sarti E. Epidemiological trends of dengue disease in Mexico (2000–2011): a systematic literature search and analysis. PLoS Negl Trop Dis 2014;8(11):e3158.

[129] Halasa YA, Shepard DS, Zeng W. Economic cost of dengue in Puerto Rico. Am J Trop Med Hyg 2012;86(5):745–52.

[130] Undurraga EA, Halasa YA, Shepard DS. Use of expansion factors to estimate the burden of dengue in Southeast Asia: a systematic analysis. PLoS Negl Trop Dis 2013;7(2): e2056.

[131] Fundacao Nacional de Saúde. Programa nacional de controle da dengue. Brasilia: Ministerio de Saúde–Fundacao Nacional de Saúde; 2002.

[132] Braga IA, Valle D. *Aedes aegypti*: histórico do controle no Brasil. Epidemiologia e serviços de saúde 2007;16(2):113–18.

[133] Oliveira L. Mosquito 'trans' promote acabar com a dengue, <http://www.diarioweb. com.br/novoportal/noticias/saude/180058,,Mosquito+trans+promete+acabar+com+a+ dengue.aspx>; 2014 [accessed 13.04.14].

[134] Nogueira ML. An overview of the circulation of dengue virus in Sao Jose do Rio Preto and in Brazil. In: Panamerican dengue research network meeting, October 19–22, 2014, Belém do Pará, Brazil; 2014.

[135] International Monetary Fund. World economic outlook database, <http://www.imf.org/external/pubs/ft/weo/2011/02/weodata/index.aspx>; 2011 [accessed January 2012].

[136] Perkins TA, Scott TW, Le Menach A, Smith DL. Heterogeneity, mixing, and the spatial scales of mosquito-borne pathogen transmission. PLoS Comput Biol 2013;9(12):e1003327.

[137] Cuong HQ. Spatiotemporal dynamics of dengue epidemics, Southern Vietnam. Emerg Infect Dis 2013;19(6):945–53.

[138] Focks DA, Daniels E, Haile DG, Keesling JE. A simulation model of the epidemiology of urban dengue fever: literature analysis, model development, preliminary validation, and samples of simulation results. Am J Trop Med Hyg 1995;53(5):489–506.

[139] World Health Organization. Choosing interventions that are cost effective (WHO–CHOICE), <http://www.who.int/choice/en/>; 2011 [accessed 04.01.11].

[140] Patil P, Alam M, Ghimire P, et al. Discussion on the proposed hypothetical risks in relation to open field release of a self-limiting transgenic *Aedes aegypti* mosquito strains to combat dengue. Asia Pac J Mol Biol Biotechnol 2010;18(2):241–6.

[141] Beech C, Quinlan M, Capurro M, Alphey L, Mumford J. Update: deployment of innovative genetic vector control strategies including an update on the MosqGuide Project. Asia Pac J Mol Biol Biotechnol 2011;19(3):101–6.

[142] Subbaraman N. Science snipes at Oxitec transgenic-mosquito trial. Nat Biotechnol 2011;29(1):9–11.

Chapter 18

Community Engagement

Danilo O. Carvalho and Margareth L. Capurro
Universidade de São Paulo, São Paulo, Brazil

STORIES FROM THE FIELD: LESSONS FROM COMMUNITY OUTREACH

Many new studies and strategies under development are targeting mosquito vector control. The resulting new information might be evaluated and shared with the local population. The local population must face the possibility that certain control strategies (i.e., genetics-based control) may result in increased numbers of mosquitoes in local houses, gardens, and streets. Moreover, the number of technicians who are new to the community will be setting and collecting traps every week to evaluate the efficiency of mosquito releases. However, public resistance to new technologies has long been recognized as one very important barrier to the application of such technologies. Public acceptance continues to depend on the understanding by the general public of the activities, of the project overall, and of the risks/benefits of using genetically modified organisms (GMOs) [1].

No technology can be isolated from its social context, and this context must be considered in any communication designed to inform the public about GMOs. Due to differences in culture, ethics, customs, and social structure between populations, even slight differences might require completely different approaches to and strategies for community engagement [2]. The size of the target area and its population may also influence the set of activities to be publicized, and the community needs to receive the proper type of information and to be informed at whatever frequency is important [3,4].

According to Tindana et al. (2007), the definition of community engagement is based on work with relevant partners involving common goals and interests are the same [5]; community contributions to protect the partners and to foster meaningful research; and collaborative partnerships with stakeholders. In 2010, Lavery et al. suggested 12 points to be considered for effective community engagement. Their work details each of those points, furnishing important

TABLE 18.1 "Points to be Consider" for Effective Community Engagement[a]

1. Rigorous site-selection procedures
2. Early initiation of community engagement activities
3. Characterize and build knowledge of the community, its diversity, and its changing needs
4. Ensure the purpose and goals of the research are clear to the community
5. Provide information
6. Establish relationship and commitments to build trust with relevant authorities in the community: formal, informal, and traditional
7. Understand community perception and attitudes about the proposed research
8. Identify, mobilize, and develop relevant community assets and capacity
9. Maximize opportunities for stewardship, ownership, and shared control by the community
10. Ensure adequate opportunities and respect for dissenting opinions
11. Secure permission/authorization from the community
12. Review, evaluate and if necessary, modify engagement strategies

[a]Complete table extracted from Ref. [6].

information regarding the basic elements needed to establish a community engagement plan [6]. The framework to be followed (Table 18.1) uses information from the community engagement plan conducted in Mexico for field cage experiments with genetically modified mosquitoes (GMMs).

In the context of the factors cited above, a brief description of community engagement will be presented based on previous well-known SIT releases in India and GMM releases previously conducted in Malaysia, Mexico, the Cayman Islands, and Brazil.

INDIA

In India, a vector control program conducted during the 1970s aimed to use classical SIT for three different species, *Culex fatigans*, *Aedes aegypti*, and *Anopheles stephensi*. According to Ref. [7], the program planned to use three types of mosquito sterilization processes: irradiation (cobalt-60), incompatibility, and chemosterilization. However, no irradiation source was available to sterilize the mosquitoes, and no research on incompatibility methods for promoting mosquito sterilization was available. The program used chemosterilization to conduct two of its five field experiments [7].

The fieldwork performed using *Ae. aegypti* never reached the intensity of that performed with *Cx. fatigans*. Moreover, the fieldwork performed with *An. stephensi* was even less advanced. These problems all resulted from the publication of a critical report claiming that all research data obtained from these experiments could be useful in "germ warfare" or that India was being used as a guinea pig for chemicals or methods not permitted in sponsoring countries, for example, in the chemosterilization project, where thiotepa was in use [8]. It was also mentioned that all the work performed in India appeared to have no relevance to malaria or filariasis control. Instead, researchers were collecting data on *Ae. aegypti*, which is the vector of yellow fever virus. The editorial mentioned the paradox that by that time, there had not been a single case of yellow fever in India for many years. It was feared that eggs of *Ae. aegypti* carrying the virus could be sent anywhere and start outbreaks all over the country. Things did not improve when a third source mentioned that the work performed using incompatibility had no effect on the mosquito population [7].

This situation had two major consequences: a program review conducted by scientists and a political investigation. The scientific review could find only minor problems with operational activities, and the reviewers offered congratulations on the achievements of the program. The level of misunderstanding reached a point at which well-founded scientific information was overlooked. For example, only bites by female mosquitoes could disseminate the virus, but the program was releasing only males. Additionally, the level of interest in *Ae. aegypti* was disclosed by the number of publications on each topic, with four papers about *Cx. fatigans* but only one regarding *Ae. aegypti*. However, from a political perspective, anything that could possibly be found to cast a negative light on the project or those involved in it was brought out. During the investigation, the United States was accused of using the program to conduct harmful experiments in India, to make preparations in case the United States ever wished to wage chemical, bacteriological, and virus warfare against India and to prepare for such warfare using India as a base [7].

The Indian government closed down the New Delhi research unit despite protests from the WHO, under whose aegis it was set up. This experience made the Indian government increasingly strict concerning the scrutiny of all projects involving foreign collaboration. A central agency that includes the defense advisor has been established to give foreign-aided projects security clearance [8]. The outcome was that the institute responsible for conducting the experiment was completely incapable of handling public relations in a way that could have prevented the escalation of a relatively small matter to substantial dimensions [7]. The institute should have pursued a policy of being open to the Indian press about its successes as well as failures.

The disastrous result of this experience was that vector control using SIT for mosquitoes was completely abandoned around the world until new genetic manipulation tools were developed using recombinant DNA instead of a conventional sterilization process. However, recent studies involving classical SIT for mosquitoes are slowly being accepted [9–12].

MALAYSIA

In Malaysia, prior to the release of a large number of GMMs, a mark-release-recapture (MRR) experiment was planned in an uninhabited area to provide basic information to establish the suppression of the controlled disease through large-scale releases. Despite substantial international involvement by Oxitec regarding this experiment, the Malaysian government is still not confident about using this technique. Accordingly, only a few MRR experiments have been conducted in Malaysia to date. However, even for MRR experiments, in this case Oxitec and its Malaysian partner (the Institute for Medical Research—IMR) were subject to the authority of the National Biosafety Board (NBB), which is the competent authority and which was followed. In turn, the NBB bases its decisions on policy recommendations by the National Genetic Modification Advisory Committee, as summarized in Figure 18.1 [13].

According to the Malaysian Biosafety Act, public consultation is mandatory in processing any transgenic release application. This consultation was implemented in the form of an newspaper announcement of the transgenic mosquito release. The invitation was extended by nongovernmental organizations. For a 30-day period after the announcement, the issuers waited for local or global reactions to the announcement of the future release to be conducted in Malaysia.

Relevant information about the proposed field trial was disseminated to the local community nearest to the release site. Material presenting this information was distributed to the community. Not only was permission from local councils and community leaders sought, but a public scientific forum was also presented at the Malaysian National Academy of Sciences, and public talks were given in selected areas.

During the public comment period established by the Malaysian team, it became known that certain individuals and groups fully supported the project, certain interests remained neutral while waiting for an expert evaluation, and a few individuals and groups expressed a negative opinion, citing alternative methods and uncertainties or even showing a lack of familiarity with the technology and its impacts [13]. All this material was collected, including the terms and conditions expressed. It was then possible for the NBB to make a decision to approve the MRR experiments. A final step was a press conference to announce the NBB decision and to provide further information to the public.

The lesson to be learned from this Malaysian example of community engagement is that there is a need to develop effective forms of communication [3,14] and to inform the findings of the field trial to the public. This example suggests that countries planning GMM releases should establish a working group to develop models for effective stakeholder or community engagement from an early stage.

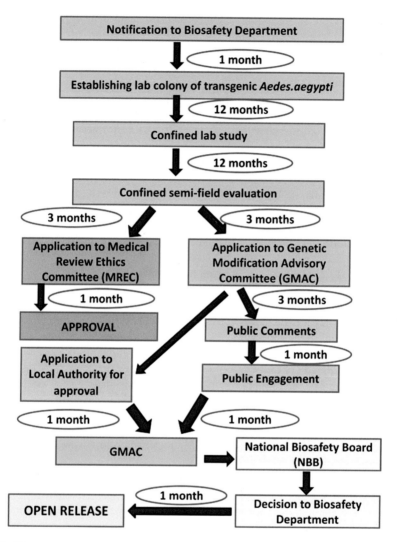

FIGURE 18.1 Malaysian biosafety regulatory process and approximate time frame. *Figure extracted from Ref. [13].*

GRAND CAYMAN ISLAND

According to Trivedi in 2011, community engagement was not properly conducted in the Grand Cayman Island program. Additionally, the scientific community raised issues regarding the releases of a GMM strain in one area of the island [15]. The Cayman Island release was the first open-field trial in the world using transgenic mosquitoes (the technical results of the project will not be discussed here); little public debate occurred. At a meeting in

Atlanta (USA) in 2010, Oxitec characterized the experiment as a "complete success" [16]. The acquisition of stakeholder permission to conduct the experiment was the only community engagement step that was performed. There were no public meetings or opportunities for residents to voice concerns about the activities and the release itself. Oxitec was then perceived as secretive. The public questioned the situation and expressed suspicion [15]. Dr. Luke Alphey (Oxitec's representative at that time) not only completely rejected the view that the program was performed secretively but also mentioned that the international community was not the target population [16]. An editorial in Nature published in 2011 also stated that the statements of Oxitec's representative added to a growing sense of unease among individuals in the field about the way in which the public was consulted and notified about such experiments [17]. The impact of this situation raised concerns and, likewise, misinformed the public. As a result of this outcome the statements of Oxitec's representative, it is conceivable that other trials could be compromised and that the same types of problems that occurred in India could once again be encountered [8].

According to Enserink in 2010 and Oxitec, there were no town hall meetings or public debates because the Cayman Islands' government did not deem them necessary. Oxitec's partner in Grand Cayman Island, the Mosquito Research and Control Unit, sent information about the study to local newspapers, and its 50 employees attended a one-time lunch meeting about the project from which information filtered out to the rest of Grand Cayman, which has approximately 50,000 inhabitants [16]. Subbaraman in 2011 reports that Alphey describes extensive and meticulous preparations for community engagement and states that people were aware that the project involved a new technology for dengue control using sterile male mosquitoes, which do not bite; furthermore, it was stated that the people were aware that not all mosquito species would be controlled [18].

A difference between the Cayman experience and the Malaysia experience is that in Malaysia, public engagement was viewed as an important tool for gaining public trust through dissemination of information about the objectives of the project, supported by the regulatory requirements for approval and experts' feedback. Subramaniam et al. [13] state: "*our experience showed that despite executing a well planned transparent public engagement process that was relevant for a release in an uninhabited site, there was still some dissatisfaction from some community groups.*" This outcome can be explained by the impossibility and impracticality of obtaining approval from every individual when the research relatively unintrusive and involved a field release in an uninhabited area.

Even so, the Cayman example cannot be considered as a complete disaster in terms of community engagement. The reason for this conclusion is that due to the Cayman experience, the social aspect of involving or engaging the community and the stakeholders is now viewed as extremely important at different levels. The Cayman example illustrates precisely how the effectiveness of community engagement should be conceptualized, even if these models have not been completely established, and the context in which they have been developed clearly influences their own nature. It is also necessary to consider the complexity of evaluating the effectiveness of community engagement in research.

MEXICO

In Mexico, the idea behind the mosquito control project was to test the competitiveness of a new transgenic mosquito line in field cages. The lack of a preexisting structure made it necessary for the scientists in the project to play critical, unbiased roles in formulating the community development pathway. Under these circumstances, it was incumbent on the researchers to identify gaps and assist in the development of regulatory norms, which included a broader regulatory environment that addressed the needs and concerns of all communities involved. After selecting the area for the field cage experiments, 3 years elapsed until the formal regulatory process was complete (Figure 18.2).

> *All of the work described here was conducted within ethical, social and cultural guidelines for community engagement activities [6]. We found that this approach helped us to develop respect and trust, basic ingredients for strong working relationships with local residents living near the field site, and for appropriate dialogue with state and national health and environmental authorities, scientists, and local and international press. Although the containment measures and communication activities taken in this work were greater than expected for research with natural strains of mosquitoes, we feel that this precautionary approach could have long-term benefits by decreasing suspicion that transgenic mosquito technology is being applied carelessly [17,19].*

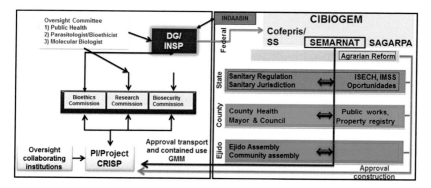

FIGURE 18.2 Schematic representation of the regulatory pathway for research involving genetically modified mosquitoes. *Figure extracted from Ref. [20].*

After all the issues raised by the experience in Grand Cayman Island and India and the precautions taken by Malaysia, the Mexican decision to take and make all efforts to support the project and its success were well planned. However, the description of the community engagement process in Mexico is more closely related to political processes and approval from different agencies to avoid any rejection. In this example, this approach was used by the Mexicans to conduct their community engagement, delegating and involving all government agencies and branches to provide enough security to all people involved.

The Mexican experiments aimed to evaluate transgenic male mosquito competitiveness in field cages, and there were no open-field releases. The involvement and activities with the local community were not well described in Ref. [20], which only mentioned attitudes and considerations that should be adopted to promote community engagement [20].

BRAZIL[1]

The Brazilian community engagement plan was based on the previous experiences using the same transgenic line used in Grand Cayman Island and considering the mistakes, considerations, and criticisms available from experts and other studies. The aim of the plan was to transmit transparent and clear information regarding Brazilian researchers' intentions to release transgenic mosquitoes in the field. In contrast to the situation in the other countries considered, in Brazil the National Biosafety Technical Commission (CTNBio) provides substrate for the National Biosafety Policy (PNB). It not only judges the biosecurity issues but also assesses the impact that a GMO may have on the community, surrounding areas, and environment. All proposed projects involving GMOs have to be submitted to the CTNBio executive secretary so that the summary can be published in the Federal Official Gazette (Diário Oficial da União—DOU), a newspaper distributed nationwide (also on the Internet) and used as the official government channel. This measure represents the first step toward a federal public consultation.

Two members of the CTNBio are assigned to evaluate the project and to prepare a final opinion to be presented to the sectorial subcommissions for deliberation before and during the plenary. To gain approval, a project needs two-thirds of the commission members' votes. The CTNBio meetings are public open hearings at which the general public is welcome. The agenda of these meetings is published ahead of time in the Federal Gazette and on the CTNBio Web site (ctnbio.gov.br). CTNBio is composed of several ministries and federal agencies, and their representatives vote, so there is no need for the project to be voted on by each organ separately. As this process also

---

1. The authors from this publication were involved in the development, execution and evaluation of this study as the project manager and the project coordinator, respectively.

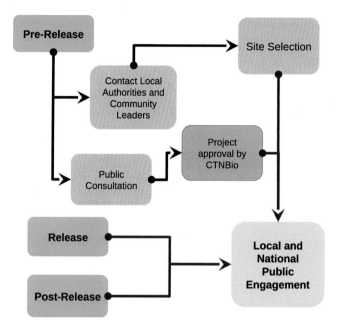

FIGURE 18.3 Brazilian community engagement process.

involves different agencies and public representatives, it can be understood as an important path for public acceptance. Accordingly, based on previous experiences, the legal framework can be defined as an element of public engagement (Figure 18.3).

The Brazilian strategy was based on transparency, as the project name shows. The project, known as PAT—*Projeto Aedes Transgênico* (*Aedes Transgenic Project*), uses the word "transgenic" in its title (Figure 18.4). Moreover, the aim is to disseminate information in a very basic and comprehensive way to the population at different levels (e.g., social, cultural, educational, and the statements of Oxitec's representative political) and also across layered levels, for example, local, regional, national, and international. The project in Brazil is always willing to receive visits from the local community, the general scientific community, public health managers, stakeholders, and the media when approached, and it also invites visitors on a regular basis to allow updating to occur. The involvement of these entities in the project was also critical to help the team to understand the local community and its culture and to decide what would be the best approach for public engagement and define the technical procedures [4,21].

The point of interest differentiating the Brazilian process of community engagement is the postrelease community engagement, where the goal was to

FIGURE 18.4   Leaflet from PAT, distribution in Brazil.

evaluate the perception of the population after the release of GMM in its neighborhood. This evaluation used a questionnaire in order to measure people's perception through multiple questions, where the positivity could be measured, for the initial evaluation there was 1,043 people involved, while the postrelease perception was conducted through 482 people's opinion. In Brazil and many other countries, it is understood that the public acceptance of particular biotechnological techniques can be high if they are perceived to provide advances of real value [22]. The strategy adopted in the studied case was that during the lifetime of the project, the public would receive constant information about various issues such as mosquito population suppression, mosquito behavior, activities and campaigns regarding the issue. When the releases of information stopped, a presentation was announced to

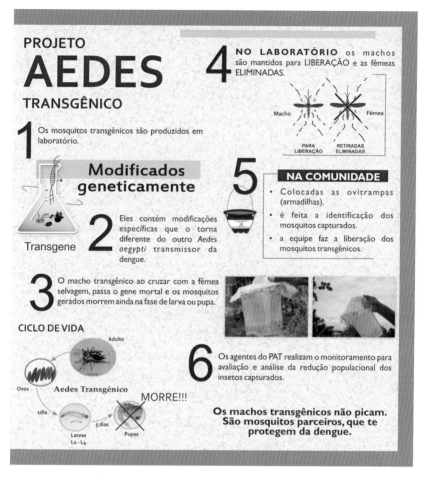

FIGURE 18.4 (*Continued*)

report all the results obtained and to evaluate people's perceptions and opinions about the releases and parallel activities (monitoring as well as public engagement itself).

Similar to Malaysia, there were positive, negative, and neutral positions. However, those considered negative were extremely low and could represent the opinions of persons unfamiliar with the technology and its impact. The positive tone of the people's perception was considered high (90%), given that the project was a pioneering project in the country and that numerous uncertainties are associated with transgenic crops. The people's understanding of the use of transgenic mosquitoes in Brazil was also evaluated, and approximately 98% of the respondents claimed to understand it. The way in

TABLE 18.2 Activities Conducted in Brazil for Disseminating Information Regarding the GMM Releases

Strategies		
Mandatory	Recommended	Suggested
Visit/interview sample/ every house in the target area	Lectures at community centers/churches-targeting adults	Action within a local event (parade, carnival, street fairs)
Meetings with local leaders, school principles, district managers	Radio spots, jingles and messages broadcasted	Driving truck with loudspeakers in the targeting area—jingle and messages
Lectures at schools— targeting kids/teens	Press releases by Moscamed journalists	Use of social media: facebook and twitter
Press coverage at local/ regional level of PAT activities: production, releases	PAT technical personnel interviewed by local/ regional/(inter)national radio stations	Press coverage at international level of PAT activities releases
	Press coverage at national level of PAT activities: production, releases	

which the community engagement process was conducted in Brazil is an example, based on previous experiences, showing that it is extremely important to involve many different spheres to guarantee the success of the project, as noted by Lavery et al. [6].

The activities conducted to implement community engagement in Brazil were split into three different stages based on what researchers considered more efficient. These activities resulted in a broad dissemination of information (Table 18.2). Mandatory activities were those that could not be omitted; recommended activities were those that demonstrated good results in disseminating the information and should be taken into consideration if a community engagement program is to be initiated. The suggested activities are those that also demonstrated a wide scope of information dissemination but were not critical to the success of implementing the technology. As mentioned previously, local aspects must be considered for each site in which the activities will be conducted. If local aspects are considered, activities considered mandatory can be interchanged with suggested activities.

A summary of the effort invested in performing this project includes the use of temporary personnel with various types of expertise. These individuals included social science students, journalists, and field or mass rearing technicians who were interested in going to the field and talking to the local population. In Brazil, a total of 20 people were directly involved in visiting all the houses included in the project, and a number of these people were also involved in additional activities that were conducted in connection with various events during the study period. The information distributed by other journalists (those not directly involved with the project) was essential to express the dialog between researchers and the various community levels using the proper amount of details and vocabulary to allow each level to fully comprehend the essential idea of the activities involved in the project.

FINAL REMARKS

It is almost a commonplace that the community presents some difficulty in understanding what occurs in science at a deep level and the misconception that only researchers know what is best, after the Brazilian experience, is faced as not completely true. The nonacademic population can understand a scientific project; it depends on the how those responsible for the project deal with it. Researchers may also be afraid to talk to the community to explain their work objectives. Note that the use of transparency is advisable. According to previous (and disastrous) experience, a lack of transparency can compromise an entire project.

After listing the most important general cultural aspects and general social aspects of an area (a country, state, village), and the decisions regarding the type of activity to be carried out to engage a local community, it is important to group similar populations to adjust the speech and the amount of information, such as the amount of details that this population must/should to know (or at least be aware of). For example, stakeholders and politicians are different from public health agents. Each of them demands a different level of information and also a different type of information. Politicians might not be interested in details of the monitoring, such as the frequency and type of trap used to collect the samples, whereas this type of information is essential for public health agents.

In summary, a commitment to ongoing review and evaluation serves to remind all participants of the complexity of community engagement and the poorly developed comprehension of what constitutes effective activities [6]. The international community needs a legal framework that harmonizes environmental safety with the humanitarian responsibility to assist nations under the social and economic burdens of disease [23]. Local communities that may be exposed during field trials (open- or contained-field trials) need to be made aware of research and risk considerations and be informed of the overall and specific context of the research. Local populations are engaged implicitly as a result of the presence of the research team among them. The social domain is a population-based regulatory system guided by cultural, political, economic, gender, equality-related, and pragmatic components [20].

REFERENCES

[1] Frewer LJ, Howard C, Aaron JI. Consumer acceptance of transgenic crops. Pestic Sci 1998;52:388−93.

[2] McNaughton D, Duong TTH. Designing a community engagement framework for a new dengue control method: a case study from central Vietnam. PLoS Negl Trop Dis 2014;8 (5):e2794. Available from: http://dx.doi.org/10.1371/journal.pntd.0002794.

[3] Lavery JV, Green SK, Bandewar SVS, et al. Addressing ethical, social, and cultural issues in global health research. PLoS Negl Trop Dis 2013;7:e2227.

[4] Lavery JV, Harrington LC, Scott TW. Ethical, social, and cultural considerations for site selection for research with genetically modified mosquitoes. Am J Trop Med Hyg 2008;79:312−18.

[5] Tindana PO, Singh JA, Tracy CS, et al. Grand challenges in global health: community engagement in research in developing countries. PLoS Med 2007;4:e273.

[6] Lavery JV, Tinadana PO, Scott TW, et al. Towards a framework for community engagement in global health research. Trends Parasitol 2010;26:279−83.

[7] Nature Editorial. Oh, New Delhi; oh, Geneva. Nature 1975;256:355−7.

[8] Jayaraman KS. Pest research centres: foreign labs shut. Nature 1982;296:104−5.

[9] Bellini R, Medici A, Puggioli A, Balestrino F, Carrieri M. Pilot field trials with *Aedes albopictus* irradiated sterile males in Italian urban areas. J Med Entomol 2013;50:317−25.

[10] Gilles JRL, Schetelig MF, Scolari F, et al. Towards mosquito sterile insect technique programmes: exploring genetic, molecular, mechanical and behavioural methods of sex separation in mosquitoes. Acta Trop 2013;1−10.

[11] Oliva CF, Jacquet M, Gilles JRL, et al. The sterile insect technique for controlling populations of *Aedes albopictus* (Diptera: Culicidae) on Reunion Island: mating vigour of sterilized males. PLoS One 2012;7:e49414.

[12] Oliva CF, Maier MJ, Gilles JRL, et al. Effects of irradiation, presence of females, and sugar supply on the longevity of sterile males *Aedes albopictus* (Skuse) under semi-field conditions on Reunion Island. Acta Trop 2013;125:287−93.

[13] Subramaniam TSS, Lee HL, Ahmad NW, Murad S. Genetically modified mosquito: the Malaysian public engagement experience. Biotechnol J 2012;7:1323−7.

[14] Dyck V, Hendrichs J, Robinson AS, et al. Public relations and political support in area-wide integrated pest management programmes that integrate the sterile insect technique. Sterile insect technique. The Netherlands: Springer; 2005. p. 547−59.

[15] Trivedi BP. The Wipeout gene. Sci Am 2011;68−75.

[16] Enserink M. Science and society: GM mosquito trial alarms opponents, strains ties in Gates-funded project. Science 2010;330:1030−1.

[17] Nature Editorial. Letting the bugs out of the bag. Nature 2011;470:2011.

[18] Subbaraman N. Science snipes at Oxitec transgenic-mosquito trial. Nat Biotechnol 2011;29:9−11.

[19] Facchinelli L, Valerio L, Ramsey JM, et al. Field cage studies and progressive evaluation of genetically-engineered mosquitoes. PLoS Negl Trop Dis 2013;7:e2001.

[20] Ramsey JM, Bond JG, Macotela ME, et al. A regulatory structure for working with genetically modified mosquitoes: lessons from Mexico. PLoS Negl Trop Dis 2014;8:e2623.

[21] Malcolm CA, El Sayed B, Babiker A, et al. Field site selection: getting it right first time around. Malar J 2009;8(Suppl. 2):S9.

[22] Reeves RG, Denton JA, Santucci F, Bryk J, Reed FA. Scientific standards and the regulation of genetically modified insects. PLoS Negl Trop Dis 2012;6:e1502.

[23] Ostera GR, Gostin LO. Biosafety concerns involving genetically modified mosquitoes to combat malaria and dengue in developing countries. JAMA 2011;305:930−1.

Chapter 19

Impact of Genetic Modification of Vector Populations on the Malaria Eradication Agenda

Vanessa Macias[1] and Anthony A. James[1,2]
[1]Department of Molecular Biology and Biochemistry, University of California, Irvine, CA, USA,
[2]Department of Microbiology & Molecular Genetics, University of California, Irvine, CA, USA

INTRODUCTION

Reports of progress on the global malaria situation are a mixed. The World Health Organization (WHO) has evidence for a continuing reduction in mortality, attributed in part to use of bed nets and combination drug therapies [1,2]. Some 3.3 million lives are estimated to have been saved since 2001. This success supports efforts to increase implementation of existing control measures with the expectation that they will continue to lower malaria incidence. However, many factors threaten these hard-won gains and these include inadequate public health infrastructures, the increasing scale over which previously successful programs must be applied, and insecticide and parasite drug resistance [1,3]. Furthermore, a number of recently recognized challenges have been identified that add to an already complex situation. These include the impact of global warming on mosquito vector distribution and the emergence of additional species of malaria parasites that can infect humans [4–6]. Thus, while there is much to celebrate about the recent reductions, we must continue to apply proven technologies while at the same time develop new disease-control tools.

The renewed call for malaria eradication stimulated cooperative planning among the malaria public health and research communities to develop agendas for reaching this goal [7]. Eradication was defined in the agenda as the reduction of transmission below a threshold level that achieves an impact on the basic reproductive rate (R_o) of the disease such that $R_o < 1$. However, it is more straightforward to express it as the complete absence of parasites in humans so that they are not able to infect mosquito vectors, and the

Genetic Control of Malaria and Dengue.
© 2016 Elsevier Inc. All rights reserved.

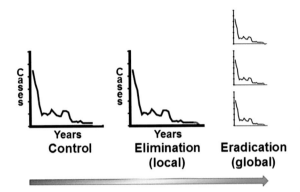

FIGURE 19.1 **Malaria eradication milestones.** Malaria eradication (right) will be achieved through a series of phases that progress (arrow) from control (left) through elimination (center). The *x*- and *y*-axes show numbers of cases and years, respectively, in arbitrary units. The red portion of the curve represents achieving elimination.

complete absence of parasites in mosquitoes so that they cannot infect humans. Recent infections of humans by parasites found previously only in non-human primates requires addressing sources of infections that originate in animal reservoirs [5,6].

Eradication is achieved through the phased operational targets of control, pre-elimination, elimination, and prevention of reintroduction [8] (Figure 19.1). The WHO defines control as less than 5% positive slides in all patients presenting with fever and elimination as no cases of locally acquired malaria for a period of 3 years as a result of deliberate control efforts. Eradication is the global elimination of malaria. This is an ambitious goal and there is a consensus that it is unlikely that any single technology will be sufficient to achieve it [7]. Contributions are needed from diagnostic, therapeutic, and prevention domains and the knowledge from a broad array of scientific disciplines must be recruited to support this effort.

It is important to ask how the goal of eradication informs the research agenda in the many contributing disciplines. This question put explicitly to vector biologists identified a number of critical needs [9]. Existing broadly applicable (insecticide treated nets, indoor residual sprays) and region-specific (environmental modification) vector-targeted prevention tools were sufficient to achieve control and elimination in many regions of the world. It is essential to use these tools where feasible and efficacious. However, there are malaria-endemic areas where it has not been possible to achieve control and elimination. This can be due to the failure to apply the currently available tools because of geographical, political, and economic difficulties, circumstances where these tools were applied but did not work (e.g., insecticide-treated nets do not impede outdoor, day-time feeding mosquitoes), and those situations where the tools worked previously but are no longer effective (e.g., the emergence of insecticide resistance) or can no longer be delivered because of failed or overwhelmed public health infrastructure. Thus, there is a clear need in the vector biology contributions to malaria eradication for better use of existing control tools and the development of novels ones to complement them.

Two of the major challenges to malaria eradication are the heterogeneity/ complexity of transmission dynamics and difficulties in sustaining control efforts [10−12]. This complexity is evident in the vector components by the large number of *Anopheles* mosquito species that have been implicated worldwide in malaria parasite transmission. There are approximately 450 described *Anopheles* species, 68 of which are known to transmit human malaria, and as many as 40 are identified as major vectors [13,14]. Indonesia alone has as many as 24 species involved in regional parasite transmission [15]. Each of these species has its own biology associated with host preferences, feeding behavior (indoor/outdoor, day/night, etc.), mating behavior, breeding-site preferences, and vector competence,[1] all of which affect their vectorial capacity. It is a significant challenge to find a single tool that accommodates all of this diversity, and this supports arguments for having multiple approaches to vector control that can be applied as needed and where effective [9].

Sustainability is a major challenge to all public health efforts and can be destabilized by both success and failure. Successful public health creates the "public health paradox;" when it is working nothing is happening. Specifically, good public health practices are characterized by the lack of disease. It is difficult under these circumstances to continue to devote resources to a problem that is perceived not to exist [8]. Withdrawal of support can lead to disease re-emergence and the ensuing costs of reasserting control are likely to be greater than those incurred by maintaining it [18].

Sustainability of vector control also is challenged by success, but has additional, intrinsic features that lead directly to failure. The most often cited is the development of insecticide resistance, and this has had a major negative impact on maintaining control in many areas of the world [19]. Additionally, migration of infected humans and mosquitoes compromises sustainability; malaria epidemics and focal outbreaks can occur in regions that have achieved elimination through the absence of the parasites but still have local competent vectors [3,20].

The prospects for success in malaria eradication will depend significantly on how well major scientific disciplines can provide tools that can address complexity and sustainability. For example, chemistry coupled with physiological insights can produce new insecticides for the vectors and prophylactic

---

1. Vector competence is a measurement of the intrinsic ability of the insect to transmit a pathogen and includes genetic components [16]. Vectorial capacity is a measurement of the efficiency of pathogen transmission, and of which vector competence is a parameter [17].

and therapeutic drugs for the parasites; these agents may have a sufficiently broad spectrum of application to be useful in managing complexity. Ecological studies can guide rational, community-wide, environmental management to remove mosquito breeding sites, and behavioral sciences can inform at-risk populations about adopting personal-protection measures (e.g., bed nets and repellents), and these also could have an impact on complexity. Immunology provides tools to probe disease progression and the basis for developing vaccines, and so contributes to sustainability at the individual level. Importantly, new tools being developed in the field of genetics can offer sustainability at a regional level.

Genetic approaches that target mosquito vectors as a means of disease control have been in consideration since the 1940s [21]. Indeed, sterile insect technologies were used to control a vector mosquito in Central America. This success was unsustainable, mostly due to civil unrest, and negative publicity in a separate effort in India decreased enthusiasm for these approaches [22,23]. However, the development of powerful molecular biological tools re-kindled enthusiasm for developing genetic control strategies. Specifically, the ability to genetically engineer specific phenotypes in mosquitoes fostered research to exploit these technologies for malaria control [24]. We anticipate a unique and important role for transgenic mosquitoes in maintaining the sustainable elimination needed for malaria eradication, but this will only be realized by careful and strategic planning.

POPULATION MODIFICATION AS A REGIONAL SOLUTION TO SUSTAINABLE MALARIA ELIMINATION

Genetic strategies for malaria control seek to eliminate vector mosquitoes or reduce their densities below thresholds needed for stable pathogen transmission (population suppression), or make them incapable of transmitting parasites (population replacement/modification) [25–30]. Transgenesis technologies were used to produce mosquito strains that carry genes that result in phenotypes that contribute to both strategies [31–35]. However, long-term, cost-effective, and sustainable malaria elimination requires the development of genetic strategies that are resilient to the immigration of parasite-infected mosquitoes and people, and the lack of such tools represents a significant unmet need in the malaria eradiation agenda. Mosquito strains for population modification carrying genes conferring parasite resistance have the appropriate design features for this purpose [25,33]. Wild, parasite-susceptible mosquitoes invading a region populated by an engineered strain will acquire the parasite resistance genes by mating with the local insects, and persons with parasites moving into the same region will not be able to infect the resident vectors, and therefore not be a source of parasites for infection of other people. Population modification also shares with other genetic control strategies the exploitation of the ability of male mosquitoes to find females, and this is expected to offer access to vector populations that would be unreachable using conventional tools [36]. Release of a population modification strain alone or in conjunction with other tools should make elimination possible in carefully selected endemic areas. Population modification strategies can be used as early as the control phase of an elimination campaign alongside other measures that will reduce disease incidence. As the efforts progress, this strategy takes on a larger role and ultimately is the mainstay of the prevention of reintroduction phase. As this elimination is achieved, the released modified mosquitoes would facilitate consolidation of this success by allowing resources to be moved to another target region with the confidence that the area just cleared will remain so. Thus, population modification offers a real chance to achieve sustainable elimination and therefore contribute significantly to malaria eradication.

Successful application of a population modification approach will depend on it being effective, that is, it achieves the goal for which it was designed, is not prohibitively more expensive than alternative approaches, and is safe for humans, animals, and the environment. Population modification strains can be generated that meet these requirements. *Anopheles* species have been engineered already with genes whose products disable *Plasmodium falciparum*,[2] and these results support the rationale for continuing to develop this approach [31,32,37]. Furthermore, population modification strains have the best design features and anticipated performance characteristics for sustainable elimination when compared to other approaches. Insecticides in all formulations and applications must be applied routinely and therefore need ongoing cost support. The same is true of proposed genetic population suppression technologies [34]. In addition as noted, the efficacy of insecticides is diminished by the emergence of resistance. Environmental modification often takes a level of infrastructure maintenance that many disease-endemic countries cannot sustain [3,38,39]. Recent work showed that introductions of exogenous symbiotic organisms into mosquitoes may increase their resistance to malaria parasites [40]. However, unlike the published reports of genetically engineered mosquitoes, these organisms have yet to be shown to completely block parasite development.

Cost-effectiveness of genetic approaches has been estimated for population suppression strains for preventing dengue virus transmission, and these are comparable initially to all other strategies [41]. However, the costs are expected to decrease as the approach drives the target population to extinction as the use of male mosquitoes to find residual populations should be less expensive and more effective than using humans to do so. Population modification should provide similar cost benefits with recurring expenses

---

2. Malaria in humans is caused primarily by four pathogen species, *Plasmodium falciparum*, *P. vivax*, *P. ovale*, and *P. malariae*. Of these, *P. falciparum* and *P. vivax* are the most significant in terms of morbidity and mortality. Recently, at least two additional *Plasmodium* species have been shown to cause disease in humans [5,6].

limited to surveillance and monitoring, which also are components of all intervention programs [42]. Furthermore, it is reasonable to expect that population modification strategies would decrease significantly the costs associated with the prevention of reintroduction phase of WHO-defined elimination [8]. It is important to acknowledge that both genetic population suppression and modification protocols most likely will only be cost-effective in areas in which a single or small number of vector species are responsible for parasite transmission. Each additional species will add to the costs, and at some point the continuous use of an unsustainable technology (e.g., insecticides) for the duration of the eradication program may be less expensive overall.

The major safety concerns of using of population modifications strains fall generally into the category of perceived risks of off-target effects that would impact other species at and outside the field-trial environment [42,43]. However, transgenic mosquito strains can be engineered with design features that make these hazards sufficiently unlikely so as to not be possible. The inclusion of specific control DNAs (e.g., those that modulate gene expression in response to a blood meal) that function in a narrow range of related species should prevent the genes from being active in beneficial insects, and inundative releases of strains that lack promiscuous (capable of spreading in many species) gene drive systems have little or no probability of moving their genes into non-target species.

The maturation pathway for new products includes distinct discovery, development, and delivery stages, and these are established for drugs, vaccines, and insecticides where rigorous industry-wide standards are used to validate performance and safety features to determine if a candidate product should advance from the laboratory to application. Such a pathway does not exist yet for genetically engineered mosquitoes [8,36]. Furthermore, the financial incentives in these approaches make it unlikely that the formulation and adoption of a consensus pathway will come from commercial interests. Therefore it is incumbent on the vector biologists, end users, and other public health stakeholders to generate such a pathway. A series of researcher-initiated efforts have taken on this challenge and guiding principles have been produced by the WHO and others [8,36]. Included in these principles are recommendations for a series of phases with "go/no-go" criteria for testing genetically engineered mosquitoes (Figure 19.2). These efforts are important but not sufficient for successful testing of a population modification approach. A strategic plan is needed that maximizes the probability of success of the first field trial of this technology while at the same time meets end user concerns about adoption of the approach.

STRATEGIC PLANNING

The primary objective of strategic planning is to ensure that the first field trial of a population modification strategy is an unqualified success. Success

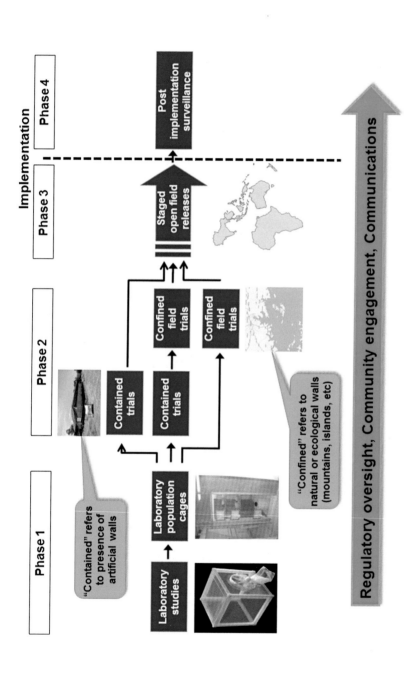

FIGURE 19.2 **Phased testing of genetically engineered mosquitoes.** A proposed scheme for the phased testing of genetically engineered mosquitoes was developed by working groups at the WHO [43]. Phase 1 is carried out entirely in the laboratory and includes the original development of the transgenic strains and small/large cage trials to estimate fitness. Phase 2 takes place in a field setting and may be either contained or confined. Phase 3 is an open field release with either or both entomological and epidemiological end points. Phase 4 is the implementation phase with the intent to achieve a sustained epidemiological impact. Regulatory oversight, community engagement, and communications should initiate early in the program. *Image adapted from Ref. [43].*

is defined here as the stable elimination of malaria at a selected trial site.[3] While this first trial will establish proof-of-principle, it is also expected that the site will remain free of malaria throughout the full duration of any ensuing eradication campaign. Thus, the outcome of the trial is expected to produce a sustained epidemiological impact.

The details of the strategic plan are meant to define major operational objectives and provide tactical guidelines that will ensure a successful trial. This trial will test the scientific features of the genetic strategy and the ability to do so in concordance with un-reproachable community engagement and an informed and transparent regulatory process [44]. Success will depend in part on meeting three operational objectives: (i) the informed selection of an optimum field site, including recruitment of the necessary personnel and resources; (ii) a well-designed and functional population modification strain; and (iii) the development of a detailed trial design and implementation plan for the release protocol.

Field-Site Selection

Site-selection criteria have been elaborated extensively in a number of publications and we highlight those relevant to a first trial of population modification [36,45−47]. Important considerations include local transmission of a single malaria parasite species, the presence of a single vector mosquito species, a limited geographical area, a thorough knowledge of the distribution and population structure of the target mosquito species and malaria epidemiology, local scientific experts with whom to collaborate, and community and government support. Political stability throughout the course of the trial also would be helpful. Importantly, there are no limitations on the persons living or moving through the experimental area.

The requirement for the malaria burden at the field site to result from a single parasite species makes it easier to monitor trial progress. *Plasmodium falciparum* or *Plasmodium vivax* offer the opportunity for the greatest impact on morbidity and mortality. The majority of existing engineered mosquitoes carry parasite resistance genes targeting *P. falciparum* and laboratory-based mosquito challenge assays are available widely for this species, therefore this makes it the best choice for the first trial [31−33,37]. This selection is not meant to diminish the significance of *P. vivax*, but allows immediate planning of trials with genes that have been proven efficacious already in the laboratory. Indeed, anti-pathogen effector genes based on altered mosquito physiology or immune enhancement or symbiont-based modification may have less-specific effects and therefore could be used in a trial targeting

---

3. The WHO certification of malaria elimination is determined at the country level [8]. We imply here a specific regional elimination that may or may not be countrywide.

multiple parasite species once the approach has been demonstrated to work for a single one.

The different levels of endemicity (hypo-, meso-, hyper-, and holoendemic [48]) are not expected to present a challenge to population modification because modification of the local vector population should suppress all parasite transmission. However, they likely will be variables that influence the amount of time that it takes to first see a positive impact of the technique and should be used as such in models. Furthermore, this technology would not be the first choice for stemming a current malaria epidemic, although once implemented, it should prevent future events.

The choice of a site in which only a single mosquito species is the vector reduces trial complexity. Genetic strategies by definition are restricted to individual interbreeding populations of a species. Complete introgression of the modification gene into a dominant vector (defined as the one contributing the most to the disease incidence in a specific area) would still leave transmission by secondary vectors, and this would prevent the trial from meeting the elimination goal. We anticipate that future applications of population modification technologies can be applied to regions with two or a small number of vectors where cost-benefit analyses provide favorable assessments for engineering strains for each target species.

Vector abundance will affect the speed at which the target population reaches fixation for the anti-pathogen effector gene. While in principle there are no constraints on the target population size, cost and logistical considerations favor a first trial in a region with low vector abundance so that sufficient mosquitoes can be reared and released, and an initial impact observed in the first or second year of a trial.

The initial trial site should maximize the geographical containment of the engineered insects [43,46]. This will assure that the release, monitoring, and surveillance activities take place on a limited scale, and therefore are not overly expensive, and contributes to meeting community engagement and regulatory considerations that we expect to be part of the first trial. This confinement is likely to be less important for future trials should the technology prove effective.

Complete and up-to-date knowledge of the vector ecology and population breeding structure is needed for site selection. This information is also important for designing the release protocol and determining those entomological parameters that can be used to monitor trial progress. Accurate epidemiological data is critical. The defined goal of the trial is to have an impact on incidence, and baseline data are needed to calibrate the success of the releases.

Local scientists, cooperative vector control facilities, and government agencies are vital to the trial process. Local scientists are nationals of the country in which the trials will take place that have the necessary expertise and authority to carry out the experiments. Sites eligible for the trial based on other criteria may lack these scientists, so strategic planning could include a training component [49]. Furthermore, it is highly likely that the laboratory scientists who developed the trial strains will have to relinquish a substantial level of control over the project as it is taken to the field [44]. Thus, success will depend largely on how well local scientists assume intellectual ownership and responsibility for the trial. This commitment of local scientists to the project goals is a significant outcome of good community engagement practices. In addition, laboratory and other facilities in which the scientists, community, and regulatory authorities have confidence are secure enough to meet the requirements for handling genetically engineered organisms are essential [42,50].

Laboratory successes in genetically engineering mosquitoes stimulated the development of community-engagement objectives and principles [44,47]. A major challenge is to obtain a form of "community consent" that is both meaningful and respects the highest ethical standards. Consent is likely to be more than an arbitrary fraction of a majority vote among community members that allows the trial to proceed. We have argued that certain cultural norms and institutions can serve as ethical surrogates for community consent [44]. These surrogates will be site specific, and what is developed for one place cannot be transferred directly to another. However, while the specifics may vary, the general considerations for developing this consent are common and sets of guidelines can be used [47]. It is worth emphasizing that trust was identified as a critical outcome of a community engagement plan. There are fears that originate from mistrust by the public of the motives of scientists. These include perceptions that the public represents a source of experimental subjects for the scientists, and that the scientists will not do any good for the trial-site community, and may actually cause harm. Here is where community engagement, including recruitment at full partnership of local scientists from the trial country, is essential. Finally, an existing statutory and regulatory structure is needed for providing trial procedure reviews and issuing authorizations to carry out the releases. A number of published documents provide useful considerations for evaluating regulatory structures [36,43,44].

Selection of the Population Modification Strain

The two major criteria for selecting the population modification strain for the trial are that it be effective and safe. Again, a strain will be effective if once released, the parasite resistance gene achieves and remains at a high-enough frequency in the local vector population so as to completely abolish parasite transmission in the target area and result in elimination. Although modeling supports the possibility of a significant effect on pathogen transmission prior to full introgression of the gene [51], it is operationally more straightforward to set complete gene fixation as a goal instead of trying to achieve some predicted sub-complete level. Elimination under these circumstances is expected to be sustainable in the presence of human and wild, parasite-susceptible mosquito immigration. Safe is defined as no significant off-target effects and no probability of vectoring a new disease agent. The off-target effects are those identified hazards and negative consequences that outweigh the benefit of having sustainably eliminated malaria [43].

We defined the optimal phenotype for a refractory gene to be one that prevents any mosquito-mediated transmission of the human-infectious forms of the parasites [33,37,52]. As far as we can determine, this means that there are *no* sporozoites (malaria parasite forms in mosquitoes that are infectious to humans) in the salivary glands of females. It is encouraging that it has been possible to produce this phenotype using different approaches [32,33].

Past vaccine and drug interventions targeting infectious agents provide many examples where resistance has been selected in pathogens thereby compromising the efficacy of the prevention or therapeutic protocol. We recommend the adoption of a "dual-transgene" approach where the population modification strain carries a compound genetic insertion comprising at least two components that disable the pathogen at different stages of its development [33]. This is functionally analogous to combined drug therapy in which the probability of a pathogen becoming resistant simultaneously to two different modes of drug action is extremely low. We expect this to be a key design feature to prevent emergence of parasite resistance in transgenic mosquitoes and sustain elimination.

The molecular targets of the anti-parasite effector gene may be polymorphic among different populations of the same species and this could affect the efficacy of the resistance phenotype. We chose targets for which there is little known variation [33], but the parasite complexity in the trial site should be characterized prior to the release to establish that it can be incapacitated by the gene products. Monitoring of the parasite population is required to mitigate the introduction of resistant parasite genotypes.

The population modification strain should also meet acceptable standards for fitness. We anticipated that the introduction of any exogenous DNA into a mosquito would necessarily come with a negative fitness cost (genetic load) because wild-type un-engineered mosquitoes have been selected in natural circumstances to be the most fit [27,53]. Any alterations to these genomes would therefore be expected to produce a fitness cost. This assumption did not take into consideration stochastic effects on population structure that could result from adaptive landscapes differing among potential interventions sites, and it is possible that wild vectors in some malaria transmission regions may not be as competitive as laboratory strains derived from them. However, the reduced-fitness rationale was used to support the need for linking gene-drive systems to parasite resistance genes. Laboratory-based empirical efforts to measure fitness produced a full range of results from showing severe negative costs in some strains to others where genes and insertions actually appear to make the transgenic mosquitoes more fit [33,37,54−58]. These results support the conclusion that engineered strains with no or minimal fitness costs can be obtained by careful selection of an anti-parasite effector gene that has been inserted into a well-characterized site in the genome. Modern genome-editing tools allow the placement of transgenes into specific regions of the host mosquito chromosomes that have been tested previously for, and insulated against, insertion-site effects [33,59].

We argued that adding an exogenous gene to the mosquito genome is likely to minimize fitness impacts when compared to manipulating transcriptional control or product abundance of endogenous genes that are involved in reproductive or immune physiology [60]. However, there are recent results that support the conclusion that targeting specific mosquito genes does not lead *a priori* to a load [61]. However, it is important to determine if laboratory performance is recapitulated in natural mosquito populations. Our experience with a population-suppression strain of the dengue vector mosquito, *Aedes aegypti*, showed that it had excellent performance characteristics in large laboratory cage trials but these could not be matched when the experiment was scaled up to much larger outdoor field cages [62].

Modification strains with some reduced fitness could still have applications in the field. Sterile insects produced for population suppression often are less fit than their wild-type counterparts because of radiation or chemical toxicity [21,63,64]. Insects with mating competitiveness lower than the targeted wild population can be effective in suppressing if released in large enough ratios [64]. Population modification strains may not share this flexibility as they are expected to remain *in situ* in the presence of immigration of wild mosquitoes, but they may be useful if they can persist at a level that is epidemiologically relevant until eradication is achieved. Modeling shows that this is possible if the anti-pathogen effector gene is linked to a chromosomal region containing genes favoring enhanced mating success [51].

An early criterion imposed on field uses of genetically engineered mosquitoes was that they should be done with male-only releases [65]. It was argued that releases of females would not be tolerated because they still could probe and feed, and therefore be a nuisance. However, there seems to be some acceptance for a relaxation of this requirement since recent trials of a *Wolbachia*-infected strain of dengue vectors could only be carried out by releasing females as the symbiont is inherited maternally [66]. Furthermore, modification-based strains rely on leaving an altered mosquito population in place and this will include reproductively active (and therefore feeding) female mosquitoes.

Safety considerations of genetically engineered mosquitoes are mainly issues about the potential for off-target effects [43,67]. These include scenarios where the transgene moves horizontally into a beneficial species in which it has a deleterious effect, potential inhalation/ingestion toxicity of transgene products, and removal of a keystone prey species from an ecological network. Here again, the design features of the strains can include components that mitigate these potential hazards. The specific gene components can be engineered such that they are functional only in the species to which they are targeted. For example, species-specific DNA control sequences can be used to direct the expression of the effector portion of the transgene, and therefore render the construct inactive if it is in a heterologous species. Inhalation/ingestion hazards also can be addressed by using products that have no inherent toxic or allergenic components, and the design feature can be such that no exogenous proteins are introduced by salivation during feeding. General allergic responses are expected to be no more frequent than human sensitivities to existing mosquito exposures [68]. Furthermore, it should be a strict requirement that all inserted transgenes in the final release strain not contain any bacterial antibiotic- or other chemical-resistance genes. This can be done easily using modern gene-editing technologies.

Population modifications strains also mitigate the issue of removing a species from an ecological network [25]. Although there are many circumstances where specific vector species are invasive and well adapted to highly artificial (not natural), human-generated ecosystems (e.g., large urban areas or agricultural regions in which the landscapes have been reshaped to favor crop production), and where complete removal of that species could be viewed as "bioremediation," there are expressed concerns about the elimination of a vector species from even badly eroded environments [67]. The modification strains leave the resident mosquitoes in place so that extant ecological dynamics remain unchanged.

Another often-expressed concern is that the specific genetic modification can produce a strain that now has the capacity to transmit a new pathogen. Fortunately, the biology of vector–pathogen interactions is complex and this represents a barrier to the transmission of new pathogen species. While there are examples of genetic changes affecting mosquito–arbovirus interactions that increased vector competence [69], these evolutionary events are likely to be rarer for protozoan parasites, including those that cause malaria. A recent cage trial study of a dengue vector in Mexico was granted permission only after laboratory experiments confirmed that the specific strain to be tested could not transmit a number of other viruses that could be expected to be found in the trial site [62]. We agree that it is good policy to test such interactions where there is a reasonable probability that a transgenic mosquito will encounter a known pathogen that it does not transmit and for which there is biological evidence that it could survive in mosquitoes. However, this should preclude unproductive research efforts for those pathogens that have never been found in mosquitoes and whose biology is not compatible with replication in mosquitoes and their cells (e.g., influenza viruses, HIV, hepatitis viruses).

Although there are a number of population modification strains under development, we are most enthusiastic about those based on single-chain antibodies (scFv) [33,37,52,70,71]. Their design features include dual-targeting components to prevent the selection of parasite resistance to the effector molecules and site-specific integration to mitigate impacts of the expression of exogenous genes on the fitness of the mosquitoes. This latter feature is also significant because it allows the remaking of a specific strain should it be lost or encounter some other difficulty that prevents the use of its original derivation. Efficacy of the scFv-based design was demonstrated in mosquitoes carrying a dual transgene that targeted the developmental stages of *P. falciparum* found in the midgut and salivary glands; no human-infectious forms were seen in the latter organs and no clear effects on fitness were observed [33].

Trial Design and Implementation

Trial implementation requires an organizational structure that maximizes cooperation and communication among all of the participants and elements of the project. The trial will require scientists who are responsible for the production, delivery, and quality control of the release strain, public health officials who will participate in monitoring and surveillance, regulatory personnel to satisfy the demands of the trial statutory conditions, individuals responsible for proactive and reactive communications and community engagement, and an administrative structure that can organize and keep the project on track.

The majority of the discovery phase of the development of population modification technologies takes place in academic settings. This is a direct result of the processes that foster creative innovation in these institutional environments. However, the subsequent stages of product development and delivery require expertise that often is not rewarded in academia, and therefore not present in the skill sets of the research scientists. We propose that a robust trial is best served by not being an academic exercise. This requires that the scientists who conceived and developed the population modification mosquito strains let go of their technology and pass it along to persons with the appropriate expertise for the next steps. This is made somewhat easier for many by the fact that these approaches are being developed for a public health benefit and not for personal or corporate enrichment, and where possible, it would be good to minimize the influence of for-profit agencies on the trials.

Good organizational practices call for the core trial team to be as small as possible while having all of the necessary expertise. Team members should be recruited from local scientists whenever possible. Participants include an on-site operational manager and persons with competency in mosquito molecular biology to ensure quality control of the product, mosquito field biologists to design and monitor release protocols, and modelers to support experimental design and define anticipated outcomes. In addition, an epidemiologist is required to track malaria incidence and prevalence, and persons familiar with regulatory criteria and community-engagement specialist with competency in the local language(s) and English are needed [36].

An explicit decision-making process is essential for the trial. All responsible persons should read and sign off on this process. Furthermore, while there are advantages to running the process as democratically as possible, ultimately there needs to be a single person who is responsible for the entire project. This person in academic settings is the principal investigator, but may have a different designation ("project leader") in a specific trial. This person needs to have the confidence of all project participants and be judged fair and knowledgeable. In addition, this person needs to be a strong advocate for the project goals while at the same time be flexible to circumstances that could demand re-evaluation of the protocols. It is important to have a person different from the project leader be the project manager. The project manager is responsible for ensuring timely coordination and communications among all project participants, and making sure that every component of the project gets what it needs, in the appropriate condition and scale, and on time.

Pre-trial efforts must include acquiring information on mating competitiveness of transgenic males compared to wild type, estimates of local vector population sizes and densities, and the expected level of adult dispersal. These data will help establish baseline entomological data for the subsequent release monitoring [72]. Furthermore, detailed information on malaria epidemiology at the site is needed. There should be enough historical data to define trends in disease incidence and prevalence so that deviations following the trial onset can be measured.

The formal design of the trial will be a challenge. Cluster-randomized trials (CRTs) have gained acceptance as the highest standard for infectious disease interventions, but these are difficult to design and could be prohibitively expensive for a population modification approach [73]. It is highly unlikely that it will be possible to find the multiple, distinct sites within reasonable geographic proximity that share enough of the matching demographic, epidemiological, and vector parameters that are needed to provide the statistical power for CRTs. Therefore, an alternative approach would be to carry out a multi-year study in a region with a previous history of malaria endemicity. Comparisons would be made between the past history of malaria prevalence and incidence at the site and what is observed subsequent to the releases.

While the frequency of the transgene in the mosquito population can be used as a surrogate marker of success, the most important end point is epidemiological. Therefore, conclusions can be confounded by other anti-malaria practices being carried out at the site during the trial, but it would be unethical to halt or withhold any other beneficial control strategy for the sake of the trial. Good trial design and statistical analyses should be able to determine that fraction of reduced incidence that results from the trial intervention. Furthermore, the ultimate outcome is conclusive, as a site that once had malaria now has achieved elimination, and this status is maintained in the relaxation or absence of other anti-malaria practices.

A successful trial requires determining the best developmental stage of the mosquito to be released. The dengue vectors in the genus *Aedes* have the remarkable property that their fertilized eggs (embryos) enter a type of diapause, called estivation, and can be stored in large numbers inactive but alive for several months [74]. These mosquitoes can be released by placing pieces of filter paper carrying as many as several thousand embryos directly into a favorable larval habitat. Unfortunately, this developmental physiology is not present in the Anopheline vectors of malaria, so an alternative stage is required. Larval and pupal stage distribution requires moving large numbers of mosquitoes in volumes of water. Weight considerations likely will make this cost-prohibitive unless the mosquitoes are reared and released locally. Furthermore, it requires placing the mosquitoes at sites in which they can complete their development. Embryo distribution is free of the need for transporting water, but these cannot be stored and also would have to be delivered physically to a place in the environment to complete development.

In the past, adults were distributed because they are have favorable weight-to-number ratios and are capable of immediate dispersal upon release [21,64]. Furthermore, the release of an engineered, blood-fed fertilized female is the numerical equivalent of a delayed release of 100−300 mosquitoes depending on the fecundity and fertility of the strain [75]. Thus, it is best to release mixed populations of transgenic adult males and blood-fed females. However, adult mosquitoes are fragile compared with the sub-adult stages, and there may be significant losses in numbers associated with packaging and transport. Therefore, this favors having a mosquito-rearing facility at or near the trial site. This would have a positive dual purpose of releasing the preferred mosquito stage while at the same time employing local vector control personnel to do the rearing and distribution. This would add significantly to the community engagement needed to engender enthusiasm for the product.

Transgenic mosquitoes are expected to be most competitive if they are similar to the wild populations at the trial site. This implies that it is necessary to put the anti-parasite effector genes into the genetic background of a strain from the trial area. This was a regulatory requirement for large-cage field trial of *Ae. aegypti* in Mexico [44]. However, there may be circumstances where releasing a laboratory-adapted strain is preferable. For example, the laboratory strain may have no or less susceptibility to insecticides than the target population. Furthermore, the laboratory strain may have been tested rigorously for their competence for other pathogens. The decision of the genetic background will be made based on the local regulatory requirements with input from the community in the trial area.

It is important to know if there are seasonal fluctuations in mosquito abundance and density. This knowledge will inform the release periods of the trial protocol to maximize survival of the transgenic adults and their subsequent progeny as well as optimize release ratios. The best time to commence releases would be at the start of the rainy season. This rain would replenish and create new oviposition and larval rearing habitats allowing the released females to deposit their fertilized eggs. The resident wild population is expected to be at their lowest size at this time, and this should maximize the impact of the released males as they should enjoy a numerical advantage over competing wild males. Habitat flooding early after trial onset could affect the initial stages of releases, so it is important that there be multiple releases over the first few weeks of the season.

Monitoring and surveillance of trial progress and outcomes should involve both entomological and epidemiological parameters. Adult trapping and molecular-detection protocols can be used to monitor the frequency of the anti-parasite gene in the mosquito population, but the capturing methods have to be adapted to the specific habits and distribution of the target species [72]. These data collecting efforts are important as they allow refining predictions of how soon we expect to see an epidemiological impact of the releases.

The specifics of evaluating the epidemiological parameters are adapted from the certification procedures developed for all elimination efforts [8]. These include surveillance mechanisms (active and passive) with full coverage of the target-site areas and reliable laboratory services to diagnose malaria. It is essential that there be full reporting of malaria cases by public and private health services with gold-standard validation of every malaria infection. This will require established and competent health services for detection, treatment, and follow-up of all possible malaria cases.

The length of the trial will be an important consideration. Since the WHO elimination certification requires at least 3 years absence of locally transmitted malaria [8], this should be the minimum trial duration. Hopefully, this will be a period long enough to calibrate stochastic factors (e.g., droughts, human migration, political instability affecting other control practices) that could confound the interpretation of the results. Limitations on the overall length depend mostly on the project showing initial success, financial support, and continued enthusiasm [36]. Two 5-year increments with annual trial review should be sufficient and definitive.

SUMMARY AND CONCLUSIONS

Population modification strategies can play a crucial role in the malaria eradication agenda. They will consolidate elimination gains by providing resistance to parasite and competent vector reintroduction and allow resources to be focused on new sites while at the same time providing confidence that treated areas will remain malaria-free. Strategic planning in three major areas, field-site selection, selection of mosquito strain, and trial design and implementation, is needed to achieve success in the first field trial. Key components of site selection include comprehensive knowledge of vectors and disease epidemiology, and local scientists who can take lead roles in trial design and implementation, and regulatory and community engagement efforts. The trial site should be geographically limited to ensure the ability to monitor it successfully and to manage costs. The preferred population modification stain should carry multiple genes for parasite resistance to assure mosquitoes have no human-infectious forms of the parasites and no probability of selecting for resistance parasites. The strain has to be sufficiently reproductively competitive to achieve gene fixation and designed such that it can be remade easily if the genes cease to function. Finally, the strain should be designed to incorporate safety features that prevent it from being a hazard to humans and the environment. Trial design and implementation should not be a strictly academic exercise, and should involve a team comprising local scientists, when possible, with specific expertise and an explicit decision-making process. The trial design is likely to be longitudinal and compare before-and-after effects on malaria epidemiology. The trial should be integrated with other malaria control efforts being conducted in the same region and requires a competent and efficient local public health system. Results should be evident and unequivocal in 3−5 years from the initial onset of releases.

ACKNOWLEDGMENTS

This work was supported in part by grants from the WM Keck Foundation and the National Institutes of Health NIAID (AI29746).

REFERENCES

[1] WHO. World malaria report, <http://www.who.int/malaria/publications/world_malaria_report_2013/report/en/>; 2013 [accessed 02.09.14].

[2] White NJ, Pukrittayakamee S, Hien TT, Faiz MA, Mokuolu OA, Dondorp AM. Malaria. Lancet 2014;383:723−35.

[3] Cohen JM, Smith DL, Cotter C, et al. Malaria resurgence: a systematic review and assessment of its causes. Malar J 2012;11:122.

[4] Sheffield PE, Landrigan PJ. Global climate change and children's health: threats and strategies for preventi. Environ Health Perspect 2011;119:291−8.

[5] Antinori S, Galimberti L, Milazzo L, Corbellino M. *Plasmodium knowlesi*: the emerging zoonotic malaria parasite. Acta Trop 2013;125:191−201.

[6] Calderaro A, Piccolo G, Gorrini C, et al. Accurate identification of the six human *Plasmodium* spp. causing imported malaria, including *Plasmodium ovale wallikeri* and *Plasmodium knowlesi*. Malar J 2013;12:321.

[7] Binka F, Brown AG, Alonso PL, et al. A research agenda for malaria eradication: cross-cutting issues for eradication. PLoS Med 2011;8:e1000404.

[8] WHO. Procedures for certification of malaria elimination. Wkly Epidemiol Rec 2014;89:321−5.

[9] Alonso PL, Besansky NJ, Burkot TR, et al. A research agenda for malaria eradication: vector control. PLoS Med 2011;8:e1000401.

[10] Spielman A, Kitron U, Pollack RJ. Time limitation and the role of research in the worldwide attempt to eradicate malaria. J Med Entomol 1993;30:6–19.

[11] Luckhart S, Lindsay SW, James AA, Scott TW. Reframing critical needs in vector biology and management of vector-borne disease. PLoS Negl Trop Dis 2010;4:e566.

[12] Cotter C, Sturrock HJW, Hsiang MS, et al. The changing epidemiology of malaria elimination: new strategies for new challenges. Lancet 2013;382:900–11.

[13] Kiszewski A, Mellinger A, Spielman A, Malaney P, Sachs SE, Sachs J. A global index representing the stability of malaria transmission. Am J Trop Med Hyg 2004;70:486–98.

[14] Hay SI, Sinka ME, Okara RM, et al. Developing global maps of the dominant anopheles vectors of human malaria. PLoS Med 2010;7:e1000209.

[15] Elyazar IRF, Sinka ME, Gething PW, et al. The distribution and bionomics of anopheles malaria vector mosquitoes in Indonesia. Adv Parasitol 2013;83:173–266.

[16] Hardy JL, Houk EJ, Kramer LD, Reeves WC. Intrinsic factors affecting vector competence of mosquitoes for Arboviruses. Annu Rev Entomol 1983;28:229–62.

[17] Smith DL, Battle KE, Hay SI, Barker CM, Scott TW, McKenzie FE. Ross, Macdonald, and a theory for the dynamics and control of mosquito-transmitted pathogens. PLoS Pathog 2012;8:e1002588.

[18] Tatarsky A, Aboobakar S, Cohen JM, et al. Preventing the reintroduction of malaria in Mauritius: a programmatic and financial assessment. PLoS One 2011;6:e23832.

[19] Sokhna C, Ndiath MO, Rogier C. The changes in mosquito vector behaviour and the emerging resistance to insecticides will challenge the decline of malaria. Clin Microbiol Infect 2013;19:902–7.

[20] Centers for Disease Control and Prevention *Plasmodium vivax* malaria – San Diego County, California, 1986. MMWR Morb Mortal Wkly Rep 1986;35:679–81.

[21] Klassen W, Curtis CF. History of the sterile insect technique. In: Dyck VA, et al., editors. The sterile insect technique: principles and practice in area-wide integrated pest management. Dordrecht, The Netherlands: Springer; 2005. p. 3–36.

[22] Oh, New Delhi, Oh, Geneva [editorial]. Nature 1975;256:355–7.

[23] Jayaraman KS. Consortium aims to revive sterile-mosquito project. Nature 1997;389:6.

[24] Report of the meeting "Prospects for malaria control by genetic manipulation of its vectors", Geneva, Switzerland. WHO TDR/BCV/MAL-ENT/91.3; 1991.

[25] Curtis CF. Possible use of translocations to fix desirable genes in insect pest populations. Nature 1968;218:368–9.

[26] Curtis CF, Graves PM. Methods for replacement of malaria vector populations. J Trop Med Hygeine 1988;91:43–8.

[27] Collins FH, James AA. Genetic modification of mosquitoes. Sci Med 1996;3:52–61.

[28] Graves PM, Curtis CF. A cage replacement experiment involving introduction of genes for refractoriness to *Plasmodium yoelii nigeriensis* into a population of *Anopheles gambiae* (Diptera: Culicidae). J Med Entomol 1982;19:127–33.

[29] James A. Control of disease transmission through genetic modification of mosquitoes. In: Handler AM, James AA, editors. Insect transgenesis: methods and applications. Boca Raton, FL: CRC Press; 2000. p. 319–32.

[30] James AA, Benedict MQ, Christophides GK, Jacobs-Lorena M, Olson KE. Evaluation of drive mechanisms (including transgenes and drivers) in different environmental conditions and genetic backgrounds. In: Knols BGJ, Louis C, editors. Bridging laboratory and field research for genetic control of disease vectors. Dordrecht, The Netherlands: Springer; 2006. p. 149–55.

[31] Ito J, Ghosh A, Moreira LA, Wimmer EA, Jacobs-Lorena M. Transgenic anopheline mosquitoes impaired in transmission of a malaria parasite. Nature 2002;417:452–5.

[32] Corby-Harris V, Drexler A, Watkins de Jong L, et al. Activation of Akt signaling reduces the prevalence and intensity of malaria parasite infection and lifespan in *Anopheles stephensi* mosquitoes. PLoS Pathog 2010;6:e1001003.

[33] Isaacs AT, Jasinskiene N, Tretiakov M, et al. Transgenic *Anopheles stephensi* coexpressing single-chain antibodies resist *Plasmodium falciparum* development. Proc Natl Acad Sci USA 2012;109:E1922−30.

[34] Marinotti O, Jasinskiene N, Fazekas A, et al. Development of a population suppression strain of the human malaria vector mosquito, *Anopheles stephensi*. Malar J 2013;26:142.

[35] Galizi R, Doyle LA, Menichelli M, et al. A synthetic sex ratio distortion system for the control of the human malaria mosquito. Nat Commun 2014;5:3977.

[36] Brown DM, Alphey LS, McKemey A, Beech C, James AA. Criteria for identifying and evaluating candidate sites for open-field trials of genetically engineered mosquitoes. Vector Borne Zoonotic Dis 2014;14:291−9.

[37] Isaacs AT, Li F, Jasinskiene N, et al. Engineered resistance to *Plasmodium falciparum* development in transgenic *Anopheles stephensi*. PLoS Pathog 2011;7:e1002017.

[38] Kitua AY. The future of malaria research and control: an African perspective. Afr J Health Sci 1998;5:58−62.

[39] Tusting LS, Thwing J, Sinclair D, et al. Mosquito larval source management for controlling malaria. Cochrane Database Syst Rev 2013;8:CD008923.

[40] Bian G, Joshi D, Dong Y, et al. *Wolbachia* invades *Anopheles stephensi* populations and induces refractoriness to *Plasmodium* infection. Science 2013;340:748−51.

[41] Alphey N, Alphey L, Bonsall MB. A model framework to estimate impact and cost of genetics-based sterile insect methods for dengue vector control. PLoS One 2011;6:e25384.

[42] Benedict M, D'Abbs P, Dobson S, et al. Guidance for contained field trials of vector mosquitoes engineered to contain a gene drive system: recommendations of a scientific working group. Vector Borne Zoonotic Dis 2008;8:127−66.

[43] WHO/TDR, FNIH. The Guidance Framework for testing genetically modified mosquitoes, <http://www.who.int/tdr/publications/year/2014/guide-fmrk-gm-mosquit/en/>; 2014.

[44] Ramsey JM, Bond JG, Macotela ME, et al. A regulatory structure for working with genetically modified mosquitoes: lessons from Mexico. PLoS Negl Trop Dis 2014;8:e2623.

[45] El Sayed BB, Malcolm CA, Babiker A, et al. Ethical, legal and social aspects of the approach in Sudan. Malar J 2009;8(Suppl. 2):S3.

[46] Malcolm CA, El Sayed B, Babiker A, et al. Field site selection: getting it right first time around. Malar J 2009;8(Suppl. 2):S9.

[47] Lavery JV, Harrington LC, Scott TW. Ethical, social, and cultural considerations for site selection for research with genetically modified mosquitoes. Am J Trop Med Hyg 2008;79:312−18.

[48] Hay SI, Smith DL, Snow RW. Measuring malaria endemicity from intense to interrupted transmission. Lancet Infect Dis 2008;8:369−78.

[49] Mnzava AP, Macdonald MB, Knox TB, Temu EA, Shiff CJ. Malaria vector control at a crossroads: public health entomology and the drive to elimination. Trans R Soc Trop Med Hyg 2014;108:550−4.

[50] Benedict M, D'Abbs P, Dobson S, et al. Arthropod containment guidelines. A project of the American Committee of Medical Entomology and American Society of Tropical Medicine and Hygiene. Vector Borne Zoonotic Dis 2003;3:61−98.

[51] Boëte C, Agusto FB, Reeves RG. Impact of mating behaviour on the success of malaria control through a single inundative release of transgenic mosquitoes. J Theor Biol 2014;347:33−43.

[52]   Jasinskiene N, Coleman J, Ashikyan A, Salampessy M, Marinotti O, James AA. Genetic control of malaria parasite transmission: threshold levels for infection in an avian model system. Am J Trop Med Hyg 2007;76:1072−8.

[53]   James AA. Gene drive systems in mosquitoes: rules of the road. Trends Parasitol 2005;21:64−7.

[54]   Catteruccia F, Godfray HCJ, Crisanti A. Impact of genetic manipulation on the fitness of *Anopheles stephensi* mosquitoes. Science 2003;299:1225−7.

[55]   Irvin N, Hoddle MS, O'Brochta DA, Carey B, Atkinson PW. Assessing fitness costs for transgenic *Aedes aegypti* expressing the GFP marker and transposase genes. Proc Natl Acad Sci U S A 2004;101:891−6.

[56]   Moreira LA, Wang J, Collins FH, Jacobs-Lorena M. Fitness of anopheline mosquitoes expressing transgenes that inhibit *Plasmodium* development. Genetics 2004;166: 1337−41.

[57]   Amenya DA, Bonizzoni M, Isaacs AT, et al. Comparative fitness assessment of *Anopheles stephensi* transgenic lines receptive to site-specific integration. Insect Mol Biol 2010;19:263−9.

[58]   Hauck ES, Antonova-Koch Y, Drexler A, et al. Overexpression of phosphatase and tensin homolog improves fitness and decreases *Plasmodium falciparum* development in *Anopheles stephensi*. Microbes Infect 2013;15:775−87.

[59]   Carballar-Lejarazú R, Jasinskiene N, James AA. Exogenous gypsy insulator sequences modulate transgene expression in the malaria vector mosquito, *Anopheles stephensi*. Proc Natl Acad Sci U S A 2013;110:7176−81.

[60]   Nirmala X, James AA. Engineering *Plasmodium*-refractory phenotypes in mosquitoes. Trends Parasitol 2003;19:384−7.

[61]   Pike A, Vadlamani A, Sandiford SL, Gacita A, Dimopoulos G. Characterization of the Rel2-regulated transcriptome and proteome of *Anopheles stephensi* identifies new anti-*Plasmodium* factors. Insect Biochem Mol Biol 2014;52:82−93.

[62]   Facchinelli L, Valerio L, Ramsey JM, et al. Field cage studies and progressive evaluation of genetically-engineered mosquitoes. PLoS Negl Trop Dis 2013;7:e2001.

[63]   Andreasen MH, Curtis CF. Optimal life stage for radiation sterilization of Anopheles males and their fitness for release. Med Vet Entomol 2005;19:238−44.

[64]   Dame DA, Curtis CF, Benedict MQ, Robinson AS, Knols BGJ. Historical applications of induced sterilisation in field populations of mosquitoes. Malar J 2009;8(Suppl. 2):S2.

[65]   Pew Initiative. *Bugs in the System: Issues in the Science and Regulation of Genetically Modified Insects.* Washington, DC, 2004.

[66]   Hoffmann AA, Montgomery BL, Popovici J, et al. Successful establishment of *Wolbachia* in *Aedes* populations to suppress dengue transmission. Nature 2011;476:454−7.

[67]   WHO. Progress and prospects for the use of genetically-modified mosquitoes to inhibit disease transmission. In: Report on planning meeting 1: technical consultation on current status and planning for future development of genetically-modified mosquitoes for malaria and dengue control, http://dx.doi.org/10.2471/TDR.10.978-924-1599238; 2010.

[68]   Peng Z, Simons FER. Advances in mosquito allergy. Curr Opin Allergy Clin Immunol 2007;7:350−4.

[69]   Higgs S. Chikungunya virus: a major emerging threat. Vector Borne Zoonotic Dis 2014;14:535−6.

[70]   James AA, Beerntsen BT, Capurro M, de L, et al. Controlling malaria transmission with genetically-engineered, *Plasmodium*-resistant mosquitoes: milestones in a model system. Parassitologia 1999;41:461−71.

[71] De Lara Capurro M, Coleman J, Beerntsen BT, et al. Virus-expressed, recombinant single-chain antibody blocks sporozoite infection of salivary glands in *Plasmodium gallinaceum*-infected *Aedes aegypti*. Am J Trop Med Hyg 2000;62:427−33.

[72] Guerra CA, Reiner RC, Perkins TA, et al. A global assembly of adult female mosquito mark-release-recapture data to inform the control of mosquito-borne pathogens. Parasit Vectors 2014;7:276.

[73] James S, Simmons CP, James AA. Field trials of modified mosquitoes present complex but manageable challenges. Science 2011;334:771−2.

[74] Christophers S. *Aëdes aegypti* (L.) the yellow fever mosquito: its life history, bionomics and structure. London: The Syndics of the Cambridge University Press; 1960.

[75] Rasgon JL, Scott TW. Impact of population age structure on *Wolbachia* transgene driver efficacy: ecologically complex factors and release of genetically modified mosquitoes. Insect Biochem Mol Biol 2004;34:707−13.

Index

Note: Page numbers followed by "*f*" and "*t*" refer to figures and tables, respectively.

DHF. *See* Dengue hemorrhagic fever (DHF)
Diário Oficial da União (DOU), 416
Disability-adjusted life years (DALYs), 381,
  396−397
Discovery Labs. *See* Hemotech system
Diseases endemic countries (DECs), 63−64
Disrupting vectorial capacity challenge, 125
  Garrett-Jones formula, 127−128
  lessons, 129
    effective delivery, 129−130
    maintenance of quality of intervention,
      130−132
    polyvalent intervention, 132−133
    sustaining effective delivery, 132
  malaria elimination and eradication, 126−127
  mosquito-dependent transmission of
    malaria, 126
  reductions in parasite transmission, 128
  Ross−MacDonald formula, 126
Dissemination rate assessment, 291−292
Docking-site-based integration, 142
  strengths, 143
  weaknesses, 143
  work, 142
Dosage compensation, 212−213
DOU. *See* Diário Oficial da União (DOU)
Doublesex (*dsx*), 209−212
Drosophila, 230
  chemosensory receptors detecting odorants
    and CO_2 in, 230
    GRs, 234−235
    IRs, 233−234
    OR, 231−233
Drosophila cinnabar gene, 150−151
Drosophila melanogaster (*D. melanogaster*),
  231−232
Drosophila simulans (*D. simulans*), 207
dsx. See Doublesex (*dsx*)
"Dual-transgene" approach, 433

E

Early warning system, 110−111
Economic burden of dengue, 377
Effect size, 126−127, 134−136
Effector gene, 281−282, 292
EIP. *See* Extrinsic incubation period (EIP)
EIR. *See* Entomological inoculation rate (EIR)
Empty neuron, 231−232
Engineering pathogen resistance, 279−280
  dissemination rate assessment, 291−292
  engineered resistance to malaria parasites,
    295*t*

evaluation, 286
  choice of parasite species, 286−287
  choice of vector species, 287−288
  infectious blood meal, 288−289
  oral challenge with malaria parasites/
    dengue viruses, 288
host factor interference, 285−286
immune activation, 281−282
immune augmentation, 283−284
incubation and analysis conditions for
  plasmodium parasites, 290*t*
infection rate assessment, 289−291
pathogen transmission rate assessment,
  292−294
RNAi, 280−281
Enterobacter sp. (*Esp_Z*), 340
Entomological inoculation rate (EIR), 86,
  126−127
Entomopathogenic fungi for disease control,
  345−347
Environmental Protection Agency (EPA), 2
EPA. *See* Environmental Protection Agency
  (EPA)
Eradication, 423−424
Estivation, 438
Ethical framework, 18
Extrinsic incubation period (EIP), 277−278

F

Feedback, 414
Female-specific release of insects carrying
  dominant lethal (fsRIDL), 206−207
FFK strains. *See* Field female killing strains
  (FFK strains)
Field female killing strains (FFK strains),
  35−36
Field-site selection, 430−432
Filter-rearing system, 34−35
Fitness, 318−320
Flavobacterium okeanokoites (*FokI*), 146
FokI. See Flavobacterium okeanokoites (*FokI*)
Forest chromosomal form, 58
Fruitless (*fru*), 209−212
fsRIDL. *See* Female-specific release of insects
  carrying dominant lethal (fsRIDL)
Fungi, 344−345

G

G-protein-coupled receptors (GPCRs), 231
Garrett-Jones formula, 127−128
GDH. *See* Glutamate dehydrogenase (GDH)

Printed in the United States
By Bookmasters